D1409870

Progress in Inorganic Chemistry

Volume 22

Advisory Board

PROGRESS IN INORGANIC CHEMISTRY

Edited by
STEPHEN J. LIPPARD

DEPARTMENT OF CHEMISTRY
COLUMBIA UNIVERSITY
NEW YORK, NEW YORK

VOLUME 22

AN INTERSCIENCE® PUBLICATION
JOHN WILEY & SONS, New York · London · Sydney · Toronto

An Interscience® Publication

Library of Congress Catalog Card Number: 59-13035
ISBN 0-471-54092-7
Printed in the United States of America
10 9 8 7 6 5 4 3 2 1

Contents

Progress in
Inorganic Chemistry
Volume 22

The Coordination and Bioinorganic Chemistry of Molybdenum

by EDWARD I. STIEFEL

C. F. Kettering Research Laboratory
Yellow Springs, Ohio
(Contribution No. 539)

ABBREVIATIONS

$acac^-$	acetylacetonate ($CH_3COCHCOCH_3^-$)
$acacen^{-2}$	1,2-bis(acetylacetoneimino)ethane ($CH_3COCHC(CH_3)NCH_2CH_2NCCH_3CHCOCH_3^{-2}$)

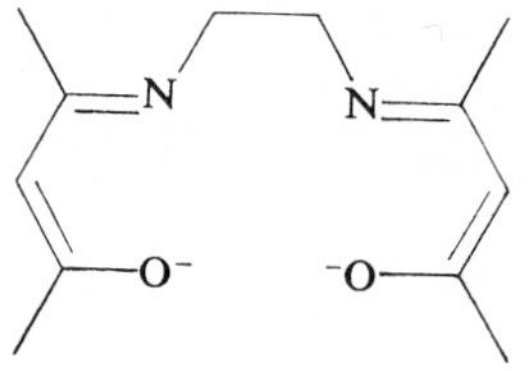

l-ala	*l*-alanine ($CH_3CH(NH_2)COOH$)
ATP	adenosine-5′-triphosphate
ADP	adenosine-5′-diphosphate
arphos	1-diphenylphosphino-2-diphenylarsinoethane ($Ph_2PCH_2CH_2AsPh_2$)
bal	2,3-dimercaptopropanol
bigH	biguanide ($NH_2C(NH)NHC(NH)NH_2$)
C_6H_5bigH	phenylbiguanide $NH_2C(NPh)NHC(NH)NH_2$
bipy	2,2′-bipyridyl $(C_5H_4N)_2$

Bu	butyl (C_4H_9)
cys	*l*-cysteine ($NSCH_2CHNH_2COOH$)
O-Mecys	*l*-cysteine methyl ester ($HSCH_2CHNH_2COOCH_3$)
cyt	cytochrome
dbm^-	dibenzoylmethide ($PhCOCHCOPh^-$)
diars	1,2-bis(dimethylarsino)benzene(C_6H_4 $(AsMe_2)_2$)
dien	diethylenetriamine ($NH_2CH_2CH_2NCH_2CH_2NH_2$)
dma	*N*,*N*-dimethylacetamide ($(CH_3)_2NCOCH_3$)
dmf	*N*,*N*-dimethylformamide ($(CH_3)_2NCHO$)
dmpe	1,2-bis(dimethylphosphino)ethane ($(CH_3)_2PCH_2CH_2P(CH_3)_2$)
dmpm	bis(dimethylphosphino)methane ($(CH_3)_2PCH_2P(CH_3)_2$)
dmso	dimethylsulfoxide (CH_3SOCH_3)
dpae	1,2-bis(diphenylarsino) ethane ($Ph_2AsCH_2CH_2AsPh_2$)
dpam	bis(diphenylarsino) methane ($Ph_2AsCH_2CH_2AsPh_2$)
dpm^-	dipivaloylmethide (t-$C_4H_9COCHCO)$-t-$C_4H_9^-$)
dppe	1,2-bis(diphenylphosphino)ethane ($Ph_2PCH_2CH_2PPh_2$)
dppm	bis(diphenylphosphino)methane ($Ph_2PCH_2PPh_2$)
dppp	1,3-bis(diphenylphosphino)propane ($Ph_2PCH_2CH_2CH_2PPh_2$)
R_2dtc^-	*N*,*N*-dialkyldithiocarbamate ($R_2NCS_2^-$)
dtd	4,7-dithiadecane ($CH_3CH_2CH_2SCH_2CH_2SCH_2CH_2CH_3$)
dth	2,5-dithiahexane ($CH_3SCH_2CH_2SCH_3$)
dtdd	5,8-dithiadodecane (n-$C_4H_9SCH_2CH_2S$-n-C_4H_9)
R_2dtp^-	dialkyldithiophosphate ($(RO)_2PS_2^-$)
$edta^{-4}$	ethylenediaminetetracetate ($(^-OOCCH_2)_2NCH_2CH_2N(CH_2COO^-)_2$)
en	ethylenediamine ($NH_2CH_2CH_2NH_2$)
EPR	electron paramagnetic resonance
Et	ethyl (C_2H_5)
FAD	flavin adenine dinucleotide
gly	glycine (CH_2NH_2COOH)
$HB(Pz)_3^-$	tris(2-pyrazolyl)hydroborate
hfa	hexafluoroacetylacetonate ($CF_3COCHCOCF_3^-$)
his	*l*-histidine ($(C_3H_3N_2)CH_2CHNH_2COOH$)
Me	methyl (CH_3)
$mida^{-2}$	methyliminodiacetate $CH_3N(CH_2COO^-)_2$
mnt^{-2}	maleonitriledithiolate (1, 2-dicyanoethylene-1, 2-dithiolate) ($S_2C_2(CN_2^{-2})$
i-mnt^{-2}	2,2-dicyanoethylene-1,1-dithiolate ($S_2C_2(CN)_2^{-2}$)
NADH	nicotinamide adenine dinucleotide
NADPH	nicotinamide adenine dinucleotide phosphate

NMR	nuclear magnetic resonance
nta^{-3}	nitrilotriacetate ($N(CH_2COO^-)_3$)
oep	octaethylporphyrin
oxineH	8-hydroxyquinoline (C_9H_7NOH)
Ph	phenyl (C_6H_5)
pbz	2-(2′-pyridyl) benzamidazole
$pdta^{-4}$	propylene-1,2-diaminetetracetate ($(^-OOCCH_2)_2$-$NCHCH_3CH_2N(CH_2COO^-)_2$
4-pic	4-picoline (4-$CH_3C_5H_4N$)
py	pyridine (C_5H_5N)
$py(anil)_2$	

C—(pyridine ring, N)—C; =N—Ph, =N—Ph

Pr	propyl (C_3H_7)
quin	quinoline (C_9H_7N)
sce	saturated calomel electrode
$salen^{-2}$	1,2-bis(salicylaldimine) ethylene

C=N, N=C, O⁻ ⁻O

N-MeSal⁻	*N*-methylsalicylaldiminate ($C_6H_4(CHN(CH_3))O^-$)
TBAP	tetra(*n*-butyl)ammonium perchlorate
tdt^{-2}	toluene-3,4-dithiolate ($CH_3C_6H_3S_2^{-2}$)
terpy	2,6-bis(2′-pyridyl)pyridine ($C_{15}H_{11}N_3$)
thf	tetrahydrofuran (C_4H_8O)
tpp	tetraphenylporphine
trias	tris(dimethylarsinomethyl)methane ($HC(CH_2AsMe_2)_3$)
triphos	bis(*o*-diphenylphosphinophenyl)phenylphosphine PPh_2-$C_6H_4P(Ph)C_6H_4PPh_2$
ttas	bis(*o*-dimethylarsinophenyl)methylarsine ($Me_2AsC_6H_4$-$As(CH_3)C_6H_4AsMe_2$)
Rxan⁻	alkylxanthate ($ROCS_2^-$)

I. INTRODUCTION

The chemistry of molybdenum has been aptly described as among the most complex of the transition elements (1). The formal oxidation states range from -2 to $+6$, with extensive chemistry within states 0, 2, 3, 4, 5, and 6. Coordination numbers span from four to eight, and mononuclear, dinuclear, and polynuclear or polymeric compounds are known in virtually all oxidation states. Some of these myriad possibilities have been selected for use by living things, and Mo has been shown to be a key component of several enzymes. Despite an intensifying interest over the last 10 years, it is our feeling that there remains a great deal to learn about the coordination chemistry and especially the biochemistry of molybdenum. Primarily, this review deals with the coordination chemistry of Mo. Emphasis is placed on recent progress, but some problems (old and new) and unsettled areas are delineated as well. A second theme falls in the area of bioinorganic chemistry, and in some cases discussions of coordination chemistry are viewed in the light of biochemical considerations.

A. Scope and Limitations

This review surveys the coordination chemistry of Mo from oxidation states 6 through 2, with a more selective treatment of lower oxidation states. A brief overview of Mo enzymes is given and, in view of the preeminence of these enzymes in nitrogen metabolism, a section on Mo-nitrogen (including dinitrogen) chemistry is included.

Structure, bonding, spectra, and (where understood) reactivity are covered. Synthetic procedures are also given for various classes of compounds. In general, solution chemistry is not covered with the notable exceptions of aquo ions and cysteine complexes. Our prediliction is to deal with stoichiometrically and, where possible, structurally well-characterized complexes. There have been hundreds of studies on complexes in solution (especially in the French and Russian literature) and most of these are not discussed. While many of these reports deal with well-defined species, others do not and it will be left to some future reviewer to assume the task of sorting them out. On the other hand, some solution studies, called "model" systems, are discussed in the appropriate sections.

The logic of organizing this review on the basis of Mo oxidation states can be questioned. After all, oxidation state is a formalism which cannot always be unequivocally applied. Uncertainty occurs when complexes contain unsaturated ligands [called "noninnocent" (2)] whose highest filled and/or lowest empty orbitals are near the energy of the metal *d* levels. In such cases (examples are found in Sections III, IV, IX, and especially VIII) the assignment of oxidation state can be quite arbitrary. Fortunately, these complexes are in the minority and the valuable procedure of assigning an oxidation state can be justified for the remainder. The oxidation state can then be identified with an electronic configuration on the metal to which we attribute distinct magnetic, spectroscopic, and reactivity patterns. These same patterns are now becoming apparent for Mo in enzymes, and the data reviewed here and organized by oxidation state may be of use to those involved in establishing the oxidation states of biological molybdenum (if indeed the enzyme ligands are innocent!).

To limit the size of this review we have arbitrarily restricted coverage of certain areas. In general, we do not separately discuss complexes containing more than two Mo's. This eliminates the extensive class of iso- and heteropoly molybdates [for Mo(VI)], the molybdenum blues [for Mo(V)-Mo(VI)], the hexanuclear $Mo_6X_8^{+2}$ clusters [for Mo(II)], and numerous polymeric oxides, sulfides, and halides. The chemistry of Mo with CN^- has not been given a full treatment, and the extensive organometallic chemistry has been covered quite selectively. The exclusion of these complexes is made easier by their general nonresemblance to other complexes discussed. Furthermore, at least at present, these complexes do not seem as relevant to Mo's biochemical role. On the other hand, when the chemistry of these compounds does interface with other studies discussed here, it is covered in that respect.

The subjectively selective treatment of organometallic compounds requires explanation. Simply stated, the literature in this area is enormous and we have had to draw the line somewhere. We recognize the artificiality of this line by realizing that, starting from low-valent compounds, cyclopentadienyl complexes have been prepared containing $Mo_2O_5^{+2}$, $Mo_2O_4^{+2}$, and $Mo_2O_2S_2^{+2}$ cores. Since these same units are routinely found in Mo(V) and Mo(VI) chemistry with very different ligands, it is clear that the core structures obtain whether or not the complex is organometallic (i.e., organometallic chemistry is indeed coordination chemistry). Likewise, there is great similarity between N_2 and CO as ligands, and while we thoroughly cover complexes of the former, complexes of the latter are treated sparingly. Since CO is an inhibitor of nitrogenase (although perhaps not at the Mo site), we feel that this omission must be explicitly noted. Nonetheless, we feel that these limitations in scope will not detract from the full sweep of Mo coordination chemistry which we wish to convey in this review.

This review was originally written based on the literature which appeared through the end of 1974. In the stage of galley proof, December 1975, brief additions were made such that important references from 1975 are also included.

B. Previous Reviews

In 1952 the first and only comprehensive and critical monograph on Mo chemistry appeared (3). In 1966 Mitchell (4, 5) reviewed much of Mo coordination chemistry in two complementary reviews, whereas Spence (6, 7) concentrated on Mo solution chemistry relevant to Mo biochemistry. Recently, Bowden (7a) has considered some chemical and enzymological aspects of Mo in biology. Spivack and Dori (7b) have recently reviewed structural aspects of Mo(IV), Mo(V) and Mo(VI) complexes while Schroeder (7c) has considered structural trends in oxo molybdenum compounds. These latter two articles supplement the present review as they contain greater structural detail than presented here. Recent broad treatments by Rollinson (8) and Kepert (8a) include a number of topics not covered here. Mo chemistry has been included in some recent reviews on other topics. Selbin (9) and, more recently, Griffith (10) reviewed oxo compounds, whereas Walton (11), centerford and Colton (12), and Ferguson (13) considered halides and oxyhalides. Cyclopentadienyl complexes have been covered by Barnett and Slocum (14). Other more specific reviews are quoted at the beginning of the appropriate section.

C. Bioinorganic Aspects

Much current research in Mo chemistry (especially in the United States and Great Britain) receives stimulation from the knowledge that Mo is an essential micronutrient for microorganisms, plants, and animals. The known and suspected Mo enzymes are a fascinating lot, possessing a molecular elegance which chemists are just beginning to fathom. Many inorganic chemists have chosen to probe the possible roles of Mo in enzymes through the use of analogy, imagination, and (not least of all) experimentation.

The experimentation sometimes takes the form of building model systems. The question arises: What constitutes a valid model for an enzyme or more precisely for the metal-ion site in this enzyme? First, we might wish our model complex to have some of the spectroscopic or magnetic properties of the enzyme. This would give us confidence that we are working with the right ligand combination, oxidation state, and geometry. Second, we might wish our model to carry out some or all of the enzyme reactions. While it is currently beyond our reasonable expectations to approach the rate and specificity of the intact enzyme, this presents an ultimate goal toward which we can strive. This ideal has not yet been reached for any enzyme. [In the case of the nonenzyme Fe-S proteins (ferredoxins), a good argument can be made that Holm's (15) recent work comes as close as any to the ideal.] However, when one does not know much about the biological system (and this was not the case for the ferredoxin models), the approach to the ideal is difficult indeed. So then, what contribution can the inorganic (ergo bioinorganic) chemist make in the case of Mo enzymes?

To begin, he can assure that any hypothesis put forward to explain some property of Mo in enzymes is not inconsistent with known chemistry. An example will illustrate the point. It has been stated in the biological literature that N *cannot* be in the first coordination sphere of Mo since no ^{14}N superhyperfine splitting is found in the Mo (V) enzyme EPR signal. However, recent work (discussed in Section IV.C) illustrates that in many simple Mo complexes, ^{14}N superhyperfine splitting is not resolved even though N is known to be a donor atom. In those cases where this splitting has been observed it is very small, so small that it would be difficult and perhaps impossible to resolve at the line widths found in the enzymes. Thus, coordination chemistry tells us that N cannot yet be eliminated as a ligand for biological molybdenum. This and other instances are discussed at greater length in the text.

Throughout the article when a particular facet of Mo chemistry seems possibly related to Mo biochemistry, that possibility is pointed out. Doubtless there are some important chemical results whose biochemical relevance we have not perceived. It is even more certain that some of the extrapolations which we have attempted bear no resemblance to real biochemical situations. Nevertheless, the lack of knowledge of details of Mo biochemistry gives us some liberty to speculate in the hope that some suggestions may be of use as working hypotheses to be tested in enzymes or models.

For Mo we feel it is proper to take a broad view of the possibilities. For example, both aqueous and nonaqueous chemistry may be relevant since even though enzymes function in water the Mo site may be hydrophobic in nature. Furthermore, the effective pH at the enzyme active site may differ vastly from that in the medium and thus aqueous studies at "nonphysiological" pH ranges could also be relevant. The lower oxidation states II and III are covered thoroughly even though their relevance to biology is uncertain. Thus the Mo oxidases seem likely to use the Mo(VI), Mo(V), and Mo(IV) states, and this seems reasonable but less certain for nitrate reductase as well. On the other hand, there is no evidence for the state of Mo in nitrogenase, and since this protein works at a very low potential, Mo(II) and (III) must be considered as possibilities. In fact, the Mo(III)/Mo(V) couple may have just the right potential for dinitrogen reduction (Section IV.E). For nitrogenase, as Chatt has succintly stated (16), ". . . all options, and there are a great number, are still open." In short, our ignorance allows us to consider a great deal of Mo coordination chemistry as potentially relevant even though model systems, in the sense described above, are clearly not present.

II. MOLYBDENUM IN BIOLOGICAL SYSTEMS

Molybdenum is an essential micronutrient for microorganisms (17),

TABLE I
Molybdenum Containing Enzymes

Enzyme	Source	Molecular weight	Mo (moles per molecular weight)	Other prosthetic groups	Refs.
Nitrogenase	Azotobacter vinelandii				
Fe-Mo Protein		226,000	2	24 Fe, 22 S	37
Fe Protein		66,000	0	4 Fe, 4 S	
Nitrogenase	Klebsiella pneumoniae				
Fe-Mo Protein		218,000	2	>30 Fe, 30 S	45, 131
Fe Protein		66,800	0	4 Fe, 4 S	45, 131
Nitrogenase	Clostridium pasteurianum				
Fe-Mo Protein		220,000	2	24 Fe, 24 S	132
Fe Protein		56,000	0	4 Fe, 4 S	133
Nitrogenase	Soybean bacteroids				
Fe-Mo Protein		200,000	1.3	29 Fe, 26 S	32, 134
Fe Protein		51,000	0		32, 134
Nitrate reductase	Neurospora crassa	228,000	1–2	cyt_b, FAD	86
Nitrate reductase	Aspergillus nidulans	190,000		haem, flavin	88
Nitrate reductase	Escherichia coli	770,000	4	(Fe, S)	85
		320,000	1.5	20 Fe, 20 S	135
Xanthine dehydrogenase	Chicken liver	300,000	2	8 Fe, 8 S, 2 Flavin,	122, 136
Xanthine dehydrogenase	Micrococcus lactilyticus	250,000	2	8 Fe, 8 S, 2 Flavin	137
Xanthine oxidase	Cow milk	275,000	2	8 Fe, 8 S, 2 FAD	107
Aldehyde oxidase	Rabbit liver	270,000	2	8 Fe, 8 S, 2 FAD (1 or 2 coenzyme Q?)	118, 119
Aldehyde oxidase	Hog liver	270,000	(0.7–1.1?)	8 Fe, 8 S, 2 FAD	119
Sulfite oxidase	Bovine liver	110,000	2	2 heme	127, 128
Sulfite oxidase	Chicken liver	(108,000)	1.6	2 heme	129

plants (18), and animals (19), with essentiality arising from its incorporation into enzymes. In microorganisms the enzymes of dinitrogen fixation, nitrate reduction, and xanthine dehydrogenation require Mo, as do the nitrate reductases found in plants. Many plants live symbiotically with microorganisms and/or depend on those in the soil to supply them with fixed nitrogen. Through this chain the production of usable nitrogen for many plants and for the animals which consume them is dependent on Mo. The mammalian enzymes, xanthine oxidase, aldehyde oxidase, and sulfite oxidase, also require Mo for activity. The last may be crucial for humans (20), and rats deficient in sulfite oxidase were shown to be extremely susceptible to poisoning by SO_2(g) or sulfite (21). The well-characterized Mo enzymes and their properties are listed in Table I. A number of recent reports describe additional Mo enzymes and as some of these are confirmed this list will grow.

We present here only a brief account of the recent work on Mo enzymes with emphasis on those points for which there seems general agreement or clearcut interpretation and which bear on the role of Mo. The juxtaposition of some biochemistry with the inorganic chemistry, which is the main theme of this review, adds some perspective to both fields. Investigation or contemplation of the intricacies of Mo enzymes must be carried out in full view of the constantly increasing knowledge of Mo coordination chemistry. Moreover, many recent studies of Mo coordination chemistry (e.g., Mo-flavin complexes, Mo-dinitrogen complexes, and Mo-cysteine complexes) have been initiated to the accompaniment of claims of biological relevance. While this may or may not be the case, some fascinating chemistry has been discovered in the process.

A. Molybdenum Enzymes

1. Nitrogenase

A large number of microorganisms "fix nitrogen," that is, they remove dinitrogen from the atmosphere or soil and reduce it to ammonia (22, 23). These include free living species (such as *Azotobacter vinelandii*) and symbiotic bacteroids (such as *Rhizobium*, which live in soybean root nodules). Recently, bacteria living in termites (24), shipworms (25), and even man (26) have been shown to fix nitrogen. The bacterial N_2 fixers span from aerobes (such as *Azotobacter*) through facultative aerobes (e.g., *Klebsiella pneumoniae*) to strictly anaerobic species (such as *Clostridium pasteurianum*). The subjects of biological dinitrogen fixation and nitrogenase action have recently been reviewed (22, 23, 27–36). The properties of some of these enzymes are listed in Table I.

Although earlier work indicated to the contrary, the most recent results

(30) reveal great similarities in the nitrogenases from the various sources. They are uniform in their requirements for reductant, using ferredoxin, or flavodoxin *in vivo*, and these or dithionite ($S_2O_4^{-2}$) *in vitro*. All nitrogenases require ATP which in the presence of Mg^{+2} is hydrolyzed to ADP and phosphate during turnover. Furthermore, as more highly purified samples have become available, the similarity in chemical composition and physical properties of the proteins from various sources has increased.

Nitrogenase is made up of two distinct and separately isolated proteins, an Fe-S protein (also called fraction 2, and azoferredoxin) and an Mo-Fe protein (also called fraction 1, and molybdoferredoxin). The former is a lower molecular-weight protein (45,000–65,000) consisting of two apparently identical subunits and containing four Fe and four acid-labile sulfides per mole. The Mo-Fe protein has two Mo atoms (although earlier claims had one Mo and some current values fall around 1.3) per mole as well as between 18 and 36 Fe's and a comparable amount of acid-labile sulfide. Its molecular weight (M.W.) is around 210,000, and the four subunits have been reported either as identical or as two different types (with two of each type). The nitrogenase enzyme is only catalytically active when both proteins, the reducing agent, ATP, and Mg^{+2} are present. Under these conditions, dinitrogen, acetylene, and other substrates (see Section II.C) are reduced, while in the absence of these substrates H_2 is evolved. In D_2O, acetylene is found to be stereospecifically reduced to *cis*(dideutero)ethylene. Hydrazine has been shown to be a substrate at high pH, where it is a neutral molecule (37).

Carbon monoxide inhibits N_2 fixation as well as the reduction of all other substrates, but it does not inhibit ATP-dependent H_2 evolution. H_2 inhibits dinitrogen reduction, however none of the other substrate reactions are inhibited. In the presence of N_2, H_2, and D_2O or N_2, D_2, and H_2O, active enzyme produces HD (previously called HD exchange). This latter reaction and the H_2 inhibition of N_2 reduction have been cogently interpreted (37) as evidence for a diimide-level intermediate which reacts with H_2 to produce $2H_2$ and N_2 (or with D_2 to produce 2HD and N_2). Thus H_2 inhibition of N_2 reduction and HD production may represent the same chemical process (interception of a bound N_2H_2 intermediate by H_2 or D_2). Both nitrogenase proteins are extremely sensitive to oxygen, with pure samples being irreversibly inactivated by O_2.

Recent intensive work (30, 38–47) has led to general conclusions on the electron transfer sequence. The evidence points to the reductant (ferredoxin or $S_2O_4^{-2}$) first reducing the Fe-S protein, which concommitant with ATP hydrolysis reduces the Mo-Fe protein. The Mo-Fe protein, in turn, reduces the substrate or evolves H_2 in the absence of substrate. EPR indicates that in a reduced state (but not in the fully reduced state capable of reducing substrate) the Mo–Fe protein shows a unique EPR signal below 15°K, with g values near 4.3, 3.7, and 2.01 (40, 41, 44, 48). Mössbauer results (41) and the slight broadening of the $g = 2.01$ (42, 48) component in the ^{57}Fe-substituted enzyme suggest that this signal arises from an Fe site. The ^{95}Mo-substituted enzyme has a signal identical to that of an enzyme containing Mo in natural abundance. This type of resonance is characteristic of a $S = 3/2$ system, which is an uncommon spin state for iron. On the other hand, monomeric Mo(III) is usually (see Section VI) a 3/2 spin system, and a study of Mo$(acac)_3$ in Al$(acac)_3$ reveals g values of 4.3, 3.5, and 1.9 with no Mo hyperfine splitting (49). Aside from this faint possibility of having observed Mo by EPR, there is no other spectroscopic evidence concerning the state or even the presence of Mo in nitrogenase. In fact, except for analytical data on the Mo-Fe protein, and the fact that nutritional studies (e.g., Ref. 50) show that Mo-free protein is inactive, there is no other direct enzymological evidence for the participation of Mo in the active site(s). Both W and V have been substituted for Mo in nitrogenase by growing organisms in appropriate media (30, 50–55). While the W-substituted enzyme is definitely inactive (52), there is controversy as to whether the V-substituted enzyme has some small activity (23, 51, 53–55). From the enzymological studies it is safe to say that not much is known about the role of Mo. It is not even certain that Mo is directly involved with substrate activation or reduction.

Model studies (56–72), on the other hand, have shown that certain Mo complexes can mimic some (but not all) of the catalytic functions of the enzyme albeit at greatly reduced levels of activity, especially for N_2 fixation. Although some of these activities have also been simulated by Fe complexes (73), generally the Mo-containing systems seem more closely analogous to the enzyme. Unfortunately, the detailed nature of the reacting species in many of these "models" is unknown and even the oxidation states are uncertain (7). The chemistry of some of these models is discussed in Section IX.

Suggestions (16, 29, 30, 36, 74) for the role of Mo in nitrogenase are therefore at a very speculative stage. Most ideas involve its use as the electron donor to substrate and relegate to Fe the function of electron transfer and/or storage. Since a total of six electrons are delivered to N_2, a molybdenum site with two Mo(III) atoms could deliver these if both were oxidized to Mo(VI) in the process. On the other hand, nitrogenase reduces a variety of substrates ($2H^+$, C_2H_2, etc.) by two-electron reactions and there is indirect evidence for the intermediacy of N_2H_2 and N_2H_4 in dinitrogen reduction. The suggestion has been offered that a dinuclear site is present with one metal involved in N_2 binding (possibly iron) and a second metal (presumably Mo) involved in transferring electrons to the substrate (28, 29, 74). The possibility has also been noted (74) that like the site in xanthine oxidase the Mo site in nitrogenase may be involved in proton transfer as well as electron transfer to substrate. In this respect, it may be significant that the Mo-Fe protein EPR signals are pH dependent (52).

The role of ATP in nitrogenase activity presents an intriguing question: How is the free energy of hydrolysis of ATP transduced into increased reducing ability (either kinetic or thermodynamic) of the Mo-Fe protein? A possible answer lies in ATP inducing a conformational change in the Fe–S protein (30, 39, 40, 75) which decreases its reduction potential to around −0.40 V. Indeed, experiments have shown (40, 76) that the ATP-Mg^{+2} complex binds to the Fe–S protein (75, 76) and lowers its reduction potential (77). Significantly, to date, no other reducing agents (no matter how strong) have been successful in reducing the Mo–Fe protein to its catalytically active state. Thus other suggestions for ATP utilization involve more direct and specific involvement with the active site. It has been suggested that ATP may aid in the removal of oxo groups from molybdenum, thus creating an open site for substrate binding or reaction (29, 30). Nonoxo molybdenum complexes (as the coordination chemistry summarized in this review shows) are characteristics of the lower oxidation states of Mo, and removal of oxo(s) could facilitate reduction. When ATP is hydrolyzed above pH 7, protons are liberated. This acid could protonate the oxo and allow it to leave the Mo coordination sphere as a hydroxo or water. Alternatively, direct phosphorylation of Mo–O by ATP could lead to an Mo–OPO_3 linkage whereupon the phosphate group could dissociate from Mo, leaving an open site. The coupling of this phosphorylation with electron transfer (78) would provide a natural explanation for the timing of ATP hydrolysis by the enzyme and would be the microscopic reverse of an oxidative phosphorylation.

Work on the model systems (60, 68) has shown that ATP enhances C_2H_2 reduction, but other studies (72, 79) have shown that contrary to the enzymic systems ATP is not hydrolyzed in the process. Rather, the acidity of ATP itself has been implicated as the factor responsible for enhanced reduction. The same effect has also been produced with sulfuric acid (72). Some recent work (65a) indicates that both the acidity of ATP and its Mo assisted hydrolysis may be involved in the stimulatory affect of ATP in the model systems.

2. *Nitrate Reductase*

Nitrate reductase has been isolated from a wide variety of microorganisms (80, 81) and higher plants (80, 82). This enzyme catalyzes the reduction of NO_3^- by NADH, NADPH, or other reductants (83). The nitrate reductases from various sources differ markedly in molecular weight [160,000 (84) to 770,000 (85)], constitution, and reactivity, but all of the more highly purified varieties contain Mo (80). Flavin nucleotides, Fe-S groups, and/or *b*-type cytochromes are commonly, but not always, present (see Table I). Although nitrate reductase has not been studied in as much mechanistic detail as either nitrogenase or xanthine oxidase, the electron transfer sequence of the enzyme from *Neurospora crassa* (M.W. = 230,000) is believed (86) to be represented by Eq. 1.

$$\mathrm{NADPH} \longrightarrow \mathrm{FAD} \longrightarrow \mathrm{Cyt}_b \longrightarrow \mathrm{Mo} \longrightarrow \mathrm{NO_3^-} \tag{1}$$

Significantly it is the Mo site of the enzyme that interacts with the substrate nitrate.

EPR studies (87) on the enzyme from *Micrococcus denitrificans* (84) reveal a resonance with $g = 1.985$ and $g = 2.045$, which is attributed to Mo(V). It is not clear how (or even if) this state of Mo fits into the catalytic sequence, but analogy with xanthine oxidase [which can reduce NO_3^- (88)] may indicate that it is an intermediate state. The g values are somewhat high, and if indeed this is an Mo signal (the Mo hyperfine interaction is not seen) then it may indicate that a number of sulfurs are coordinated to Mo. Tungsten has been substituted for Mo by growing plants in high-W/low-Mo media (89, 90), but the tungsten-substituted enzyme is inactive. Vanadium could not be incorporated (90).

Model studies (91–96) have shown that aqueous molybdenum complexes are capable of catalyzing the reduction of nitrate (although unlike the isolated enzymic system, the reactions often proceed further than NO_2^- yielding NO, N_2O, or NH_4^+, depending on conditions). The reduction of NO_3^- by Sn(II), catalyzed by molybdate in acid solution, has been interpretated (92, 93) in terms of Mo(IV) as the active reductant, as have polarographic experiments (91). These and other studies (95, 96) have also been interpreted (94–96) in terms of monomeric Mo(V) being the active reductant. Any conclusions drawn from these studies are limited by the lack of detailed knowledge of the species present in the solutions.

In very recent work (96a, 96b) monomeric mono oxo Mo(V) species have been used in nonaqueous solvents to reduce NO_3^- to NO_2 as the initial product. The NO_2 formed in these systems disproportionates to NO_2^- and NO_3^- if water is present. The NO_2^- so formed is reducible to NO by the same Mo(V) systems (96b). It would thus appear that monomeric Mo(V) in a nonaqueous environment is required to prevent the further reduction of NO_2 formed. In another recent study (96c) it was shown that Mo(III) in the form of $Mo(H_2O)_6^{+3}$ in p-toluenesulfonic acid reduces NO_3^- to NO_2^- in a quantitative reaction. These results may harbinger future progress as more detailed data become available from both biological and inorganic systems.

The complexation and reactions of Mo with flavins have also been studied in solution (97–101). This area (as well as nitrate reductions) has been reviewed by Spence (7) and will not be discussed here. Recently (102), the isolation of Mo(IV) and Mo(V) complexes of flavin derivatives has been reported, although the detailed structures remain uncertain (102a).

3. *Xanthine Oxidase and Xanthine Dehydrogenase*

Xanthine oxidase and dehydrogenase catalyze the oxidation of xanthine (structure **1**) to uric acid (structure **2**) (103–105).

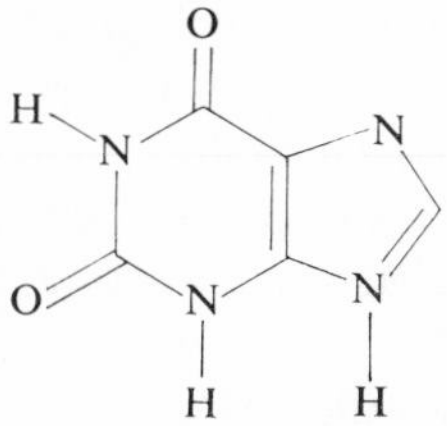

O H N N O N N H H

1

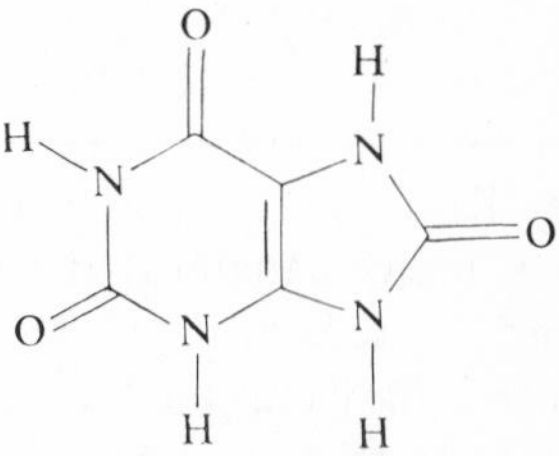

O H H N N O O N N H H

2

While the oxidase uses O_2 as oxidant, the dehydrogenases use NAD^+ or ferredoxin as electron acceptor. Both can use other oxidants as well (e.g., cyt_c, $Fe(CN)_6^{-3}$). The oxygen incorporated in the product derives from H_2O and not O_2 (106), even when O_2 is the oxidant (i.e., these are not oxygenases). Xanthine oxidase is by some margin the best characterized of the Mo enzymes, both with respect to composition and mechanism. The subject has recently been reviewed (104, 105).

The oxidase enzyme is readily isolated from cows' milk as well as from other mammalian sources (such as liver). It contains two Mo atoms, two FAD moieties, two Fe_4S_4 clusters, and has a molecular weight of 275,000 (104, 105). The linear electron transfer sequence (Eq. 2)

$$\text{xanthine} \longrightarrow \text{Mo} \longrightarrow \text{FAD} \longrightarrow Fe_4S_4 \longrightarrow O_2 \qquad (2)$$

was favored by earlier work, but more recently (106–110) it has become apparent that the actual situation is more complex. It appears that this enzyme (and perhaps the others discussed here) has multiple sites for electron ingress and egress. Furthermore, there seems to be an equilibrium distribution of electrons between the various carriers within the enzyme, and intramolecular electron transfer is rapid and not rate limiting. In view of these refinements, the xanthine binding and reduction site remains strongly implicated as the Mo site. Cyanide is known to inactivate xanthine oxidase

(111), and it has been shown that this involves abstraction of S from the protein (as SCN^-), leading Massey (109) to suggest a persulfide nucleophile for xanthine and Bray (112) to suggest a persulfide ligand on Mo. The molecule alloxanthine (structure **3**) (formed by xanthine-oxidase oxidation of allopurinol) is a potent inhibitor of xanthine oxidase (113) and binds

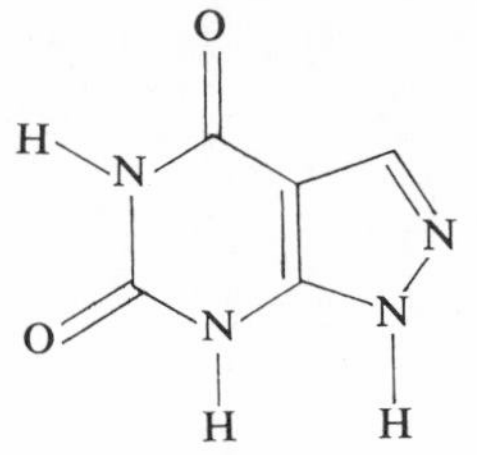

3

to the fully reduced form of the enzyme. An Mo(IV) state has been implicated in the binding of this inhibitor (113).

The Mo oxidation states have received much attention, and an Mo(IV), Mo(V), Mo(VI) set seems to be consistent with all verified data on active enzyme. The resting enzyme appears to contain Mo(VI), which is reduced to Mo(IV) during substrate oxidation. A sequence of two one-electron transfers through the FAD and/or Fe_4S_4 components to oxidant serves to reactivate (i.e., reoxidize) the Mo site to Mo(VI). The intermediate Mo(V) state is identified by its EPR as determined using rapid-freezing techniques (105). The g and A values (see Table II) have been used to suggest sulfur binding of Mo in an octahedral coordination (but see Section IV for comments). Of key importance is the observation of superhyperfine splitting from a single proton in the Mo(V) signal (105, 114–116). The proton in question is labile, but if 8-deuteroxanthine is used as a substrate the signal initially appears in its deutero form (105). This and other evidence is consistent with a coupled transfer of two electrons and one proton to the Mo site.

The question of the position of the proton on the enzyme remains open, and a model in which it is attached to an Mo ligand (74, 78, 108) has been suggested. The possibility of direct hydride transfer to Mo has also been considered (108, 109). However, the ^1H splitting observed is nearly isotropic, which is inconsistent with the presence of an Mo–H bond unless, by coincidence, the true isotropic splitting is exactly one-half the smaller component of the anisotropic splitting and of the same sign (109). In this case the a_H values would appear to be isotropic, although actually the components would differ in sign. (Incidentally, solution spectra are of no help in this case due to slow protein tumbling.) Furthermore, direct hydride transfer to Mo(VI)

TABLE II
Electron Paramagnetic Resonance of Molybdenum in Enzymes[a]

	$g_\perp$		$g_\parallel$		$A_\perp(^{95}Mo)$		$A_\parallel(^{95}Mo)$		$a_\perp(^1H)$		$a_\parallel(^1H)$		
Enzyme	g_x	g_y	g_z	g_{ave}	A_x	A_y	A_z	A_{ave} (^{95}Mo)	a_x	a_y	a_z	a_{ave} (^1H)	Refs.
Xanthine oxidase (very rapid)	1.951	1.956	2.025	1.977	37	24	41	34					105, 157
Xanthine oxidase (rapid, complex I)	1.964	1.969	1.989	1.974			64		12	12	12	12	105, 115, 116
Xanthine oxidase (CH_3OH inhibited)	1.953	1.977	1.989	1.973	57(?)	25	57(?)	46.3	5.6	3.9	4.4	4.6	105, 156
Aldehyde oxidase				1.97									122
Sulfite oxidase (pH = 7.5)	1.968		2.000	1.979	45.8		62.5	51.4	10		12	11	128
Sulfite oxidase (pH = 10)	1.984	1.961	1.950	1.965			55						128
NADH dehydrogenase	1.948	1.977	2.018	1.981	48	47	52	49					138
Nitrate reductase	1.985		2.045	2.005									87

[a]Hyperfine splittings (A and a) are expressed in gauss.

requires an Mo(VI) hydride bond which remains intact upon one-electron oxidation. Since this leaves the single unpaired electron in the Mo–H bond, a much larger a_H would be predicted. In light of these considerations an Mo hydride is highly unlikely.

A mechanism (74) involving proton transfer to a ligand coordinated to Mo is strongly supported by the evidence (104, 105, 108). In this model the timing of proton transfer to accompany electron transfer is attributed to the increased basicity of coordinated ligands in the Mo(IV) state. The Mo(VI) state would have a deprotonated N (or O) ligand that, upon reduction of Mo from (VI) to (IV), becomes basic and aids the extraction of the 8-proton from xanthine. This process may occur in conjunction with nucleophilic attack on the 8-carbon by an enzyme persulfide as suggested by Massey et al. (108, 109). The process is pictured in Fig. 1. Additional support for this model

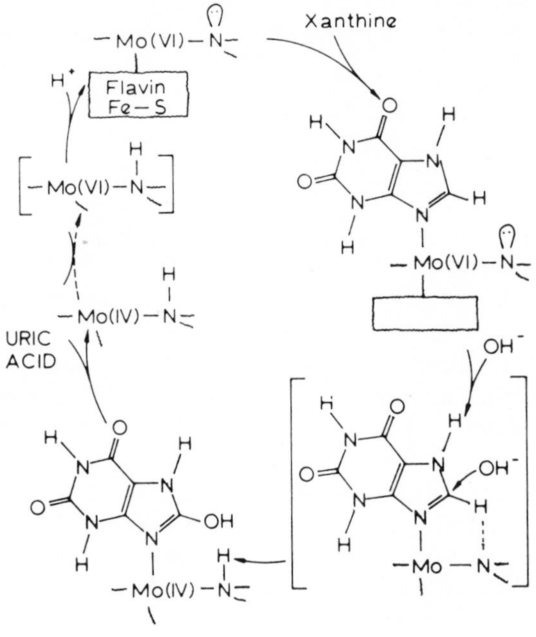

Fig. 1. Proposed cycle for the action of xanthine oxidase (74, 78). (In this scheme, the nucleophile is OH^-, but evidence has been presented (111) that a persulfide on the protein may be the initial nucleophile, with subsequent attack by OH^- producing the final product.)

comes from the strong binding of alloxanthine to the Mo(IV) state of the enzyme. In this state the ligand coordinated to Mo in the enzyme is protonated, and alloxanthine, having a nitrogen in the 8-position, would be expected to strongly bind to this site using both Mo–N and hydrogen bonding (structure **4**).

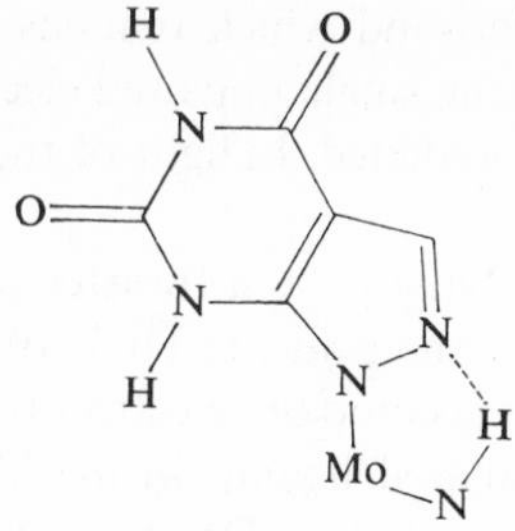

H O N O N H N N H Mo N

4

This arrangement may strongly resemble the transition state for the coupled proton—electron transfer, as many enzyme-inhibitor complexes resemble the proposed transition state for the catalyzed reaction where enzyme substrate binding is believed to be tightest (117).

To date, model systems using simple Mo complexes have not achieved success in the oxidation of xanthine. EPR studies of Mo-cysteine and thiol complexes that show EPR properties similar to those of xanthine oxidase are discussed in Section IV.

4. Aldehyde Oxidase

Aldehyde oxidase can be isolated and purified from chicken liver and other sources (104, 118, 119). Although it catalyzes the oxidation of RCHO to RCOOH by O_2, the enzyme also accepts a variety of pyrimidines as oxidizable substrates (120). Since aldehyde oxidation is done more efficiently by other (non-Mo) enzymes, it is more likely that pyrimidine oxidation is the physiological function of the enzyme. Thus the very similar enzymes xanthine and aldehyde oxidase may together constitute a set of purine and pyrimidine oxidases.

Like xanthine oxidase, aldehyde oxidase contains two Mo atoms, two FAD units, and two Fe_4S_4 clusters per molecular weight of around 300,000, with a similar electron transfer chain. A potential difference lies in the possible presence of coenzyme *Q* as an additional electron carrier in aldehyde oxidase (104). Although it has not been studied as much as milk xanthine oxidase, most work indicates gross similarities in mechanistic detail (121–125). The unhydrated form of the aldehyde (126) has been shown to be the preferred substrate for xanthine oxidase (which also oxidizes aldehydes).

5. Sulfite Oxidase

Sulfite oxidase has been purified from bovine (127, 128) and chicken liver (129) and identified in human liver, wheat germ, and *Thiobacillus thioparus* (129). The reaction catalyzed is the oxidation of SO_3^{-2} to SO_4^{-2}, although the source of O is H_2O and not O_2. Sulfite oxidase is the smallest known Mo protein at M.W. = 110,000 and contains two Mo's and two *b*-type cytochromes (127, 128). It has two identical (M.W. = 55,000) subunits. The electron transfer sequence $SO_3^{-2} \rightarrow Mo \rightarrow heme \rightarrow O_2$ is consistent with EPR studies (128). As in xanthine oxidase an Mo(V) EPR signal showing proton superhyperfine splitting is seen (see Table II). A pK_a of 8.2 has been determined for the proton responsible for the splitting (128).

A tungsten-substituted sulfite oxidase has been prepared (130) by feeding rats tungsten instead of Mo in their diets. The tungstoprotein is immunologically identical to the molybdoenzyme, but it is catalytically inactive. However, in the presence of the strong reductant $S_2O_4^{-2}$, the tungstoprotein does show an EPR signal unequivocally assignable to W(V) (130), which unlike the Mo(V) signal in this enzyme does not disappear when $S_2O_4^{-2}$ is present in excess. This observation is consistent with the greater difficulty in reducing tungsten, as compared to Mo complexes. The combined W and Mo results are consistent with Mo(VI), Mo(V), and Mo(IV) being accessible to sulfite oxidase, whereas only the higher two states are possible for the catalytically ineffective W protein. It is apparent that the three known Mo oxidases have many similarities in mechanistic detail, at least insofar as Mo is concerned.

6. *Other Enzymes*

A molybdenum-containing NADH dehydrogenase was isolated (138) from *Azotobacter vinelandii* grown in low-Fe medium, and the Mo was identified by EPR (Table II). A CO_2 reductase from *E. coli* (139) has been partially purified and the presence of Mo inferred by nutritional studies. Similar studies (140) reveal the production of formate dehydrogenase to be enhanced by Mo, and in some cases Se is also found to have a stimulatory effect. An Mo-activated hydrogenase from *Clostridium pasteurianum* has been discussed (141), but an active enzyme free of Mo has been isolated from the same source (142). EPR signals attributable to Mo have been found in complex I isolated from beef heart mitochondria (143). As some of these findings are confirmed further information on the role of molybdenum in biology will be forthcoming.

B. The Mo Cofactor

The idea of a molybdenum cofactor common to all Mo enzymes has gained considerable support in recent years. The evidence for this comes both from genetics and enzymology. The genetic evidence (144–146) will not be discussed here, but the enzymology will be briefly considered.

The work is largely that of Nason and co-workers (147–150), who utilize a mutant strain of *Neurospora crassa* which does not produce an active form of nitrate reductase. This mutant, designated nit-1, does, however, produce an inactive form of nitrate reductase when grown in the presence of NO_3^- (i.e., nitrate continues to induce synthesis of nitrate reductase albeit in an inactive form). Nason's key discovery was that this inactive nitrate reductase could be made active by mixing with acid-hydrolysis products of *any* Mo-containing enzyme. Simple Mo complexes did not suffice nor did hydrolysis products from other (non Mo) enzymes have any effect. The acid-hydrolysis products of the Mo-containing enzymes become more effective in activating the mutant enzyme if MoO_4^{-2} is added to the solutions. Furthermore, when $^{99}MoO_4^{-2}$ is used it is incorporated into the newly activated nitrate reductase only in the presence of the hydrolyzed Mo enzymes. These data are consistent with a labile Mo complex being one of the hydrolysis products of all Mo enzymes. This Mo complex (ergo cofactor) combines with the incomplete nitrate reductase of the mutant to produce a fully active enzyme. Partial isolations of the cofactor have been briefly reported (148, 150–152) and it would appear to be a low molecular-weight (<1000) complex possibly containing a polypeptide ligand.

The presence of a single Mo binding group in proteins may indicate a common or at least similar mechanism for the Mo site of various enzymes. This may involve the ability of complexed Mo to transfer protons and electrons, its proclivity to form dissociable dimers, or its ability to engage in O-atom reactions. Some of these possibilities are discussed in the next section. The question of whether mononuclear or dinuclear Mo is found in enzymes is discussed in Section IV.

C. Substrate Half Reactions

Although each of the Mo enzymes is unique, all feature internal electron transfer chains as part of their natural catalytic cycles. Being oxidoreductases, they catalyze full redox reactions, for example, the oxidation of SO_3^{-2} by O_2 or the reduction of NO_3^- by NADH. By design, however, the Mo enzymes do not cause close approach and direct-electron or atom transfer reaction to occur between these substrates, Rather, a low-activation energy pathway is provided by which electrons can flow through the enzyme from reductant to oxidant. In a sense, each enzyme is like an electrochemical cell that allows separation of anodic and cathodic reactions. In this crude analogy the internal electron transport chain of the enzyme is like the wire in the electrochemical cell (whereas the solution external to the protein is like the salt bridge of the electrochemical cell).

The molybdenum site of the enzyme is concerned externally only with a half reaction, that is, it is either the anode or the cathode. In fact, although the evidence is not quite conclusive for nitrogenase, the Mo site on each protein is the electron carrier that interacts directly with the named substrate and *not* [except in special cases (88, 100, 153)] with the physiological oxidant or reductant. In order to focus on the role of Mo we look directly at the substrate half reactions. These are listed in Table III. Nitrogenase (28, 29), xanthine oxidase (104), and aldehyde oxidase (104, 120) each have an ex-

TABLE III
Representative Substrate Half Reactions for Molybdenum Enzymes

Nitrogenase (28, 29, 158, 159):

$$N_2 + 6H^+ + 6e^- \longrightarrow 2NH_3$$
$$N_2H_4 + 2H^+ + 2e^- \longrightarrow 2NH_3$$
$$HCN + 6H^+ + 6e^- \longrightarrow CH_4 + NH_3$$
$$N_2O + 2H^+ + 2e^- \longrightarrow N_2 + H_2O$$
$$HN_3 + 2H^+ + 2e^- \longrightarrow N_2 + NH_3$$
$$RNC + 6H^+ + 6e^- \longrightarrow RNH_2 + CH_4$$
$$RCN + 6H^+ + 6e^- \longrightarrow RCH_3 + NH_3$$
$$C_2H_2 + 2H^+ + 2e^- \longrightarrow C_2H_4$$
$$2H^+ + 2e^- \longrightarrow H_2$$
$$[ATP + H_2O \xrightarrow{Mg^{+2}} ADP + P_i]$$

Nitrate reductase (83):

$$NO_3^- + 2H^+ + 2e^- \longrightarrow NO_2^- + H_2O$$

Xanthine oxidase (104, 120):

$$\text{Xanthine } (\mathbf{1}) + H_2O \longrightarrow \text{uric acid } (\mathbf{2}) + 2H^+ + 2e^-$$

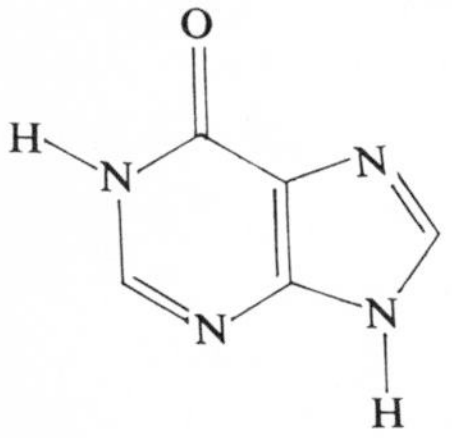

$$+ H_2O \longrightarrow \text{xanthine } (\mathbf{1}) + 2H^+ + 2e^-$$

Aldehyde oxidase (104, 120):

$$RCHO + H_2O \longrightarrow RCOOH + 2H^+ + 2e^-$$

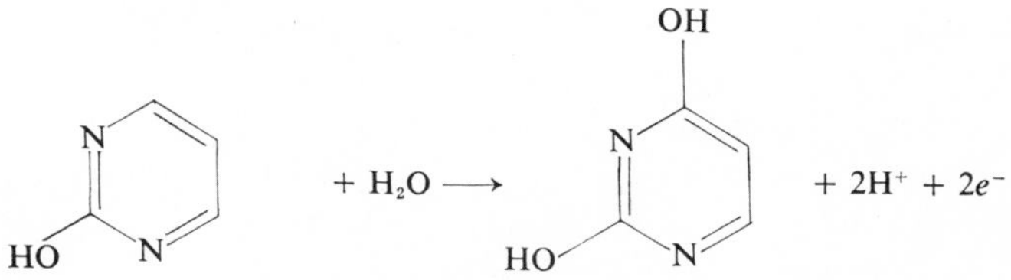

$$+ H_2O \longrightarrow \quad + 2H^+ + 2e^-$$

Sulfite oxidase (127):

$$SO_3^{-2} + H_2O \longrightarrow SO_4^{-2} + 2H^+ + 2e^-$$

tremely broad range of substrate specificity (unusual for enzymes). All of the substrate half reactions proceed formally with transfer of two electrons (or some multiple thereof) and one or two protons (or some multiple thereof) in the same direction. In addition, H_2O is sometimes added or removed. It has been suggested (74, 78) that Mo enzymes may be intimately involved with proton transfer as well as electron transfer between enzyme and substrate. The viability of this suggestion receives support from considerations of coordination chemistry.

The chemistry of Mo and other elements demonstrates that the acidity (pk_a) of coordinated ligands is strongly dependent on the oxidation state of the metal. For example, in acid solutions Mo(VI) exists as $MoO_2(H_2O)_4^{+2}$ (Section III), whereas Mo(III) is in the form $Mo(H_2O)_6^{+3}$ (Section VI). If an Mo complex or enzyme site has ligands capable of proton transfer, an increase in oxidation state drastically lowers the pK_a of those ligands. In xanthine oxidase and sulfite oxidase there is direct evidence from EPR that such an exchangeable proton is present in the Mo(V) state of the enzyme. Furthermore, the pK_a of this proton is found to be close to 8 (i.e., close to the physiological pH). The known effect of oxidation state on pK_a thus requires that in the lower oxidation state [e.g., Mo(IV)], the pK_a of this site will be very high and the site will be very basic. Similarly, in the Mo(VI) state, the pK_a must be very low and the site must be strongly acidic By this effect, proton transfer between substrate and enzyme can accompany electron transfer. This process has been termed "coupled proton-electron transfer" (74, 78), and for xanthine oxidase (and to a lesser extent the other oxidases) much evidence is compatible with the idea (108) (see discussion in Section II.D.3). Although the experimental data bearing on the Mo states in reductases are much sketchier, the same process (or more precisely, its microscopic reverse) has been suggested here (74). If these hypotheses are verified, they suggest that the chemical basis for the biological role of Mo involves its ability to couple proton and electron transfer reactions.

There are, of course, potential alternative mechanisms for all of the Mo enzymes. Except for nitrogenase, each enzyme can formally be viewed as adding an O atom to (or removing an O atom from) substrate. It is also well known that Mo complexes can transfer an oxo group [e.g., $MoO_2(R_2dtc)_2 \rightarrow MoO(R_2dtc)_2$; see Section III]. The postulation of this activity for Mo enzymes is thus an attractive alternative hypothesis (106, 154, 155). However, at present the evidence for O-atom transfer is not strong, although the process may occur in the special case of pyridine *N*-oxide oxidation of xanthine catalyzed by xanthine oxidase (106). Since the evidence for proton transfer is strong for xanthine oxidase, it is difficult (but not impossible) to imagine an Mo site both donating an O atom and accepting a proton. This becomes especially difficult if the idea of an active nucleophilic persulfide group is adopted (109). Future work should provide data to sort out these possibilities.

III. THE CHEMISTRY OF MOLYBDENUM(VI)

Molybdenum occurs naturally in the hexavalent state in $PbMoO_4$ (1), $MgMoO_4$ (1), and in sea water (160). The coordination chemistry of Mo (VI) is almost entirely that of oxo complexes. Nonoxo complexes, although rare, include such interesting compounds as MoS_4^{-2} (161), $Mo(O_2)_2^{-2}$ (162), $[Mo(O_2C_6H_4)_3]_2$ (162), MoF_7^- (164), and MoF_8^{-2} (164), as well as the simple octahedral species MoF_6 (3) and $MoCl_6$ (165). Some dithiolene and related complexes which have formal Mo(VI) oxidation states are discussed in Section VIII, and nitrido compounds are considered in Section IX.

It seems clear that the Mo(VI) state is involved in the activity of biological molybdenum. It is in the form of molybdate salts that Mo is usually added to culture media (17), soils (18), or animal feeds (19). Tungstate, added to these nutrients, is an effective antagonist for Mo, with W having been substituted for Mo in some enzymes (52, 89, 90, 130). These tungstoproteins are structurally intact and (where studied) immunologically identical to the corresponding molybdoenzymes (130), but display no catalytic activity. In view of the greater difficulty in reducing WO_4^{-2} it seems clear that W(VI) is present in the tungstoproteins, and its strong affinity for the Mo site suggests that this site binds Mo(VI) in the native enyzme. Furthermore, Mo(V) (105), and in one case W(V) (130), has been identified as the state in the Mo oxidases that appears (in most cases) upon *reduction* of the resting enzyme with very small amounts of substrate (105) or reductant (130). Thus Mo(VI) is a very likely participating state for at least some of the Mo enzymes.

In this section we start with the structural systematics of Mo(VI) complexes and follow with a discussion of simple molybdates and their thio and seleno derivatives. A treatment of halides and oxyhalides is followed by a discussion of other complexes containing the cis-dioxo grouping. A final section serves as a potpourri covering some reactions, cis-trioxo complexes, and polydentate ligands. Some compounds which are not discussed explicitly in the text are listed in Table V.

A. Structural Studies

The structural chemistry of Mo(VI) is dominated by the presence of oxo groups bound to Mo. These oxygen atoms utilize sigma and pi donation to produce a strong (multiple) Mo–O bond. In our discussion an oxo group which is not used to bridge to other Mo's is designated O_t (for terminal), whereas bridging oxo is designated O_b. A list of Mo(VI) complexes whose structures have been investigated crystallographically is given in Table IV.

TABLE IV
X-ray Crystallographic Structural Studies on Mo(VI) Complexes

Complex	Refs.
K_2MoO_4	165
$MoO_3(dien)$	168
$Na_4[MoO_3(edta)MoO_3]\cdot 8H_2O$	169
$K_2[Mo_2O_5(C_2O_4)_2(H_2O)_2]$	170
$(NH_4)_2[Mo_2O_5(dmf)_4Cl_2]\cdot 2H_2O$	171
$(NH_4)_2[Mo_2O_5(O_2C_6H_4)_2]\cdot 2H_2O$	172
$Mo_2(O_2C_6Cl_4)_6$	163
$MoO_2(PhCOCHCOPh)_2$	173
$MoO_2(acac)_2$	174
$MoO_2(oxine)_2$	175
$MoO_2(Et_2dtc)_2$	176
$MoO_2(Pr_2dtc)_2$	177
$K_2[MoO_2(O_2C_6H_4)_2]$	177
$Rb_2[MoO_2F_4]$	178
$K_2[MoO_2F_4]\cdot H_2O$	179
$MoO_2Cl_2(dmf)_2$	180
$H[MoO_2((OCH_2CH_2)_3N)]$	181
$MoO_2(bipy)\,Br_2$	184
$K_2[MoO(O_2)F_4]\cdot H_2O$	183
$(NH_4)_3[MoO(O_2)F_4]F$	184
$K_2[MoO(O_2)_2(C_2O_4)]$	185
$MoO(O_2)_2(OP(NMe_2)_3)(H_2O)$	186
$MoO(O_2)_2(OP(NMe_2)_3)(py)$	186
$Zn(NH_3)_4[Mo(O_2)_4]$	161
$K_2[Mo_2O_3(O_2)_4(H_2O)_2]\cdot 2H_2O$	187
$(pyH)_2[Mo_2O_3(O_2)_4(H_2O)_2]$	188, 189
$(pyH)_2[Mo_2O_2(OOH)_2(O_2)_2]$	188, 189

The well-known molybdate ion MoO_4^{-2} containing four terminal oxo groups is the dominant form in Mo(VI) solutions above pH 7. It has a T_d structure, with Mo–O_t = 1.76 (1) Å found in K_2MoO_4 (166). Related chalcogenide derivatives MoS_4^{-2} and $MoSe_4^{-2}$ also have T_d structures, and mixed oxo-thio, oxo-seleno molybdates have structures based on tetrahedral coordination of Mo (161). The tetrahedral 4 coordination is a characteristic of Mo species isolated from *basic* solutions. As the pH is lowered, an increase in coordination number from 4 to 6 occurs and octahedral Mo(VI) predominates. In the absence of other ligands isopolymolybdates are formed and can be isolated. In the presence of a variety of cations (e.g., Ce^{+4}, Al^{+3}, and Co^{+3}) or anions (e.g., PO_4^{-3}, AsO_4^{-3}, and SO_4^{-4}) the heteropoly molybdates are obtained. Both iso- and heteropolymolybdate anions are found to contain exclusively 6-coordinate MoO_6 polyhedra joined by the sharing of corners or edges (but not faces) (1, 167). Their structural aspects have been reviewed (1, 167) and will not be discussed here.

1. Monomeric Complexes

In oxo complexes of hexavalent molybdenum, MoO^{+4}, MoO_2^{+2} and MoO_3 structural units are found. By far, the largest number are 6-coordinate complexes containing the MoO_2^{+2} group and many have been structurally investigated (173–182). Invariably, the basic octahedral structure contains two short $Mo–O_t$ bonds in cis positions. The $Mo–O_t$ distances with but few (perhaps understandable) exceptions lie at 1.67 ± 0.05 Å, and O–Mo–O angles range from 102 to 114° (with the exception of the $MoO_2F_4^{-2}$ structures discussed below). The large deviation from the nominal 90° angle is attributable to bond—bond repulsion between the multiply bound oxo ligands. Despite this deviation the $O_t - O_t$ distance is short at 2.66 ± 0.07 Å, which is less than the O–O Van der Waals contact distance of 2.80 Å.

The structure of the $MoO_2F_4^{-2}$ ion has been determined in both Rb^+ (178) and K^+ (179) salts. In the $Rb_2[MoO_2F_4]$ structure (178) a disorder between O and F is present, and quite atypical Mo–O bond lengths are reported. $K_2[MoO_2F_4] \cdot H_2O$ has the cis-dioxo octahedral structure, with deviations from C_{2v} symmetry probably caused by H bonding to the water of crystallization (179). The $Mo–O_t$ averages 1.71 Å, and while the Mo–F distance trans to $Mo–O_t$ averages 1.97 Å those cis to $Mo–O_t$ (and trans to each other) average 1.93 Å. The small structural trans effect is not significant in this study, but it is probably real and is unmistakable in other structural studies discussed below.

The complex $MoO_2Cl_2(dmf)_2$ also contains solely monodentate ligands and is found to have cis oxo's, cis dmf's (O bound), and trans chlorines (180). The Cl–Mo–Cl angle at 161.3° is far removed from the nominal 180°, and this is attributed to repulsion between $Mo–O_t$ and Mo–Cl groups. The complex $MoO_2(bipy)Br_2$ also has "trans" halides, and again the nonlinear "trans" Br–Mo–Br angle of 159.7° is attributable to repulsive interaction with the oxo groups (182).

The structural trans effect is well illustrated by those MoO_2L_2 complexes in which two symmetrical bidentate ligands are present. For example, in $MoO_2(PhCOCHCOPh)_2$, illustrated in Fig. 2 (173), the Mo–O bonds trans to $Mo–O_t$ average 2.17 Å, whereas those cis to $Mo–O_t$ (and trans to each other) average 1.99 Å. In $K_2[MoO_2(O_2C_6H_4)_2] \cdot 2H_2O$ (171) the corresponding distances are 2.15 and 2.05 Å, whereas in $MoO_2(Et_2dtc)_2$ (175) the

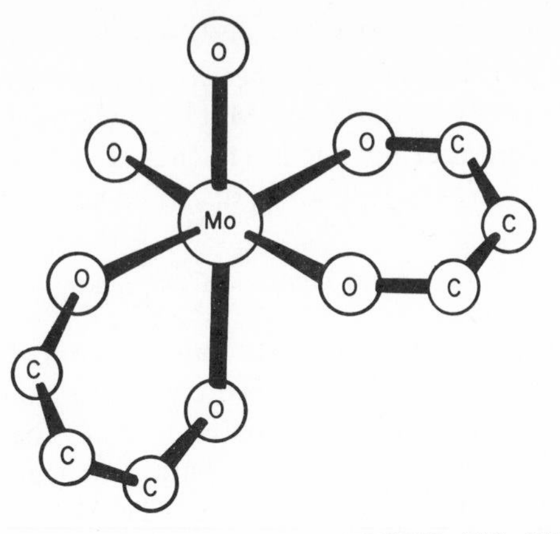

Fig. 2. The structure of $MoO_2(RCOCHCOR)_2$ (173, 174).

Mo–S trans to Mo–O_t is 2.63 Å and that cis to Mo–O_t is 2.44 Å. The trans effect observed in these structures is a general property of the Mo–O_t linkage, and this trans-bond weakening will be noted many times in the chemistry of Mo(VI), Mo(V), and Mo(IV).

With the unsymmetrical bidentate ligands 8-hydroxyquinoline (oxine H) the complex $MoO_2(oxine)_2$ (175) displays a structure with C_2 symmetry having quinoline nitrogens trans to Mo–O_t. With the tripod ligand tris (hydroxyethyl)amine a cis-dioxo structure obtains (181) for $H[MoO_2((OCH_2CH_2)_3N)]$, with N trans to one of the O_t's. The Mo–O bond length trans to the second Mo–O_t at 2.34 Å is much longer than the averaged value of the other nonoxo Mo–O's which lie trans to each other at Mo–O = 1.94 Å. In this structure the Mo–O_t length averages at the atypically long value of 1.78 Å, and the position of the proton is not specified (181).

A second structural unit found in a monomeric oxo complex is MoO_3. The complex MoO_3(dien) has the 6-coordinate octahedral structure (168), shown in Fig. 3a, with the cis-trioxo arrangement, an Mo–O_t distance averaging 1.74 Å, and O–Mo–O angles averaging 106°. This unit has not proved to be common, and, in fact, the complex MoO_3(dien) shows little stability in solution (see Section III.E). A second 6-coordinate monomeric structure containing the MoO_3 unit has been found for $MoO_3(nta)^{-3}$ where one of the ligand carboxylates is uncoordinated (168a).

The MoO^{+4} structural unit has recently been found by Weiss and coworkers (177a) in the complex ion $MoO(Et_2dtc)_3^+$ and in the neutral complexes $MoOCl_2(Et_2dtc)_2$ and $MoOBr_2(Et_2dtc)_2$. The complexes each have seven-coordinate pentagonal-bipyramidal structures with oxygen occupying one apical position. In the case of the halo complexes the halogen occupies the other apical site. The Mo–O_t bond lengths of 1.684 (6), 1.701 (4) and 1.656

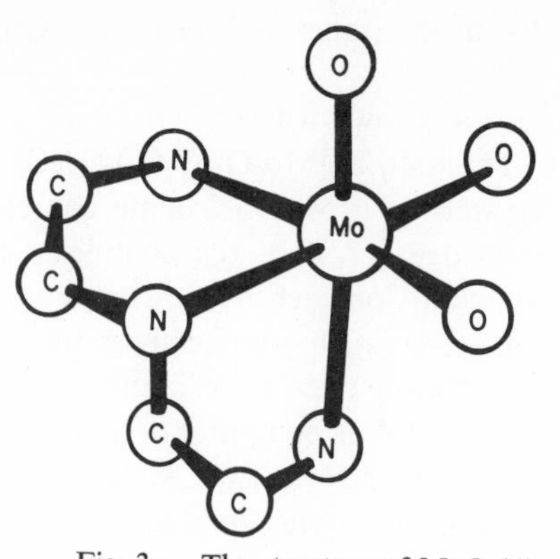

Fig. 3*a*. The structure of MoO_3(dien) (168).

(5) in the three structures are not significantly different from those of the 6-coordinate dioxo complexes. The halo complexes are formed either by reaction of hydrohalic acid with $MoO_2(Et_2dtc)_2$ or by the oxidative addition of Cl_2 or Br_2 to the Mo(IV) complex $MoO(Et_2dtc)_2$ (177a, 177b).

2. *Dimeric Complexes*

The *cis*-MoO_3 unit with virtually identical dimensions to that in MoO_3 (dien) is found in the dimeric complex $Na_4(O_3Mo(edta)MoO_3)\cdot 8H_2O$ (169). Here $edta^{4-}$ binds to each MoO_3 group through an amino nitrogen and two carboxylate oxygens, with the ethylenediamine bridging the two halves of the molecule. As shown in Fig. 3b, no direct Mo–O–Mo or Mo–Mo

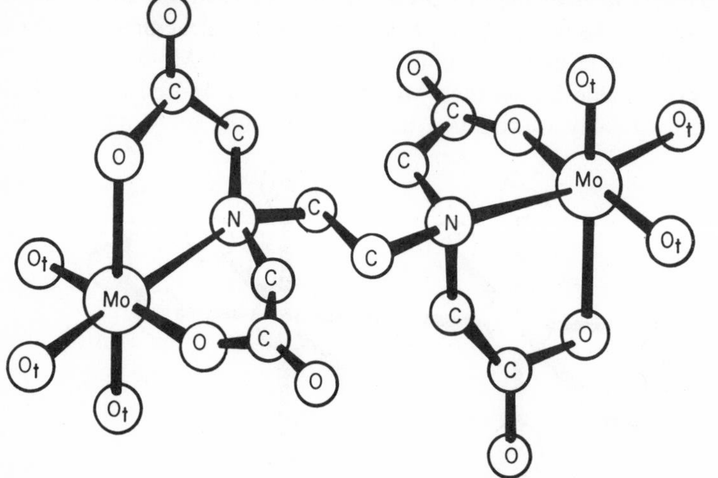

Fig. 3*b*. The structure of $MoO_3(edta)MoO_3^{-4}$ in $Na_4[MoO_3(edta)MoO_3]\cdot 8H_2O$ (169).

connection exists between the metal atoms, and the overall symmetry is quite close to C_{2h}.

Other dimeric structures are known where a linear or near-linear Mo–O–Mo linkage is present. The salt $K_2[Mo_2O_5(C_2O_4)_2(H_2O)_2]$, crystallized during attempts to grow crystals of a Mo(V) oxalate complex, contains a perfectly linear Mo–O–Mo bridge (170). In the complex anion shown in Fig. 4 the two terminal oxo ligands on each Mo are mutually cis and cis to the bridging oxo. The complex is crystallographically centrosymmetric, with one oxalate and one water completing the 6-coordination sphere on each Mo atom. The Mo–O_t distance at 1.69 Å is comparable to those found in cis-dioxo complexes. The Mo–O_b distance is 1.88 Å, whereas the Mo–OH_2 distance (trans to Mo–O_t) at 2.33 Å is quite long, again illustrating the large structural trans effect of the O_t ligand. The two Mo–O distances to the oxalate ligand at 2.09 Å (trans to O_b) and 2.19 Å (trans to O_t) indicate that O_t has a larger trans-bond weakening effect than O_b. The largest O–Mo–O bond angles are found (170) to correspond to the shortest Mo–O bond distances, which is in agreement with the notion that interligand (and/or interbond) repulsions determine the more subtle aspects of these coordination spheres.

A near-linear bridge is also found for the complex $[MoO_2Cl(dmf)_2]_2O$, obtained unexpectedly in the attempted crystallization of $MoO_2Cl_2(dmf)_2$ (171). Here the overall symmetry is C_2 with an Mo–O_b–Mo angle of 171°. The terminal Mo–O_t's with a distance of 1.68 Å are disposed cis to each other on each Mo and cis to Mo–O_b, but the $Mo(O_t)_2$ groups lie on opposite sides of the molecule and are related by the C_2 axis. The Cl's lie trans to the bridging oxygen, whereas the two O-bound dmf's on each Mo are cis to each other but trans to the two dmf's on the other Mo. The structure is similar to the binuclear oxalate and is simply related to that of $MoO_2Cl_2(dmf)_2$ by replacement of a chloride on each Mo by bridging O. Indeed the dinuclear complex is reasonably viewed as a hydrolysis product of $MoO_2Cl_2(dmf)_2$.

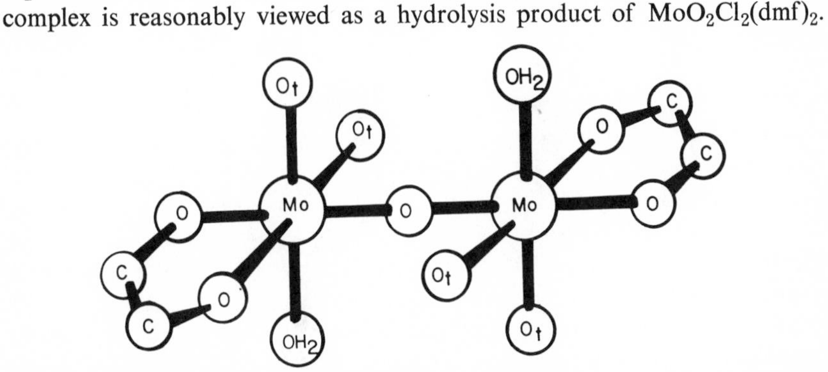

Fig. 4, The structure of $Mo_2O_5(C_2O_4)_2(H_2O)_2^{-2}$ in $Ba[Mo_2O_5(C_2O_4)_2(H_2O)_2]$ (170)

Thus in both instances in which linear Mo–O–Mo dimers have been found, the preparations were fortuitous and as yet there is no case for Mo(VI) of this linkage having been purposefully prepared and studied.

The complex $(NH_4)_2[Mo_2O_5(O_2C_6H_4)_2] \cdot 2H_2O$ was found (172) to have a triple bridge consisting of a single nonlinear oxo bridge and two bridges formed through the sharing of catechol oxygens of different ligands. The structure can be roughly considered as two octahedra sharing a face with the three bridging oxygen atoms at average distances of 1.92, 2.16, and 2.37 Å from each Mo. The catechol oxygens do not bridge symmetrically, and each catechol is more closely associated with a single Mo. The two 6-coordination spheres are completed by two terminal oxo's and one nonbridging catechol oxygen on each Mo. The Mo–Mo distance is 3.13 Å, but, of course, no Mo–Mo bonding is possible for this $4d^0$ system. Recently (172a) a similar triple bridge has been found in the complex $Mo_2O_5(PQ)_2$ (where PQ is the 9,10-phenanthrenequinone radical anion). A triple bridged structure has also been briefly reported (172b) for $(NH_4)\,[Mo_2O_5(C_6H_{11}O_6)] \cdot H_2O$ where the organic ligand is a polyanion of the sugar mannitol. Here the three bridging O atoms each symmetrically span the two Mo atoms with different Mo–O distances.

The nonoxo dimeric complex $Mo_2(O_2C_6Cl_4)_6$, prepared from the reaction of $O_2C_6Cl_4$ (*o*-tetrachlorobenzoquinone) with $Mo(CO)_6$, also has bridging catecholate ligands (163). The distorted octahedral coordination sphere of each Mo is made up of two O atoms from bridging catechols and four O atoms from two nonbridging ligands. In this case, the bridging is symmetrical and each bridging ligand has one O in each Mo coordination sphere. Remarkably, the formally similar complex $Mo(PQ)_3$ shows a mononuclear structure containing near trigonal-prismatic coordination about Mo (163a). We note that these complexes can be alternatively formulated as containing Mo(O) and quinone ligands and as such are related to the dithiolene ligands discussed in section VIII.

3. *Peroxo Complexes*

A substantial number of Mo(VI) peroxo complexes have been structurally investigated by Weiss (183, 186, 188, 189) and Stomberg (161, 184, 185, 187) and their collaborators. These complexes have side-on-bound peroxide and are found as mononuclear, dinuclear, tetranuclear (190), and heptanuclear (190) compounds. The latter two complexes are related to the isopolymolybdates and are not discussed here. The designation of coordination number in these complexes is equivocal insofar as the side-on-bound peroxide can be considered to occupy either one or two coordination sites. As has been done by the workers in the field and for the purpose of our discussion we adopt the convention that O_2^{n-} occupies two coordination sites when bound side-on.

Structurally characterized mononuclear peroxy compounds include $[Zn(NH_3)_4][MOo(_2)_4]$ (161), $(NH_4)_3[MoO(O_2)F_4]F$ (184), $K_2[MoO(O_2)F_4] \cdot H_2O$ (183), $K_2[MoO(O_2)_2C_2O_4]$ (185), $MoO(O_2)_2(OP(N(CH_3)_2)_3)H_2O$ (186), and $MoO(O_2)_2(OP(N(CH_3)_2)_3)py$ (186). The Mo in $Mo(O_2)_4^{-2}$ is dodecahedrally coordinated by eight oxygen atoms. The O–O distance is 1.55 Å, and the O–O bonds span the *m* edges (between *A* and *B* sites) of a triangular dodecahedron. The overall symmetry is close to D_{2d} making this complex isostructural to the well-known $Cr(O_2)_4^{-3}$ ion (191). Other mononuclear peroxo complexes have one $Mo–O_t$ in addition to one or two peroxo ligands. The $Mo–O_t$ bond lengths fall in the narrow range 1.63 to 1.69 Å, whereas the Mo–O (peroxo) distances span from 1.91 to 1.96 Å. The peroxo ligand is always found cis to the oxo group, and the plane determined by the Mo–O–O ring is nearly perpendicular to the $Mo–O_t$ direction. The overall coordination geometry is in each case describable as a pentagonal bipyramid with the oxo ligand in an apical position and peroxo groups in the equatorial plane. The structure of $MoO(O_2)_2(C_2O_4)^{-2}$ is illustrated in Fig. 5.

Notwithstanding the utility of the 7-coordinate designation for descriptive purposes, the structural systematics of these peroxo complexes are nicely understood by considering the O_2^{-2} ligand to take the place of an oxo ligand. For example, if the four oxo groups in MoO_4^{-2} are replaced by properly oriented peroxo ligands, the $Mo(O_2)_4^{-2}$ structure quite nearly obtains. In fact, the actual dodecahedral structure of $Mo(O_2)_4^{-2}$ is such that the midpoints of the O–O bonds form a near tetrahedral array about Mo. In the monooxo, monoperoxo compounds the cis arrangement of the oxo and peroxo ligands is reminiscent of the cis-dioxo structure. Moreover, the orienta-

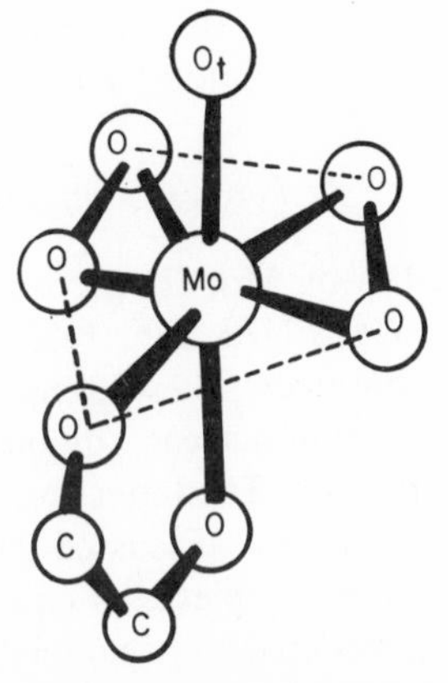

Fig. 5. The structure of $MoO(O_2)_2(C_2O_4)_2^{-2}$ (185).

tion of the O–O bond is such as to allow it to be a pi donor much in the manner of the terminal Mo–O_t. This analogy breaks down somewhat for the oxo-diperoxo complexes. Thus, although the oxo and peroxo groups are mutually cis as in MoO_3 complexes, there are only two additional coordination sites filled, making the structures more like trigonal bipyramidal rather than octahedral MoO_3 compounds.

The dinuclear complex ion $(H_2O)(O_2)_2OMoOMoO(O_2)_2(H_2O)^{-2}$ was structurally investigated in its pyH^+ (188, 189) and potassium salts (187). The monooxo bridge is nonlinear (132 and 148° in the two salts), and it is found to be cis to the Mo–O_t groups on each Mo. The two peroxo ligands on each Mo are cis to the Mo–O_t's, as in the monomeric compounds. The overall structure is describable as two pentagonal bipyramids bridged through an oxygen atom shared by their equatorial planes. The oxo and water ligands occupy apical positions (with O_t groups on the two Mo atoms oriented trans to each other), whereas the peroxides occupy the remaining equatorial sites.

The structure of $(pyH)_2[O(O_2)_2Mo(OOH)_2MoO(O_2)_2]$ shows dimers (188, 189) with two bridging hydroperoxo ligands. Each Mo is again heptacoordinate, and the structure is viewed as two pentagonal pyramids sharing an axial-equatorial edge to give a centrosymmetric structure. The peroxo ligands occupy the remaining equatorial sites while the axial position not involved in the bridge contains the terminal oxo. The Mo–O_b trans to O_t is 2.39 Å, whereas the other bridging Mo–O_b distance is 2.05 Å. The bridging and nonbridging peroxo groups both have distances near 1.47 Å, which is consistent with their peroxide formulation.

When O_2^{n-} is bound to a transition metal, the question of its formal oxidation state (and that of the metal) arises. In the compounds discussed above the Mo(VI)-peroxide formulation is always given, but in the absence of structural data an Mo(IV)-dioxygen formulation is possible. We can consider the hypothetical reaction of Mo(IV) and O_2 as an oxidative addition, with the extent of charge transfer determining the proper formulation. In the complexes discussed here the O–O distance lies in the range 1.44 to 1.55 Å (with the exception of 1.36 (3) Å being found for $(NH_4)_3[MoO(O_2)F_4]$). Comparison of these distances with 1.44 Å in H_2O_2 and 1.21 Å in O_2 indicates the Mo(VI) peroxide formulation seems quite appropriate for these compounds.

The peroxo complexes are of interest as reagents in organic synthesis (192) and as possible intermediates in molybdate-catalized peroxidation and oxidation reactions (193, 194). There is, however, no evidence that dioxygen or peroxides interact with Mo in enzymes. In fact, quite the contrary, the evidence is usually strong that in Mo oxidases the dioxygen interacts with the Fe–S, heme, or flavin parts of the protein and not the Mo site (104, 105, 128).

Recently, Weiss and coworkers (189a) have reacted $MoO_2((n—Pr)_2dtc)_2$ with H_2S and obtained the blue "disulfur" complex $MoOS_2((n—Pr)_2dtc)_2$. This complex has a 7-coordinate pentagonal-bipyramidal structure totally analogous to that found for $MoO(O_2)_4^{-2}$ Fwith the two bidentate dtc ligands replacing the four F donors. The S–S distance is 2.018 (8) Å and while this is not a sufficient basis on which to formulate the charge on S_2, an Mo(VI)-S_2^{-2} formulation is reasonable by analogy with the peroxo complexes.

4. Trends

The structural trends in oxo and peroxo Mo(VI) complexes involve the uniform tendency toward cis disposition of oxo (and peroxo) groups. This is explicable in terms of the maximum utilization of $d\pi$- (d_{xz}, d_{yz}, d_{xy} with ligands along x, y, and z) acceptor orbitals by the strongly pi-donating oxo groups. For example, two trans-disposed oxo's would be required by symmetry to share the same two $d\pi$ orbitals, whereas two cis-oxo atoms can utilize all three $d\pi$ orbitals (sharing one and each having exclusive use of another) (170).

The increase in $Mo–O_t$ bond length from near 1.68 to 1.74 Å as the number of terminal oxo ligands increases from two to three is understood in a similar manner (168). In the dioxo complex 1.5 $d\pi$ orbitals are available for bonding with each $O(p\pi)$. For the trioxo complex only one net $d\pi$ orbital is available for each oxo, since each of the three $d\pi$ levels is shared by two oxos, with each oxo sharing two $d\pi$ orbitals.

After accounting for the cis-oxo tendency the more subtle details of the coordination spheres can usually be understood in terms of interligand (and/or interbond) repulsion. Here the short metal-oxo linkage again exerts a dominant effect, with the largest bond angles uniformly found between the shortest bonds.

B. Molybdates and Thiomolybdates

In basic solution the well-known tetrahedral molybdate ion MoO_4^{-2} is present. Protonation of MoO_4^{-2} as the pH drops below 7 leads rapidly (195) to an increase in coordination number from 4 to 6. If chelating ligands are present, complex formation will often occur upon acidification. In the absence of such ligands isopolymolybdates are formed, which have exclusively 6-coordinate Mo. The increase in coordination number can be viewed as a consequence of the tendency of Mo to achieve near neutrality in charge [i.e., the electroneutrality principle (196)]. In MoO_4^{-2} four strongly sigma- and pi-donating oxo ligands suffice to neutralize the positive charge on Mo(VI). When one (or more) of these is protonated it is less effective as a donor, leaving Mo with more residual positive charge whereupon it takes up additional ligands to reapproach neutrality.

Below pH 7 the isopolymolybdates are formed according to Eqs. 3 and 4

$$7MoO_4^{-2} + 8H^+ \longrightarrow Mo_7O_{24}^{-6} + 4H_2O \quad (3)$$
$$Mo_7O_{24}^{-6} + MoO_4^{-2} + 4H^+ \longrightarrow Mo_8O_{26}^{-4} + 2H_2O \quad (4)$$

and remarkably there is no evidence for an intermediate form between the molybdate and heptamolybdate stages (195). An interesting explanation for the lack of intermediates has been advanced by Cruywagen and Rohwer (195a) based upon their conclusion that it is the *second* protonation of molybdate which causes the increase in coordination number from 4 to 6. The equilibria (197) are rapid (0.2–200 msec range) (195), and as the solution becomes more acidic, still higher polymeric forms may be present (197). Eventually, with sufficient acid ($pH \leq O$) depolymerization occurs, and in strong acids the MoO_2^{+2} ion is found appropriately coordinated by water or the anion of the acid to complete its 6-coordination sphere.

The Mo species in very dilute solutions are, however, quite different in the pH range 1 to 6. Thus while the formation constant for the heptamolybdate ion is very large (197), monomeric species may still be the stable form at very low Mo concentrations. Working at 10^{-6} to 10^{-8} *M*, Burclova et al. (198) used the ^{99}Mo radionucleide and electrophoretic migration and dialysis techniques to deduce the species present in $HClO_4$. Above pH 6.0 strongly anionic species exist (presumably MoO_4^{-2}), whereas between pH 2.5 and 6.0 $HMoO_4^-$ (or its hydrated form) is postulated as present. From pH = 1 to 2.5 a neutral species is indicated by zero-electrophoretic mobility (possibly hydrated H_2MoO_4), whereas below pH = 1 the cationic species predominate, culminating in the formation of MoO_2^{+2} in 6M perchloric acid solutions (198).

These solution studies may have biological implications. At present there is no evidence for the participation in a biological system of anything resembling an isopoly- or heteropolymolybdate. The presence of only two Mo atoms per enzyme rules against the presence of polymeric species in the molybdoenzymes. However, at this point virtually nothing is known about the absorption, transport, and storage of Mo in biological systems. In the biochemistry of Fe, where absorption, transport, and storage are better understood, the storage form (ferritin) has a hydrous iron-oxide-phosphate inner core which is a polymeric Fe complex (199). Although physiological pH is usually considered to be about 7.4, local fluctuations can occur within the organized structures of cells or tissue. Since the polymolybdates form at pH 7 and below (not far from the biological range), they remain a possibility as a storage form of Mo. On the other hand, unless Mo is concentrated from its normal level in cells or culture media ($\sim 10^{-7}M$), the formation of the polymolybdates in biological systems is unlikely.

The thio derivatives of MoO_4^{-2} have been known since the 19th century and have received a great deal of attention in the last few years, largely through the efforts of Müller and co-workers. Diemann and Müller thoroughly reviewed the field in 1972 (161), and bonding and spectroscopic aspects were considered by Müller, Diemann, and Jørgensen (200).

The ions MoO_3S^{-2}, $MoO_2S_2^{-2}$, $MoOS_3^{-2}$ and MoS_4^{-2} have been prepared from MoO_4^{-2} and H_2S in basic solutions, with the detailed stoichiometry determined by manipulating concentrations, reaction times and conditions of isolation (161, 201–204). Analogous selenium compounds such as $MoOSe_3^{-2}$ and $MoSe_4^{-2}$, mixed sulfur–selenium species, and mixed sulfur—selenium—oxygen compounds have all been prepared (161, 205). Compounds isolated with a given cation (e.q., K^+, Rb^+, Cs^+, and Ph_4As^+) are often isomorphous to each other and to the corresponding sulfate salt and are found to contain discrete tetrahedral dianions (201, 205, 206). The acid thiomolybdate $(Ph_4As)(HMoO_2S_2)$ has also been isolated (202).

While the reaction chemistry of these species has not been thoroughly studied, the anions MoS_4^{-2}, MoO_3^{-2}, and $MoO_2S_2^{-2}$ have been shown to act as bidentate ligands (207–209). Each of these anions has at least two *cis*-sulfur atoms, and the S–S distance makes these ligands quite comparable to other 1,1-dithiolates (210) (especially those containing a PS_2 donor set). Complex anions of the formula $Zn(MoS_4)_2^{-2}$, $Ni(MoS_4)_2^{-2}$, $Fe(MoS_4)_2^{-2}$, $Zn(MoOS_3)_2^{-2}$, $Ni(MoOS_3)_2^{-2}$, $Ni(MoO_2S_2)_2^{-2}$, and $Co(MoO_2S_2)_2^{-2}$ (161, 208, 209) have been isolated as tetraphenylphosphonium or tetraphenylarsonium salts. The diamagnetic nickel complexes have electronic spectra consistent with a square-planar NiS_4 chromophore, and the MoS_2 ligand has a position quite similar to dithiophosphinate in the spectrochemical series (209).

K_2MoS_4 reacts with H_2K_2edta in aqueous solution at pH 6 to give the ion $Mo_2O_4S(edta)^{-2}$, which is isolated as a red crystalline compound $K_2Mo_2O_4S(edta) \cdot H_2O$ (211). The dianion can also be generated by bubbling H_2S through a water solution containing $Mo_2O_6(edta)^{-4}$ at pH 7.0 to 8.0. A structure involving a bridging S, two terminal O groups on each Mo, and a bridging edta molecule seems likely (211).

The bonding in the $4d^0$ tetrahedral molybdate systems has been described by molecular orbital (MO) treatments (200). The appropriate MO correlation diagram in T_d is shown in Fig. 6. The oxo (or thio or seleno) ligands are involved in both sigma and pi bonding with Mo. The difference between the $3t_2$ and $2e$ levels represents the Δ_t value for the system. Since there are no d electrons this splitting is not attainable from analysis of d-d transitions. It can, however, be estimated from the difference in energy of

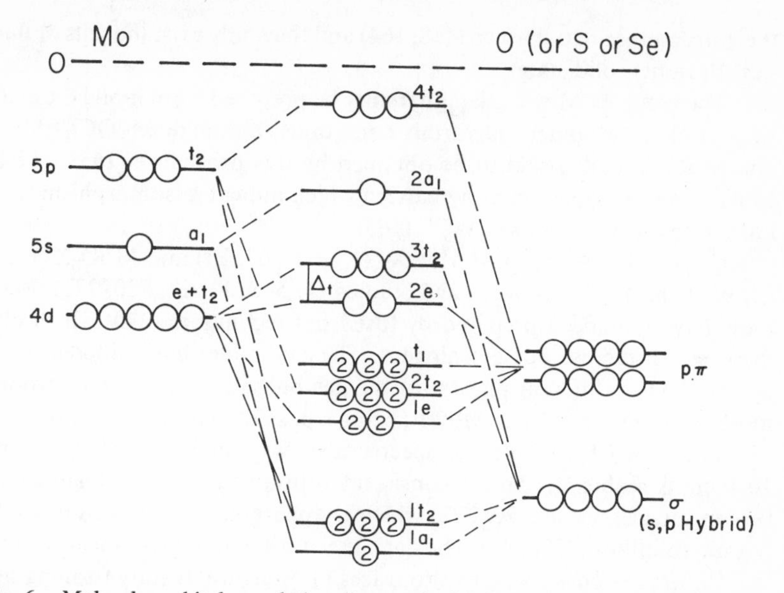

Fig. 6. Molecular orbital correlation diagram for MoX_4^{-2} tetrahedral complexes.

charge transfer bands, provided they can be properly assigned. Although the charge transfer spectra in tetrahedral oxanions has been a controversial area (200), the three lowest charge transfer transitions in MoS_4^{-2} seem reasonably assigned as $\nu_1(t_1 \rightarrow 2e)$, $\nu_2\,(2t_2 \rightarrow 2e)$ and somewhat less certainly, $\nu_3(t_1 \rightarrow 3t_2)$. Thus the difference between ν_3 and ν_1 may correspond closely to the Δ_t value. Electronic spectra have been studied in some detail (200, 212-214) and are roughly consistent with this scheme.

The infrared and Raman spectra of the tetrahedral molybdates, thiomolybdates, and selenomolybdates have also received study (215–218). For the T_d, tetrahedral dianion MoO_4^{-2}, the two stretching modes (A_1 and F_2) are Raman active and one (F_2) is infrared active as well. Bending modes are of E and F_2 symmetry, and while both are Raman active only F_2 is infrared active. Bands at 897 and 837 cm^{-1} have been assigned as stretches, whereas those at 325 and 317 cm^{-1} are assigned to bends. Studies on MoO_3S^{-2} and $MoO_2S_2^{-2}$ reveal the ν(Mo–O) bands expected for C_{3v} and C_{2v} structures, respectively (16). Selected complexes with their electronic and infrared absorptions are listed in Table V.

C. Halo and Oxohalo Compounds

The octahedral MoF_6 complex is well known (3) and has been used as a fluorinating agent (219). The ions MoF_7^- and MoF_8^{-2} have been claimed, but their structures are unknown (4, 8, 164) and they only exist in melts of fluoride salts or in the solid state.

The complex $MoCl_6$ can apparently be prepared from molybdic acid and thionyl chloride under anhydrous conditions, although $MoOCl_4$, $MoOCl_3$, and $MoCl_5$ are also said to be obtained by this procedure (165). The black $MoCl_6$ powder shows an x-ray pattern which indicates isomorphism with the known octahedral complex WCl_6 (165).

Oxyhalides of the form $MoOX_4$ (X = F, Cl, Br) and MoO_2X_2 (X = F, Cl, Br, I) have been isolated and studied (4, 5, 8, 11, 12, 220, 221). Many of these have now been prepared by fused-salt techniques (220) which obviate the need for the use of free halogens. The oxyhalides have different states of aggregation in different phases. In the gas phase the evidence is strong that most are monomeric [with $MoOF_4$ being a possible exception (220)].

For $MoOCl_4$ the infrared spectrum shows a single $\nu(Mo\text{–}O_t)$ at around 1000 cm^{-1} (Table V). This is consistent with a C_{4v}, square pyramidal structure in the gas phase (220, 222) which is apparently maintained in CCl_4 and hexane solutions (222, 223). The dark-green solid and its red-brown solutions in CCl_4 are extremely sensitive to traces of moisture, readily forming molybdenum blue [the general name for nonstoichiometric hydrous oxides of oxidation state between Mo(V) and Mo(VI)]. $MoOCl_4$ is photolytically unstable (223) to ultraviolet radiation according to Eq. 5:

$$MoOCl_4 \xrightarrow{h\nu} MoOCl_3 + 1/2\,Cl_2 \tag{5}$$

It reacts with pyridine, acetone, or acetylacetone to give (223) the Mo(V) products $MoOCl_3(py)_2$, $MoOCl_3(acetone)_2$, and $MoOCl_3(acacH)$, respectively. Recently, $MoOCl_4$ was shown to cause dimerization of acacH to a tetraoxa-adamantane molecule (224). These reactions illustrate the strongly oxidizing and reactive nature of $MoOCl_4$ which precludes its being a valuable starting material for the preparation of Mo(VI) complexes.

The gas-phase structure of $MoOF_4$ is uncertain and may be polymeric (220). In the solid state the polymeric structure (225) contains asymmetric fluorine bridges spanning from the cis position of one $MoOF_5$ octahedron to the trans position of the adjacent octahedron. The complex $Rb[Mo_2O_2F_9]$ is formed by reaction of excess $MoOF_4$ with RbF in liquid SO_2 under N_2 at $-20°C$ (226). The complex anion $Mo_2O_2F_9^-$ shows ^{19}F NMR (8:1 intensity ratio), infrared, and Raman data consistent with the presence of a bridging fluoride in the formulation $F_4OMo\text{-}F\text{-}MoOF_4^-$. The same ion is apparently generated when acetylacetone is reacted with $MoOF_4$ in CH_3CN (227).

For MoO_2X_2 the gas-phase infrared data strongly support a C_{2v} structure (*vide infra*). MoO_2Cl_2 has been most thoroughly studied, and in $CHCl_3$ solution the data are consistent with the presence of dioxo-bridged dimers, whereas a polymeric species containing no Mo–O_t is believed present in the solid state (228). MoO_2Cl_2 has much more hydrolytic and redox stability than $MoOCl_4$, and derivatives of the form $MoO_2Cl_2L_2$, L = acetone, dmf, dmac, dmso, Ph_3PO, Ph_3AsO, pyO (223, 229, 230), can be prepared by direct reaction of MoO_2Cl_2 with L. These complexes as well as MoO_2Cl_2 (in contrast to $MoOCl_4$) are good starting materials for the preparation of other Mo(VI) complexes [e.g., $MoO_2(acac)_2$, $MoO_2(oxine)_2$ (223)]. Alternatively, the complexes $MoO_2Cl_2L_2$ can be prepared by oxidation of lower valent halides or oxyhalides. For example, $MoO_2X_2(Ph_3AsO)_2$(X = Cl, Br) is formed when MoX_3 reacts with molten Ph_3AsO (231).

The complex anions $MoO_2X_4^{-2}$ (X = F, Cl) have been studied (232–235) in solution and isolated, although in concentrated HCl there is some doubt as to whether $MoO_2Cl_3^-$ or $MoO_2Cl_4^{-2}$ is the dominant anion present (233, 235). Analogous complexes of thiocyanate of the form $(NEt_4)_2MoO_2(NCS)_4$ have been prepared by acidification of $MoO_4^{-2}-NCS^-$ solutions and rapid

TABLE V
Properties of Some Mo(VI) Complexes

Complex	Infrared absorptions, cm^{-1} (R for Raman), $\nu(Mo–O_t)$	Electronic absorptions cm^{-1}	$(\varepsilon)^a$	Refs.
MoO_4^{-2}(aq.)	897(R), 838	43,200	(2,800)	161
		48,000	(8,400)	200
				215
				217
				270
MoO_3S^{-2}	882, 833	25,400		161
		34,700		200
		44,100		272
$(NH_4)_2MoO_2S_2$	867, 842	25,400	(300)	200
		31,400	(6,000)	271
		34,700	(3,000)	272
$MoOS_3^{-2}$	862	21,500	(2,300)	200
		25,500	(8,700)	273
		32,600	(6,600)	174
		38,500		
$(NH_4)_2MoS_4$	—	[19,000][b]		200
		21,400	(13,000)	270
		31,500	(17,000)	275
		41,300	(24,000)	
$MoSe_4^{-2}$	—	[16,000][b]		200
		18,000		274
		27,800		
		37,200		
MoO_2F_2 (gas)	1009, 987			220
MoO_2Cl_2 (gas)	994, 972			220
MoO_2Br_2 (gas)	991, 969			220

Table V (continued)

MoO_2I_2 (gas)	972, 950			220
MoO_2Cl_2 (solid)	905			228
$MoOF_4$ (gas)	1030			220
$MoOCl_4$ (CCl_4)	1010	14,430	(315)	220
		20,750	(2,820)	222
		37,170	(2,920)	223
$MoOBr_4$	998			220
$Na_2[MoO_2F_4]\cdot H_2O$	948, 912			264
	951, 920 (R)			
$Na_2[MoO_2Cl_4]$	964, 925			235
$MoO_2Cl_4^{-2}$ in HCl	960, 922	32,300	(5,000)	232
		44,200	(7,000)	235
$(NEt_4)_2$ $[MoO_2(NCS)_4]$	925, 890			236
				237
$Rb[F_4Mo(O)FMo(O)F_4]$	1033, 1022			226
$MoO_2Cl_2(dmf)_2$	939, 905			223
$MoO_2Cl_2(dma)_2$	948, 909			223
$(C_5H_5)MoO_2Cl$	920, 887			276
$(NEt_4)MoO_2Cl_2(acac)$	931, 897			282
$MoO_2(acac)_2$	935, 905	31,200	(3,200)	239
		37,000	(22,000)	
$MoO_2(oxine)_2$	926, 899	27,000		239
		39,100		277
		41,400		
$MoO_2(N\text{-}MeSal)_2$	922, 904			242
$MoO_2(salen)$	920, 885			278
$MoO_2(SCH_2CH_2NH_2)_2$	894, 874	28,300	(6,130)	279
		36,200	(6,180)	
		39,725	(7,250)	
$MoO_2(O\text{-}MeCys)_2$	912, 884	28,500	(5,300)	279
$Na_2[MoO_2(cys)_2]\cdot dmf$	922, 892	32,700	(12,000)	279
$MoO_2(Me_2dtc)_2$	909, 875	26,500	(630)	252
		34,500	(6,300)	277
		39,700	(15,000)	
$MoO_2(Et_2dtc)_2$	905, 877	26,320	(3,700)	252
$MoO_2((n\text{-}C_4H_9)_2dtc)_2$	909, 877	25,320	(1,020)	252
$MoO_2(S_2C(4\text{-}morph))_2$		26,700	(3,680)	254
$MoO_2(S_2PPh_2)_2$	932, 901	38,000		277
$MoO_3(dien)$	835			262
	892, 839 (R)			264
				281
$Na_4[(MoO_3)_2edta]\cdot 8H_2O$	893, 840			263
				280
				285
$(C_5H_5)_2Mo_2O_5$	930, 920, 890			276
	(Mo-O-Mo 770?)			

[a]Molar extinction coefficient in unets l mole^{-1} cm^{-1}.
[b]Numbers in brackets reprecent shoulders or uncertain bands.

precipitation (236, 237). Infrared data indicates N-bound thiocyanate. In stronger acid solutions excess thiocyanate produces Mo(V) as do HI and HBr (slowly) (238).

The cis-dioxo grouping is characterized by two infrared- and Raman-active stretching modes of symmetry a_1 (symmetric) and b_2 (antisymmetric) in C_{2v}. Trans-dioxo complexes would show a single infrared-active stretch (antisymmetric) and a single Raman-active stretch (symmetric) at a different frequency. The presence in the infrared of two Mo–O_t bands in virtually all 6-coordinate MoO_2^{+2} compounds discussed here (see Table V) is strongly indicative of the cis-dioxo structure. In no case have *trans*-MoO_2 complexes been substantiated for Mo(VI), although this arrangement is found in the Mo(IV) complex, $MoO_2(CN)_4^{-4}$, and the reasons for this are discussed in Section V. The splitting between the two stretching modes ranges from 20 to 50 cm^{-1}, and vibrational analysis (216) indicates the symmetric mode to be higher in frequency.

The C_{2v}, tetrahedral MoO_2X_2 complexes in the gas phase show $\nu(Mo–O_t)$ at much higher values than related octahedral complexes (220). In these 4-coordinate complexes the opportunity exists for further O → Mo pi bonding without unduly increasing the accumulation of negative charge on Mo. The splitting between the two bands, uniformly close to 22 cm^{-1} (220), is smaller than for the corresponding 6-coordinate complexes. The decrease in $\nu(Mo–O_t)$ as halogen (X) changes from F to I is consistent with the increasing pi donor ability of the halo ligands in this same order.

D. MoO_2L_2 Complexes

The MoO_2(β-diketonate)$_2$ complexes (173, 174, 239–241) are prepared by acidification of aqueous MoO_4^{-2}-β-diketonate solutions or, by reaction of β-diketones, with MoO_2Cl_2. The well-studied complex $MoO_2(acac)_2$ (239–241) has found use in preparation of additional MoO_2^{+2}-containing compounds (242).

The NMR spectra of MoO_2(β-diketonate)$_2$ complexes below room temperature are consistent with the cis-dioxo structure found in the solid state (173). Since each β-diketone ligand is unsymmetrically located with respect to the two oxo's (see Fig. 2), there is inequivalence of the ligand substituents. For example, at room temperature $MoO_2(C_4H_9COCHCOC_4H_9)_2$, ($MoO_2$ (dpm)$_2$), shows two sets of signals from the *t*-butyl protons (243), whereas below room temperature $MoO_2(acac)_2$ shows methyl and methine signals in 3:3:1 ratio (244). In each case as the temperature is raised the (C_4H_9 or CH_3) lines broaden and coalesce to a single line. The process is first order in complex, and averaging mechanisms involving complete β-diketonate dissociation can be ruled out by the lack of sufficiently rapid ligand exchange. The activation energy for the process is estimated at 17 kcal mole^{-1} for $MoO_2(dpm)_2$ in CH_2Cl_2, 17 kcal mole^{-1} for $MoO_2(acac)_2$ in C_6H_6, and 13 kcal mole^{-1} for $MoO_2(acac)_2$ in $CHCl_3$ (243, 244). Intramolecular processes involving either bond breaking to form a 5-coordinate intermediate or twisting of the 6-coordinate polyhedron cannot be distinguished from the data at hand, although some argument in favor of the twisting mechanism has been offered (243).

The yellow-orange complex $MoO_2(oxine)_2$ is readily formed (239–245, 246) and is, in fact, used in the gravimetric determination of Mo(VI) (247). Its cis-dioxo structure has been confirmed crystallographically (175), and the oxine nitrogens are found to lie trans to the oxo groups. The proton NMR spectrum is consistent with the maintenance of that structure in dmso and CH_2Cl_2 (248) solutions. Investigations of oxine complexes of Mo appear to be stimulated by the resemblance of the oxine ligand to one of the potential binding sites of riboflavin (249). Since three of the Mo enzymes contain flavin nucleotide prosthetic groups, this area has received some study. However, there is, as yet, no evidence for direct Mo-flavin interactions in enzymes nor have any Mo(VI)-flavin complexes been isolated (but see section II. B and Ref. 7). The electrochemical reduction of $MoO_2(oxine)_2$ in dmso was studied using cyclic voltammetry and two one-electron steps (at − 1.07 and − 1.12 V sce) were found. The possible relevance of this observation to Mo enzymes was discussed (249).

Complexes of 5-chloro-8-quinolinate and 2-methyl-8-quinolinate of the form MoO_2L_2 have been prepared. The proton NMR of the 2-methyl derivative in dmso has been interpreted as indicating no Mo–N bonding in that solvent. The CH_3 group in the 2 position of this ligand can be seen from models to produce steric strain by contacting the oxo ligand (248).

Complexes of the form $MoO_2(R_2dtc)_2$ were prepared by Malatesta (210, 250, 251) by acidification of aqueous molybdate solutions of the appropriate ligand. Moore and Larson (252) and others (253, 254) assigned octahedral structures with cis-dioxo ligands based on the observation of two strong Mo–O_t absorption bands in the infrared. This was confirmed crystallographically by Kopwillen (175) and Weiss (177). The complexes $MoO_2(R_2dtc)_2$ show large dipole moments (9.51 D for R = Et, 8.05 D for R = n-C_3H_7, and 7.60 D for R = n-C_4H_9) whichare consistent with the presence of the cis structure in solution (252).

Colton and Rose (255) prepared complexes which they formulated as β-$MoO_2(R_2dtc)_2$ (R = Me, Et, propyl and benzyl) by aerial oxidation of $Mo(CO)_2(R_2dtc)_2$ or $Mo(CO)_3(R_2dtc)_2$ in CCl_4, benzene, or a mixture thereof. These yellow complexes, designated β, are not identical to those of the same stoichiometry studied by Moore and Larson (252), Jowitt and Mitchell (253), and Newton et al. (254) and designated α-$MoO_2(R_2dtc)_2$. The β complexes show an unmistakable single $\nu(Mo{-}O_t)$ around 945 cm^{-1} ($\pm$ 5 cm^{-1}) as compared to, for example, 912 cm^{-1} and 880 cm^{-1} for the α-$MoO_2(Et_2dtc)_2$. Mass spectral comparison (255) reveals the α form to have a parent ion peak corresponding to $MoO_2(Et_2dtc)_2^+$, whereas the β form shows many peaks at mass numbers higher than the monomer (but no apparent parent). It is suggested (255) that a di-μ-oxo-bridged structure is present and that the β form is a dimer in solid and gas phase. However, the yellow complex β-$MoO_2(Et_2dtc)_2$ upon dissolution in CH_2Cl_2 or $CHCl_3$ turns orange, is monomeric by molecular weight, and still displays a single "slightly shifted" $\nu(Mo{-}O_t)$ band. It is suggested that in solution β-$MoO_2(R_2dtc)_2$ is present as a monomeric trans complex, but this does not reconcile with the high value of $\nu(Mo{-}O_t)$. The system certainly deserves further study.

The reactions of α-$MoO_2(R_2dtc)_2$ with phosphines lead to Mo(IV) complexes according to (254, 256) reaction 6:

$$MoO_2(R_2dtc)_2 + PPh_3 \longrightarrow MoO(R_2dtc)_2 + OPPh_3 \tag{6}$$

Reaction with 0.5 mole of phophine leads to the Mo(V) dimer $Mo_2O_3(R_2dtc)_4$, presumably by direct combination of the Mo (IV) complex $MoO(R_2dtc)_2$ with the Mo(VI) species. In fact, the equilibrium shown in Eq. 7 was observed by Barral et al. (256) and Newton et al. (254), and an equilibrium constant was estimated. This is further discussed in Section IV.

$$MoO_2(Et_2dtc)_2 + MoO(Et_2dtc)_2 \rightleftharpoons Mo_2O_3(Et_2dtc)_4 \tag{7}$$

The complex $MoO(Et_2dtc)_2Br_2$ reported by Nieuwpoort (257) appears to be a 7-coordinate Mo(VI) complex. It is formed as a side product in the one-electron oxidation of $Mo(Et_2dtc)_4$ by Br_2. It can also be formed directly by reaction of $MoO(Et_2dtc)_2$ with Br_2 in an oxidative addition process (258) or by treatment of MoO_2 $(Et_2dtc)_2$ with HBr (177b).

The cis-dioxo structure for all MoO_2^{+2} -containing complexes is established by infrared spectroscopy and ocassionally confirmed by NMR, dipole moment, and crystallographic studies. Table V lists some of the complexes with their infrared and electronic spectral absorptions.

E. Reactions and Other Complexes

The reactions of Mo(VI) complexes where studied have proved to be quite rapid, requiring application of the techniques of fast reactions (stopped flow and relaxation) for kinetic measurements. The dominant feature of all complex-forming reactions which start with MoO_4^{-2} appears to be a rapid change of coordination number from 4 to 6 as the pH is lowered. Diebler et al. (259, 260) studied the equilibria and kinetics of complex formation between MoO_4^{-2} and 8-hydroxyquinoline or 8-hydroxyquinoline-5-sulfonic acid in the pH range 7.5 to 9.8. The temperature-jump method at 20.0°C (Δt = 5°C) was employed. The protonated form $HMoO_4^-$ is considered to react rapidly by addition of either 8-hydroxyquinoline or its anion (or the sulphonated derivatives), with reactions of MoO_4^{-2} being several orders of magnitude slower. The maximum bimolecular rate constant for the reaction of $HMoO_4^-$ and oxine is reported as 1.5×10^8 l mole^{-1} s^{-1}. The formation of 1:1 and 2:1 catechol—Mo complexes was studied by Kustin and Liu (261). The two rate constants are similar, and it is felt that both may represent substitution on 6-coordinate Mo(VI) ions. This would favor a mechanism in which the 4- to 6-coordination shift is a precursor rather than a result of complex formation (261). No matter which of the possibilities is correct, it is clear, nevertheless, that the shift from 4- to 6-, tetrahedral to octahedral coordination is a key step in Mo(VI) chemistry.

Complexes containing MoO_3 group are far less numerous than those containing MoO_2^{+2}. Thus only the edta and dien complexes have been isolated as crystalline compounds (168, 169, 262, 263). The MoO_3 group, when of C_{3v} symmetry, has two stretching modes, a_1 and e, both of which are infrared and Raman active. For MoO_3(dien) in the solid state these have been assigned to the bands that occur at 879 and 835 cm^{-1} (infrared and Raman in solid samples), with the OMoO bend assigned to a band at 317 cm^{-1} (in the Raman) (264). However, other studies using only infrared implicate the 839 cm^{-1} band (168, 262) as containing both a_1 and e modes.

The isolated complex MoO_3(dien) in solution reacts with oxine in a manner almost identical to the reaction of Na_2MoO_4 with oxine (265). This led Taylor et al. (265) to suspect the presence of MoO_4^{-2} in the MoO_3(dien) solutions. The identity of the solution Raman and infrared spectra of MoO_4^{-2} and MoO_3(dien) would seem to indicate extensive dissociation of the complex below pH 10. The fact that ethanol precipitates MoO_3(dien) is indicative of a rapid equilibrium (Eq. 8),

$$MoO_3(dien) + H_2O \rightleftharpoons MoO_4^{-2} + dienH_2^{+2} \tag{8}$$

which lies far to the right but which can be forced to the left by addition of alcohol and precipitation of the complex. The complex is insoluble in dmso, $CHCl_3$, acetone, and dioxane and is only soluble in water in which it hydrolyses. Its solution stability is clearly not extensive, and its occurrence in the solid state may be secured to a large extent by the H-bonded network in which it is involved (168).

On the other hand, the edta complex $(MoO_3)_2edta^{-4}$ appears to have distinct solution stability in the pH range 3 to 8. In fact, its structure was correctly deduced by NMR studies (266, 267) prior to the crystallographic determination (169). In solution a 1:1 complex, probably of the form MoO_3-$edta^{-4}$, has also been identified (268). The rapid kinetics of complex formation have been investigated by temperature-jump relaxation techniques in the pH range 7.5 to 8.25 (268) and are again consistent with the protonated form $HMoO_4^-$ being reactive with ligand. A two-step process leading first to MoO_3 $edtaH^{-3}$ and subsequently to $(MoO_3)edta(MoO_3)^{-4}$ is found and the rate constants reported. Mo(VI) and nta^{-3} produce a 1:1 complex with equivalent methylene groups (267) in the NMR. A 7-coordinate C_{3v} structure, $(nta)MoO_3^{-3}$, containing a long Mo–N bond fits both the NMR data (267) and the stereochemical requirements of the nta^{-3} ligand and MoO_3 moiety (169). However, a recent brief report (168a) of the crystal structure of K_3-$[MoO_3(nta)] \cdot H_2O$ shows distinct 6-coordination with one of the nta carboxylate groups uncoordinated to Mo. The nmr and x-ray structural results could be reconciled either by assigning different structures in the solid state and solution or by invoking rapid exchange of free and coordinated carboxylate groups of the nta ligand. Equilibrium solution structures of complexes formed by $edta^{-4}$ and related ligands with Mo(VI) have recently been reviewed [269].

Dimeric complexes have beeen found in crystallographic studies to contain linear (170, 171) or bent (187, 188, 189) O_b groups in monooxo-bridged complexes. Complexes containing nonoxo (oxygen) ligands as bridges (162, 187, 188, 189) have also been crystallographically identified. However, the existence of many of these species in solution is uncertain, and some complexes may be artifacts of the crystallization procedure [e.g., MoO_3dien, $Mo_2O_5(dmf)_2Cl_2$].

IV. THE CHEMISTRY OF MOLYBDENUM(V)

The chemistry of Mo(V), like that of Mo(VI), is dominated by the presence of strong Mo–O bonding. However, unlike the Mo(VI) state where monomeric and polymeric complexes are dominant, dimeric species play a major role in the Mo(V) state. The monomeric complexes of Mo(V) can be isolated as monooxo species from strongly acid solutions (e.g., $MoOCl_5^{-2}$, $MoO(NCS)_5^{-2}$) or as nonoxo species with certain ligands (e.g., $Mo(R_2dtc)_4^+$, $Mo(CN)_8^{-3}$). Dinuclear compounds are found bridged by either single oxo groups or by two oxo, two sulfido, or two chloro ligands. Most recently, bridging by both O and S (258) and triple-bridged structures (285) have also been found. Most of the dimeric compounds also contain a single terminal Mo–O_t linkage on each Mo. Exceptions to this include Mo_2Cl_{10} and some of the sulfido-bridged compounds.

Although Mo(V) is certainly a participating state in at least some of the Mo enzymes, it is not known whether biological Mo is mononuclear or dinuclear. All Mo enzymes contain two Mo's, and in this section we show that, especially near pH 7, dimers pervade the coordination chemistry of Mo(V). The chemical and biological presence of paired Mo is indeed suggestive of a dinuclear Mo site in enzymes. However, the evidence for a mononuclear site in enzymes is at least equally suggestive. Thus, except for nitrogenase, all Mo enzymes show an EPR signal characteristic of monomeric Mo(V) (see Section II.B and Section IV.B) during turnover. This signal shows an unpaired electron associated with *one* Mo(V), and (although a mixed IV-V or V-VI state cannot be firmly eliminated) there is no evidence for any Mo–Mo interaction. Furthermore, all Mo enzymes (including nitrogenase) also have at least two of all other prosthetic groups (e.g., flavins, Fe_4S_4 clusters, hemes) and have subunits that come in sets of two or multiples thereof. Therefore, it is also reasonable to assume that there are two active sites per enzyme, each replete with the full complement of catalytic machinery. Thus at present both mononuclear and dinuclear Mo complexes can be considered as potentially relevant to the role of Mo in enzymes.

We begin this section with a survey of the common structural types encountered. Oxo and nonoxo mononuclear complexes are then presented, and a separate discussion of electron spin resonance (ESR) results is included in view of the extensive use of this technique in the study of Mo enzymes. The various classes of dimeric complexes are considered, as are the reactions of chloride and thiocyanate complexes. For these simple ligands there is some understanding of the manner of interconversion of monomeric and various dimeric species. Finally, the Mo(V) cysteine and glutathione systems are discussed in light of their possible relevance to Mo enzymes.

A. Structural Studies

Although over 25 structural studies have been completed in the last 10 years, only few structural types have been found. Nonoxo, nonsulfido structures are scantily represented by $(NBu_4)_3[Mo(CN)_8]$, $[MoCl_4(diars)_2]I_3$, and Mo_2Cl_{10} and the complex $[Cl(PMe_2Ph)_4ReN_2MoCl_4(OCH_3)]CH_3OH \cdot HCl$ (306) is discussed in Section IX. Almost all other structures, be they monomeric, dimeric, or tetrameric, contain one terminal oxo per molybdenum. The exceptions occur where the electronically similar terminal sulfido or alkylimido take the place of the terminal oxo. For dimeric complexes the $Mo_2O_3^{+4}$, $Mo_2O_4^{+2}$, and $Mo_2O_2S_2^{+2}$ units are strongly represented. A single triple-bridged structure and an unusual tetrameric species have been reported which may foreshadow future directions. A list of those complexes whose structures have been determined is given in Table VI.

TABLE VI
Crystal Structures of Mo(V) Complexes

Complex	Refs.	Complex	Refs.
$K_2[MoOF_5]\cdot H_2O$	284	$[Mo_2O_4(his)_2]\cdot 3H_2O$	295
$K_2[MoOBr_5]$	285	$Mo_2O_4(O\text{-}Etcys)_2$	296
$[Ph_4As]\ [MoOBr_4(H_2O)]$	286	$Mo_2O_4((i\text{-}C_3H_8)_2dtc)_2$	305
$MoOCl_3(SPPh_3)$	287	$Mo_2O_3(\mu\text{-}(S,\ O)\text{-}SCH_2CH_2O)(oxine)_2$	283
$MoOCl_3(OP(N(CH_3)_2)_3)_2$	287	$Mo_2O_2S_2(C_5H_5)_2$	297
$(NB\mu_4)_3[Mo(CN)_8]$	288	$Mo_2O_2S_2(his)_2$	298
$[MoCl_4(diars)_2^+](I_3^-)$	289	$Cs_2[Mo_2O_2S_2(edta)]\cdot 2H_2O$	299
Mo_2Cl_{10}	290	$Na_2[Mo_2O_2S_2(cys)]_2\cdot H_2O$	300
$Mo_2O_3(Etxan)_4$	291	$MoO_2S_2(O\text{-}Mecys)_2$	301
$Mo_2O_3(Et_2dtp)_4$	292	$(N(n\text{-}C_4H_9)_4)_2[Mo_2O_2S_2(i\text{-}mnt)_2]$	302
$Mo_2O_3((i\text{-}C_3H_8)_2dtc)_4$	177	$Mo_2(N\text{-}t\text{-}C_4H_9)_2S_2(C_5H_5)_2$	303
$Ba[Mo_2O_4(H_2O)_2(C_2O_4)_2]\cdot 5H_2O$	293	$Mo_2S_4((n\text{-}C_4H_9)_2dtc)_2$	304
$Na_2[Mo_2O_4(cys)_2]\cdot 5H_2O$	294	$[Cl(PMe_2Ph)_4ReN_2MoCl_4(OCH_3)]\ CH_3OH\cdot HCl$	306

1. Mononuclear Complexes

Structurally investigated mononuclear compounds listed in Table VI include oxo and nonoxo species. In the oxo class the complexes contain halo ligands in addition to oxo. Displaying the same trends as Mo(VI) complexes, the C_{4v} anions $MoOBr_4(H_2O)^-$ (286), $MoOBr_5^{-2}$ (285), and $MoOF_5^{-2}$ (284) have substituted octahedral coordination spheres characterized by very short Mo–O_t bonds and long bonds trans to O_t. For example, in $MoOF_5^{-2}$ (284), shown in Fig. 7, the Mo–O_t distance is 1.66 Å, the four *cis*-Mo–F bonds average 1.88 Å, and the *trans*-Mo–F is 2.02 Å. In each case the molybdenum is displaced out of the equatorial plane in the direction of the oxo ligand. These three features, the short Mo–O_t bond, the longer bond to the atom trans to O_t, and the displacement of the Mo toward the oxo are common features of most of the oxo-Mo complexes discussed in this section.

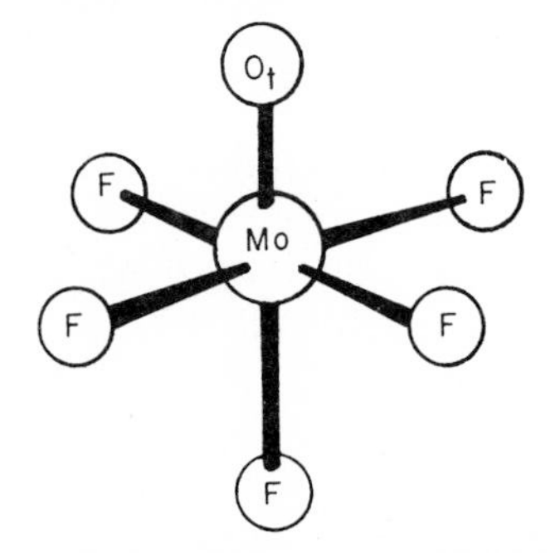

Fig. 7. The structure of $MoOF_5^{-2}$ in $K_2[MoOF_5]$ (284).

The complex $MoOCl_3(SPPh_3)$, prepared by reaction of $MoOCl_3$ and $SPPh_3$ in CH_2Cl_2, was recently (287) shown to have a monooxo structure containing 5-coordinate Mo. The approximate square-pyramidal geometry has O_t at the apex with the position trans to it vacant. The Mo is 0.65 Å above the plane described by the S and three Cl's, and the Mo–O_t bond length is at 1.647(3) Å. This distance is shorter than the Mo–O_t bond length found in other monomeric and dimeric structures and is consistent with the observation of ν(Mo–O_t) at the high value of 1008 cm^{-1} in this complex. The bond length and infrared data indicate a very strong Mo–O_t bond in this case where there is no other oxo group in the coordination sphere and no group trans to oxygen. Thus, in the extreme, the trans-bond-weakening effect of Mo–O_t can produce an open-coordination position and an even stronger Mo–O_t.

The crystal structure of $NH_4[MoO_2(C_8H_7O_7N_3)]H_2O$ has been briefly reported (168a). The uramil-N,N-diacetate ligand is tetradentate and a unique (for Mo(V)) MoO_2^{+1} grouping is said to be present.

For nonoxo monomeric species the two complexes studied reveal 8-coordination spheres. The paramagnetic salt $(N(C_4H_9)_4)_3[Mo(CN_8]$, prepared by oxidation of $Mo(CN)_8^{-4}$, was found (288) to have a near-regular, D_{2d} triangular-dodecahedral geometry. However, the salt $K_3[Mo(CN)_8]\cdot H_2O$ is found to be isomorphous to $K_3W(CN)_8\cdot H_2O$, where W is known to have a D_{4d} square-antiprismatic arrangement (305). The probable stereochemical lability of the $M(CN)_8^{-3}$ anions is discussed in Section IV. C.

The complex $[MoCl_4(diars)_2]I_3$ was prepared quite accidentally, and not reproducibly, during the crystallization of $[Mo(CO)_2(diars)_2Cl]I_3$ in $CHCl_3$ (289). Its dodecahedral structure shows the diarsine ligands in the *A* positions (elongated tetrahedron) and the Cl ligands in the *B* positions (flattended tetrahedron), as found for $TiCl_4$ $(diars)_2$.

2. Dinuclear Complexes

a. Mo_2Cl_{10}. $MoCl_5$ is an often used starting material in the preparation of pentavalent complexes. Its structure (290) in the solid is dimeric and, as shown in Fig. 8, contains bridging Cl atoms. The long Mo–Mo distance of 3.84 Å and the magnetic moment of 1.64 B.M. per Mo at 293°K are consistent with the absence of Mo–Mo bond formation. Significantly, $MoCl_5$ does not remain dimeric in the gas phase where a monomeric 5-coordinate pentagonal bipyramidal structure is found (307). The dimeric structure reveals significant Cl–Cl repulsions that may be responsible for the breaking of the dimer (290) upon dissolution or vaporization.

b. Structures Containing the Linear Mo-O-Mo Bridge. The three monooxo-bridged species that have been described each contain 1,1-dithiolate ligands (210). In 1964 Blake, Cotton, and Wood (291) reported the structure

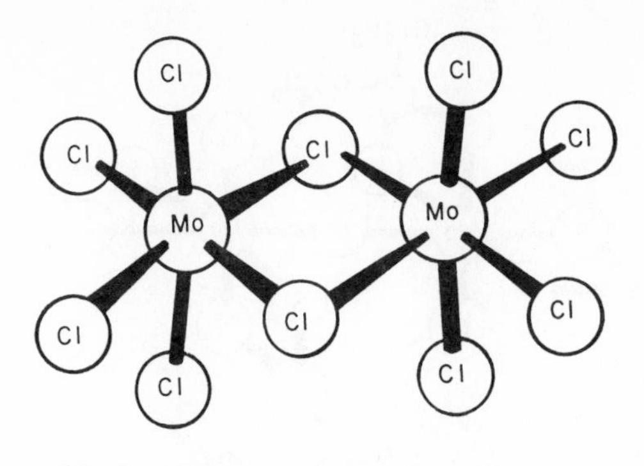

Fig. 8. The structure of Mo_2Cl_{10} (290).

of $Mo_2O_3(Etxan)_4$ while in 1969 Knox and Prout published the structure of $Mo_2O_3(Et_2dtp)_4$ (292). These two structures are notable for their gross similarities, and although they have some important differences, we find it convenient to discuss them together.

The complexes each display 6-coordinate, distorted octahedral Mo atoms joined by a rigorously linear (in the dtp case) or nearly linear [178 (4)° for the Etxan complex] Mo–O–Mo linkage. The structures are displayed in Figs. 9 and 10, with important distances shown on the Etxan structure. The key difference between the structures is the presence of trans-directed terminal oxygens in the dtp case and cis-directed (nearly eclipsed) terminal oxygens in the Etxan case. The dtp complex has a crystallographically required center of inversion, whereas the xan complex, although possessing no crystallographically required symmetry, has an approximate twofold axis. In the latter complex the dihedral angle between the planes defined by $O_1Mo_1O_2$ and $O_1Mo_2O_3$ is 4.5 (0.5)°. Significantly, both complexes show substantial variation in the Mo–S distances, with the order Mo–S (trans

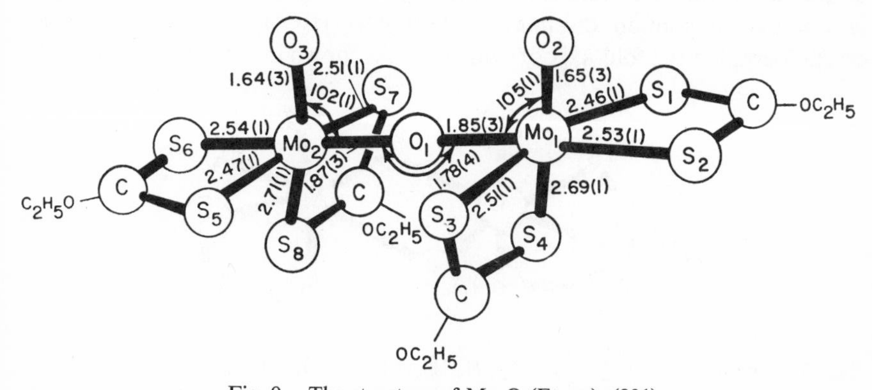

Fig. 9. The structure of $Mo_2O_3(Etxan)_4$ (291).

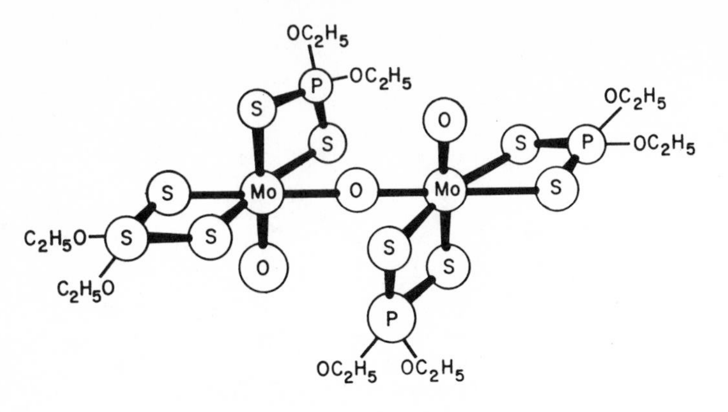

Fig. 10. The structure of $Mo_2O_3(Et_2dtp)_4$ (292).

to S) < Mo–S (trans to O_b) < Mo–S (trans to O_t) found in each case. This rather dramatic trans influence of oxo is typical of complexes which display strong M–O bonding, as are the short Mo–O_t distances in the two crystals.

Both of the above compounds are strictly diamagnetic, indicating some form of pairing of the spins on the two Mo(V) ions. As is discussed in Section IV. C, either the cis or trans, but not the perpendicular, orientation of the Mo–O_t bonds favors pi-bonding across the bridge, which is responsible for the diamagnetism.

The structure of $Mo_2O_3((C_3H_7)_2dtc)_4$ was recently solved by Ricard, Weiss, and co-workers (177) and is found to be closely similar to that of the xanthate complex. In fact, within the $Mo_2O_3S_8$ skeleton the distances and angles for the dtc and xan complexes are virtually identical.

c. Structures Containing the Mo_2O_2Bridge. The first di-μ-oxo structure was reported by Cotton and Morehouse in 1965 (293) and illustrates many of the key features of this class. The salt $Ba[Mo_2O_4(C_2O_4)_2(H_2O)_2] \cdot 5H_2O$ was found to contain discrete $Mo_2O_4(C_2O_4)_2(H_2O)_2^{-2}$ ions lying on a crystallographic twofold axis of symmetry. As shown in Fig. 11, the structure

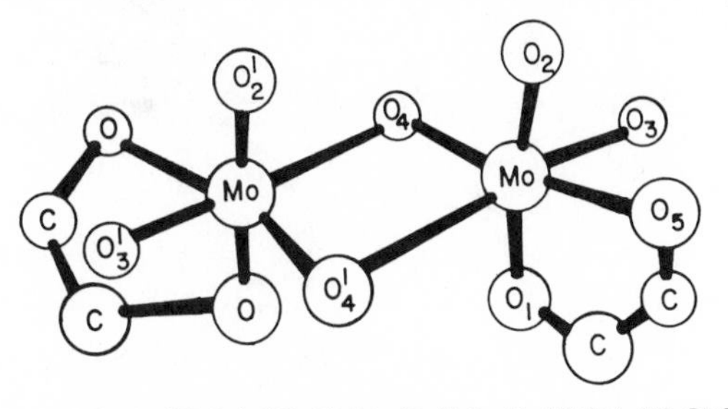

Fig. 11. The structure of $Mo_2O_4(C_2O_4)_2(H_2O)_2^{-2}$ in $Ba[Mo_2O_4(C_2O_4)_2(H_2O)_2] \cdot 5H_2O$ (293).

consists of two distorted MoO_6 octahedra sharing an O–O edge. Important distances are $Mo–O_2 = 1.70(3)$, $Mo–O_4 = 1.93(3)$, $Mo–O'_4 = 1.88(3)$, $Mo–O_3(H_2O) = 2.24(4)$, $Mo–O_5 = 2.14(4)$, and $Mo–O_1 = 2.11(3)$ Å. The Mo–Mo distance of 2.541 Å suggests direct Mo–Mo bonding, and Cotton and Morehouse argue that a Mo–Mo bond order close to 1 is reasonable for this complex. Significantly, the Mo_2O_2 bridge is not planar with the center of the $O_4–O'_4$ line lying on the C_2 axis at a point 0.32(3) Å from the center of the Mo–Mo line (also on the C_2). This corresponds to a dihedral angle of 151° between the two MoO_2 planes. Although not explicitly stated, it is seen from bond angle data that the Mo is significantly out of the plane approximated by O_4, O'_4, O_3, and O_5 in the direction of the O_t atom.

The structure of $Na_2[Mo_2O_4(cys)_2]\cdot 5H_2O$, which is shown in Fig. 12*a* (294), is fundamentally similar to that of the oxalate complex. The overall symmetry is close to C_2, although no crystallographically required symmetry is present. The $Mo–O_t = 1.71$, $Mo–O_b = 1.95, 1.91$, Mo–Mo = 2.57 Å are comparable to those in the oxalate structure, as is the dihedral angle of 151° between MoO_2 planes. The Mo atoms lie 0.38 Å out of the $O_b–O'_b$–S–N planes toward the terminal oxo group, a situation seen to be common in both mononuclear and polynuclear complexes of the oxo type.

The structure of $Mo_2O_4(his)_2$ is very close to that of the cysteine analog (295). Again, with no crystallographic symmetry imposed, the molecular symmetry is close to C_2. The $Mo–O_t$ at 1.71, $Mo–O_b$ at 1.92, the Mo–Mo bond at 2.55 Å, and the displacement of Mo by 0.34 Å out of the O_2N_2 plane serve to illustrate the extreme closeness of these parameters in the $Mo_2O_4^{+2}$ structures. A notable feature of this complex is the shorter value for Mo–O (carboxyl) compared to the cysteine derivative [2.21 vs. 2.30 (1)]. This simply reflects the greater stereochemical flexibility in the six-membered chelate ring containing Mo, N, and N compared to the five-membered ring containing Mo, N, and S in the cysteine structure. This permits the carbon bearing the carboxylate group to dip farther below the coordination plane, thereby

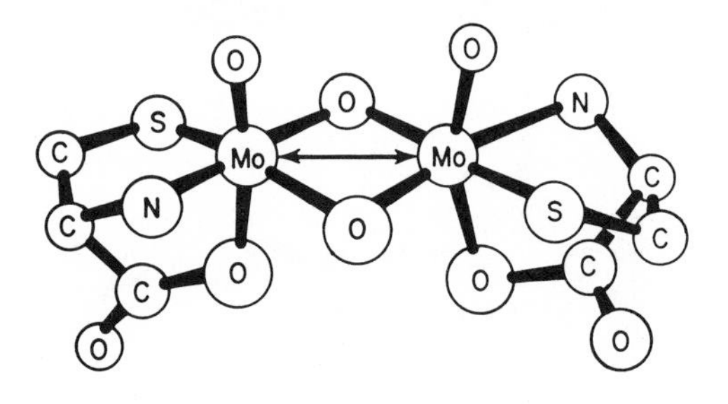

Fig. 12*a*. The structure of $Mo_2O_4(cys)_2^{-2}$ in $Na_2[Mo_2O_4(cys)_2]\cdot 5H_2O$ (294).

allowing Mo–O (carboxylate) to assume the smaller value in the histidine structure.

Cotton et al. (293) pointed out that the systematic variation in O–Mo–O bond angles correlates inversely with the lengths of the Mo–O bonds which define the angle, that is, the shorter Mo–O bonds have larger angles between them. This is attributed to ligand-ligand (or bond-bond) repulsion within the coordination sphere. Donahue (308) pointed out that the combined bond angle and bond distance trends were such as to define O—O distances in a narrow range around 2.78 Å, providing a neat confirmation of Cotton's suggestion (309) that bond angles are controlled by O–O repulsions

A side view of an Mo_2O_4 complex, seen perpendicular to the plane defined by the C_2 axis and the Mo–Mo bond, is shown in Fig. 12*b*. The nonplanarity of the bridge and the overall shape of the molecule can be explained by incorporating a number of factors. First, each Mo lies out of the equatorial plane of its four tetragonal donor atoms, with the strong $Mo–O_t$ bond defining the local tetragonal axis and responsible for the effect. Second, the Mo–Mo distance of 2.55 Å allows Mo–Mo bonding, and the bent bridge permits closer approach of the Mo atoms without unduly enlarging the O–Mo–O or contracting the Mo–O–Mo angle. Finally, the terminal oxo groups on the two Mo atoms clearly bend back away from each other. If they were not bent back, their distance would equal 2.55 Å (the Mo–Mo distance), which is considerably shorter than the van der Waals contact distance of 2.80 Å. These three effects, tetragonal distortion, Mo–Mo bonding, and $O_t–O_t$ repulsion, apparently contribute with various weights to produce the overall geometry of the $Mo_2O_4^{+2}$ grouping.

The structure of Mo_2O_4(*O*-Etcys)$_2$ (296) differs from the other di-μ-oxo species in that (neglecting any Mo–Mo bonding as part of the coordination sphere) the Mo atoms are 5-coordinate. Although the geometry about the Mo is clearly distorted, the authors prefer to refer their discussion to a trigonal bipyramidal structure. A particular bridging oxygen atom is then found in the axial position of one of the Mo-coordination spheres, but in an equatorial position in the sphere of the other Mo. An idealized drawing of this geometry

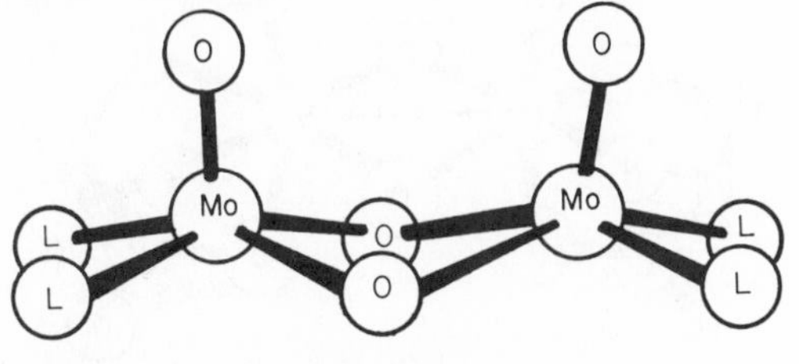

Fig. 12*b*. The basic $Mo_2O_4^{+4}$ core.

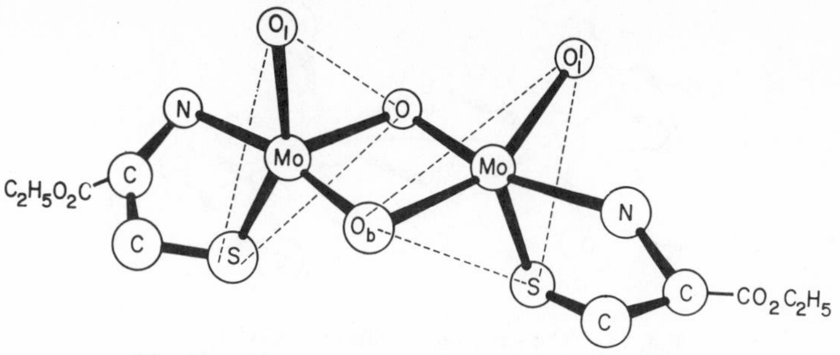

Fig. 13. The structure of $Mo_2O_4(O\text{-Etcys})_2$ (296).

is shown in Fig. 13, where an approximate (noncrystallographic) twofold axis is present bisecting the O_b–O_b and Mo–Mo lines. The two "equatorial" planes shown by the dotted lines in the figure subtend a dihedral angle of 5.3°, indicating their approximate coplanarity. As in the other di-μ-oxo dimers, the bridge is nonplanar and the Mo_2O_2 dihedral angle equals 144°. The deviation from idealized geometry at the metal is clearly seen by the fact that the O_b–Mo–N angles average 154° compared to the idealized 180°. The Mo–Mo distance of 2.562(3) Å is again consistent with metal-metal interaction and thus with the observed diamagnetism of the complex. Finally, the Mo–S bond distance of 2.38 Å is significantly shorter than that of the other complexes. It is interesting that no group lies strictly trans to this sulfur, and thus the bond length trend would appear to be Mo–S (trans to O_t) > Mo–S (trans to O_b) > Mo–S (trans to S) > Mo–S (no trans group).

Structural studies on a salt of $Mo_2O_4(\text{edta})^{-2}$ have apparently been carried out (310), and while the overall structure is said to be similar to the others discussed here, problems with disorder have apparently precluded complete solution.

Very recently the structures of $Mo_2O_4(Et_2dtc)_2$ (310a) and $Mo_2O_4(\text{bipy})_2$-$(O_2PH_2)_2$ (310b) have been reported. Both contain Mo_2O_4 bridges of the now familiar type (Fig. 12a). The Mo geometry in the Et_2dtc complex like that of $Mo_2O_2S_2(\text{i—mnt})_2^{-2}$ lends itself to a square pyramidal description. The second complex has one monodentate hypophosphito ligand on each Mo in a position trans to O_b and one bidentate bipy on each Mo completing roughly octahedral coordination spheres.

d. Structures Containing the MoS_2Mo Bridge. The first of the eight di-μ-sulfido structures was solved by Stevenson and Dahl (297) for $Mo_2O_2S_2$-$(C_5H_5)_2$, prepared along with other compounds by the reaction of cyclohexene sulfide with $((C_5H_5)Mo(CO)_3)_2$. The molecular structure shown in Fig. 14 displays the di-μ-sulfido complex lying on a crystallographic center

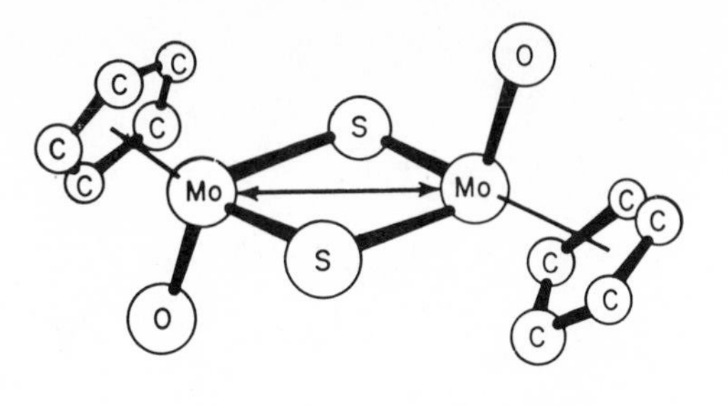

Fig. 14. The structure of $Mo_2O_2S_2(C_5H_5)_2$ (297).

of symmetry with trans-oxo and trans-cyclopentadienyl groups. The structure is close to C_{2h} symmetry if the cyclopentadienyl groups are assumed to be cylindrically symmetrical. (The C_2 axis is along the S–S line, whereas the mirror plane bisects this line and the cyclopentadienyl group and contains the metal-metal axis.) The Mo–Mo distance is 2.89 (1) Å, and while this is again consistent with some Mo–Mo interaction, it is considerably longer than the Mo–Mo distance of 2.55 Å found in the di-μ-oxo dimers. However, unlike the four di-μ-oxo dimers and most of the other di-μ-sulfido complexes (*vide infra*), the Mo_2S_2 bridge in this case is rigorously planar and the terminal oxo groups are trans. Recently, Dahl et al. (303) solved the structure of an analogous compound in which the terminal oxo groups are replaced by *t*-butylimido ($NC(CH_3)_3$) groups. In this similarly centrosymmetric complex a near linear [176.3(3)°] Mo–N–C angle occurs and detailed resemblance to the oxo structure is apparent.

The $Mo_2O_2S_2^{+2}$ structures containing histidine (298), cysteine (300), and edta (298, 311) contain cis Mo–O_t's and are remarkably similar to each other insofar as the $MoO_2S_2^{+2}$ unit is concerned. The Mo–O_t distance is 1.68 ± 0.01, Mo–S_b is 2.30 ± 0.03, and Mo–Mo is found at 2.82 ± 0.03 Å. Besides the MoS_2Mo core the overall structures are quite similar to their di-μ-oxo analogs. Moreover, the MoS_2Mo bridge is also nonplanar with a MoS_2–MoS_2 dihedral angle of 150 to 160°. Thus even the shape of the bridge is very similar to that found in the oxo analogs (i.e., Fig. 12*b*). The edta analog displays the feature of an additional ethylenediamine bridge spanning the Mo atoms and connecting them at positions trans to O_t (Fig. 15). In short, except for the two complexes that contain cyclopentadienyl groups, the Mo_2S_2 bridge is bent, the O_t's are cis, and the Mo–Mo distance lies in a narrow range around 2.80 Å. The two cyclopentadienyl structures, on the other hand, are centrosymmetric, contain a planar Mo_2S_2 grouping and trans O_t or $NC(CH_3)_3$ groups, and have Mo–Mo distances near 2.90 Å.

The di-μ-sulfido cis-dioxo structures as a class are remarkably similar *structurally* to the di-μ-oxo cis-dioxo compounds, with the only major dif-

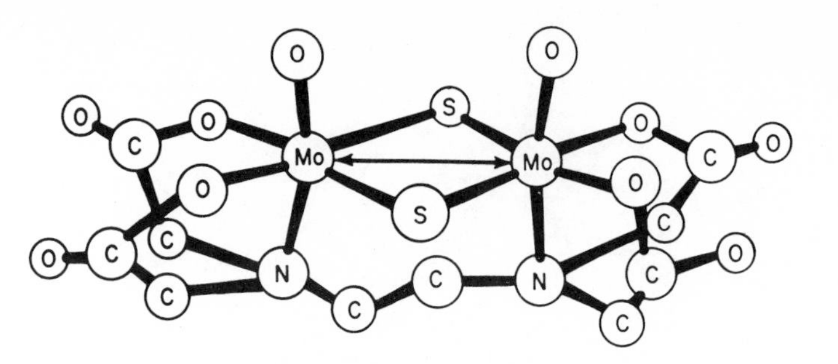

Fig. 15. The structure of $Mo_2O_2S_2(edta)^{-2}$ in $Cs_2[Mo_2O_2S_2(edta)]\cdot 2H_2O$ (299).

ference being the size of the bridge (while the shape of the bridge remains quite similar). This difference is fully attributable to the larger size of sulfur compared to that of oxygen. The bridge in cis $Mo_2O_4^{+2}$ and $Mo_2O_2S_2^{+2}$ species seems to represent a single structural type, with oxygen and sulfur substituting interchangeably. This notion is reinforced by the recent isolation of a mixed OS-bridging system (258).

Finally, recent work of Dori et al. (304) has established the presence of an $Mo_2S_4^{+2}$ grouping in $Mo_2S_4((n\text{-}C_4H_9)_2dtc)_2$. This structure is similar to $Mo_2O_4^{+2}$ structures, with sulfurs replacing the bridging *and terminal* oxos. The Mo–S_t distance of 1.94 Å is extremely short, showing this to be a multiply bound sulfur, and the Mo–Mo distance of 2.80 Å is comparable to that in $MoO_2S_2^{+2}$ compounds. The combined results establish a range of isostructural compounds containing both terminal and bridging oxo and sulfido groups.

The complexes $MoO_2S_2(O\text{-Mecys})_2$ (301) and $[NBu_4]_2[MoO_2S_2(S_2CC(CN)_2)_2]$ (302) are found to have 5-coordinate Mo (again, not including the Mo–Mo bond). The di-μ-sulfido cysteine methyl ester complex is very similar to the di-μ-oxo cysteine ethyl ester complex (Fig. 13), and the Mo coordination geometry is also describable in terms of trigonal-bipyramidal structure. On the other hand, in the anion $MoO_2S_2(S_2CC(CN)_2)_2^{-2}$ each nominally 5-coordinate Mo has a coordination sphere that is clearly described as a square pyramid (302).

e. Triply Bridged Structure. One new type of structure has recently been found in studies by Enemark, Haight, and co-workers (283) on the compound prepared according to reaction 9:

$$Mo_2O_4(oxine)_2(py)_2 + HSCH_2CH_2OH \longrightarrow Mo_2O_3(SCH_2CH_2O)(oxine)_2 + 2py + H_2O \qquad (9)$$

This complex has the unique triply bridged structure shown in Fig. 16. There is a bent μ-oxo linkage, and the 2-thiolatoethoxo ligand symmetrically

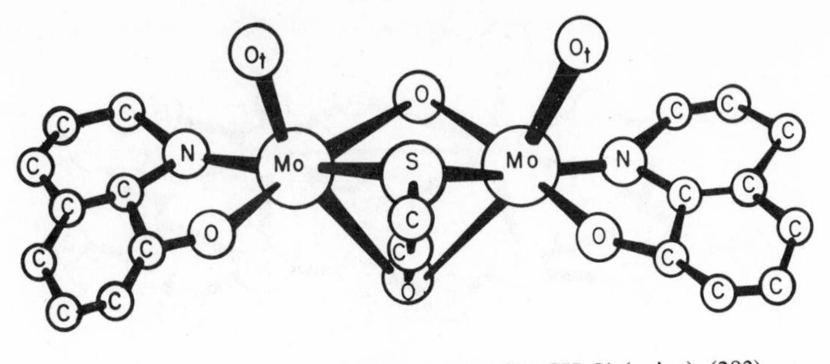

Fig. 16. The structure of Mo_2O_3(u-(S, O)-SCH_2CH_2O) $(oxine)_2$ (283).

bridges the two Mo's via its S and O atoms. This triply bridged structure suggests intriguing possibilities for interconversion of mono-, di-, and tri-bridged structures which may occur in chemical or biological systems. This is discussed at greater length in Sections IV. D and IV. E.

f. Tetrameric Mixed-Valence Structure. There is one crystal structure study of a mixed Mo(V)-Mo(VI) complex (312) which is discussed because of its novelty and relation to dimeric species. The complex $Mo_4Cl_4O_6$-(O-n-$C_3H_8)_6$ was initially prepared accidentally during attempts to grow crystals of CH_3MoCl_2(O-n-$C_3H_8)_2$. The procedure involves reacting $MoCl_5$ and dry n-C_3H_8OH at 80°C to give $MoCl_3(OC_3H_8)_2$, which is treated with $Zn(CH_3)_2$ in slight excess to give $CH_3MoCl_2(OC_3H_8)_2$. When dissolved in pentane and allowed to stand, light-red crystals of $Mo_4Cl_4O_6(OC_3H_8)_6$ precipitate. Although there is no rationale or simple equation for this reaction, it is apparently reproducible (312). As shown in Fig. 17, the centrosymmetric tetrameric structure contains both terminal and bridging oxo and propoxide groups and can be viewed in terms of two dimeric halves. The upper (labelled) part of the tetramer in many ways resembles the di-μ-oxo dimers

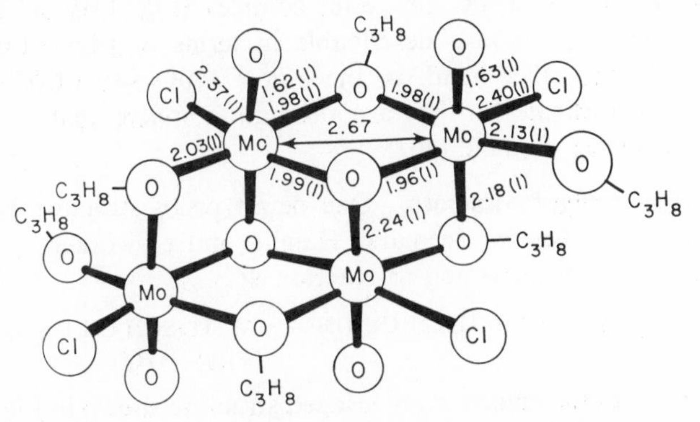

Fig. 17. The structure of $Mo_4Cl_4O_6$(O-n-$C_3H_8)_6$ (310).

having an MoO_2–MoO_2 dihedral angle of 157°. The Mo–O_t distances of 1.62 and 1.63 Å are among the shortest known, and despite the fact that one of the bridging atoms is a propoxide oxygen, the bridge is remarkably symmetrical. Comparison with the bridges found in di-μ-oxo dimers reveals a slightly enlarged bridge with a longer Mo–Mo distance of 2.67 Å. This difference would appear to be due to the difference in oxidation state between the Mo(V) dimers and this mixed Mo(V)-Mo(VI) complex, where the Mo–Mo bond would be weaker as only one electron is involved in it rather than two. Looking at the binding between the dimeric portions we find that the two bridging oxo ligands within the dimeric fragment also bridge to the other dimer at the longer distance of 2.24 Å. Additionally, propoxide oxygens which are terminal on each dimer also bridge to a molybdenum of the second dimer. The Mo_2O_2 bridges in this case are nearly or exactly planar, and in each case the bridging atom is trans to a terminal oxo group. Furthermore, the Mo–Mo distances between dimers at 3.43 Å and at 3.36 Å preclude any Mo–Mo bonding. *It would thus appear that the bonding between the dimeric portions of the molecule is weaker than that within the dimers.* This observation causes one to wonder whether the tetramer would remain in solution (no solution molecular weights are given) or split up into dimers, each having formally one Mo(V) and one Mo(VI) and thus displaying paramagnetism and potentially interesting ESR properties.

Recently, Sawyer and coworkers have postulated the presence of mixed valence tetrameric species in solution following electrochemical reduction of $Mo_2O_2S_2(Et_2dtc)_2$ (312a).

B. Mononuclear Complexes

The mononuclear complexes logically divide into monooxo and nonoxo subclasses. Our consideration of the monooxo complexes includes a discussion of bonding which is of further use as a starting point in the treatment of bonding in dimeric compounds.

1. Monomers Containing the MoO^{+3} Unit

The Mo–O_t bond is clearly one of great strength. Its short distance, high stretching frequency (and corresponding high force constant), and general chemical persistence attest to this point. Only a few reactions are known which remove Mo–O_t in a nonredox process. These include reactions which form tris(dithiolene)-like complexes (see Section VIII), and others in which sulfur replaces oxygen atoms.

The electronic structure of complexes containing the oxo-molybdenum (V) group (MoO^{+3}, molybdenyl) were considered by Jørgensen (313) and

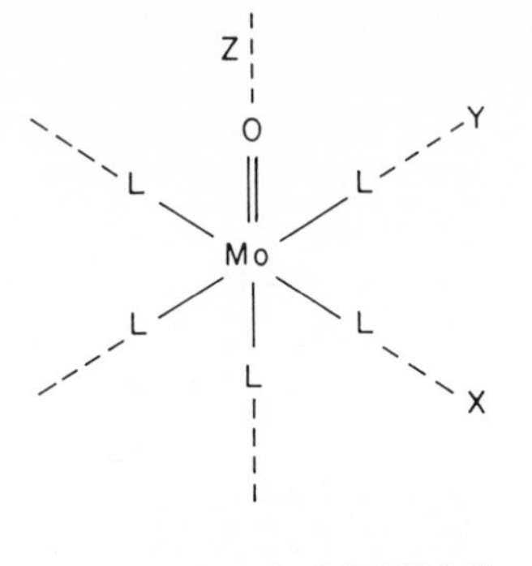

Fig. 18. Coordinate system for $MoOX_5^{-2}$ C_{4v} complexes.

Gray and Hare (314). The latter study followed closely the treatment of the vanadyl ion by Ballhausen and Gray (315). The dominant feature in a molecular orbital treatment of the $MoOX_5^{-2}$ ion is the strong interaction between Mo and O, with the oxo group engaging in strong sigma- and pi-donor interactions.

The coordinate system used in this discussion is shown in Fig. 18. In C_{4v} symmetry the d orbitals transform according to a_1 (d_{z^2}), b_1 ($d_{x^2-y^2}$), $b_2(d_{xy})$, and $e(d_{xz}, d_{yz})$ representations, and the d-orbital (ligand field) portion of the MO diagram is shown in Fig. 19. The a_1 level is strongly sigma antibonding due to interaction with a $p\sigma$ orbital on O and (a weaker interaction) with a $p\sigma$ orbital on the trans group. The b_1 level (mostly $d_{x^2-y^2}$) is also sigma antibonding, but does not lie as high in energy as a_1 because the four halogen ligands in the equatorial plane are not as tightly bound as the oxo group. The d_{xz} and d_{yz} orbitals are involved in strong pi bonding with O p_x and p_y orbitals and as a result the e level is significantly raised in energy above the $b_2(d_{xy})$ level. Assuming Mo–X pi bonding to be negligibly small, the $b_2(d_{xy})$ level is essentially nonbonding. Comparison with the octahedral scheme is shown in Fig. 19, where it is seen that the orbitally nondegenerate b_2 level lies lowest and bears the single unpaired electron. The $^2B_2$ ground

Fig. 19, Ligand-field splitting Diagram for C_{4v} $MoOCl_5^{-2}$ complexes.

state should be characterized by a near spin-only magnetic moment and perusal of Table VII reveals this to be the case. Three "*d-d*" transitions $(b_2)^1 \rightarrow (e)^1$ $(^2B_2 \rightarrow {}^2E)$, $(b_2)^1 \rightarrow (b_1)^1$ $(^2B_2 \rightarrow {}^2B_1)$, and $(b_2)^1 \rightarrow (a_1)^1$ $(^2B_2 \rightarrow {}^1A_1)$ are predicted by this scheme. In most complexes two low-energy, low-intensity bands are seen in the electronic spectrum (at around 13,000 and 22,000 cm^{-1}) and are assigned to the first two of these transitions. The third *d-d* band is often obscured by higher intensity peaks attributable to charge transfer (316). Very recently Garner et al. (316a) have suggested (based on polarization and variable temperature studies) that the second low energy band (at ~ 22,000 cm^{-1}) is probably due to a $^2B_2 \rightarrow {}^2E$ transition which may involve an Mo–O(π) → Mo–O(σ*) or Mo–O(π) → Mo(d_{xy}) excitation or some admixture thereof.

There is still uncertainty as to the detailed assignments of the charge transfer transitions. Gray and Hare (314) assigned the lowest charge transfer as L→M (O(π_b)→d_{xy}) from the pi-bonding oxygen orbitals to the half-filled b_2 level. However, the large increase in the energy in this band as X goes from Br to Cl to F is difficult to reconcile with this notion and would seem to implicate a X(π)→d_{xy} transition at least for the Br and Cl cases (317–320). A great deal of charge transfer data has been summarized by So and Pope (317), who use the optical electronegativity system of Jørgensen (321) to argue persuasively for the halogen(π)→b_2 assignment for the lowest charge transfer band.

The monooxo complexes in addition to possessing near spin-only magnetic moments and characteristic electronic spectra show an extremely intense ν(Mo–O_t) in the infrared with the range 940 to 1020 cm^{-1}. This peak is often the strongest in the spectrum and is a potent diagnostic for the presence of Mo–O_t. However, similar Mo–O_t vibrations are seen in dimeric complexes, and therefore magnetic susceptibility and especially EPR spectroscopy (discussed later in this section) are more useful in establishing the presence of monomeric Mo(V).

The weakness of the bond in the coordination site trans to O_t is emphasized by the isolation of complexes where this site is empty. The x-ray crystallographic study of $MoOCl_3(SPPh_3)$ displays this structural type (287). The salts M[$MoOCl_4$] and M[$MoOBr_4$] have been characterized by a number of workers (319, 322–324), and while their mononuclear nature has not been proved unequivocally, it is generally assumed to be the case. M[$MoOCl_4$] and M[$MoOBr_4$] were first prepared by reaction of $MoCl_5$ and MCl in liquid SO_2 (M = Rb, Cs, pyH, quinH, Et_2NH_2) (324). Preparations in concentrated HCl or HBr, on the other hand, usually lead to $MoOCl_5^{-2}$ or $MoOBr_5^{-2}$ salts being isolated (324). However, Piovesana and Furlani (323) found that when large cations such as Ph_4As^+, $PhCH_2N(CH_3)_3^+$, $N(CH_3)_4^+$, or NBu_4^+ were used, the $MoOCl_4^-$ salts could be produced from $MoCl_5$ and MCl in concentrated

HCl, by reduction of MoO_3 in HCl or by oxidation $M_3[MoCl_6]$ in CH_3OH. The latter reaction even occurs in a N_2 atmosphere, showing the extreme ease with which some $MoCl_6^{-3}$ salts are oxidized (either by CH_3OH or by traces of O_2).

Equilibria between 5- and 6-coordinate monomeric complexes have recently been studied (324a) in CH_2Cl_2 solution. The complexes $MoOCl_3L$ and $MoOCl_3L_2$ readily interconvert with the position of the equilibrium strongly dependent on L. For L = Ph_3PO, $(Me_2N)_3PO$ or thf no $MoOCl_3L$ is found but for L = $SPPh_3$, $(Me_2N)_2CS$ or $(CH_3)_2S$ the 5-coordinate complexes predominate. Kinetic studies (324b) on ligand substitution by Cl^- or Br^- in $MoOCl_3(OPPh_3)_2$ reveal an inverse dependence of the rate on $OPPh_3$ concentration and no dependence on halide, consistent with a dissociative mechanism for this reaction.

Furlani and Piovesana (323) studied the solvent dependence of the electronic spectrum of (NBu_4) $[MoOCl_4]$. They find that the two lowest energy transitions at 14,100 and 22,800 cm^{-1} remain the same in energy, but that the second transition increases greatly in intensity ($\sim$10-fold) when CH_3OH is used as the solvent (as compared to conc. HCl or CH_3CN). Their preferred interpretation of this effect calls on coordination of CH_3OH in the position trans to $Mo–O_t$. This interpretation would then differ from that of Haight (325) and Gray and Hare (314) for *aqueous* solutions of $MoOCl_5^{-2}$, which upon lowering the acidity also show an enhanced absorption in this band. In this case, discussed later, dimerization is considered to be the cause of the enhancement.

Scane and Stephens (322) compared the infrared spectra of solid $R[MoOBr_4]$ and $R[MoOBr_4(H_2O)]$. When R = $(Ph)_4As^+$ they found $\nu(Mo–O_t)$ at 1007 cm^{-1} for the complex without coordinated H_2O, which shifts to 981 cm^{-1} when water is coordinated. Structural studies show the position trans to $Mo–O_t$ to contain the H_2O in (Ph_4As) $[MoOBr_4(H_2O)]$ (286). Dehydration can be effected by heating in vacuum and the hydrate regenerated by exposure to atmospheric moisture. Similar behavior is noted for $[(n\text{-}C_4H_9)_4N]$ $[MoOCl_4]$ and $[(n\text{-}C_4H_9)_4N]$ $[MoOCl_4(H_2O)]$ wherein $\nu(Mo–O_t)$ values of 1011 and 980 cm^{-1} are found, respectively (322). If the retention of monomeric Mo(V) is assumed (as is tacitly done in all of these studies), then it is reasonable to assume that a C_{4v} structure is present for $MoOCl_4^-$. This then illustrates the effect that the presence or absence of a group trans to $Mo–O_t$ has on the $\nu(Mo–O_t)$ as well as the chemically important labilization effect that the oxo group has on the trans position.

A possible formulation of $MoOX_4^-$ compounds as containing 6-coordinate dimeric Mo with halide bridges was suggested by Selbin (9) and by Allen and Neuman (238). This suggestion has precedent in the known structures of $MoOCl_3$ (326) and Mo_2Cl_{10} (290), both of which show chloro bridges and near spin-only moments per Mo. The idea is not inconsistent with most of the data and even the high ν(Mo–O) found by Scane (322) fits, as polymeric $MoOCl_3$ has a (Mo–O) at 1020 cm^{-1}. Since no conductivity or x-ray structural studies have been published, it remains difficult to totally eliminate this possibility, although the EPR parameters for (NBu_4) $[MoOCl_4]$ are characteristic of monomeric species (327).

Complexes of the form $MoOCl_3L$ and especially $MoOCl_3L_2$ are quite common and some are listed in Table VII. These can be formed by direct reaction of $MoOCl_3$ and L (11, 223, 328–330), or by reaction of L with $MoOCl_4$, $(NH_4)_2[MoOCl_5]$, $MoCl_5$, $MoCl_4$, or $MoCl_3$ under appropriate conditions (11, 330–333). $MoOCl_3$(bipy) can be prepared from $(bipyH_2)$-$[MoOCl_5]$, as discussed in Section IV. D. $MoOCl_3(Ph_2POCH_2CH_2POPh_2)$ and related compounds are prepared by oxidation of the zerovalent carbonyl-phosphine complex with excess halogen (334, 335). The affinity of Mo in its high oxidation states for oxo ligands is further illustrated by reaction of $MoCl_5$ with Ph_3AsO, which produces (Ph_3AsCl) $[MoOCl_4]$ (333) and not the simple adduct $MoCl_5(Ph_3AsO)$(330). The reaction of $MoOCl_3$ with potentially bridging ligands (336) gives complexes $MoOCl_3L_2$, with L = $NCCH_2CN$, $NCCH_2CH_2CN$, 4,4′bipyridyl, pyrazine. The expected bridged structure is undoubtedly present, but the complexes show magnetic and spectroscopic properties of mononuclear $MoOCl_3L_2$ complexes. While the ligands probably bridge two $MoOCl_3$ centers, the physical and, to a large extent, the chemical properties are not altered from the mononuclear analogs (336). $MoOBr_3$ reacts similarly to $MoOCl_3$. The reactions of Mo halides and oxyhalides have been reviewed in detail by Walton (11).

TABLE VII
Properties of Mononuclear Mo(V) Complexes

Complex	Infrared ν(Mo–O_t)	μeff (T/°C)	*Electronic spectra*[a] cm^{-1}	$(\varepsilon)^b$	Refs.
$M_2[MoOCl_5]$ in HCl			14,100	(11)	324
[M = NH_4, K, Rb, Cs,			22,470	(10)	and
pyH, quinH, $(CH_3)_3NH$,			28,010	(570)	refs.
$(CH_3)_2NH_2$, $(C_2H_5)_2NH_2$]			32,260	(5,300)	therein
			40,000	(3,600)	
$M_2[MoOBr_5]$ in HBr			14,290	(7)	324
			21,280	(560)	
[M = NH_4, K, Rb, Cs,			24,100	(3,200)	
pyH, quin]			26,530	(2,500)	
$(bipyH_2)$ $[MoOCl_5]$	985	1.73 (303)	14,710	(27)	378
			23,260	(49)	
			28,250	(620)	
$Cs_2[MoOCl_5]$	952	1.72			379

TABLE VII (continued)

Complex	Infrared ν(Mo–O$_t$)	μeff(T/°C)	*Electronic spectra*[a] cm^{-1}	(ε)[b]	Refs.
					380 324
(bipyH$_2$) [MoOBr$_5$]	982	1.85 (295)	14,400 21,500 24,300 26,600	(24) (606) (3,670) (2,920)	381
Cs$_2$[MoOBr$_5$]	948	1.73			379 382 324
(NBμ_4) [MoOCl$_4$]	1,000–1,015		14,100 22,800 (in MeOH)	(26) (330)	323
(PhCH$_2$NMe$_3$) [MoOCl$_4$]	975–995		13,900 22,600 (in CH$_3$CN)	(11) (11)	323
			14,300 22,500 (in HCl)	(13) (14)	323
			14,100 22,800 (in MeOH)	(23) (250)	323
[N(C$_3$H$_7$)$_4$] [MoOCl$_4$]	990	1.71 (297)	14,800 [22,700] [26,700] (solid)		348
(NEt$_4$)$_2$ [MoOCl$_4$Br]			13,250 22,500	(16) (15)	383
(Ph$_4$As) [MoOCl$_4$]	1,012 987	1.73 (301)	13,300 [14,100] 22,500 26,700		322 319
(NEt$_4$)$_2$[MoO(NCS)$_5$]	946	1.67 (293.5)	12,500 [18,400] 21,600 [24,400] 29,800	(104) (22,240) (24,060)	384
(pyH)$_2$[MoO(NCS)$_5$]	950 [970]	1.69 (297)	12,700 [18,600] 21,500 [24,500] 29,800	(76) (17,800) (18,370)	384

TABLE VII (continued)

Complex	Infrared $\nu(Mo{-}O_t)$	μeff(T/°C)	*Electronic spectra*[a] cm^{-1}	(ε)[b]	Refs.
$(NEt_4)_2[MoO(NCSe)_5]$	958				385
$MoOCl_3$	1,007				283
$MoOCl_3(py)_2$	966		13,800		223
			20,800		328
			26,670		
$MoOCl_3(dmso)_2$	1,032, 999	1.64			330
$MoOCl_3(thf)_2$	1,000, 985	1.73	13,400	(40)	328
			19,050		
$MoOCl_3(C_4H_8OS)_2$	990	1.70	13,510		328
			17,540		
$MoOCl_3(Ph_3PO)_2$		1.72	13,700	(38)	335
			22,600	(46)	330
$MoOCl_3(CH_3CN)_2$	980	1.70 (299)	13,700		329
			19,000		
			26,000		
$MoOCl_3(PPh_3)_2$	950	1.72 (298)			329
					332
$MoOCl_3(SPPh_3)$					287
$MoOCl_3(acacH)$	990		14,290		223
			27,780		
$MoOBr_3(bipy)$	968	1.80 (297)	13,900	(40)	381
			21,300	(8,450)	
$MoOCl_3(diars)$	956, 932	1.4–1.5			352
$(NEt_4)[MoOCl_3(ttfa)]$	965	1.87	13,700	(35)	320
			18,900	(300)	
			20,100	(400)	
			[23,800]		
			24,900	(20,000)	
$MoOCl_3(dppe)$ (red)	941	1.73			332
$MoOCl_3(dppe)$ (brown)	952	1.69			332
$MoO(OH)(oxine)_2$	935	1.83	18,200	(2,720)	386
			23,000	(4,300)	340
			25,800	(5,700)	
$MoOCl(acac)_2$	962		[13,510]	(30)	
			[16,810]	(83)	
			26,320	(3,300)	
MoO(OH)(tpp)	901	1.70 (79)			341
MoO(OOH)(tpp)	941, 901	1.72 (79)			341

TABLE VII (continued)

Complex	Infrared $\nu(Mo{-}O_t)$	μeff(T/°C)	Electronic spectra[a] cm^{-1}	(ε)[b]	Refs.
MoOCl(tpp)	990	1.68 (79)			341
MoO(OH)(oep)					342
$MoO(OCH_3)(oep)$	896				387
$[Mo(Et_2dtc)_4]I_3$		1.70			257
$K_3[Mo(CN)_8]$		1.66	25,000	(1,100)	4
			25,900	(1,280)	
			26,500	(1,200)	
			37,700	(2,550)	
			40,500	(2,960)	
$(NEt_4)[MoCl_6]$		1.55 (302)	16,700	(300)	319
			21,700	(950)	
			27,000	(2,300)	
			31,900	(6,700)	
$(Ph_4As)[MoCl_6]$			[16,900]		319
			22,100	solid	
			27,000		
$Cs[MoF_6]$			24,000		388
			29,000		
			35,000		
$(NEt_4)[Mo(OCH_3)_2Cl_4]$		1.71 (297)	11,700	(16) (CH_3NO_2)	348
			21,800	(27)	
$(NEt)[Mo(OEt)_2Cl_4]$		1.73 (297)	12,100	(19)	348
			21,500 (CH_3NO_2 or CH_3OH, HCl)	(29)	
$MoCl_5(N(CH_3)_3)$		1.80			389
$MoCl_5(C_4H_8O_2)$		1.36			333

[a]Numbers in brackets represent shoulders.
[b]Molar extinction coefficients in units l mole^{-1} cm^{-1}.

A report (337) has appeared of a monomeric Mo(V) dithiocarbamate complex formulated as $MoOCl_2$((5-phenylpyrazoline)dtc). The green complex was prepared by reaction of Mo(V) in 7*N* HCl with ligand and extraction into $CHCl_3$-isopentanol. The analytical data and reported characterization (337) leave this a tentative formulation for this interesting species. EPR studies (338) indicate the presence of a number of halo-oxo-dithiocarbamato complexes in concentrated acid solutions of Mo(V) and R_2dtc^- salts.

The unusual complex $(C_5H_5)MoOCl_2$ has been prepared (276) by the aerobic photolysis of $[(C_5H_5)Mo(CO)_3]_2$ in $CHCl_3$ or by treating (C_5H_5)-

MoO_2Cl, $[(C_5H_5)MoO_2]_2O$, or $[(C_5H_5)MoO_2]_2$ with HCl under a variety of conditions. This complex displays $\nu(Mo–O_t)$ at 949 cm^{-1}, has a magnetic moment of 1.74 B.M. (at 297.5°C), and shows a broad EPR signal at $g < 2$. The analogous bromo complex $(C_5H_5)MoOBr_2$ has also been reported (276).

The complex of empirical formula $Mo_2Cl_6O \cdot 6CH_3CN$ was shown to be $[MoCl_2(CH_3CN)_4]$ $[MoOCl_4(CH_3CN)] \cdot CH_3CN$, a mixed Mo(III)-Mo(V) salt (339).

Reactions of $R_2(MoOCl_5)$ with freshly prepared concentrated solutions of thiocyanic acid lead to compounds of the form $R_2[MoO(NCS)_5]$ [R = NH_4^+, pyH^+, and $N(CH_3)_4^+$] (379). Magnetic susceptibility (340, 406) and magnetic resonance (406) studies firmly established the monomeric nature of the complexes and show their close similarity to $MoOX_5^{-2}$ species (where X = F, Cl, Br, or I). The observation of $\nu(CN)$ at 2000 to 2100 cm^{-1}, $\nu(C–S)$ at 870 cm^{-1}, and most importantly $\delta(NCS)$ at 470 cm^{-1} firmly establishes N coordination for all five coordinated thiocyanates.

There are very few examples of MoO^{+3} complexes wherein no halide or pseudohalide ligands are present in the coordination sphere. The complex $MoO(OH)(oxine)_2$ reported by Mitchell and Williams (340) is one potential example, as are the porphyrin complexes MoO(OH)(porphyrin) (341, 342).

Molybdenum porphyrin compounds have not been studied in great detail, but both mono- and dinuclear complexes have been reported. $Mo(CO)_6$ and tetraphenylporphine in refluxing decalin upon workup (including Al_2O_3 chromatography) give the complex MoO(OH) (tpp) (344). Upon dissolution in HCl the complex MoOCl(tpp) is formed and can be isolated in crystalline form as MoOCl(tpp)·HCl. A dimeric form $[MoO(OH)(tpp)]_2$ and a hydro peroxy complex were also claimed (341). The monomeric complexes display magnetic moments and EPR absorption consistent with their formulations (Tables VII and VIII). The complex MoO(OH)(oep) was studied polarographically and was found to be reducible at −0.21 V [in dmso vs. sce with TBAP 0.01*M*]. The reduction is assigned to the Mo(V)-Mo(IV) couple (342), There is no evidence for Mo porphyrins in any biological system.

2. *Nonoxo Mononuclear Mo(V) Complexes*

Monomeric nonoxo Mo(V) complexes are known for a number of ligands. Thus MoF_6^-, $MoCl_6^-$, $MoCl_4(OR)_2^-$, $Mo(CN)_8^{-3}$, and $Mo(R_2dtc)_4^+$ are well characterized ions, and while the 6-coordinate complexes are hydrolytically quite unstable, the 8-coordinate cyanide and dithiocarbamate complexes can be handled in aqueous media. The complex Mo_2Cl_{10}, while dimeric in the solid state, is a trigonal-bipyramidal monomer in the gas phase (307, 343) and is believed to be dissociated in benzene and CCl_4 solutions as well (11).

Salts of $MoCl_6^-$ can be prepared by reaction of Mo_2Cl_{10} with the appropriate chloride (318, 319) and are extremely unstable toward hydrolysis (319). The $MoCl_6^-$ ion in solidified KCl–$MoCl_5$ melt shows an absorption band at 24,100 cm^{-1} (318), assignable to the $(t_{2g})^1 \rightarrow (e_g)^1$ $[^2T_{2g} \rightarrow {}^2E_g]$ transition of an octahedral complex (see Fig. 19). Brisden, Walton, and co-workers (319, 344) found a vastly different spectrum for (Et_4N) $[MoCl_6]$ which showed "low-" intensity bands at 16,700 cm^{-1} (300) and 21,700 cm^{-1} (950). A splitting of the octahedral $^2T_{2g}$ level by symmetry lowering and/or spin-orbit coupling is invoked to explain the spectrum. Single-crystal and temperature-dependence studies on $MoCl_6^-$ doped in Cs_2ZrCl_6 reveal a single low-energy absorption from 23,830 to 25,697 cm^{-1} with considerable vibronic detail (345). In this study no evidence is found for a static Jahn-Teller effect or distortion. On the other hand, the observation of EPR at room temperature for solid (NEt_4) $[MoCl_6]$ would seem to confirm a highly distorted nature for $MoCl_6^-$ in this salt (346).

The more intense bands commencing at 27,000 cm^{-1} in all $MoCl_6^-$ salts are assigned to $L(Cl(\pi) \rightarrow M)$ charge transfer (318, 319, 345). This compares with $\sim$28,000 cm^{-1} for the first intense band in $MoOCl_5^{-2}$ (319), and the proximity of the two numbers lends support to the assignment of this band to $Cl \rightarrow M$ [rather than $O(\pi) \rightarrow M$] charge transfer in the oxo complex as well (318, 320).

The chloro alkoxide complexes $MoCl_4(OR)_2^-$ [R = CH_3, C_2H_5] have been isolated (347) and studied by Brubaker and co-workers (348, 349). The complexes are prepared (348) by low-temperature reaction of $MoCl_5$ with dry ROH and isolated as NMe_4^+, pyH^+, or $quinH^+$ salts; their properties are summarized in Tables VII and VIII. The low-energy absorption bands and infrared and EPR spectra are all consistent with a trans-octahedral structure for these compounds. The alkoxide displays some lability as the ethoxide complex can be prepared by reaction of the methoxide complex with ethanol (349).

The octacyanide ion (4) $Mo(CN)_8^{-3}$ can be prepared by MnO_4^- or Ce^{+4} oxidation of $Mo(CN)_8^{-4}$. Its structural and EPR aspects are considered in other parts of this section. The unique nature of the octacyanide complexes makes it difficult to extrapolate from their interesting chemistry to make any generalizations about the chemistry of Mo.

The same may be said of the complexes $[Mo(R_2dtc)_4]X$ [X = Cl^-, Br^-, I_3^-], also prepared by the oxidation of the corresponding Mo(IV) compounds. Thus the voltammetric reduction potential for the $Mo(R_2dtc)_4^+$ / $Mo(R_2dtc)_4$ couple is -0.55 V (vs. sce in CH_3COCH_3, $NaClO_4$, R = Et) (351) or -0.51 V (vs. sce in CH_2Cl_2, NBu_4ClO_4, R = Et) (257), and iodine or bromine can be used as chemical oxidants. While the compounds are almost certainly 8-coordinate (no monodentate dtc's are seen in the infrared), their structural details are uncertain. The EPR parameters $g = 1.979$ and $A = 39$ G (257) are not typical of 6-coordiate MoS_6 and are discussed later.

The 8-coordinate complex $[MoCl_4(diars)_2]I_3$ has been structurally investigated (289), but except for its fortuitous preparation (289) and various unsuccessful attempts at its preparation (352), there is nothing known of its chemistry. The thiocyanate complexes formulated as $MoCl(NCS)_4 \cdot 2CH_3COCH_3$ and $Mo(NCS)_5 \cdot 2CH_3COCH_2CH_3$ have been reported (353), but are not fully characterized. Various other addition and substitution products of $MoCl_5$ are known, but they are structurally uncharacterized (4, 8, 11).

C. EPR Studies

The Mo(V) oxidation state has been detected by EPR in four of the five well-studied Mo enzymes (Section II. B). For the three oxidases, the similarities in g and A values are such as to suggest a similar but not identical environment for Mo in these enzymes. The EPR signals show rhombic or axially symmetric g and A tensors, and under certain conditions superhyperfine splitting is seen from a single exchangeable proton (Table II). There are at present no well-characterized, isolated Mo(V) complexes which have the g values and hyperfine couplings characteristic of Mo in enzymes. Depite this fact, "model" studies in solution have been of value in establishing something of the character of the Mo site.

Meriwether et al. (354) first implicated thiolate ligands in the binding of Mo in enzymes. Thus Na_2MoO_4 and excess thioglycolic acid in acetate or phosphate buffer (pH 3–7) reveal up to 10% of the Mo as an EPR-active complex with $\langle g \rangle = 1.978$ and $\langle A \rangle = 37$ G. These solutions are unstable and give way within minutes to hours to a species with $g = 2.006$ and finally to one with $g = 1.987$. Thiolactic acid, bal, ethane-1,2-dithiol, and propane-1, 2-dithiol are also able to reduce Mo(VI) to Mo(V) in water, ethanol, acetone, or dmso solutions and show similar EPR signals (354). The g and A components are quite similar to those seen in xanthine oxidase (105, 157) and Meriwether et al. concluded that S is a likely ligand for biological Mo.

Subsequently, Huang and Haight (355–357) found that *l*-cysteine, 2-mercaptopropionic acid, 2-aminoethanethiol, glutathione, and the apoenzyme of putidaredoxin all display similar EPR signals in Mo(V) solutions, whereas the ligands *l*-ala, *l*-his, 1-propanethiol, and edta gave no EPR signal (355–357). Apparently a thiolate donor incorporated into a potentially chelating ligand is required to give an EPR-active species. [The details of the studies on the Mo cysteine system are discussed in the last part of this section.]

Unfortunately, our ignorance of the full donor—atom set and detailed coordination geometry in these model systems is only slightly less than our ignorance of these parameters for the enzyme-bound Mo. While sulfur donors of some sort seem to be implicated, the detailed nature of the coordination spheres and, in fact, whether the complexes are mononuclear or dinuclear is not known with certainty. While diamagnetic dimeric compounds can be eliminated as the source of the EPR signal, complexes of mixed oxidation state [e.g., Mo(IV)-Mo(V) or Mo(V)-Mo(VI)] cannot be eliminated either for enzymes or for these "models."

Recently, Mitchell and Scarle found that proton hyperfine splitting of the magnitude found in Mo oxidases can be observed in Mo complexes containing a pyrrole ligand (376). However, again the nature of the complexes producing this interesting EPR signal is not yet clear. Very recently, proton hyperfine splitting has been seen in some well characterized 6-coordinate nonoxo Mo(V) compounds (237a). For example, the complex $Mo(Et_2dtc)(NHSC_6H_4)_2$ shows a proton hyperfine splitting of 7.4 gauss and an ^{14}N splitting of 2.4 gauss. These findings establish the feasibility of assignment of the proton hyperfine splittings in Mo enzymes to N–H groups bound directly to Mo (237a).

A further example of the promise and pitfalls of the model approach is again found in the original study of Meriwether et al. (354). Here, the inverse linear relation between g values and hyperfine coupling was pointed out, and a plot of g versus A revealed most data points to lie on a straight line. Interestingly, at that time the only complex whose values were found to lie significantly off the line was $Mo(CN)_8^{-3}$, whose structure differs vastly from the near octahedral ones known (or assumed) for other compounds in the study. The fact that the $\langle g \rangle$ and $\langle A \rangle$ values for xanthine oxidase fall on this linear correlation was taken to imply that octahedral Mo (V) is responsible for the Mo signal in the EPR-active enzyme. While this may indeed be correct, more recent work shows that the conclusion may still be premature. For example, $MoOBr_5^{-2}$ falls far off the linear correlation, whereas $Mo(Et_2dtc)_4^+$ falls squarely on it. The latter complex, like $Mo(CN)_8^{-3}$, has a nonoctahedral 8-coordinate structure and, in fact, shows $\langle g \rangle$ and $\langle A \rangle$ values quite close to those of the enzyme. While one can be reasonably certain that xanthine oxidase does not have an MoS_8 core, this result, nevertheless, indicates that isotropic g and A values cannot alone definitively identify either the coordination geometry or the donor—atom set of Mo in enzymes.

Despite these caveats the early model studies (354) served as a potent stimulus to further studies on sulfur-donor complexes of Mo(V). Unfortunately, as discussed later in this section, the vast majority of these studies have found dimeric, diamagnetic complexes possessing MoOMo, MoO_2Mo, or MoS_2Mo bridges. A clear need in this area is for some isolated monomeric Mo(V) complexes whose structures can be ascertained and whose EPR parameters can be correlated with enzyme values. Only then will the enzyme EPR reach its true potential in the definition of the coordination sphere of Mo in enzymes.

The remainder of our EPR discussion largely involves complexes which are at least stoichiometrically well characterized. Most of these are isolated species, which are observed in single crystals, polycrystalline solids, or in liquid or frozen solutions (glasses). We discuss some results of solution studies, but we do not cover this area in detail.

The most intensively studied compounds, and, in fact, the only ones for which single-crystal data are available, are the $MoOX_5^{-2}$ complexes (X = F, Cl, Br). These compounds show axially symmetric g and A tensors with parameters given in Table VIII. The g values for $MoOF_5^{-2}$ are interpreted in a relatively straightforward manner in terms of a conventional second-order perturbation theory treatment (358). In this approach, spin-orbit-interaction acts as a perturbation on the ground state $^2B_2$ wave function and allows other levels to mix in with the ground state causing orbital angular momentum to change the g factor from its spin only value of 2.0023. In the simplest model, only the d levels (i.e., the conventional ligand field levels) are considered to mix with the ground state. Since the d-orbital excitations (which appear in the expressions for g shifts) are available from spectroscopic data, it is a simple matter to show that this treatment predicts $2.0023 > g_{\perp} > g_{\parallel}$, and this is, in fact, found for $MoOF_5^{-2}$ (358). However, for $MoOCl_5^{-2}$ (359) and $MoOBr_5^{-2}$ (360), the experimental finding is $g_{\parallel} > g_{\perp}$, and in $MoOBr_5^{-2}$, $g_{\parallel}$ is greater than the free-electron value of 2.0023. These observations are impossible to reconcile with the simple d-orbital model.

Kon and Sharpless (360) offer the explanation that low-energy charge transfer excitations (terminating in b_2) contribute strongly to the g shift. These excitations from a filled level to b_2 (d_{xy}) will cause a positive g shift for $g_{\parallel}$ (as opposed to d excitations which must give a negative shift in these d^1 compounds). Thus the $b_2 \rightarrow b_1(d\text{-}d)$ transition lowers $g_{\parallel}$, but the $b_1(L(\pi)) \rightarrow b_2)$ transition raises $g_{\parallel}$. In this model the $g_{\parallel} > 2$ for the bromide complex is explained as due to a lower excitation energy which combines with a greater degree of covalent bonding to produce a large positive g shift.

DeArmond (361) and Manoharan and Rogers (358) prefer an alternate explanation in which ligand spin-orbit coupling contributes substantially to g shifts in these highly covalent compounds containing Cl and expecially Br. These halogens have large values for the spin-orbit coupling constants and can make substantial contributions toward raising $g_{\parallel}$. This model, like that of Kon and Sharpless, requires substantial equatorial pi bonding and covalency. The presence of such bonding is substantiated by the observation (358) of significant and anisotropic superhyperfine coupling with the Cl or Br nuclei. The superhyperfine structure is indicative of substantial spin-density in the equatorial in-plane $p\pi$ orbitals of the halogen. The ground b_2 (d_{xy}) level is adapted by symmetry for spin transfer to these orbitals.

Marov et al. (362) considered a large number of compounds and suggested that both charge-transfer excitations and ligand spin-orbit interactions must be considered as additional contributors to $g_{\parallel}$. No matter what the

TABLE VIII
Electron Spin Resonance Data for Mo(V) Complexes

Complex	Medium	$\langle g \rangle$	$\langle A \rangle$	$g_{\parallel}$	$g_{\perp}$	$A^*_{\parallel}$	$A^*_{\perp}$	Remarks	Refs.
$MoOF_5^{-2}$	35% HF	1.913	69.					$\langle A \rangle_F = 13 \pm 1$ G for four equivalent fluorine	183
K_2MoOF_5	Single crystals in $K_2[NbOF_5] \cdot KHF_2$	1.905	69.	1.874	1.911	106.2	50.6	$\langle A \rangle_F = 11.1$ for equatorial F's	358
$MoOCl_5^{-2}$	Single crystals of $(NH_4)[InCl_5(H_2O)]$	1.947	51.2	1.9632	1.9400	81.5	36.0		361
$MoOCl_5^{-2}$	Single crystals of $(NH_4)_2[InCl_5(H_2O)]$	1.949	49.5	1.970	1.938	81.1	33.7		358
$MoOCl_5^{-2}$	10–12M HCl	1.947	50						359
$MoOBr_5^{-2}$	HBr, frozen glass	1.993	44.8	2.090	(1.945)	67.6	(33)		360
$MoOBr_5^{-2}$	Single crystals	1.993	41.7	2.090	1.945	66.0	30.0		390
$MoOI_5^{-2}$	HI solution	2.508						Overlapping I and Mo hfs prevent resolution of either	390
$MoO(NCS)_5^{-2}$	2M $HClO_4$ 0.5M NCS^-	1.940	49					$A_N = 2.2$ G at $-25°C$ for four equatorial N's	391
$(NEt_4)_2[MoO(NCS)_5]$	Acetone-benzene (1 : 9)	1.938	49						384
$(NEt_4)_2[MoO(NCS)_5]$	Acetone	1.935	50	1.928	1.944	76	38		385
$(NEt_4)_2[MoO(NCSe)_5]$	Acetone	1.937	49	1.923	1.947	74	33		385
$(NBu_4)[MoOCl_4]$	CH_3NO_2, 78°K	1.943	54.7	1.968	1.930	82.2	40.9		327
$(NBu_4)[MoOCl_4]$	CH_2Cl_2	1.956	51.5						327
$(C_{12}H_{19}SN)MoOCl_4$	CH_3CN, $CHCl_3$ (1 : 40)	1.950	59					Also shows phenothiazine radical cation at $g = 2.0067$	392
$(NEt_4)[MoOCl_3(hfa)]$		1.947							320
$(NEt_4)[MoOCl_3(dbm)]$		1.945							320
$(NEt_4)[MoOBr_3(dbm)]$		1.982							320

MoO(OH)(tpp)	Benzene	1.960	48							341
MoOCl(tpp)	$CHCl_3$	1.969	50							341
MoO(OOH)(tpp)	Benzene	1.965	50							341
$(NEt_4)[MoCl_6]$	Polycrystalline solid	1.949		1.977	1.935			The similarity of this spectrum to $MoOCl_5^{-2}$ makes this a tentative assignment		346
$(NEt_4)[MoCl_4(OCH_3)_2]$	Powder			1.970	1.920					327
$(NEt_4)[MoCl_4(OCH_3)_2]$	CH_3NO_2, 297°K	1.939	47.8							327
	CH_3NO_2, 78°K	1.939	48.7	1.971	1.923	77.0	34.5			
$Mo(S_2C_2Ph_2)_3^-$	Diglyme	2.011	11.4							701
$(Ph_4As)Mo(S_2C_2(CF_3)_2)_3$	$CHCl_3$	2.0097	12.2							704
$Mo(tdt)_3^-$	Acetone	2.003	29.1							712
$Mo(NHSC_6H_4)_3^-$	thf	1.988	38							720
$Mo(Et_2dtc)_4^+$		1.9794	37.1							257
$K_3Mo(CN)_8$	H_2O, 298°K	1.9915	32.5					$\langle A \rangle_{C^{13}} = 11.7$		350
	Glycerine	1.9920	33.0	1.9981	1.9889	15	42.1			350
	$K_4Mo(CN)_8$	1.9848		1.9779	1.9882					350
$Mo(CN)_8^{-3}$	MeOH glass	1.989	32	1.997	1.985	12	42			363
$(NBu_4)_3Mo(CN)_8$	CH_3CN powder	1.991								288
	(298 or 100°K)	1.991								
$MoOCl_5^{-2}$+oxine	dmf	1.950	48.8	1.935	1.968					277
$MoCl_5^{-2}$ + thiooxine	dmf	1.971	40.3	2.002	1.950					277
$MoCl_5^{-2}$ + 8-mercapto-quinoline	dmf	1.945	50.0	1.969	1.933					277
$MoOCl_5^{-2}$ + $tdtH_2$	dmf	1.992	30.6	1.975	2.004					277
				g_1	g_2	g_3	A_1*	A_2*	A_3*	
Mo(V)-cysteine	Aqueous pH=8	1.975	37	1.931	1.972	2.029	34	54	24	355
Mo(V)-cysteine	BH_4^- treatment of $Mo_2O_4(cys)_2^{-2}$	1.995	41	1.975	1.990	2.019	57	29	37	446

*Hyperfine splitting from $^{95,97}Mo$ in gauss.

detailed contributions are, it is clear that the EPR studies confirm the results of optical studies (317–320) which implicate the in-plane halide ligands in nonnegligible pi bonding.

Remarkably, $MoCl_6^-$ in $NEt_4(MoCl_6)$ is said to show resonance at room temperature (346). This is an unlikely occurrence unless the $MoCl_6^-$ ion is highly distorted such as to produce a large splitting of the degenerate $^3T_{2g}$ ground state. In view of the reported closeness of the g values to those of $MoOCl_5^{-2}$ and the extreme hydrolytic instability of $MoCl_6^-$ (319), this result must be viewed with some skepticism. However, the optical spectrum of $NEt_4(MoCl_6)$ has been interpreted in terms of a highly distorted $MoCl_6^-$ structure (319).

The complex ions $MoCl_4(OR)_2^-$ have been studied by Brubaker and co-workers (327, 348), and their axially symmetric g and A tensors have been nicely interpreted in terms of a trans (D_{4h}, $MoCl_4O_2$) coordination geometry.

The 8-coordinate ion $Mo(CN)_8^{-3}$ has been assigned a square-antiprismatic structure from the analysis of its g tensor in frozen solutions (350, 363). However, when doped in $K_4Mo(CN)_8 \cdot 2H_2O$ the g values change and become consistent with the $Mo(CN)_8^{-3}$ ion adopting the triangular dodecahedral structure of the $Mo(CN)_8^{-4}$ host (350). X-ray studies (288) reveal a triangular dodecahedral arrangement for the anion in $(NBu_4)_3[Mo(CN)_8]$, but the complex $K_3Mo(CN)_8 \cdot 2H_2O$ is isomorphous to $K_3W(CN)_8 \cdot 2H_2O$ which contains square antiprismatic W (305). The combined EPR and structural results may implicate a nonrigid 8-coordination sphere for $Mo(CN)_8^{-3}$. The observation of only one ^{13}C hyperfine splitting (350) from the eight cyanides, although consistent with a square-antiprismatic structure, is also consonant with other explanations. For example, a dodecahedral structure may be present, where the ^{13}C coupling constants for the two types of CN are roughly equal or a rapidly fluxional molecule with a dodecahedral ground state may cause averaging of the splittings. The one-electron reduction product, $Mo(CN)_8^{-4}$, is fluxional to such an extent that even at $-150°C$ its ^{13}C NMR spectrum consists of a single peak [$((n\text{-}C_3H_7)_4N)_4(Mo(CN)_8)$ in $CHClF_2$-CH_2Cl_2 (90:10) as solvent (364, 365)]. It seems that the $Mo(CN)_8^{-3}$ ion is quite possibly fluxional on the EPR time scale.

Russian workers (338, 366–374) have performed a large number of sometimes detailed EPR studies on solutions of Mo(V). These are invariably performed either in strongly acidic media or in an organic solvent which is used to extract the EPR-active complex from the acidic Mo(V) solutions. Some of the earlier work has been reviewed by Marov (371). In virtually all of these studies complexes are not isolated, but in some cases sufficient hyperfine and superhyperfine structure is resolved to definitively identify the stoichiometry and (sometimes) the structure of the species undergoing EPR absorption.

For example, Garifiyanov et al. (366) studied Mo(V) in 35% HF solutions. They found that when the temperature was lowered to 245°K both $^{95,97}Mo$ and ^{19}F hyperfine splittings were resolved. At 217 ± 3°K optimal resolution is achieved, whereas further cooling produces broadening and an anisotropic spectrum appears below 180°K. The hyperfine splitting from four equivalent ^{19}F's establishes $MoOF_5^{-2}$ or $MoOF_4^-$ (or both) as the absorbing species. The parameters found are similar to those of isolated $MoOF_5^{-2}$ salts (358).

There are numerous hints in the Russian work to indicate that Mo(V) EPR signals may be extremely temperature dependent, especially insofar as line width is concerned. For example, Ryabchikov, Marov, and co-workers studied the EPR of acidic Mo(V) thiocyanate solutions (374). At room temperature, the Mo(V) signal due to even isotopes of Mo is unsplit, but at lower T the line (still in liquid solution) is split into nine equidistant superhyperfine components. At these temperatures the line width has decreased to 1.1 G, and the pattern corresponding to splitting by four equivalent nitrogens with $a(^{14}N) = 2.2$ G has appeared. Garifiyanov and Kamenev (373) found a similar temperature dependence and splitting in ether extracts of acidic Mo(V) thiocyanate solutions. Recent results (237) indicate that the isolated complex $(NEt_4)_2$ $[MoO(NCS)_5]$ in CH_2Cl_2 shows a single line for Mo-even isotopes at room temperature, which at low (but above freezing) temperature sharpens to reveal the 9-line superhyperfine pattern.

The observation of ^{14}N hyperfine splitting and its magnitude have some implications for enzymological studies. Several studies on the Mo enzymes have argued that N can be eliminated as a ligand for Mo since no ^{14}N hyperfine splitting is seen in the observed Mo(V) EPR signal (105, 128). This argument is fallaceous. The largest ^{14}N hyperfine in an Mo complex is the value of 4.0 G found for $MoNO(CN)_5^{-3}$ (375). This formal Mo(I) complex is a unique case where strong Mo–NO sigma and pi bonding is present and a very short Mo–N distance is likely. In Mo(V) compounds of thiocyanate and some aminethiolate complexes (237, 272, 274) ^{14}N splittings of less than 3 G are found. [The ligand trans to O_t in the $MoO(NCS)_5^{-2}$ complex while apparently N bound (by infrared studies) gives no resolvable superhyperfine splitting.] The small values found for ^{14}N splitting indicate that for many compounds in which the line width exceeds 3 G this splitting would not be resolved. Furthermore, even in compounds where this superhyperfine splitting is resolved in liquid solution, upon freezing the hyperfine pattern is often no longer seen (237, 373). The majority of enzyme studies has been done at low temperatures in frozen solutions, and the results display line widths which would preclude the observation of ^{14}N splitting, even if N were one of the ligands in the first coordination sphere of biological molybdenum, Clearly the observation of ^{14}N hyperfine splitting would establish its presence as a ligand, but the nonobservance of such splitting does not eliminate ^{14}N as a possibility (74).

Lee and Spence (377) studied complex formation in anhydrous dmf using $(NH_4)_2[MoOCl_5]$ as a source of Mo(V) and the ligands oxine, thio-oxine, 8-aminoquinoline, and 3,4-dimercaptotoluene(tdt). Complexes are shown by Job's method (using optical data) to be 1:1 [ligand to Mo(V)], but whether dmf or chloride (or both) occupy the remaining coordination positions is not clear. Significantly, for the three 8-substituted quinoline ligands the formation constants decrease in the order $S > O > NH_2$. The g and A values for the complexes are shown in Table VIII, but their detailed structures remain uncertain.

While it is undoubtedly true that studies with isolated complexes often confirm solution studies, it is also true that isolation of complexes occasionally illustrates a fallacy in the solution work. A possible example is found in the work of Larin et al. (338) which deals with Et_2dtc^- complexes of Mo(V) in halo-acid solutions. In excess R_2dtc^- the species $MoO(Et_2dtc)_3$ is claimed to be present based on the lack of hyperfine splitting from halogens. However, the reported g and A values for this complex (338) are virtually identical to those of $Mo(Et_2dtc)_4^+$ (257). While it is barely conceivable that $MoO(Et_2dtc)_3$ and $Mo(Et_2dtc)_4^+$ have the same g and A values, it is more likely that $Mo(Et_2dtc)_4^+$ was produced in these studies (338).

D. Dinuclear Complexes

Dinuclear complexes are common in the Mo(V) state with μ-oxo, di-μ-oxo, di-μ-sulfido, di-μ-chloro, and triple bridges represented. The di-μ-chloro arrangement is found for Mo_2Cl_{10}, polymeric $MoOCl_3$, and perhaps a few other compounds (11), and these differ from other bridged species in that they are paramagnetic, displaying near spin-only magnetic moments. In contrast, all other bridges are characterized by diamagnetism, and their chemistry and bonding are discussed below.

1. Complexes Containing the $Mo_2O_3^{+4}$ *Unit*

The structural aspects of $Mo_2O_3^{+4}$ complexes have been discussed in Section IV. A. There are many possible arrangements of the Mo_2O_3 unit, and four of these are displayed in Fig. 20. Arrangements *a* and *b* have been structurally identified by x-ray techniques. The linear structure *d* is known in Re chemistry [for Re(V), $5d^2$ systems] and has been found for the dinuclear Mo prophyrin complex $Mo_2O_3(porphyrin)_2$ (387, 387a). Structure *c* is unknown to date, but has been suggested as a possibility (241) for a reported paramagnetic dimeric complex.

Bonding considerations for the linear $Mo_2O_3^{+4}$ bridge were given by Cotton and co-workers (291) and others (393, 394). We first consider the arrangement where the O_t groups are cis to each other and adopt the same

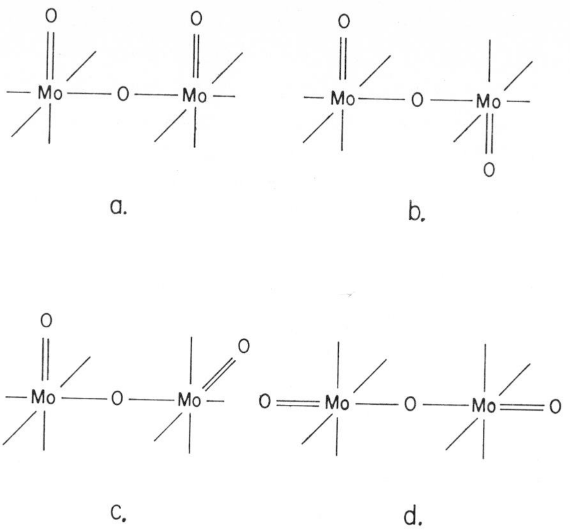

Fig. 20. Structures containing the linear-bridged $Mo_2O_3^{+4}$ unit.

coordinate system as for the mononuclear oxo compounds with Mo–O_t along the Z axis and Mo–O–Mo in the X direction for each metal (Fig. 18). The d orbital splitting on each Mo is again dominated by strong Mo–O_t bonding (see Fig. 19). The $4s$, $4p_x$, $4p_y$, $4p_z$, $3d_{x^2-y^2}$ and $3d_{z^2}$ orbitals on each metal are used in sigma bonding, whereas the $3d_{xz}$ and $3d_{yz}$ orbitals are used [in the conventional (315) manner] in strong pi bonding, with the terminal oxo on each metal. In the mononuclear oxo complexes the d_{xy} orbital is nonbonding, lowest in energy, and contains the unpaired electron. In the dimeric compounds these d_{xy} orbitals are in proper alignment with the p_x orbital on the O_b to allow formation of three-center delocalized molecular orbitals. The interaction produces a bonding, a nonbonding, and an antibonding level. The four electrons [two from O($2p$) and one from each metal] fill the bonding and nonbonding levels and produce a delocalized three-center pi bond over the Mo–O–Mo linkage. The overlap in the bonding M. O. is shown in Fig. 21. This scheme predicts an Mo–O_b bond order of 1.5 and neatly explains the observed diamagnetism of the compounds. It is valid when either cis or trans

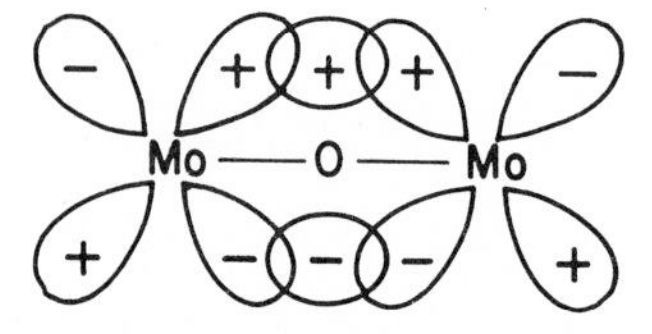

Fig. 21. Overlap in the three-center bonding molecular orbital of Mo-O-Mo.

(Fig. 20*a* or *b*) O_t groups are present. For the case where Mo–O_t (O_b–Mo–O_t dihedral angles) are at 90° to each other, this three-center bond cannot form, and individual pi bonds occur between each Mo d_{xy} and different $O_b(p\pi)$ orbitals. However, now each of the degenerate antibonding levels contains one unpaired electron producing individual Mo–O_b pi-bond orders of 0.5. In this case, although the Mo–O_b bonding is substantially the same in strength a paramagnetic complex with $S = 1$ ground state would be obtained.

The available structural evidence seems to implicate the diamagnetic cis (Fig. 20*a*) or trans (Fig. 20*b*) forms of $Mo_2O_3^{+4}$ dimers as being lower in energy than the paramagnetic skew (Fig. 20*c*) form, with the latter not having been confirmed in any structural study. However, the possibility of a small barrier to rotation associated with a small singlet—triplet energy difference cannot be eliminated. Thus, in solution, an equilibrium between paramagnetic and diamagnetic forms may be present or the paramagnetic form may occur as an intermediate in the interconversion of cis and trans structures.

The assignments of the infrared vibrational bands for complexes containing the $Mo_2O_3^{+4}$ unit have been in some doubt. Thus, conflicting assignments have been published both for ν(Mo–O_t) and (Mo–O_b–Mo) bands (216, 254, 281, 380, 295, 296). However, the latest papers (254, 395, 396) seem generally but not unanamously (380) to agree on the assignments. For the cis structure the Mo–O_t bonds should give rise to two infrared-active stretching bands (symmetric, ν^s, and antisymmetric, ν^a), whereas only one band (antisymmetric, ν^a) should be infrared active for the centrosymmetric trans complexes. In fact, as is seen in Table IX, only one infrared band assignable to ν(Mo–O_t) is usually seen in the appropriate region ~ 920 to 980 cm^{-1}. For the cis complexes this observation may be ascribed either to a very small coupling of the two Mo–O_t oscillators causing overlap of the two bands or to a low intensity of the antisymmetric mode (396). If the latter explanation is correct, then both modes may have been observed in at least one case, and the antisymmetric ν(Mo–O_t) may lie some 50 cm^{-1} higher than the symmetric ν(Mo–O_t) mode (396). O^{18} isotope studies would clarify the situation.

The assignment for the bridge vibrations is somewhat clearer. In the 730 to 785 cm^{-1} region a (usually fairly weak) band is often observed and assigned to the antisymmetric stretch. The band of moderate intensity from 430 to 450 cm^{-1} is considered to be the symmetric stretch. These bands are lower in frequency than the Mo–O_t bands due to weaker Mo–O bonding across the bridge (and concommitant smaller force constants). The much lower frequency of the symmetric vibration is due to its comprising a motion of the Mo atoms, whereas the antisymmetric stretch involves mostly a motion of the lighter O atom. Further support for this assignment comes from the nonobservance of the lower energy band in the centrosymmetric Mo_2O_3-$(S_2P(OR)_2)_4$ complexes where the symmetric stretch is infrared inactive (396).

A further distinctive feature of the $Mo_2O_3^{+4}$ complexes is the presence in their electronic spectrum of an intense absorption at $\sim 19{,}000\ cm^{-1}$ which causes their characteristic deep (usually purple) coloration. This band does not appear in either monomeric, di-μ-oxo, or di-μ-sulfido complexes, and it is not ligand dependent to any great (or systematic) extent. While $L \rightarrow M$ charge transfer has been suggested as the source of the absorption, the lack of variation with ligand and the absence of this band in similar compounds lacking the linear Mo–O–Mo bridge make this unlikely. Rather, it would seem to be characteristic of the linear bridge and could be due to a transition terminating in the antibonding component of the three-center Mo–O–Mo bond. This peak is so distinctive of this class of compounds that those few complexes formulated as containing linear $Mo_2O_3^{+4}$ but not absorbing in this region [e.g., $(NEt_4)_4Mo_2O_3Cl_8$ (397) and $Mo_2O_3(SCH_2CH_2OH)_4$ (279)] should be regarded with skepticism. Of course complexes such as $Mo_2O_3(SO_4)(NCS)_6^{-4}$ which have other probable bridging groups besides O_b will not have a linear bridge, and their lack of absorption at $19{,}000\ cm^{-1}$ is consistent with their proposed structure (380).

The $Mo_2O_3^{+4}$ unit is found with a large number of different ligands occupying the nonoxo sites. All eight remaining sites in the two octahedral Mo coordination spheres are occupied, and donor atom sets of S_8, N_8, O_8, O_4N_4, N_4S_4, and perhaps O_4S_4 and Cl_8 have been established. Some of the known $Mo_2O_3^{+4}$ compounds are tabulated in Table IX. The preparation of $Mo_2O_3Cl_8^{-4}$ will be discussed in Section IV. E, and Mitchell's review (5) covers some of the early work. Here we discuss some of the more recent development and their implications.

By far, the largest number of well-characterized $Mo_2O_3^{+4}$ complexes (and, in fact, the only ones for which structural information is available) are those of the 1,1-dithiolate ligands R_2dtc^-, $Rxan^-$, and $(RO)_2PS_2^-$.

The $Mo_2O_3(R_2dtc)_4$ complexes were first prepared by Malatesta using SO_2 or $S_2O_4^{-2}$ reduction of Mo(VI)-R_2dtc^- solutions (210, 250), but difficulty with the preparations and low yields appear common (252, 254). Moore and Larson (254) used $(NH_4)_2[MoOCl_5]$ as a source of Mo(V), whereas Newton et al. (252) preferred the "the anaerobic hydrolysis product of $MoCl_5$" at ice-bath temperatures with excess ligand. The latter workers also found that the complexes $Mo_2O_3(R_2dtc)_4$ can be prepared by stirring 1:1 $CHCl_3$-methanol solutions of $Mo_2O_4(R_2dtc)_2$ with the appropriate dtc^{-1} ligand. Colton and Scollary (395) found $Mo_2O_3(R_2dtc)_4$ as an oxidation product of $Mo(CO)_3(R_2dtc)_4$ or $Mo(CO)_2(R_2dtc)_4$ by O_2. Mass spectral analyses have found use in the characterization of these compounds (395, 398).

The complexes $Mo_2O_3(S_2P(OR)_2)_4$ (R = Et, Ph) were isolated from the reaction of dithiodiethyl- or dithiodiphenylphosphoric acid in Na_2CO_3 solution with $(NH_4)_2[MoOCl_5]$ (253, 399). The xanthate complexes are conveniently prepared (400) by the acidification of Na_2MoO_4-$Rxan^-$ solutions. This method utilizes $Etxan^-$ as both ligand and reductant and is superior to those which use $S_2O_4^{-2}$ as a reductant, as the latter is capable of reducing Mo to the tetravalent state.

Some interesting reactions have been found for xanthate and dithiocarbamate complexes (254,400). In the presence of alcohols reaction 10 occurs

$$Mo_2O_3(Rxan)_4 + ROH \longrightarrow Mo_2O_2S_2(Rxan)_2 + 2(RO)_2CS + H_2O \qquad (10)$$

whereas excess ligand causes the transformation, Eq. 11, to obtain.

$$Mo_2O_3(Rxan)_4 + (NEt_4)(Rxan) \longrightarrow (NEt_4)_2[Mo_2O_2S_2(S_2CO)_2] \qquad (11)$$

Thus the xanthate complexes are capable of internally extracting S from coordinated ligands. These reactions illustrate the instability of the Mo–O–Mo bridge with respect to the very stable di-μ-sulfido bridge. Additional examples of this are given below.

The complexes $Mo_2O_3(R_2dtc)_4$ have been found (254, 256) to display an unusual and potentially significant property in solution. Thus the disproportionation equilibrium, Eq. 12.

$$Mo_2O_3(R_2dtc)_4 \rightleftharpoons MoO_2(R_2dtc)_2 + MoO(R_2dtc)_2 \qquad (12)$$

has been postulated (254) to explain the noncompliance of the absorption band at 19,700 cm^{-1} with Beer's law. The Mo(IV) and Mo(VI) complexes can be mixed together to generate the $Mo_2O_3(R_2dtc)_4$ species and an equilibrium constant,

$$K_{eq} = \frac{[MoO(R_2dtc)_2]\,[MoO_2(R_2dtc)_2]}{[Mo_2O_3(R_2dtc)_4]}$$
$$= 4 \times 10^{-3} \text{ mole } l^{-1} \; (R = n\text{-}C_3H_7, 41°C)$$

has been determined (256). Solutions of $Mo_2O_3(R_2dtc)_4$ react with azodicarboxylic esters according to the overall reaction 13.

$$Mo_2O_3(R_2dtc)_4 + R'OOCNNCOOR' \longrightarrow MoO_2(R_2dtc)_2 + MoO(R_2dtc)_2(R'OOCNNCOOR') \qquad (13)$$

Since the Mo(IV) complex had previously been shown (401) to react with the azodicarboxylate ester, the reaction undoubtedly illustrates a shift of equilibrium (Eq. 12) to the right as the Mo(IV) product $MoO(R_2dtc)_2$ is removed by complexation.

Newton et al. (254) have speculated that this equilibrium between the monooxo-bridged Mo(V) dimer and monomeric Mo(IV) and Mo(VI) species could occur in Mo enzymes. For example, in the Mo reductases an Mo(V) dimer could split up to produce one active Mo(IV) center which would be involved in reduction of substrate. In Mo oxidases the Mo(VI) center so formed could be the oxidant for the substrate. The internal disproportionation of the dimer could provide the enzyme with a means of concentrating its reducing and/or oxidizing power into a single center capable of undergoing a two-electron redox process.

While the above studies (254) have been useful in delimiting some of the solution properties of $Mo_2O_3(R_2dtc)_4$ compounds, our understanding of their solution chemistry remains incomplete. Thus it is invariably found that even freshly prepared solutions of "analytically pure" $Mo_2O_3(R_2dtc)_4$ show EPR signals (237, 402). The g and A values indicate a rhombic symmetry for the paramagnetic complex (402), but until a fully paramagnetic substance is isolated, discussion of its structure must be considered speculative.

The complexes $Mo_2O_3(oxine)_4$ (340, 403) and $Mo_2O_3(acac)_4$ (404) have been reinvestigated (241, 248, 386). Andruchow and Archer (386) prepared complexes of formula $Mo_2O_3(oxine)_4$ from reactions of the Mo(III) compounds $MoCl_3(py)_3$ or K_3MoCl_6 with oxine. It is of interest that the Mo(V) complex obtains even though precautions are taken to keep the Mo(III) reactions anaerobic. Remarkably, four, presumably isomeric, complexes are claimed with one being paramagnetic to the extent of 1.83 B.M. Andruchow and Archer noted that when the unsymmetrically bidentate oxine ligands are allowed to chelate the remaining positions on the four $Mo_2O_3^{+4}$ frameworks (Figs. 20*a–d*), a total of 21 isomers becomes possible. Furthermore, intermediate configurations in which the $Mo–O_t$'s are skewed are also possible. Perhaps in that perspective the reported four isomers are not so surprising and, as discussed previously, the paramagnetic isomer could be assigned the structure of Fig. 20*c*. On the other hand, the complexes are not reported as being isolated in crystalline form, and various mixtures of dimers or even mixtures of monomers and dimers cannot be eliminated at this time. Other workers (405) have claimed that two different forms of $Mo_2O_3(oxine)_4$ exist which differ in their ability to hold solvent of crystallization. These forms are interconvertible by properly choosing the solvent, and it is possible that geometrical isomers have also been detected in this work.

These results must be compared with the NMR work of Amos and Sawyer (248). Here, the proton NMR of solutions of $Mo_2O_3(oxine)_4$ in dmso was found to display ligand chemical shifts virtually identical to those of

$MoO_2(oxine)_2$. In $MoO_2(oxine)_2$ the ligands are equivalent and the oxine N's are found trans to O_t (175) in the solid state. This caused Amos and Sawyer to propose structure **5**,

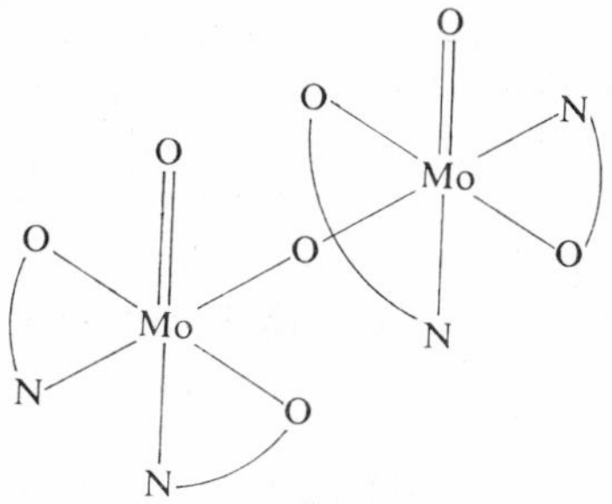

5

with the N's trans to either O_t or O_b. This would suggest that the trans effects of O_b and O_t are comparable at least insofar as chemical shifts in these ligands are concerned. Structurally, both O_t and O_b have substantially larger trans effects than nonoxo ligands, but that of O_t is larger than O_b. In their study there is no hint of the presence of different isomers.

Work with oxine complexes is considered important since the oxine ligand (structure **6**) has a binding structure similar to a potential binding site of flavin (structure **7**) in one of its tautomeric forms. Since three of the Mo enzymes contain flavin, this work may be helpful in assessing Mo flavin binding in chemical and perhaps in enzymological systems.

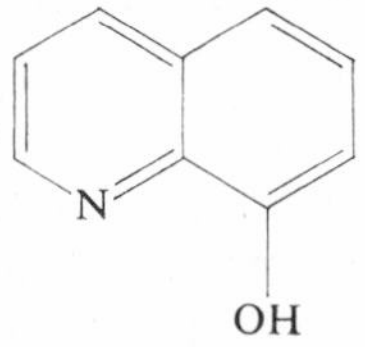

6 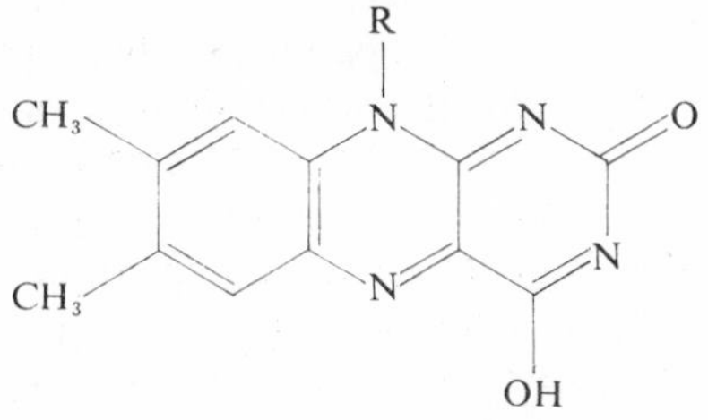 **7**

The only isolated Mo flavin complexes were reported recently by Selbin et al. (102) for Mo(IV) and (one) Mo(V) systems, although a substantial number of solution studies have been performed (97–101). The nature of the isolated species is uncertain (102a).

Gehrke and Veal (241) reinvestigated $Mo_2O_3(acac)_4$ and discussed some of the probable impurities found in earlier preparations. They favor reaction of $MoO_2(acac)_2$ with acacH in a sealed tube for preparative purposes.

TABLE IX

Some Complexes Containing the $Mo_2O_3^{+4}$ Unit

Complex	*Infrared spectra cm*$^{-1a}$		*Electronic spectra*^{a}		Refs.
	$\nu(Mo{-}O_t)^a$	ν(Mo–O–Mo)	cm^{-1}	(ε max)	
$(NEt_4)_4[Mo_2O_3Cl_8]$	983, 958	735, 516			397
$(pyH)_4[Mo_2O_3(NCS)_8]$	955	[868] (873)			380
$(NEt_4)_4[Mo_2O_3(NCS)_8]$	937	[760]	12,700	(930)	
			19,500	(16,000)	237
$[Mo_2O_3(SC(NH_2)_2)_8]Cl_4$			Green		407
$[Mo_2O_3(SC(NH_2)_2)_8]Br_4$			Green		407
$Mo_2O_3(acac)_4$	944 (958)	(435)	20,400	(1,435)	24
			24,500	(3,862)	404
			30,500	(15,000)	408
$Mo_2O_3(oep)_2$		630			387
$Mo_2O_3(oxine)_4$	933		18,200	(10,200)	340
			24,600	(14,200)	386
Mo_2O_3(*N*-Mesal)$_4$	931, 910		14,500		278
			19,300		
Mo_2O_3(*O*-Mecys)$_4$	925	[425]	[14,000]		409
	(932)		19,000		279
			30,000		
Mo_2O_3(*O*-Etcys)$_4$	925	[425]	19,840	(7,550)	279
			30,300	(13,500)	
$Mo_2O_3(SCH_2CH_2OH)_4$	950		[13,300]		279
			14,300	(180)	
			[15,600]		
			[16,700]		
			29,300	(3,870)	
			31,000	(4,140)	
			45,500	(7,120)	
$Mo_2O_3(Me_2dtc)_4$	938	758, 435			254
	(930)	(433)			252
$Mo_2O_3(Et_2dtc)_4$	398 (930)	750,438	19,500	(see text)	254
		(810, 431)	26,450	(5,200)	398
					395
					252
Mo_2O_3((*i*-C_3H_8)$_2$dtc)$_4$	943	758, 438			254
Mo_2O_3((*n*-C_3H_8)$_2$dtc)$_4$	975, 930	770, 447			396
Mo_2O_3(S_2C(*i*-pip))$_4$	938	750, 440	19,500	(see text)	254
			26,500	(5,880)	
$Mo_2O_3(S_2C(4\text{-morph}))_4$	940	735, 435	19,500	(see text)	254
			26,500	(5,500)	
$Mo_2O_3((PhCH_2)_2dtc)_4$	940	427	19,300	(7,000)	398
			26,700	(8,000)	
			32,300	(27,000)	
			38,200	(75,000)	
Mo_2O_3(5-phenylpyr-azolinedtc)$_4$	975, 945	770	19,800	(4,500)	337

TABLE IX (continued)

$Mo_2O_3(Mexan)_4$	952 (945)	766, 432	19,700	(5,020)	254
		(420)			252
$Mo_2O_3(Etxan)_4$	948	766, 431	19,500		254
	(954)	(760, 438)	26,450	(3,430)	252
					396
$Mo_2O_3((i\text{-}C_3H_7)xan)_4$	943	[748], 432	19,680	(11,800)	254
$Mo_2O_3((i\text{-}C_4H_9)xan)_4$	953 [945]	[750], 432	19,680	(13,200)	254
$Mo_2O_3((n\text{-}C_4H_9)xan)_4$	953	[747], 427	19,720	(12,000)	254
$Mo_2O_3(S_2P(OPh)_2)_4$	972	785	15,400	(700)	253
			20,100	(1,000)	
$Mo_2O_3(S_2P(OEt)_2)_4$	967	782			253
$Mo_2O_3Cl_4(bipy)_2$	966		14,200		341
			19,500		
			24,200		
$Mo_2O_3Br_4(bipy)_2$	960		14,200		381
			19,600	(intense)	
			24,200		
			27,100		
$Mo_2O_3Br_4(phen)_2$	965		13,400		412
			18,000		
			23,200		
$[Mo_2O_3(C_2O_4)_2(phen)_2] \cdot 2H_2O$	965	720	[27,800]	(1,460)	410
			36,400	(20,900)	
$Na_2[Mo_2O_3(C_2O_4)_2(oxine)_2] \cdot 5H_2O$	970	710	27,170	(6,540)	410
			32,260	(9,700)	
			39,840	(71,500)	
$Na_2[Mo_2O_3(C_2O_4)_2(pic)_2] \cdot 3H_2O$	973	720	[20,800]	(700)	410
			31,900	(4,100)	
			37,600	(9,200)	
$(pyH)_4[Mo_2O_3(SO_4)(NCS)_6]$	956				380
$(pyH)_4[Mo_2O_3(C_2O_4)(NCS)_6]$	963				380
$(pyH)_3[Mo_2O_3(ClO_4)(NCS)_6]$	971, 981				380

[a]Numbers in brackets represent shoulders or very weak absorptions.

$Mo_2O_3L_4$ complexes, with L = *O*-ethylcysteine, *O*-methylcysteine, and *N*-methyl-salicylaldimine, have been reported (Table IX).

Salts of the form $R_4(Mo_2O_3(NCS)_8)$ [R = $(CH_3)_2NH^+$, $(CH_3)_4N^+$] were prepared by James and Wardlaw (433) from $R_2(MoOCl_5)$ and NH_4-NCS in concentrated aqueous solutions. The NCS vibrations are consistent with the presence of N-bound thiocyanate.

The dimeric complex $O[MoO(oep)_2]_2$ is of interest in that recent structural work (387a) shows a linear O=Mo–O–Mo=O linkage with Mo–O_t = 1.71 and Mo–O_b = 1.94 Å. It has an infrared band at 630 cm^{-1} assigned to the asymmetric stretch of the OMoOMoO unit. The low frequency is consistent with this assignment (387).

2. *Complexes Containing the* $Mo_2O_4^{+2}$ *Unit*

The structure of the $Mo_2O_4^{+2}$ group has been discussed in Section IV. A. In all known cases the MoO_2Mo bridge has a bent arrangement with the $Mo–O_t$ groups on the same side of the bridge (i.e., cis). A chemically significant point is that (not including the Mo–Mo bond) the Mo's can achieve either 5 or 6 coordination. In 5 coordination both square pyramidal and distorted trigonal-bipyramidal coordinations are possible about Mo. The terminal oxo group has the ability to weaken the bond trans to it, and in the extreme this trans ligand is absent. The special lability of the site trans to oxo is illustrated by isotope exchange studies (413) on $Mo_2O_4(NCS)_6^{-4}$, which show that only two of the six thiocyanate ligands exchange rapidly with free ^{14}C-labeled NCS^- in H_2O or methanol. The labile NCS^- ligands presumably lie trans to $Mo–O_t$.

The electronic spectra of $Mo_2O_4^{+2}$ species do not, in general, display the low-energy transitions which are characteristics of both MoO^{+3} monomers and $Mo_2O_3^{+4}$ dimers. The bonding in the $Mo_2O_4^{+2}$ compounds has been considered both qualitatively and quantitatively (380, 394, 414). There is general agreement that a direct Mo–Mo bond is present, and this is strongly supported by structural (Section IV.A.), magnetic (415, 416), and spectral studies (380, 394).

For a qualitative description of the bonding we adopt the same coordinate systems about Mo which were used for MoO^{+3} and $Mo_2O_3^{+4}$ species. After sigma and pi $Mo–O_t$ bonding are accounted for, each metal is left with a single unpaired electron in its d_{xy} orbital. These d_{xy} orbitals are oriented to allow direct overlap and Mo–Mo bonding (Fig. 22). The bonding level of this interaction accommodates the two electrons in accord with the diamagnetism of the compounds in this class.

The $Mo_2O_4^{+2}$ grouping is characterized by terminal [$\nu(Mo–O_t)$] stretches between 925 and 985 cm^{-1}. Although generally only one intense absorption is observed, a second band has been reported in some cases (see Table X), and this would agree with the prediction of two infrared-active modes (antisymmetric and symmetric) for the *cis*-$Mo–O_t$ grouping. Vibrational

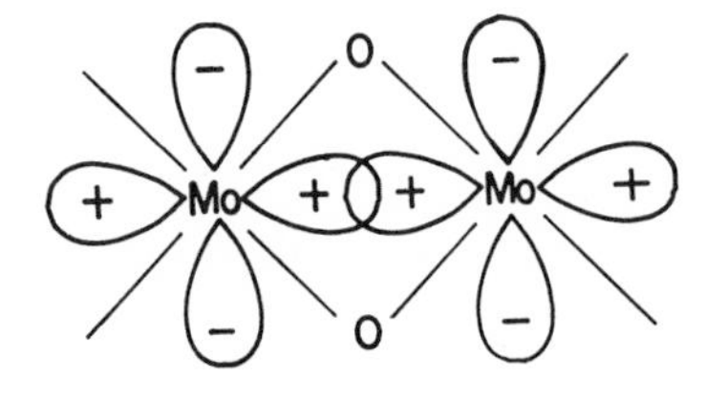

Fig. 22. Overlap in the Mo-Mo bond across the Mo_2O_2 Bridge viewed down the *Z* axis.

TABLE X
Some Complexes Containing the $Mo_2O_4^{+2}$ Unit

Complex	Infrared spectrum cm^{-1} $\nu(MoO)$	$\nu(MoO_2Mo)$	Electronic spectrum cm^{-1}	(ε)	Refs.
$(NEt_4)_4$ $[Mo_2O_4(NCS_6)]$	943				281
$(pyH)_4$ $[Mo_2O_4(NCS)_6]$					331
					430
$(pyH)_2[Mo_2O_4Cl_4(H_2O)_2]$	978				281
$Li_2[Mo_2O_4Br_4(H_2O)_2]$					281
$(NH_4)_2[Mo_2O_4F_4(H_2O)_2]$	980, 940				431
$Mo_2O_4(bipy)_2Cl_2$	952		13,900		411
			20,200		
			24,400		
$Mo_2O_4(phen)_2Br_2$	952		13,600		412
			28,400		
			23,400		
$Mo_2O_4(phen)_2Cl_2$	955				432
$Mo_2O_4(bipy)_2Br_2$	955		13,600		381
			18,000		
			22,600		
$Ba[Mo_2O_4(C_2O_4)_2(H_2O)_2]$	975				281
$(pyH)_2[Mo_2O_4(C_2O_4)_2$ $(H_2O)_2]$	975	725	[20,000]		410
			26,000	(1,180)	
			32,800	(3,420)	
$(pyH)_2[Mo_2O_4(C_2O_4)_2$ $(py)_2]$	960	743	[20,000]		410
			26,000	(187)	
			32,800	(3,580)	
$K_2[Mo_2O_4(ox)_2(H_2O)_2]\cdot 3H_2O$			20,400	(82)	281
			26,000	(340)	
			32,800	(7,040)	
			38,900	(7,040)	
$Na_2[Mo_2O_4(edta)]\cdot nH_2O$	943		25,640	(336)	263
			33,670	(8,980)	426
					427
$Ba[Mo_2O_4(edta)]\cdot 3.5H_2O$	945		23,800		426
			30,800		
$Mo_2O_4^{+2}$ (aq.)			26,040	(103)	427
			33,900	(3,500)	418
			39,370	(4,000)	
$Mo_2O_4(his)_2$	941	750, 475	33,700	(4,000)	419
$Mo_2O_4(C_5H_5)_2$	925	710, 462			276
					424
$Na_2[Mo_2O_4(cys)_2]\cdot 5H_2O$	955	(735)	32,700	(12,000)	279
	970, 945, 925	730, 425	48,800	(19,000)	409
			32,600	(11,800)	
$Mo_2O_4(O\text{-}Mecys)_2$	978 (950)	739	34,700	(8,900)	279
			36,200	(9,050)	

TABLE X (continued)

$Mo_2O_4(SCH_2CH_2OH)_2$	956	702	31,600 45,700	(3,200) (6,200)	279
$Mo_2O_4(O\text{-Etcys})_2$	980 (945)	735	34,700 36,200	(8,900) (9,050)	409
$Mo_2O_4(Me_2dtc)_2$	978 (960)	736, 480			254
$Mo_2O_4(Et_2dtc)_2$	977 (960)	733, 480			254
$Mo_2O_4((1\text{-pip})\ dtc)_2$	980 (965)	732, 472			254
$Mo_2O_4((4\text{-morph})dtc)_2$	982 (965)	734, 470			254

analyses of the MoO_2Mo bridging modes have been given (380, 417). Experimentally, two bands are usually associated with this grouping. The first band, occurring from 700 to 750 cm^{-1}, is of moderate to strong intensity, whereas the second band, occurring from 460 to 490 cm^{-1}, is more variable in strength. While the positions of these bands are somewhat similar to those of the $Mo_2O_3^{+4}$ grouping, the 700 to 750 cm^{-1} band is much more intense and, in general, serves to diagnose the presence of the $Mo_2O_4^{+2}$ linkage. The data for a number of complexes are displayed in Table X.

The $Mo_2O_4^{+2}$ (aq.) ion is apparently the main species present in acid aqueous solutions below $[H^+] = 2M$ (418). The high pH limit for its existence and its fate at the hands of the increasing hydroxide concentrations are unknown, but ultimately $MoO(OH)_3$ precipitates. By carrying out preparations in acidic (but not too acidic) solutions, a large number of $Mo_2O_4^{+2}$-containing complexes have been isolated, with precipitation often affected by raising the pH or adding alcohol. In a number of cases the di-μ-oxo species has been prepared by hydrolysis of an $Mo_2O_3^{+4}$-containing complex. For example (409), the purple mono-μ-oxo-bridged *O*-methylcysteine complex can be cleanly converted to the yellow dioxo-bridged species (Eq. 14) and a similar preparation of $Mo_2O_4(R_2dtc)_2$ complexes has been reported (252).

$$Mo_2O_3(O\text{-Mecys})_4 + H_2O \xrightarrow{CH_3OH} Mo_2O_4(O\text{-MeCys})_2 \quad (14)$$

Amino-acid complexes of the $Mo_2O_4^{+2}$ core have been isolated for *l*-cys (279, 409) and *l*-his (419), and others have been detected by their ORD spectra in solution (420, 421) in the pH range 2.5 to 8.5. Complexes with saccharides and other naturally occurring ligands have also been investigated in solution (421–423).

A cyclopentadienyl complex having the $Mo_2O_4^{+2}$ core has been prepared by Cousins and Green (276). The complex $Mo_2O_4(C_5H_5)_2$ is obtained from the air oxidation of $(C_5H_5)Mo(CO)_3H$ or from the photochemical oxidation of $[(C_5H_5Mo(CO)_3]_2$ in $CHCl_3$ (276). It can also be formed by hydrolysis of $(C_5H_5)MoCl_4$, with the tetrameric species $Mo_4O_8(C_5H_5)_4$ isolated as an intermediate (424). The di-μ-oxo complex shows absorptions at 925, 910, and 710 cm^{-1}, which may be attributable to terminal, terminal, and bridging Mo–O virbrations, respectively. If these assignments are correct, they imply a structure with *cis*-O_t atoms. This would mean that $Mo_2O_4(C_5H_5)_2$ and $Mo_2O_2S_2(C_5H_5)_2$ are not isostructural since the latter has a trans centrosymmetric structure.

Salts of the anion $Mo_2O_4(NCS)_6^{-4}$ were first prepared by James and Wardlaw (433) by reacting $MoOCl_5^{-2}$ with NH_4SCN in a warm solution [more dilute than for the $Mo_2O_3(NCS)_8^{-4}$ preparation]. The isolated complexes $R_4[Mo_2O_4(NCS)_6]$ ($R = pyH^+$, $quinH^+$) are diamagnetic [340] and once formed are stable in the nonaqueous solvents in which they dissolve. Infrared evidence is consistent with N-bound thiocyanate.

The complex $Na_2[Mo_2O_4(edta)] \cdot H_2O$ contains the dinuclear di-μ-oxo dimolybdenum group which is further bridged by the edta ligand. Preparation involves neutralization of Mo(V)-edta solutions [with Mo(V) prepared by reduction of Mo(VI) with Hg in 3*N* HCl]. A similar procedure leads to isolation of the pdta complex (263). The proton NMR of these complexes in aqueous solutions indicate the diamagnetic dimeric structures to be retained from pH 0 to 9.0 (425). The barium salt $Ba[Mo_2O_4(edta)] \cdot 3.5H_2O$ was isolated by Hruskova et al. (426) and serves as a useful precursor for the preparation of other salts. Reaction of $Ba[Mo_2O_4(edta)] \cdot 3.5H_2O$ with M_2SO_4 leaves a precipitate of $BaSO_4$ and allows for isolation of $M_2[Mo_2O_4(edta)] \cdot xH_2O$ (M = H, Li, Na, K, Rb, Cs, NH_4) from the resulting solutions (426). Complexes of Mo(V) with the related ligands nta and mida (425) were studied in solution and found to be stable between pH 1 and 8 (nta) and between pH 4 and 9 (mida). However, the structures of these complexes remain uncertain.

The equilibrium and kinetics of dissociation of $Mo_2O_4(edta)^{-2}$ to form $Mo_2O_4^{+2}$ at $H^+ = 0.5$ to $2.0M$ ($HClO_4$) have been studied by Sasaki and Sykes (427). The dissociation and reassociation are H^+ dependent in a complex way, and the structural requirements of the $Mo_2O_4(edta)^{-2}$ complex are probably responsible for the complication. The stereochemistry of the complex ion requires the edta nitrogens to be trans to the O_t groups in order for edta to be hexadentate on the $Mo_2O_4^{+2}$ core (see Fig. 14 for the di-μ-sulfido structure). However, in the acid solution the nitrogens of the free edta are fully protonated and initial attack on Mo must occur through carboxylate oxygen. If the COO^- binds trans to O_t, the stable complex cannot form, but it is the trans site which is the more labile. Thus unstable complexes form and must dissociate before the stable entity is produced, and this may be the cause of the atypical behavior of this edta complexation reaction when compared to those of other metals (427).

The ion $Mo_2O_4(edta)^{-2}$ is reducible in a four-electron, four-proton, apparently reversible process to the dimeric Mo(III) complex formulated as

$Mo_2O_2(H_2O)_2(edta)^{-2}$ (428). The formal reduction potential for this couple (vs. hydrogen electrode) is very pH dependent with E° = 0.10 V at pH 2, but −0.50 V at pH 7. In basic solution the Mo(III) complex is an exceedingly strong reductant capable, for example, of reducing $Cd(NH_3)_4^{-2}$ to Cd metal (478). This is the only system where a reduction potential for the Mo(III)/Mo(V) couple is experimentally known at pH = 7. The value of −0.50 V is remarkably close to what is required for the Mo–Fe protein of nitrogenase (75–77). Detailed voltammetric studies have confirmed the presence of a quasi-reversible four-electron four-proton reduction process (428a). Significantly, the nature and concentration of oxyanions present as buffers or electrolytes has profound effects on the voltammetric characteristics.

The reactions of $Mo_2O_4^{+2}$ species which involve breaking one or both of the bridges have not been extensively studied, but there are some results. $Mo_2O_4^{+2}$ complexes can be transformed to $Mo_2O_2S_2^{+2}$ or $Mo_2S_4^{+2}$ as discussed in Section IV.D.3. Mitchell (410) used $Mo_2O_4(C_2O_4)_2(H_2O)_2^{-2}$ as a starting material to make a number of $Mo_2O_4^{+2}$ and $Mo_2O_3^{+4}$ complexes. Newton et al. (400) formed the $Mo_2O_3(R_2dtc)_4$ complexes from $Mo_2O_4(R_2dtc)$ by reaction with excess R_2dtc^-. Reactions of $Mo_2O_4(cys)_2$, halide, and thiocyanate complexes are discussed below, as are reactions which lead to triple bridges.

Recently $Mo_2O_4(R_2dtc)_2$ and $Mo_2O_4(S_2PR_2)$ complexes were shown to react (237a, 258b) with certain thiolate containing bidentate ligands to produce monomeric complexes. With o-hydroxybenzenethiol the complex $MoO(S_2P(i\text{–}C_3H_7)_2)(OSC_6H_4)$ ($g=1.971$, $A_{Mo}=39$, $a_P=38$ gauss) can be formed while o-aminobenzenthiol and 1,2-benezenedithiol yield the nonoxo complexes $Mo(Et_2dtc)(NHSC_6H_4)_2$ ($g=1.991$, $A_{Mo}=38$, $a_N=2.4$, $a_H=7.4$ gauss) and $Mo(Et_2dtc)(S_2C_6H_4)_2$ ($g=2.003$, $A_{Mo}=31.8$ gauss). These results indicate that with proper choice of ligand it may be possible to overcome the necessity for strong acid in the preparation of mononuclear Mo(V) complexes.

3. *Complexes Containing the μ-Sulfido Linkage*

The complexes containing the $MoO_2S_2^{+2}$ grouping have either planar or bent MoS_2Mo bridges (297–303) and their structures have been discussed in Section IV.A. Complexes with planar bridges contain *trans*-$Mo\text{–}O_t$ groups and (at least to date) have only been found with cyclopentadienyl ligands (297, 303). Although two $\nu(Mo\text{–}O_t)$ stretching modes are predicted to be infrared active for the *cis*-$Mo\text{–}O_t$ structure, the complexes listed in Table XI show only one $\nu(Mo\text{–}O_t)$ vibration appearing from 920 to 970 cm^{-1}. The electronic absorption spectra appear to have a characteristic band at 27,000 cm^{-1}.

In general, the disulfido bridge can be formed by reaction of dioxo-bridged complexes with H_2S (211, 279, 434), and the complex anions $Mo_2O_2S_2(cys)_2^{-2}$ and $Mo_2O_2S_2(his)_2^{-2}$ are prepared by bubbling H_2S into aqueous

TABLE XI
Complexes Containing the $Mo_2O_2S_2^{+2}$ Group

Complex	Infrared spectrum $\nu(Mo\text{–}O_t)$ cm^{-1}	*Electronic absorptions* cm^{-1}	(ε)	Refs.
$Na_2MoO_2S_2(cys)_2 \cdot 3H_2O$	948	13,000	(160)	279
		23,200	(820)	
		28,600	(2,800)	
		36,000	(7,000)	
$Na_2Mo_2O_2S_2(cys)_2 \cdot 5H_2O$	955	30,300		211
		35,200	(10,000)	
		44,000	(25,000)	
$Na_2Mo_2O_2S_2(his)_2 \cdot 5H_2O$		27,000		211
		32,200		
		35,800	(8,000)	
		40,000	(sh)	
$Mo_2O_2S_2(C_5H_5)_2$	920			435
$Mo_2O_2S_2(edta)^{-2}$				299
$Mo_2O_2S_2^{+2}$ (aq.)		27,000	(sh)	436
		32,000		
		34,500		
		49,000		
$Mo_2O_2S_2(O\text{-}Etcys)_2$	966	34,500	(5,800)	279
		39,500	(7,000)	
		44,200	(10,600)	
$Mo_2O_2S_2(Mexan)_2$	825[a]			
$Mo_2O_2S_2(Etxan)_2$	830[a]	26,800	(1,360)	400
$Mo_2O_2S_2(i\text{-}C_3H_7xan)_2$		27,000	(1,560)	400
$Mo_2O_2S_2(i\text{-}C_4H_9xan)_2$		26,950	(730)	400
$Mo_2O_2S_2(i\text{-}C_4H_9xan)_2$	968 (900)	26,880	(970)	400
$(NEt_4)_2Mo_2O_2S_2(SOCS)_2$	968 (955)	28,800	(4,250)	400

[a]In KBr discs, $\nu(Mo\text{–}O_t)$ increases to ~ 980 cm^{-1} on dissolution indicating probable Mo=O–Mo bridging in the solid.

solutions of the di-μ-oxo analogs (211, 279). The substitution of the two bridging oxo's is characterized by disappearance of the ~ 740 cm^{-1} absorption characteristic of the dioxo bridge.

$Mo_2O_2S_2[(n-C_4H_9)_2dtc]_2$ was prepared by Moore and Larson (252) by bubbling H_2S into solutions of the Mo(VI) complex $MoO_2[(n\text{-}C_4H_9)_2dtc]_2$ in benzene. The preparation of $Mo_2O_2S_2(Rxan)_2$ from $Mo_2O_3(Rxan)_4$ was discussed previously (400). The complex $Mo_2O_2S_2(C_5H_5)_2$ is isolated through air oxidation of the crude products obtained from the reaction of $[(C_5H_5)$-$Mo(CO)_3]_2$ with cyclohexenesulfide. It is extremely stable to both air and moisture (435).

The ion $Mo_2O_2S_2(cys)_2^{-2}$ can be hydrolyzed free of its cysteine ligands using 1.0*N* HCl (436). Under these extreme conditions the resultant Mo_2O_2-S_2^{+2} ion is stable and can be purified by passage through a Sephadex G-10 column. Remarkably, no H_2S is evolved even when the complex is left in 10*N* acid solutions for extended periods. The great inertness of the MoS_2Mo bridge would seem to eliminate its presence in biological systems *if* bridge cleavage of an Mo(V) system is postulated as part of a catalytic cycle. The $Mo_2O_2S_2^{+2}$ ion shows a distinctive visible spectrum, with maxima shown in Table XI.

Substitution of the terminal oxygens by sulfurs has also been affected, but it requires more vigorous conditions. For example, the reaction of $Mo_2O_4(R_2dtc)_2$ with P_4S_{10} in xylene (437) leads to $Mo_2S_4(R_2dtc)_2$. Crystallographic studies (304) on the R = *n*-butyl complex indicate a multiply bonded terminal Mo–S analogous to Mo–O_t. The complex $[MoS_2(Etxan)]_x$ (where x is greater than 1 but otherwise unknown) is possibly of similar structure, but insolubility has hampered its characterization. It is prepared by reaction of $Mo_2O_3(Extan)_4$ with H_2S in benzene-ethanol (434). However, in $CHCl_3$ solution reaction of $Mo_2O_3(Etxan)_4$ and H_2S gives $Mo_2O_2S_2(Etxan)_2$ (400). Two isomeric compounds of composition $Mo_2S_4(C_5H_5)_2$ have been reported, but their structures are unknown (439).

The complexes $Mo_2O_3S(R_2dtc)_2$ have been isolated in which one of the bridging oxos in $Mo_2O_4(R_2dtc)_2$ has been replaced by bridging, S, whereas the second oxo bridge remains intact (258). These μ-oxo-μ-sulfido compounds are part of the series $Mo_2O_4^{+2}$, $Mo_2O_3S^{+2}$, $Mo_2O_2S_2^{+2}$, $[Mo_2OS_3^{+2}]$, $Mo_2S_4^{+2}$, where only the species in brackets has yet to be found in an isolated compound. Apparently, S and O can be interchangeably substituted in the Mo(V) coordination sphere without affecting gross structural change.

The reaction of $Mo_2O_3(R_2dtc)_4$ or $MoO_2(R_2dtc)_2$ with excess H_2S leads to the isolation of compounds of formula $Mo_2O_2S_3(R_2dtc)_2$ (258). These compounds have one additional sulfur atom per dimer, and in analogy with results on Zn(II) and Ni(II) 1,1-dithiolate complexes (438) are designated as sulfur-rich species. Significantly, CN^- or Ph_3P can abstract the sulfur (as SCN^- or Ph_3PS) to produce the known $Mo_2O_2S_2(R_2dtc)_2$ complexes. The sulfur is perhaps present as part of a persulfido linkage, and a characteristic band at 550 cm^{-1} is present in the sulfur-rich, but not in the $Mo_2O_2S_2(R_2dtc)_2$, complexes. Whether the persulfido linkage is in the bridge or in the ligand (as in the Ni and Zn cases) is not known.

Significantly, a persulfido linkage has been suggested as a key participant in xanthine and aldehyde oxidase activity. Edmondson et al. (111) have postulated an RSS^- group as a protein-bound nucleophile which attacks the 8-position of xanthine (or the aldehyde carbon) in the course of the catalytic cycle. Cyanide can abstract a sulfur from xanthine and aldehyde oxidases, and this has been suggested (111) as the cause of cyanide inactivation of the enzyme. On the other hand, Bray (112) has suggested that the persulfide linkage may serve as a ligand in the Mo coordination sphere. The above results on the dithiocarbamate complexes illustrate that persulfido Mo complexes may exist and show reactivity comparable to the enzyme persulfide. While they add some weight to the postulation of a persulfide ligand to Mo (112), they do not in any way rule out the persulfide nucleophile hypothesis. In fact, the two suggestions (111, 112) are not incompatible. Furthermore, the possibility must be carefully considered as to whether terminal Mo–S linkages are present in these cases.

4. *Other Bridged Structures*

Haight, Enemark, and co-workers (283) have found that Mo_2O_4-(oxine)$_2$(py)$_2$ (429) reacts with $HSCH_2CH_2OH$ to give the triply (O, S, O) bridged structure $Mo_2O_3(\mu\text{-}SCH_2CH_2O)(oxine)_2$, which is shown in Fig. 16.

R′SH reacts with $Mo_2O_4(R_2dtc)_2$ to form compounds of the form $Mo_2O_3(SR')_2(R_2dtc)_2$ (254), which are also formulated as being triply bridged (by one oxo and two alkyl thiolates).

The interesting compounds $(pyH)_4[Mo_2O_3(SO_4)(NCS)_6]$, $(pyH)_3$-$[Mo_2O_3(ClO_4)(NCS)_6]$, and $(pyH)_4[Mo_2O_3C_2O_4(NCS)_6]$ were prepared (380) by reaction of $(pyH)_4[Mo_2O_4(NCS)_6]$ with sulfuric, perchloric, and oxalic acid, respectively. The oxyanion ligand was implicated as bridging by the change in its infrared spectrum upon coordination. For example, in the sulfate case the intense ν_3 band is split into a triplet, indicating a local C_{2v} symmetry about S. Unsymmetrical monodentate sulfate is eliminated by the mode of preparation of the compounds from $Mo_2O_4(NCS)_6^{-4}$, with no evidence for NCS dissociation, and by the Mo–O–Mo vibrations. Since the ligands appear to be bridging, two modes of bonding, structures **8** and **9**, can be considered

8

9

Although the structural studies might appear to support the triple bridge (structure **9**), this would require 7-coordinate Mo. Structural characterization of these compounds is clearly desirable in view of the role of Mo enzymes in catalyzing reactions of oxyanions (NO_3^-/NO_2^-, SO_4^{-2}/SO_3^{-2}, ATP/ADP + P_i).

Very recently (380a) the complex anion $Mo_2O_4(NCS)_4(CH_3CO_2)^{-3}$ was isolated and structurally characterized in its pyridinium salt. It has structure **9a** where a conventional $Mo_2O_4^{-2}$ type bridge is supplemented by bridging of the acetate ligand. A similar bridging acetate (380b) has been found for a Mo(III) dimer having an $Mo(OH)_2Mo$ bridge and these results may be indicative of a common type of supplemental bridging in Mo_2X_2 bridged systems (X=O, OH).

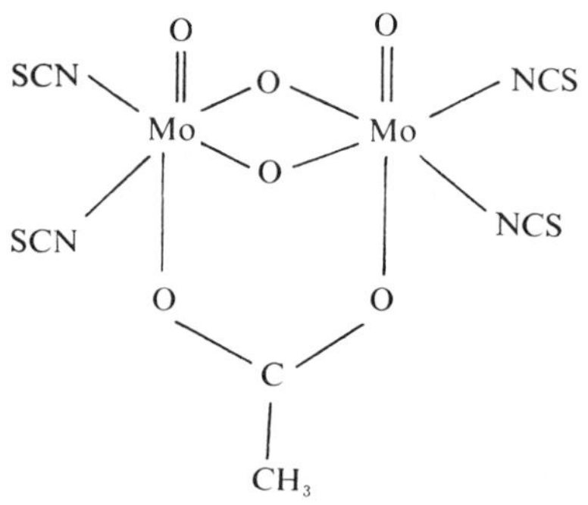

SCN SCN O O Mo Mo O O NCS NCS O O C CH_3

9a

E. Reactions and Interconversions

The study of reactions of Mo complexes is at a primitive stage. Kinetic and equilibrium parameters are virtually unknown, and we are at a point where reaction conditions which lead to the various classes of compounds are just becoming comprehensible. In this subsection we discuss the chemistry of chloride and thiocyanate complexes where there is some notion of the manner of interconversion of structural types. In addition, the reactions of cysteine and glutathione Mo(V) complexes are considered.

1. Reactions of Halides and Pseudohalides

In concentrated hydrochloric acid (>6*M*) Mo(V) exists as the green monomeric species $MoOCl_5^{-2}$ which shows spectral and magnetic properties essentially identical to those of the salts $M_2(MoOCl)_5$ [M = NH_4^+, pyH^+, Cs^+, etc. (379)]. When the acidity of these solutions is decreased, the color changes to brown, orange, or yellow (depending on the concentration of Mo), corresponding to a spectral change first analyzed by Haight (325). Concommitant with the spectral change there is a decrease in both the intensity of the EPR signal (359) and in the magnetic susceptibility of the solution (439). In brief, the monomeric form dominates at higher acid, but dimeric forms appear as the acidity is decreased. The monomer-dimer equilbrium is temperature dependent (325), and the endothermicity of dimer formation is illustrated by the thermochromic nature of the solutions. At room temperature in 4 to 5*M* HCl yellow solutions can be prepared which turn green upon heating (440). The process is reversible and thermodynamic parameters have been estimated (325, 440). There are at least two dimeric forms present, one being diamagnetic. This behavior is inferred from the observation that the EPR signal disappears more rapidly than the decrease in susceptibility would predict as acidity is lowered (359) and is consistent with spectral studies (325). Haight (325) showed that two H^+ are consumed as two monomers are formed from a dimer and formulated one of the dimers as structure **10**.

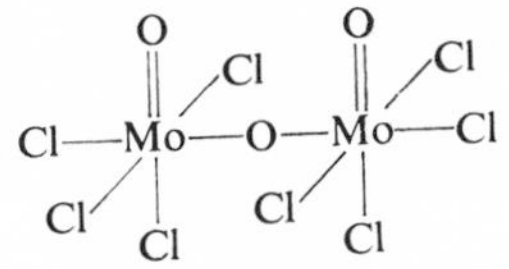

```
      O           O
      ‖  Cl       ‖  Cl
      ‖ /         ‖ /
Cl—Mo———O———Mo——Cl
   /|        /|
 Cl  Cl    Cl  Cl
```

10

Support for this formulation came from Colton and Rose (397), who isolated the diamagnetic salts $Cs_4[Mo_2O_3Cl_8]$ and $(NEt_4)_4[Mo_2O_3Cl_8]$ by adding base to concentrated solutions formed by $MoCl_5$ and H_2O (although other workers (396) found this preparation difficult to reproduce). The visible spectra of the isolated salts are said to resemble Haight's dimer in 1*M* acid, and the infrared spectrum (Table IX) is said to be consistent with the formulation. Thus in scheme 15 the compound of Colton and Rose would resemble Haight's second dimer.

$$MoOCl_5^{-2} \rightleftharpoons [\text{dimer I}] \rightleftharpoons [\text{dimer II}] \qquad 15$$

Curiously, the spectra of the monomeric form of $MoOCl_5^{-2}$ and the first dimer do not differ greatly in their absorption maxima, but they do differ drastically in the intensity observed at the various wavelengths. Thus $MoOCl_5^{-2}$ ($[HCl] = 12M$) shows absorption at 13,900 cm^{-1} ($\varepsilon = 13$) and 22,200 cm^{-1} ($\varepsilon = 12$), corresponding to ${}^2B_2 \rightarrow {}^2E$ ($d_{xy} \rightarrow d_{xz}, d_{yz}$) and ${}^2B_2 \rightarrow {}^2B_1$ ($d_{xy} \rightarrow d_{x^2-y^2}$) transitions in C_{4v} symmetry (314). The first dimeric species shows absorption intensity similar to the monomer at 13,900 cm^{-1}, but shows greatly enhanced absorption around 22,000 cm^{-1} with ε estimated at 2200 per dimer (325). The intensity enhancement was interpreted by Hare et al. (359) as due to substitution of one of the equatorial chlorides since the affected transition is localized within the equatorial plane. This substitution would place a bridging oxo cis to a terminal oxo group. The interesting possibility (291) is that dimer I is of the form illustrated in Fig. 20*c* where a paramagnetic triplet state would occur. Thus both of Haight's dimers could be monooxo bridged with the diamagnetic form having C_{2v} or D_{2h} symmetry (Figs. 20*a* and *b*) and the paramagnetic form having D_{2d} symmetry (Fig. 20*c*).

The nature of the species present in more dilute HCl (and in $HClO_4$)

was studied by Ardon and Pernick (418) using ion-exchange techniques. They prepared Mo(V) solutions by dissolving $MoO(OH)_3$ in concentrated HCl and diluting to 1*M* HCl. Yellow-orange solutions are obtained from which a cation can be absorbed on a cation-exchange resin. The ion exchange and freezing-point depression behavior are consistent with the formation of $Mo_2O_4^{+2}$. The similarity of the absorption spectrum to that of known $Mo_2O_4^{+2}$ containing complexes (Table X) is supportive of this formulation. It appears that as acidity is decreased sufficiently, the mono-μ-oxo dimers give way to di-μ-oxo species.

These observations are further supported by a number of preparative studies. For example, the Mo(V)-bipyridyl-chloride system was investigated by Mitchell (411) and later in greater detail by Saha and Haldar (378). The general scheme for interconversion of these complexes is illustrated in Eq. 16.

$$(bipyH_2)MoOCl_5 \xrightarrow{-HCl} MoOCl_3(bipy) \xrightarrow[-HCl]{H_2O} Mo_2O_3Cl_4(bipy)_2 \xrightarrow[-HCl]{H_2O} Mo_2O_4Cl_2(bipy)_2 \qquad (16)$$

The properties of individual complexes in the scheme are given in appropriate tables (see Tables VII through X). The first complex, formed in concentrated HCl, is the bipyridinium salt of the well-known $MoOCl_5^{-2}$ ion. This compound can be made to lose HCl by, for example, refluxing in aqueous ethanol to yield the complex $MoOCl_3(bipy)$. The latter complex can also be prepared from $MoCl_5$ and bipy in CCl_4 containing a trace of moisture. Although Mitchell (411) noted only a khaki complex, Saha and Haldar (378) reported pink and green as well as khaki isomers, depending on the detailed mode of preparation. The dimeric compounds containing $Mo_2O_3^{+4}$ or $Mo_2O_4^{+2}$ units are produced by varying the amount of H_2O allowed to contact the $MoOCl_5^{-2}$. Saha and co-workers found similar interconversions for Mo-phen-Cl (432), Mo-bipy-Br (412), and Mo-phen-Br (381) systems. By working in weakly acidic solutions, the compounds $(pyH_2)_2[Mo_2O_4Cl_4(H_2O)_2]$ (380, 417) and $((CH_3)_4N)_2[Mo_2O_4Cl_4(OH_2)_2]$ (380) have also been isolated.

A simplified general scheme for reaction of chloro complexes as the pH of the solution is raised is then as follows (Eqs. 17–20):

$$2MoOCl_5^{-2} + H_2O \rightleftharpoons [\text{paramagnetic dimer}] + 2H^+ \qquad (17)$$

$$[\text{paramagnetic dimer}] \rightleftharpoons Mo_2O_3Cl_8^{-4} \qquad (18)$$

$$Mo_2O_3Cl_8^{-4} + H_2O \rightleftharpoons Mo_2O_4Cl_6^{-4} + 2Cl^- + 2H^+ \qquad (19)$$

$$Mo_2O_4Cl_6^{-4} \rightleftharpoons Mo_2O_4^{+2} + 6Cl^- \qquad (20)$$

We feel that only $MoOCl_5^{-2}$ and $Mo_2O_4^{+2}$ at the extrema of this scheme are reasonably certain to be present while intermediate complexes exist in various states of aquation or ligation, depending on the conditions. The nature of the paramagnetic dimer is uncertain, and while it could be a D_{2d} or D_{4h} isomer of $Mo_2O_3Cl_8^{-4}$ (Fig. 20*c* or *d*), it could also be hydroxo or chloro bridged or have a nonlinear Mo–O–Mo linkage.

Despite some understanding of the reactions our knowledge of the situation is far from satisfactory. We note first that the magnetic susceptibility (439), EPR (359), and spectrophotometric (325) studies, which together hold the key to our understanding of this system, were all done at different concentration ranges (0.3, 0.03, and 0.003*M*, respectively) and thus cannot be quantitatively compared. Furthermore, at low acidity Haight reports vastly decreased visible absorption, which apparently was confirmed by Colton and Rose and assigned to the $Mo_2O_3Cl_8^{-4}$ ion. This low absorption, however, is usually associated with di-μ-oxo dimers and not mono-μ-oxo species which in all well-documented cases are highly colored with absorptions around 500 nm (ε = 10,000). However, the known monooxo bridged species all have linear bridges and $Mo_2O_3Cl_8^{-4}$ could have a bent Mo–O–Mo group (417). Thus while the gross features may be delineated by Eqs. 17 to 20, the structural details of the species present and their relative concentrations as functions of [Mo], [H^+], [Cl^-], and temperature are not yet clear. Furthermore, some workers have postulated tetrameric species either in solution or in isolated complexes (394), and while their existence has not yet been proved, neither have they been eliminated by definitive experiments.

The situation for the thiocyanate complexes is similar. Here the monomer $MoO(NCS)_5^{-2}$ (379), mono-μ-oxo dimer $Mo_2O_3(NCS)_8^{-4}$ (433), and di-μ-oxo dimer $Mo_2O_4(NCS)_6^{-4}$ (433) are well-characterized species (Table VII–X). Their methods of preparation differ largely in concentration of acid and ligand used. This is consistent with interconversions 21 and 22,

$$2MoO(NCS)_5^{-2} + H_2O \rightleftharpoons 2H^+ + 2NCS^- + Mo_2O_3(NCS)_8^{-4} \quad (21)$$

$$Mo_2O_3(NCS)_8^{-4} + H_2O \rightleftharpoons 2H^+ + 2NCS^- + Mo_2O_4(NCS)_6^{-4} \quad (22)$$

where strong acid and excess ligand favor the monomer.

Allen and Neuman (238) investigated Mo(V)-HBr solutions by spectrophotometry and concluded that even at the highest HBr concentrations dimeric forms predominate with no evidence for $MoOBr_5^{-2}$. However, EPR studies by Dowling and Gibson (343) showed that $MoOBr_5^-$ is almost certainly present in quantity in concentrated HBr solutions of Mo(V).

Unfortunately, the *lowest* acid concentration used in all of the above studies is ~0.1*M*, which is, of course, still quite acidic. As the pH is raised past 7, a brown compound usually formulated as $MoO(OH)_3$ invariably precipitates. The question remains: what intermediate states, which are detectable and structurally distinct, exist between $Mo_2O_4^{+2}$ and $MoO(OH)_3$? That is, in the general scheme that occurs as pH is raised (scheme 23),

$$MoO^{+3} \rightleftharpoons Mo_2O_3^{+4} \rightleftharpoons Mo_2O_4^{+2} \rightleftharpoons [?] \rightleftharpoons MoO(OH)_3 \quad (23)$$

what belongs in the brackets? The answer (as is seen in the next subsection) depends strongly on the additional ligands present and may have much to say about the nature of biological molybdenum.

2. *Interaction of Cysteine and Glutathione with Mo(V)*

a. Cysteine with Mo(V). The observation of EPR signals attributable to Mo(V) in four of the Mo enzymes has had much influence on the course of model studies. The observed g and A values are compatible with Mo(V) bound to sulfur (354) and this has led workers seeking model systems to look at Mo(V) reactions with the one sulfhydryl-containing amino acid, *l*-cysteine.

In 1963 Spence and Chang studied (452) the complexation of Mo(V) by cysteine in solution between pH 3 and 6. They concluded, using Job's method, that a 1:1 complex was formed and estimated the equilibrium constant,

$$K_{eq} = \frac{[\mathrm{MoL}]}{[\mathrm{Mo}][\mathrm{L}]} = 10^6 \qquad (\mathrm{L} = \mathrm{cys})$$

Spence and Chang noted some problems with full understanding of this system, due to inherent complications and while much progress has been made since this initial study, many questions remain unanswered.

A key step in the Mo-cysteine studies was the isolation of a crystalline sample of the complex anion $Mo_2O_4(cys)_2^{-2}$. The complex salt $Na_2[Mo_2O_4(cys)_2]\cdot 5H_2O$ was initially reported by Kay and Mitchell (443) and, independently, by Melby (409). Mitchell's preparations involved the reduction of aqueous solutions made from Na_2MoO_4 and cys·HCl with hydrazine hydrate, sodium dithionite, or excess cys·HCl (with the last reductant giving the poorest yield). Mitchell also isolated the barium salt (279) of the anion, which is said to be insoluble in H_2O. Melby's preparation consisted of reacting $MoCl_5$ "dissolved" in 3*N* HCl with a 3:1 excess of cys·HCl. The solution was neutralized with NaOH and ethanol was added to promote crystallization. Melby also prepared the guanidinium salt by reaction of the sodium salt with guanidinium hydrochloride (409).

The structure of the dimer has already been discussed and is shown in Fig. 12. The complex has a magnetic moment of 0.53 B.M. at 295°, indicating strong coupling between the Mo's and approximate diamagnetism. In agreement with this, a normal NMR spectrum is recorded in D_2O (444) with no apparent broadening attributable to paramagnetic species.

The reactivity of this complex was first noted by Mitchell, who reported that in the presence of O_2 the 32,700 cm^{-1} peak decreased in energy to 31,650 cm^{-1} and in intensity from 12,800 to ~ 5000 l $mole^{-1}$ cm^{-1} (443). From concentrated aqueous solutions of $Mo_2O_4(cys)_2^{-2}$ in the presence of air at pH = 9.2,

Mitchell (443) claims to have isolated both cystine (cys-cys) and an Mo(VI) complex said to be $MoO_2(cys)_2$. Clearly other products containing a less than 1:1, cysteine to Mo ratio must be present to quantitatively account for this reaction. The air oxidation is said to be slow but accelerated by increased pH, excess cys, or Fe salts (444). The mechanism of this reaction has been studied (445), but the details are not yet available (445).

Bayer and Krauss (446) reacted $Mo_2O_4(cys)_2^{-2}$ and BH_4^-, and although the reaction product is not fully characterized, it appears to maintain the 1:1, Mo:cys ratio. Significantly, it gives rise to a strong EPR signal (and it is implied that 100% of the Mo is in an EPR-active form). The *g*-values shown in Table VIII differ from those reported by Haight and Spence (vide infra), but they are nonetheless similar to the values for xanthine oxidase (Table II). The complex is reported to be brown and, in view of the brown color of Schrauzer's Mo-thiol C_2H_2 reducing system (59–65), would appear to warrant further investigation.

Huang and Haight (355, 356) pointed to some of the complexities in the chemistry of the Mo(V)-cysteine system when they measured the EPR properties of solutions of Mo(V) ($\sim 10^{-3}M$) and cys ($\sim 10^{-2}M$) in 0.2*M* phosphate buffers over the pH range 6 to 12. An EPR signal appears above pH 6, increases in intensity up to pH 10, and eventually disappears when the pH is increased to 12. Identical results were obtained using $Na_2Mo_2O_4(cys)_2 \cdot 5H_2O$ and cysteine in phosphate buffer or in unbuffered water at pH = 8 (adjusted with NaOH). The EPR signals appear to be due to a single monomeric species, and its intensity is proportional to the square root of the dimer concentration which is consistent with an equilibrium crudely expressed as Eq 24.:

$$(MoL)_2 \rightleftharpoons 2MoL \tag{24}$$

The quantitation of the EPR signal shows less than 2% of the monomer to be present at all times. This percentage increases reversibly with increasing temperature which is consistent with an endothermic formation of monomer. The *g* and *A* values are given in Table VIII.

Huang and Haight also reported the presence of a band at 17,200 cm^{-1} (580 nm) (ε = 120) at pH = 10 in these same solutions, and they associated the blue color with the paramagnetic species. One cannot help note the strong correlation of these EPR parameters and absorption maxima with those observed in the molybdenum enzymes [Table II (104, 105)]. Huang and Haight also noted that heating of the solid $Na_2Mo_2O_4(cys)_2 \cdot 5H_2O$ produced similar Mo(V) EPR signals (357).

Kroneck and Spence (447, 448) and, independently, Stephenson and Schultz (449) investigated the Mo(V) cysteine system in more detail at pH

7 to 12. While most of Haight's observations were reproduced by these workers, the additional data presented showed that the initial interpretations were at best oversimplifications of a complex situation. Both studies (447, 449) agree that excess free cysteine has profound effects on the course of the reactions in basic solution. Thus both the rate of formation and the amount of absorbing species were increased by the presence of free cysteine (or other thiols). Furthermore, both studies agreed that the species responsible for the EPR spectrum and the 580nm absorption (the blue color) were not the same. Kroneck and Spence concluded (447, 448) that when the reaction to form the optically absorbing species was complete, 2 moles of OH^- were consumed per mole of Mo(V) dimer, whereas Stephenson and Schultz (449) found 1 mole of OH^- per dimer.

Both groups did kinetic studies of the cysteine-catalyzed disappearance of $Mo_2O_2(cys)^{-2}$. Their combined results are consistent with the rate law

$$\frac{-d[Mo_2O_4(cys)_2^{-2}]}{dt} = k_1[Mo_2O_4(cys)_2^{-2}] + k_2[cys][Mo_2O_4(cys)_2^{-2}]$$

where

$k_1 = 1.48 \times 10^{-5}\ sec^{-1}$ [at 25°C, pH = 10.0, 0.50M borate (447, 448)]

$k_2 = 2.06 \times 10^{-2}\ M^{-1}sec^{-1}$ [at 25°C, pH = 10.0, 0.50M borate (447, 448)]

{[$k_2 = 3.75 \times 10^{-2}\ M^{-1}sec^{-1}$ [at 30°C, pH = 9.2, μ = 2.0 borate or phosphate buffer (449)]}

Although their formulations differ as to detail, both groups consider the blue reaction product to be a mono-μ-oxo dimeric species. The evidence for this structure is said to be the spectral similarities to known μ-oxo species. We note, however, that while known μ-oxo species (which all contain linear bridges) do indeed show lower energy absorption than related di-μ-oxo species, these bands came from 500 to 510 nm (19,000–20,000 cm^{-1}) and not from 580 to 650 nm (15,400–17,200 cm^{-1}) as observed for the complex in question. More importantly, the visible band in the known μ-oxo complexes is two orders of magnitude more intense (ε = 10–15,000 vs. 100–200) than that in the postulated complex. Thus although a μ-oxo bridge may be present, its existence has not been proved. There are numerous outstanding questions. If there is a μ-oxo bridge, is it linear [as implied in (449)] or bent [as implied in (447, 448)]? The latter seems more likely in view of the spectral differences between the complex and the known linear μ-oxo complexes. Have other potential bridging groups been eliminated? Which of the three cysteine donor atoms remain coordinated to Mo? Are the Mo's equivalent? Could the cysteine-amino group become deprotonated as the pH is raised? What is the relationship between the blue absorbing species and the EPR-active species?

These and other questions remain unanswered, and it is our feeling that until the species responsible for the blue color, the EPR signal, and the chemical reactivity of this system are isolated and structurally characterized, this situation may continue. A possible key observation (449) is that only 75 ± 15% of the Mo present gives a positive Mo(V) test (either in conc. HCl or by ethanolic oxine). Should this result be confirmed, it may indicate the occurrence of internal redox reactions, and the presence of small amounts of mixed valence species must be carefully considered.

Kay and Mitchell (444) have looked into other aspects of the reactivity of $Mo_2O_4(cys)_2^{-2}$. At lower pH's the coordinated carboxylates are protonated and displaced by water from the coordination spheres of Mo (Eq. 26).

$$Mo_2O_4(cys)_2^{-2} + 2H^+ + 2H_2O \longrightarrow \text{HOOC—} \; \text{S, NH}_2\text{, O, O, Mo, O, O, Mo, NH}_2\text{, S, H}_2\text{O, H}_2\text{O} \; \text{—COOH} \quad (25)$$

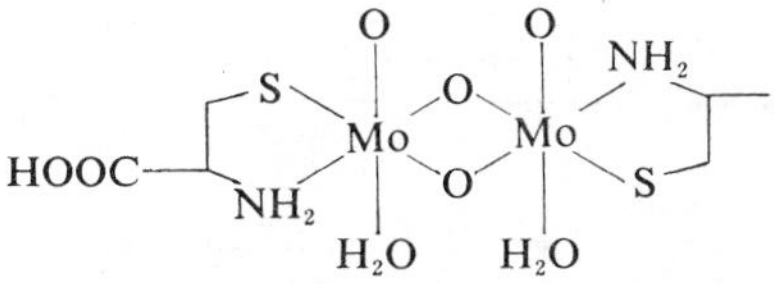

This claim is based on pH titration and NMR studies from which pK_a values of 2.03 and 4.07 are estimated for the removal of two protons from the di-aquo complex. These are reasonable values in view of the pK of around 2 for the carboxylate ionization in free cysteine (450, 451). The di-aquo complex shows absorption at 33,900 cm^{-1} as compared to the 32,070-cm^{-1} peak in the starting material. As H_2O is higher than carboxylate in the spectrochemical series, this result is compatible with the above transition having some "*d-d*" character. Working at various pH's Mitchell found that the spectrum did not change with time in acetate, phthalate, borate, or carbonate buffers, but did change in phosphate or tartrate. With phosphate buffer a new peak at 21,000 to 22,000 cm^{-1} appeared. These results, coupled with Marov's work (371) on phosphate Mo(V) systems and Spence's work on tartrate Mo(V) systems (452), may indicate buffer participation in bridge-opening reactions that lead to paramagnetic species.

$Mo_2O_4(cys)_2^{-2}$ has been used by Mitchell (444) in the preparation of the known complexes Mo_2O_3(*O*-Etcys)$_4$, Mo_2O_3(oxine)$_4$, and $Mo_2O_4Cl_2$(phen)$_2$ and in the preparation of some new complexes formulated as $Mo_2O_4(SO_4)$-(phen)$_2$, Mo_3O_8(phen)$_2 \cdot 2H_2O$, and Mo_3O_4(*O*-Etcys)$_4$(phen)$_2$. The last complex is said to be readily soluble in organic solvents, giving a blue color. This interesting complex bears further investigations. Finally, $Mo_2O_4(cys)_2^{-2}$ was reacted by Kay and Mitchell (444) with diazomethane in attempts to form a dinitrogen complex. The isolated products, however, were hydrated Mo(V) hydroxide, *S*-methyl-*l*-cysteine, and unreacted starting materials. It is noted that *S*-methyl-cysteine does not form stable complexes with Mo(V), reinforcing the notion that thiolate and not thioether groups are required for stable complex formation.

The use of $Mo_2O_4(cys)_2^{-2}$ in a catalytic system that reduces acetylene is discussed in Section IX.

Some general conclusions as to the reactivity of the $Mo_2O_4(cys)_4^{-2}$ complex can be made. First, the coordinated carboxylate group, trans to the Mo–O_t bond, appears to be the site of initial lability in the Mo coordination sphere. In the presence of additional ligands the cysteines can be fully replaced. The di-μ-oxo bridge can be converted to a di-μ-sulfido bridge in the presence of H_2S and to a mono-μ-oxo bridge in the presence of certain ligands. When the pH is high (especially when other thiol ligands, including excess cysteine, are present), a reaction ensues that produces a blue product and a small amount of EPR-active species which are not identical. The reaction clearly involves some alteration of the MoO_2Mo bridge and may involve the formation of a mono-oxo-bridged species.

b. Complex of Mo and Glutathione. Glutathione is the biologically significant (453) tripeptide δ-glutamylcysteinyl-glycine (structure **11**):

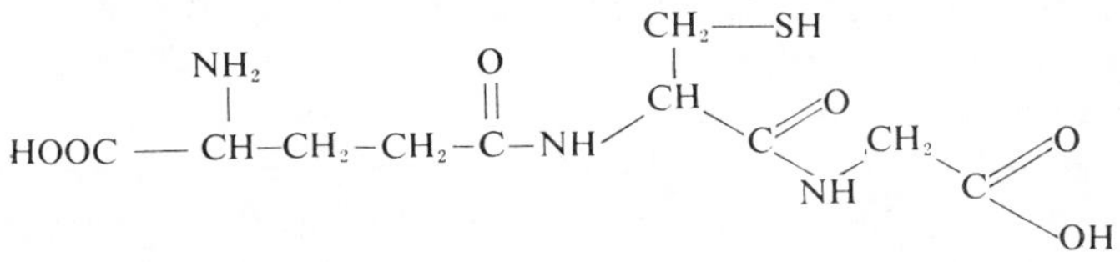

```
                                      CH2—SH
          NH2              O           |
          |                ||          CH       O
HOOC — CH–CH2–CH2–C–NH/       \C//    CH2      O
                                         \  /      \  //
                                          NH         C
                                                      \
                                                       OH
```

11

Huang and Haight (357) isolated a complex of Mo(V) and glutathione under argon. The complex, light brown and diamagnetic, is formulated as Na-$[Mo_2O_4(glutathione)]\cdot 4H_2O$. Like other di-$\mu$-oxo complexes it possesses infrared bands at 965, 730, and 470 cm^{-1} due to terminal (965 cm^{-1}) and bridging (730, 470 cm^{-1}) metal oxygen vibrations. The ligand vibrations are broadened in the complex with respect to the free ligand, as has been found in the Mo(V)-cysteine complex. The complex is formulated according to the structure shown in Fig. 23, with the unusual (454) feature of an unionized coordinated amide nitrogen. This represents the first and, as of this date, the only isolated Mo-peptide complex reported in the literature. A similar compound has been isolated by Evans et al. (66) and used with BH_4^- in the catalytic reduction of acetylene and hydrazine.

In basic solution a paramagnetic species is formed in low concentration never exceeding 4%, which nonetheless, is a higher concentration than the paramagnetic species formed from the cysteine dimer. However, the nature of the EPR signal, in this case, differs vastly from that in the cys case. Using

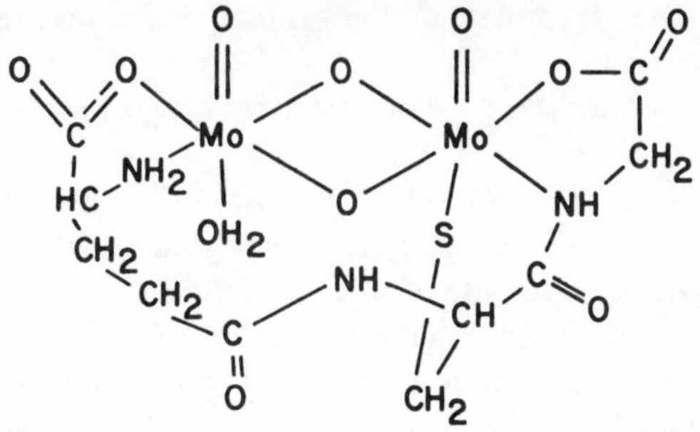

Fig.23. Proposed structure for the Mo(V) Glutathione complex (357).

isotopically enriched ^{95}Mo ($I = 5/2$) an 11-line EPR signal is seen at room temperature in solution indicating that the unpaired electron (electrons) interacts with two magnetically similar Mo's. Huang and Haight assigned this spectrum to a dimeric triplet-state complex containing two unpaired electrons. However, the expected low-intensity half-field transition was not observed and this assignment must be regarded as tentative. At 77°K the spectrum is complex and broadened, precluding a complete analysis. Various assumptions were made allowing a partial assignment and evaluation of the zero-field splitting constant (D) from which Mo–Mo distances of either 6.0 or 4.8 Å are calculated. In view of the complexity of the situation, these calculations must be regarded with caution. Significantly, Huang and Haight again correlated the appearance of absorption at 580 nm with the presence of the EPR signal.

V. THE CHEMISTRY OF MOLYBDENUM (IV)

Complexes in the (IV) oxidation state have now become common in Mo chemistry. Although Mo(IV) was thought to be prone to disproportionation in aqueous solution, this notion has been made suspect by recent work. Octacyanide, hexaisothiocyanate, and tetrachloride compounds have been known for some time, and these have been used recently as starting materials in the preparation of other Mo(IV) complexes. Both oxo and nonoxo compounds are found, but the oxo grouping does not dominate the chemistry of Mo(IV) as it does that of Mo(V) and Mo(VI).

Since all Mo enzymes catalyze two-electron redox processes, and since Mo(VI) is a strong candidate for one of the reacting oxidation states, the participation of the $4d^2$, Mo(IV) state seems reasonable for a monomeric Mo site. In fact, in the case of xanthine oxidase, direct evidence shows that the

Mo(IV) state can be attained and suggests its participation in the catalytic cycle (see Section II and Ref. 104).

Chemical studies of Mo(IV), in addition to illustrating its viability as an oxidation state in both oxo and nonoxo systems, demonstrate a range of coordination numbers (4 through 8) which is greater than for any other oxidation state.

In this section we depart from our procedure of first covering structural aspects. Crystallography has revealed the complexes of Mo(IV) to (in most cases) have monomeric structures, and the coordination numbers 5, 6, 7, and 8 are represented by square-pyramidal, octahedral, capped octahedral (or pentagonal bipyramidal), and dodecahedral geometries, respectively. In Table XII we list the structures that have been solved, but find it more convenient to consider their details in the individual parts of this section.

TABLE XII
X-Ray Structural Studies on Mo(IV) Complexes

Complexes	Refs.
$NaK_3[MoO_2(CN)_4]\cdot 6H_2O$	458
$Na_3[MoO(OH)(CN)_4]$	459
$[MoOCl(dppe)_2]\ [ZnCl_3OC(CH_3)_2]$	460
$MoOCl_2(PMe_2Ph)_3$ (blue)	461, 462
$MoOCl_2(PEt_2Ph)_3$ (green)	461, 463
$MoO((n\text{-}C_3H_7)_2dtc)_2$	177a
$MoO((n\text{-}C_3H_7)_2dtc)_2(TCNE)$	464
$K_2[MoCl_6]$	465
$MoCl_2(N\text{-}Mesal)_2$	466
$MoCl_4(PMe_2Ph)_3$	467
$MoBr_4(PMe_2Ph)_3$	468
$K_4[Mo(CN)_8]\cdot 2H_2O$	469, 470
$Mo(CH_3NC)_4(CN)_4$	471, 471a
$Mo(PMe_2Ph)_4H_4$	472
$Mo(S_2CPh)_4$	473, 474
$Mo_2S_2(SCN(n\text{-}C_3H_7)_2)_2\ ((n\text{-}C_3H_7)_2dtc)_2$	475, 476
$[Mo_3S_4(C_5H_5)_3]\ [Sn(CH_3)_3Cl_2]$	477
$(C_5H_5)_2MoH_2$	478
$(C_5H_5)_2MoCl_2$	479
$(C_5H_5)_2MoCl(Et)$	479
$[(C_5H_5)_2Mo(OH)(NH_2CH_3)]\ (PF_6)$	479
$[(C_5H_5)_2Mo(cys)]Cl$	480
$[(C_5H_5)_2Mo(cys)]PF_6$	480
$[(C_5H_5)_2Mo(sarcosine)]Cl$	480
$[(C_5H_5)_2Mo(gly)]Cl$	480
$(C_5H_5)_2Mo(tdt)$	481
$[(C_5H_5)_2Mo(SCH_2CH_2NH_2)]I$	482
$(C_5H_5)_2Mo(\mu\text{-}SC_4H_9)_2FeCl_2$	483, 484

A. Oxo Complexes

Oxo complexes of Mo(IV) contain the MoO^{+2} group and have structures derived from an octahedron with one very strong Mo–O_t bond and a weak bond or an open position trans to the strong Mo–O. Unlike the Mo(V) state where monomeric MoO^{+3} complexes usually contain halide ligands and require strong acid for preparation, the MoO^{+2} complexes have been found with phosphine (485), 1,1-dithiolate (399), 1,2-dithiolate (486), Schiff base (487), isonitrile (488), and phthalocyanine (489) ligands and do not, in general, require acid for preparation. The MoO_2 grouping is known only in the $MoO_2(CN)_4^{-4}$ ion (490, 491). The complexes and their properties are displayed in Table XIII.

For the monooxo complexes, whether 5 or 6 coordinate, a $(d_{xy})^2$ [$(b_2)^2$ or $^1A_{1g}$] ground state is expected in C_{4v} symmetry (Fig. 19). While most compounds have been reported as diamagnetic, magnetic moments as high as 0.6 to 0.9 B. M. have been observed, and these have been attributed to temperature-independent paramagnetism (399, 492, 487). The NMR spectra, where reported, are consistent with the presence of diamagnetism and spin pairing in the $(b_2)^2$ configuration, and this seems likely for all complexes in this class.

Spectroscopically, three *d-d* bands are possible for one-electron excitations from b_2 to e, a_1, or b_1 levels, but interelectronic repulsions, two-electron excitations, low-energy charge transfer, and splitting of the e level due to lower symmetry cause the situation to be complex, and therefore detailed assignments have not been attempted. The infrared spectra are characterized by an intense ν(Mo–O_t) from 930 to 1010 cm^{-1}. In general, for complexes of a given ligand the ν(Mo–O_t)'s follow the trend Mo(IV) > Mo(V) > Mo(VI), and this has been attributed (399) to increased pi bonding when only a single oxo is present, as in the Mo(IV) complexes.

A number of oxomolybdenum(IV) complexes contain a 5-coordinate $MoOS_4$ coordination sphere. The red-purple complex $MoO(S_2P(O\text{-}i\text{-}C_3H_7)_2)_2$ was prepared (492) when K_3MoCl_6 and $KS_2P(O\text{-}i\text{-}C_3H_7)_2$ were mixed in dilute acid solution. Jowitt and Mitchell (399) prepared $MoO(S_2P(OR)_2)_2$ and $MoO(R_2dtc)_2$ complexes by reduction of the Mo(V) species $Mo_2O_3(LL)_4$ using zinc dust or thiophenol, or alternatively by reduction of aqueous solutions of sodium molybdate and sodium dialkyldithiocarbamate with $S_2O_4^{-2}$. It is notable that dithionite affects the reduction of Mo to the IV state when sulfur ligands are present. The same reductant is often used to obtain the reduced state in Mo enzymes, where sulfur has also been postulated as a ligand to Mo. $MoO(S_2PF_2)_2$ was prepared by Cavell and Sanger (494) according to (Eq. 26):

TABLE XIII
Properties of Oxo Mo(IV) Complexes

Complex	Infrared absorption $\nu(Mo–O_t)$, cm^{-1}	Electronic absorption cm^{-1} (or color)	(ε)	Refs.
$MoO(S_2P(O\text{-}i\text{-}C_3H_7)_2)_2$	975			492
$MoO(S_2P(OPh)_2)_2$	998	15,100	(5)	399
		19,800	(100)	
		24,600	(52)	
		32,000	(470)	
$MoO(S_2P(O\text{-}Et)_2)$	1,000	15,250	(20)	399
		19,150	(80)	
		24,650	(55)	
		32,000		
$MoO(Et_2dtc)_2$	962	13,350		399
		16,900		
		19,200		
		25,800		
		32,000		
$MoO(Me_2dtc)_2$	975			401
$MoO(S_2PF_2)_2$	1,006			494
$MoO(oxine)_2$	930	14,500		399
		17,100		
		19,050		
$MoO(py(anil)_2)Cl_2$	950	8,000	(700)	487
		16,800	(5,770)	
		27,800	(7,700)	
$MoO(py(anil)_2)Br_2$	950	8,200	(790)	487
		16,130	(6,450)	
		27,780	(8,110)	
$MoO(C_{32}H_{16}N_8)$	972	13,330	(6,450)	489
($C_{32}H_{16}N_8$ = phthalocyanine)		14,160	(87,100)	
		15,700	(18,600)	
		27,170	(31,600)	
$MoO(tpp)(py)_2$	990			341
$[MoOCl(CNCH_3)_4](I_3)$	946	18,280		488
$[MoOBr(CNCH_3)_4]$ (BrI_2)	950			488
$MoOX_2$ L_3				
X = Cl, L = $MePPh_2$	945	Green		485
= $MePPh_2$	951	Blue		
= Me_2PPh	943	Green		

TABLE XIII (continued)

Complex	Infrared absorption $\nu(M\text{–}O_t)$, cm^{-1}	Electronic absorption cm^{-1} (or color)	(ε)	Ref.
$= EtPPh_2$	941	Green		485
$= Et_2PPh$	940	Green		485
$= (n\text{-}C_3H_7)PPh_2$	941	Green		485
$= (n\text{-}C_3H_7)_2PPh$	940	Green		485
$= Me_2AsPh$	952	Blue green		485
$X = Br; L = Me_2PPh$	955	Blue green		485
$X = I; L = Me_2PPh$	946	Green		485
$X = NCO; L = Me_2PPh$	940	Blue		485
$X = NCS: L = Me_2PPh$	948	Blue		485
$MoOCl_2(triphos)$	949	Green		485
$[MoOCl(dppe)]\ BF_4$	940–956	Blue green		493
$K_4[MoO_2(CN)_4] \cdot 6H_2O$	800 broad	19,050		490
$K_3[MoO(OH)(CN)_4] \cdot 2H_2O$	921	16,900	(38.7)	490
		44,450	(19,900)	
		[46,500]	(17,000)	
$[Mo(OH)_2(CN)_4^{-2}]$		16,400		
$MoO(Me_2dtc)_2(MeOOCCCCOOMe)$	933			401
$MoO(Me_2dtc)_2(TCNE)$	935			401
$MoO(Et_2dtc)_2(EtOOCNNCOOEt)$	935	25,600	(4,000)	401
		28,700	(3,250)	
$(NB\mu_4)_2MoO(mnt)_2$	930			497
$(PPh_4)_2MoO(mnt)_2$	940	15,100	(120)	486
		16,100	(170)	
		20,100	(260)	
		27,100	(4,430)	

$$MoOCl_4 + 4F_2PS_2H \longrightarrow MoO(S_2PF_2)_2 + (F_2PS_2)_2 + 4HCl \qquad (26)$$

An interesting mode of preparation of the $MoO(Et_2dtc)_2$ species is by the reaction (495) of triphenylphosphine with the Mo(VI) species $MoO_2(Et_2dtc)_2$ (Eq.6):

$$MoO_2(Et_2dtc)_2 + (C_6H_5)_3P \longrightarrow MoO(Et_2dtc)_2 + (C_6H_5)_3PO \qquad (6)$$

This is an effective oxygen-atom transfer reaction, analogs of which could be involved in certain aspects of the function of Mo in enzymes (154, 155) and in other catalytic schemes (256).

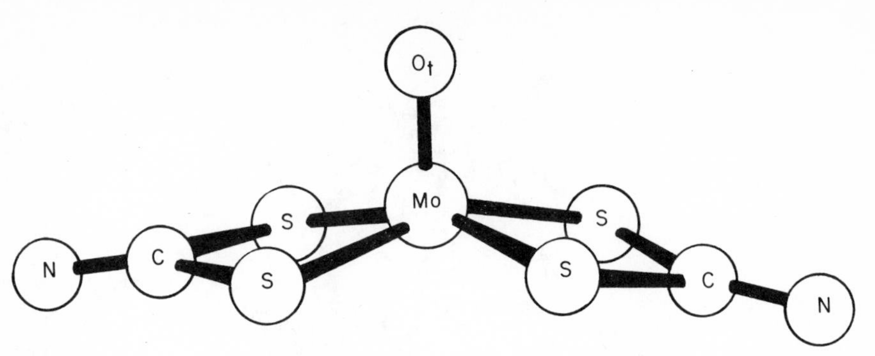

Fig. 24. The structure of $MoO((n\text{-}C_3H_7)_2dtc)_2$ (177). (The *n*-propyl groups are omitted for clarity.)

The x-ray structure of $MoO((n\text{-}C_3H_7)_2dtc)_2$ (177) is shown in Fig. 24. The Mo–O_t distance is 1.695 Å, and the molecule displays a square-pyramidal geometry with C_{2v} symmetry and the Mo lying 0.7 Å above the plane of the four sulfurs. These parameters do not differ greatly from those found in oxo-Mo(V) complexes, although the Mo(IV) is out of the equatorial plane by a slightly greater distance.

Schneider, Newton, and co-workers (401) demonstrated that these 5-coordinate complexes are coordinately unsaturated and capable of undergoing (oxidative) addition-type reactions such as reaction 27.

$$MoO(Et_2dtc)_2 + EtOOCN{=}NCOOEt \longrightarrow MoO(Et_2dtc)_2(EtOOCNNCOOEt) \quad (27)$$

Here, the diethyl azodicarboxylate reacts with the pink Mo(IV) complex to produce a yellow adduct [roughly comparable to the adducts formed by Vaska's complex (496)]. In moist $CHCl_3$, this complex is hydrolyzed to EtOOCNHNHCOOEt and $MoO_2(Et_2dtc)$, making the overall process possibly analogous to one of the steps in nitrogenase activity. A tetracyanoethylene adduct of $MoO(Et_2dtc)_2$ was also prepared, as were diethylazodicarboxylate, dimethylacetylenedicarboxylate, and tetracyanoethylene adducts of $MoO((CH_3)_2dtc)_2$ (401).

The structure of the tetracyanoethylene adduct of $MoO((n\text{-}C_3H_7)_2dtc)_2$ was solved by Ricard and Weiss (646) and the result is shown in Fig. 25. The TCNE is bound to Mo confirming adduct-formation reaction. If TCNE is considered a bidentate ligand, then the coordination geometry is roughly pentagonal-bipyramidal with the terminal oxygen and a sulfur of one of the ligands occupying the apical positions and the TCNE, and three sulfurs in the equatorial plane. The TCNE ligand occupies coordination sites cis to the Mo–O_t bond, and this enables it to act effectively as a pi-acceptor ligand for

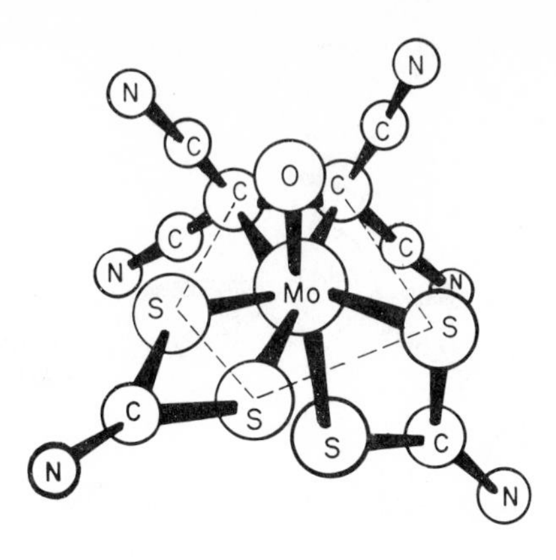

Fig. 25. The structure of $MoO((n\text{-}C_3H_7)_2dtc)_2(TCNE)$ (464). (The *n*-propyl groups are omitted for clarity.)

the electrons in the d_{xy} metal orbital. The similarity of this structure to that of the peroxo-Mo(VI) complexes discussed in Section III is notable (464). The formulation of the Mo oxidation state is equivocal and the discussion of Section III.A.3 similarly applies. The Mo–O_t bond length at 1.682 Å is not significantly different from that of the starting material at 1.695 Å, or from 1.664 Å found for $MoO_2(n\text{-}C_3H_7)_2dtc)_2$ (177).

Oxidative addition of Cl_2 or Br_2 to $MoO(R_2dtc)_2$ leads to the 7-coordinate Mo(VI) compounds $MoOX_2(R_2dtc)_2$ (X = Cl, Br) (258a). However, the reaction of $MoO(R_2dtc)_2$ with HCl or HBr does not involve oxidative addition and leads to $MoX_2(R_2dtc)_2$ (258a) according to Eq. 28:

$$MoO(R_2dtc)_2 + 2HX \longrightarrow MoX_2(R_2dtc)_2 + H_2O \qquad (28)$$

Atkinson et al. (493) prepared complexes of the form $MoOCl_2L_3$, $MoOCl_2L(LL)$, and $[MoOCl(LL)_2]X$ [where L = tertiary phosphine, LL = ditertiary phosphine, and X = BF_4^- or $B(C_6H_5)_4^-$] by reaction of sodium molybdate and the alkylarylphosphines, L, in ethanol containing a small amount of HCl. An alternate preparative procedure involves the combination of $MoOCl_3(thf)_2$ and L in thf to yield red solutions, presumably containing $MoOCl_3L_2$, which are readily reduced by Zn(Hg) to yield $MoOCl_2L_3$. As shown in Table XIII, the Mo(IV) complexes in this class display the characteristic ν(Mo–O_t) in the 940 to 956 cm^{-1} range.

The crystal structure (460) of the purple complex $[MoOCl(dppe)_2^+]$ $[ZnCl_3(OC(CH_3)_2^-]\cdot OC(CH_3)_2$ shows a near-octahedral geometry for the molybdenum(IV) cation. The Mo–O_t length is 1.69 Å, whereas the Mo–Cl trans to Mo–O_t has a bond length of 2.46 Å, somewhat longer than the Mo–Cl distance found when Mo–Cl is not trans to Mo-O_t.

Butcher and Chatt (485) prepared a large series of complexes of the form $MoOX_2L_3$(where X = I, Br, NCS, or NCO, and L = tertiary phosphine). The complexes $MoOCl_2L_3$ are prepared by reacting $MoCl_4(NCC_2H_5)_2$ with the appropriate phosphine in ethanol. (In $CHCl_3$ the complexes *trans*-$MoCl_4L_2$ are formed.) Metathesis reactions then allow replacement of chloride by other halides or pseudohalides. Butcher and Chatt (485) noted that some of the $MoOX_2L_3$ complexes were blue while others were green. Furthermore (except for X = NCO), the green complexes show $\nu(Mo–O_t)$ below 946 cm^{-1}, whereas the blue complexes show $\nu(Mo–O_t)$ at frequencies greater than 946 cm^{-1}. For example, $MoOCl_2(P(CH_3)_2C_6H_5)_3$ was isolated as a green solid with $\nu(Mo–O_t)$ = 943 cm^{-1} and as a more stable blue solid with $\nu(Mo–O_t)$ = 954 cm^{-1}. Both isomers were assigned a meridional arrangement for the phosphines, and the difference between them was said to be that the green form had L trans to O, whereas the blue form had Cl trans to O as shown in structures **12** and **13**.

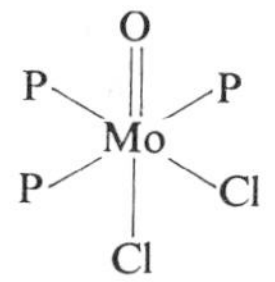

cis-mer-$MoOCl_2L_3$
(12)

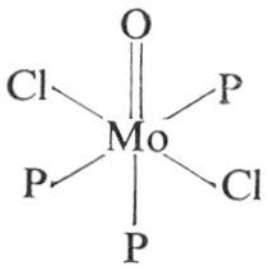

trans-mer-$MoOCl_2L_3$
(13)

However, structural studies indicate a different origin for the blue and green isomers. Thus, Chatt, Manojlovic-Muir, and Muir (461–463) determined the structures of the green form of $MoOCl_2(P(C_2H_5)_2C_6H_5)_3$ [$\nu(Mo–O_t)$ = 940 cm^{-1}] and the blue form of $MoOCl_2(P(CH_3)_2C_6H_5)_3$ [$\nu(Mo–O_t)$ = 954 cm^{-1}], and they found both compounds to display the cis-mer arrangements with Cl's cis and phosphines mer to each other (structure **12**). The difference between the structures is subtle, with the blue complex having bond lengths $Mo–O_t$ = 1.676(7) Å and Mo–Cl = 2.551(3) Å trans to O, while the green complex has $Mo–O_t$ = 1.803(11) Å and trans Mo–Cl = 2.476(6) Å. These changes in bond length are coupled with differently distorted coordination polyhedra in the two compounds. Both polyhedra adopt arrangements of ligand atoms (describable as distorted octahedral) which apparently minimize repulsions of the ligands. However, there seem to be (at least) two ligand arrangements within the cis-mer configuration which are at least local minima in the potential energy surface describing the intramolecular isomerization of these molecules. Chatt (485) suggests that the green and blue isomers for a given complex be called distortional isomers, a still more subtle form of structural isomerism.

The Schiff-base ligand produced from 2,6-diacetylpyridine and aniline is tridentate and forms complexes of the form $MoO(py(anil)_2)X_2$, where X = Cl or Br. These are obtained (487) by reacting Mo(III) starting materials with aniline and 2,6-diacetylpyridine in boiling *n*-butanol. The blue-indigo complexes are very air sensitive in solution and somewhat so in the solid state. The compounds are isomorphous to $VO(py(anil)_2)Cl_2$ and $VO(py(anil)_2)$-Br_2 and thus their possession of the strong M–O bond and octahedral structure characteristic of these species is likely.

$MoO(C_{32}H_{10}N_8)$ is obtained by reaction of phthalonitrile with MoO_2Cl_2 in dmf (489). The phthalocyanine formulation for this complex is strongly supported by its electronic and mass spectra. The complex is diamagnetic and a weak ESR signal (g = 1.97) at 77°K is attributed to a small amount of [presumably Mo(V)] impurity. Novotny and Lippard prepared a series of $[MoO(CNR)_4X]^+$ complex ions by the reaction of MoX_4 with the appropriate isocyanide in methanol (488). The dianionic complex $MoO(mnt)_2^{-2}$ was prepared by reaction of Mo(VI) and mnt^{-2} and isolated as Ph_4P, Et_4N, Ph_3-PCH_2Ph, and $N(C_4H_9)_4$ salts (486, 497). Selbin et al. (102) recently reported the isolation of monooxo Mo(IV) complexes of flavin derivatives (102) but a more recent study casts some doubt on this formulation (102a).

The oxo Mo(IV) complexes which have been discussed so far contain a single strong Mo–O_t bond. However, in Mo(IV) cyanide chemistry the D_{4h} species *trans*-$MoO_2(CN)_4^{-4}$ has been well characterized chemically (490, 491) and structurally (458, 498, 499). The trans Mo–O bonds are 1.83 Å in length, somewhat longer than the values near 1.70 Å found in monooxo Mo(IV) complexes. The isolation of this complex requires strongly basic conditions, and the complex is readily protonated in solution. Salts of $MoO(OH)(CN)_4^{-3}$ are isolated at pH = 7, and structural studies (459, 499) reveal Mo–O_t = 1.75 and Mo–OH = 2.10 Å trans to Mo–O_t in the C_{4v} anion. The compound designated $Mo(OH)_2(CN)_4^{-2}$ (491) seems more likely to be $MoO(OH_2)(CN)_4^{-2}$ (500). It displays pK_a values at 25°C of 9.96 and 12.6, the latter coorresponding to deprotonation of $MoO(OH)(CN)_4^{-3}$. Recently, Robinson et al. (499a) reported the crystal structure of $[Pt(en)_2][MoO(H_2O)(CN)_4]$. $2H_2O$ wherein the anion, $MoO(OH_2)(CN)_4^{-2}$, is found to have trans Mo–O_t and Mo–OH_2 groups with Mo–O distances of 1.67 and 2.27 Å, respectively. Despite its facile protonation, $MoO_2(CN)_4^{-4}$ is a genuine example of a dioxomolybdenum-(IV) species.

Why is it that the dioxomolybdenum(IV) species is only viable when the other four ligands in the coordination sphere are cyanides? The answer lies in the strong pi-acceptor capacity of the CN^- ligands. The strong Mo–O multiple bonding that dominates the chemistry of Mo^{+5} and Mo^{+6} can be viewed as satisfying the necessity to neutralize the large formal positive charges on Mo [Compounds in the (V) and (VI) states only exist with pi-donor ligands, i.e., the electroneutrality principle (196) is operative]. The Mo–O_t bond puts sufficient electron density on Mo through sigma plus pi bonding to neutralize the large formal positive charge. For Mo(VI) this requires at least two terminal oxo's in a 6-coordinate complex, whereas for most Mo(IV) complexes only one oxo is required because of the lower oxidation state. However, in the $MoO_2(CN)_4^{-4}$ ion the CN ligands in the xy plane withdraw charge via the $(d_{xy})^2$ system and therefore effectively permit more pi donation by the other ligands, and in this complex the presence of two oxo's becomes possible. The facile protonation of one of the oxygens is, however, a sign that the oxygen does not donate its charge as fully as do the oxo groups in Mo(VI) complexes which are more difficult to protonate. The dioxotetracyano species thus illustrates the control that ligands in the coordination sphere can have on the acidity of other ligands in that same sphere. This could be a factor of the choice of ligands by Mo enzymes.

Relevant points about oxo-Mo(IV) complexes include their variability in coordination number and their ability to undergo oxidative-addition reactions.

B. Mo(IV) in Aqueous Solution

In 1968, in a little noted (457) but potentially significant paper, LaMache (456) reported that Mo(IV) was indeed a viable species in aqueous solution. Thus it was shown that the salts of both Mo(V) and Mo(IV) could be isolated from electrolysis of Mo(VI) in acid media. Furthermore, the reaction of equimolar Mo(III) and Mo(V) in hot $<4.5M$ HCl was reported to produce Mo(IV). Although Mo(III) and Mo(V) do not produce Mo(IV) in strong acid, once the Mo(IV) ion is formed in the weaker acid solution it is stable at any acidity. LaMache formulated the species as $MoO(OH)^+$ and noted its color as rose.

More recently, Ardon and Pernick (455) described the existence of a Mo(IV) ion in aqueous solutions of *p*-toluenesulfonic acid. In these solutions, as in those of Mo(II) and Mo(III) in the same medium, the *p*-toluenesulfonate anion is assumed to have little coordinating ability, and the species present are thus aquo (and/or hydroxo or oxo) complexes. Ardon and Pernick noted that the "almost colorless' solutions of $Mo(H_2O)_6^{+3}$ in *p*-toluenesulfonic acid acquired a pinkish tint upon prolonged storage. Ion-exchange chromatography led to the separation of a dark-red band. A rational preparative procedure for this ion showed it to be formed (but not quantitatively!) from an equimolar solution of Mo(III) and Mo(V) in $1.0M$ HPTS when the solution was heated at 90° for 1 hr in an inert atmosphere. The ion-exchange and redox behavior of the ion indicated a Mo(IV) complex with an ionic charge +4. Ardon and Pernick formulated the ion as structure **14**

$$(H_2O)_4Mo\langle O_2 \rangle Mo(H_2O)_4^{+4}$$

14

but other possible formulations include complexes with hydroxobridges and hydroxo-terminal ligands and species with mixed oxidation states. Recently, Ramasami et al. (455a) have presented evidence from kinetics and ion exchange behavior that supports a monomeric formulation, presumably MoO-$(H_2O)_5^{+2}$, for this species. Nevertheless, the gross formulation of this ion appears sound, and it is especially significant that the Mo(IV) species does not rapidly disproportionate to Mo(III) and Mo(V) as had often been considered the fate of aqueous solutions of Mo(IV). In contrast, previous studies invariably had coordinating anions (Cl^-, Br^-, etc.) or reducible anions (ClO_4^-, NO_3^-) present in the solution either of which could compromise the stability of Mo(IV).

Taken together, the above studies present interesting evidence for the existence of Mo(IV) in aqueous solution and, coupled with the polarographic detection of Mo(IV) and its postulation as a kinetic intermediate (48–50) in a number of reactions, lend weight to its postulation as a participant in Mo enzymes.

C. Halide and Pseudohalide Complexes

The halide chemistry of Mo(IV) has been reviewed (13, 414, 501) and is not considered in detail here. For our purpose the polymeric Mo(IV) halides are of interest as starting materials in the production of complexes of Mo-(IV) and other oxidation states.

Molybdenum tetrafluoride (13) is obtained along with MoF_5 from the disproportionation of Mo_2F_9, prepared from $Mo(CO)_6$ and F_2 at $-75°C$ or from Mo and SF_4. The light-green polymeric solid has not been investigated in great detail.

$MoCl_4$ exists in two polymorphs, differing apparently in the Mo–Mo distance and interaction. Alpha-$MoCl_4$, originally prepared by simply refluxing $MoCl_5$ in benzene (502), is now most often prepared by reaction 29,

$$2MoCl_5 + Cl_2C{=}CCl_2 \longrightarrow 2MoCl_4 + Cl_3CCCl_3 \qquad (29)$$

in which Cl is abstracted from $MoCl_5$ by the tetrachloroethylene, and the α-$MoCl_4$ thus produced is isomorphous with the metal-metal bonded $NbCl_4$. The structure consists of infinite chains of octahedra sharing opposite edges with alternating long and short M–M distances. The magnetic moment of α-$MoCl_4$ is 0.85 B.M. which is consistent with metal-metal interaction. *Beta*-$MoCl_4$, on the other hand, has a 3.50-Å Mo–Mo distance and magnetic behavior characteristic of two unpaired electrons. Structurally, it has Mo in octahedral coordination with edges shared by adjacent Mo's. Lately, the α-form of $MoCl_4$ has been used in many procedures owing to its relatively simple preparation. Thermally, $MoCl_4$ is unstable to disproportionation to $MoCl_3$ and $MoCl_5$ above 170° (13).

Molybdenum tetrabromide (503) is prepared by bromination of Mo metal or by reaction of Br_2 and $Mo(CO)_6$ or $MoBr_3$ (13). The $MoBr_4$ is quite unstable, readily giving $MoBr_3$ and Br_2.

The tetrahalides are, in general, reactive to air and water, but with proper precautions are useful as starting materials for preparation of some of the complexes that are shown in Table XIV and discussed below.

Potassium salts of $MoCl_6^{-2}$ can be prepared at 200° according to Eq. 30.

$$2KCl + MoCl_5 \longrightarrow K_2MoCl_6 + 1/2Cl_2 \qquad (30)$$

TABLE XIV
Properties of Nonoxo Mo(IV) Complexes

Complex	Magnetic moment, B.M.	(T, °K)	Electronic absorption[a] cm^{-1}	(ε)[b]	Refs.
$K_2[MoCl_6]$	2.28	(300)	24,100		465
	(θ = 54°)		27,800		
			31,750		
			35,700		
			40,800		
$(NMe_4)_2[MoCl_6]\cdot 0.5CH_3CN$	2.44		22,000		504
			25,800		
$(pyH)_2[MoCl_6]$	2.36		21,510		505
			25,640		
			27,780		
$Rb_2[MoBr_6]$	2.18	(300)	16,100		465
	(θ = 140°)		19,600		
			24,100		
			28,000		
			41,700		
$K_2[Mo(NCS)_6]$	3.02				506
$(pyH)_2[Mo(NCS)_6]$	2.45	(297)	17,700	(25,000)	340
			[23,800]	(53,000)	
			24,400	(62,500)	

TABLE XIV (continued)

Complex	Magnetic moment B.M.	(T, °K)	Electronic absorption[a] cm^{-1}	$(\varepsilon)^b$	Ref.
$(N(C_4H_9)_4)_2[Mo(NCS)_6]$	2.80				506
$MoCl_4(CH_3CN)_2$	2.48		20,200	(75)	507
	(θ = 40°)		[25,000]	(400)	
			[27,800]	(900)	
			31,800	(2,900)	
			35,100	(2,400)	
$MoCl_4 \cdot (C_2H_5CN)_2$	2.50		19,800	(75)	507
	(θ = 40°C)		25,300	(1,800)	
			32,300	(5,000)	
			34,500	(5,000)	
$MoCl_4(py)_2$	2.71		21,200		508
			24,700		
$MoCl_4(thf)_2$	2.52		[20,800]		508
			26,000–29,000		
$MoCl_4(PPh_3)_2$	2.43		15,600		508
			20,000		
			25,000		
			27,800		
$MoCl_4(PEt_2Ph)_2$	2.61	(300)	Maroon		485
$MoCl_4(PMePh_2)_2$	2.48	(300)	Maroon		485
$MoCl_4(AsPh_3)_2$	2.36		15,600		508
			20,000		
			24,400		
			27,000		
$MoCl_4(dppe)$	2.18	(300)	Red brown		485
$MoCl_4(bipy)$	2.36	(300)	19,600		508
			25,000		
			27,000		
$MoCl_4(phen)$	2.28	(300)			508
$MoCl_4(diars)$	1.96	(293)			509
$MoCl_2(acac)_2$	2.63	(296.3)	19,690	(1,943)	510
			25,910	(3,450)	
			29,240	(4,600)	
$MoCl_2(\beta\text{-diketonate})_2$	Paramagnetic				511
$Mo(N\text{-Mesal})_2Cl_2$	2.74	(298.3)	18,870		510
			23,800		
			28,570		
			31,750		

TABLE XIV (continued)

Complex	Magnetic moment B.M.	(T, K°)	Electronic absorption cm⁻¹	(ε)	Ref.
$Mo(p\text{-}ClC_6H_4Nsal)_2Cl_2$	2.50		17,390		510
	(θ = 10°)				
$Mo(salen)Cl_2$	2.61	(292)	10,050		510
			17,390		
			42,550		
$Mo(acacen)Cl_2$			17,540		510
			25,000		
$K_4Mo(CN)_8$			[24,400]	(90)	346
			27,000	(158)	
			42,000	(12,000)	
$(NH_4)_4[Mo(C_2O_4)_4]\cdot 8H_2O$	diam		15,380	(15)	512
			18,940	(50)	
			23,200	(60)	
$K_4[Mo(CO_3)_4]\cdot K_2CO_3\cdot 2H_2O$	diam				513,514
$Mo(Me_2dtc)_4$	1.0				515
$Mo(Et_2dtc)_4$	0.68				516
	(θ = 71°)				
$Mo((CH_2)_4dtc)_4$	2.11		18,900		517
			[21,300]		
			22,200		
			28,200		
			37,000		
$Mo(S_2CCH_2Ph)_4$	diam		13,500	(1,900)	518
	(0.5–0.6)		17,100	(1,900)	
			22,300	(2,400)	
			[25,000]		
$Mo(S_2CPh)_4$	diam		11,600	(14,100)	518
	(0.5–0.6)		[13,000]		
			17,600	(6,700)	
			20,800	(4,790)	
			[21,500]		
$Mo(SC(Ph)CHC(Ph)O)_4$	diam				519
$Mo(CN)_4(CNCH_3)_4$	diam		(26,000)	(130)	
			(30,000)	(350)	471,471a,
			41,500	(15,700)	471b
			43,800	(11,000)	

[a]Numbers in bracketts represent shoulders or poorly resolved bands.
[b]Molar extinction coeffecients in l $mole^{-1}$ cm^{-1}.

probably through the intermediacy of $MoCl_6^{-1}$ which at that temperature (in the melt) is unstable to the internal redox. Alternatively, the preparation can be effected at 150° by reacting $MoCl_5$ with an excess of iodine monochloride. Other salts of $MoCl_6^{-2}$ can be prepared by the ICl method. The K_2MoCl_6 salt is isomorphous to K_2PtCl_6 and, the Mo–Cl distance is reported as 2.31 Å (465) from powder studies. The hexabromomolybdates(IV) can be prepared from $MoBr_3$, MBr, and IBr (where M = Rb, Cs). Complex salts of Mo(IV) and iodide could not be prepared (465). The tetravalent hexahalides hydrolyze slowly in moist air.

The tetrahalides are reactive toward addition of ligands which break the polymer structures to form 6-coordinate monomeric species. MoF_4, $MoCl_4$, and $MoBr_4$ are found (11, 508, 520, 521) to undergo addition reactions to yield compounds of the form MoX_4L_2 (where L_2 is two neutral monodentate ligands or one neutral bidentate ligand). For many of the MoX_4L_2 complexes L is simply the donor solvent in which MoX_4 dissolves. A list of known adducts is shown in Table XIV, and although no x-ray structural information is available, there is little doubt that these represent 6-coordinate octahedral Mo(IV) complexes. For the $MoCl_4L_2$ complexes the question of whether the cis or trans isomer is present is an open one, although where L is a phosphine, a trans structure has been assigned (485) from infrared data.

MoX_4L_2 complexes can also be prepared by the reaction of $Mo(CO)_4L_2$ complexes with excess X_2 [X = Cl, Br; L = $P(CH_3)_2C_6H_5$, $P(C_2H_5)_2C_6H_5$] (522). The complexes so prepared show a single ν(Mo–Cl) and are also assigned octahedral structures with trans phosphines.

Moss and Shaw (522) further found that the complex $MoBr_4(P(CH_3)_2C_6H_5)_3$ could be prepared by bromine oxidation of $Mo(CO)_3(P(CH_3)_2C_6H_5)_3$. This complex and its tetrachloride analog have been structurally characterized. As illustrated in Fig. 26, the $MoX_4(P(CH_3)_2C_6H_5)_3$ (X = Cl, Br)

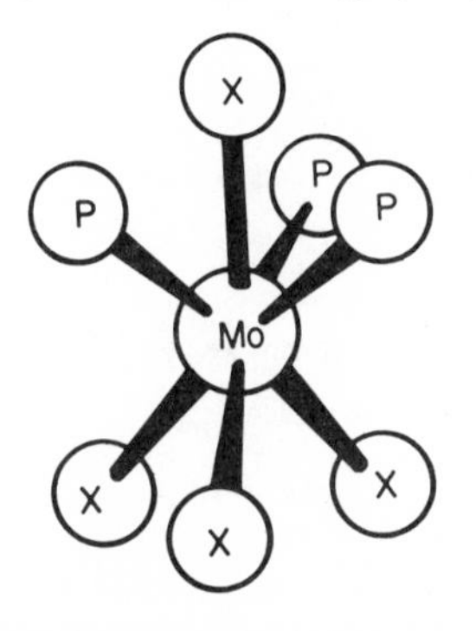

Fig. 26. The structure of $Mo(PMe_2Ph)_3X_4$ (467, 468). (Only the P donor atom of the phosphine ligand is shown.)

complexes have a C_{3v} capped-octahedral 7-coordination. The capped face consists of the three P atoms, while the cap and the remaining three vertices are occupied by Cl or Br. These structures contain seven large donor atoms about Mo(IV) and illustrate that nonoxo Mo(IV) complexes can expand their coordination shell beyond 6.

In one case a probable 5-coordinate adduct, $MoF_4(C_6H_5N(CH_3)_2)$, is isolated presumably because of the steric bulkiness of the ligand (521). The compound described as $MoCl_4((C_6H_5)_3AsO)_4$, an 8-coordinate Mo(IV) complex, has been shown to be a complex of Mo(V) (519).

Westland and Uzelac (520) have measured the heats of solution ΔH_1 and ΔH_2 for processes 31 and 32, respectively.

$$MoCl_4(s) + 2L(l) \longrightarrow MoCl_4L_2 \text{ (dissolved in } x\text{-2 moles of L)} \quad \Delta H_1 \tag{31}$$

$$MoCl_4L_2(s) + L(l) \longrightarrow MoCl_4L_2 \text{ (dissolved in } x \text{ moles of L)} \quad \Delta H_2 \tag{32}$$

They then calculated the ΔH for process 33 as $\Delta H_1 - \Delta H_2$,

$$MoCl_4 + 2L(l) \longrightarrow MoCl_4L_2 \tag{33}$$

and values of -53.9 ± 4, -40.4 ± 0.6, and -31.3 ± 0.8 kcal mole^{-1} were reported for L = pyridine, tetrahydrofuran, and tetrahydrothiophene, respectively. Comparison of the O-donor, thf, and the S-donor tetrahydrothiophene show that more heat is evolved when the former binds to $MoCl_4$. This would seem to implicate $MoCl_4$ as a class A (523), hard (524) acceptor. However, it would be dangerous to weight these findings too heavily as they are for solid compounds and not gas-(or solution-) phase species, and the effects of changing the crystal structure can not be clearly accounted for in light of the missing thermodynamic data for heats of crystallization.

The use of anionic ligands leads to the replacement of chloride in $MoCl_4$. Complexes of the form $Mo(LL)_2Cl_2$ (where LL is a β-diketonate) have been prepared (see Table XIV). The red-purple $Mo(acac)_2Cl_2$ complex was first prepared by Larson and Moore (502) from $MoCl_4$ and acacH, and in view of its general insolubility, it was characterized as polymeric. More recently, Doyle (511) showed that a variety of β-diketones reacted with either $MoCl_4$ or $MoCl_5$ to give the MoL_2Cl_2 product. With other β-diketonates (e.g., dibm, hpd, and dpm) Doyle obtained soluble, monomeric complexes. The observation of increasing solubility with increasing size of the ligand suggests that the observed insolubility of the $Mo(acac)_2Cl_2$ is in line with existing trends, without requiring a polymeric structure. The compounds are paramagnetic, and the only structural information so far available is the apparent existence of two Mo–Cl stretches at 342 and 312 cm^{-1} for $Mo(acac)_2Cl_2$, consistent with nonlinear (i.e., *cis*) chlorides. However, as this spectrum is of an insoluble solid, the geometrical assignment must be considered highly tentative.

In addition to the β-diketonate ligands, the monoanionic 8-hydroxyquinolinate and *N*-alkylsalicylaldiminate ligands have been found to form complexes of the form $Mo(LL)_2Cl_2$ (510), whereas the tetradentate ligands (LLLL = salen, and acacen) (510) produce complexes of the form $Mo(LLLL)X_2$. The complexes are prepared (510) by reaction of $MoCl_4(NCCH_3)_2$, or $MoCl_3(py)_3$ with excess of the Schiff base. They are paramagnetic, consistent with roughly octahedral structures and a ground state derived from the $^3T_{2g}$ state in octahedral symmetry. The structure of a representative of this class has been solved by diffraction techniques, and Davies and Gatehouse(466) found that $Mo(N\text{-}Mesal)_2Cl_2$ had a centrosymmetric structure with trans chlorides. The Mo–Cl, Mo–O, and Mo–N distances are 2.39, 1.95, and 2.14 Å, respectively, and the bond angles about the metal do not differ substantially from 90°. It should be noted that both cis and trans complexes of the general from $M(LL)_2Cl_2$ have been identified. Thus $M(acac)_2X_2$ (M = Ti, Sn, Zr, Hf) (525, 526) and $M(oxine)_2Cl_2$ (M = Ti, Sn) (527) all have cis arrangements, whereas $Re(acac)_2Cl_2$ (528) and $Mo(N\text{-}Mesal)_2Cl_2$ (466) both have trans structures. While it is too early to proclaim a trend, the observation is made that the d^0 or d^{10} species, in general, have cis geometries, whereas the d^3(Re) and d^2(Mo) systems have trans arrangements. However, the adoption of the trans geometry by $Ti(salen)Cl_2$ shows that this arrangement is also possible for d^0 species when the ligand (in this case, salen) displays a preference for such an arrangement. Thus the energy difference between the cis and trans forms may, in fact, be quite small.

The only (nonoxo) pseudohalide complexes of Mo(IV) are those of thiocyanate. The $Mo(NCS)_6^{-2}$ ion can be prepared by oxidation of $Mo(NCS)_6^{-3}$ with $Fe(CN)_6^{-3}$ (529), by reaction of $MoCl_4(CH_3CN)_2$ with KSCN in CH_3CN (506), or by reaction of SCN^- with "Mo(IV)" produced *in situ* by mixing 2 moles of Mo(VI) with 1 mole of Mo(III) and prepared electrolytically (340). The tetrabutylammonium (506), tetraphenylarsonium (506), and pyridinium (529) salts (340) have also been isolated and characterized. The $Mo(NCS)_6^{-2}$ is assigned N-bonded thiocyanate from infrared data, a conclusion that is quite reasonable in view of the established Mo–N bonds in $Mo(NCS)_6^{-3}$ and the expected increase in class A character (hardness) in going from Mo(III) to Mo(IV).

The 6-coordinate halo and pseudohalo complexes are all found to be paramagnetic with moments ranging from 1.9 to 2.8 B. M. The temperature dependence of the magnetic susceptibility of these complexes agrees with a configuration derived from a $^3T_{1g}$ ground state (in O_h) appropriately split by spin-orbit coupling and fields of lower symmetry. The lowest transitions for the $(t_{2g})^2$ configuration are $^3T_{1g} \rightarrow {}^3T_{2g}$ and $^3T_{1g}(F) \rightarrow {}^3T_{1g}(P)$, and while these assignments have been made in some cases (510, 530), the detailed assignments remain tentative.

The coordination number 7 has also been found in $MoCl(R_2dtc)_3$, described in Section IX, and the apparently similar compound $MoCl(SC(Ph)$-$CHC(Ph)O)_3$ (where the ligand is monothiodibenzoylmethide) has been prepared from the reaction of $MoCl_5$ with thiodibenzoylmethane. These, along with $MoX_4(PMe_2Ph)_3$ (X = Cl, Br), illustrate that 7 is a viable coordination number for Mo(IV) and that 6-coordinate Mo(IV) compounds need not be considered coordinately saturated. This notion is reinforced in the next section.

D. Eight-Coordinate Complexes

The dodecahedral octacyanide ion has been known for many years and has been well studied in terms of synthesis (531), molecular structure (469, 470), electronic structure (530), and photochemical and thermal reactivity (4, 8). This first 8-coordinate Mo complex has been joined quite recently by a host of others (289, 471–474, 512–517), mostly of the form $Mo(LL)_4$, where LL is a dithiocarbamate, dithiocarboxylate, or monothio-β-diketonate ligand. Jowitt and Mitchell (515) prepared $Mo(Me_2dtc)_4$ by reaction 34.

$$Mo(CO)_6 + 2Me_2NC(S)SSC(S)NMe_2 \longrightarrow Mo(Me_2dtc)_4 + 6CO \tag{34}$$

A similar reaction (516) forms the ethyl derivative. Bradley and Chisholm (531) reported the preparation of $Mo(NR_2)_4$ (R = CH_3 and C_2H_5) and demonstrated that a "CS_2-insertion" reaction of the form shown in reaction 35 could be used to prepare the tetrakis complex.

$$Mo(N(CH_3)_2)_4 + 4CS_2 \longrightarrow Mo(S_2CN(CH_3)_2)_4 \tag{35}$$

Brown and Smith (517) were able to prepare the tetramethylene-dithiocarbamate complexes by the direct reaction shown in reaction 36.

$$Na^+S_2CN(CH_2)_4^- + MoCl_4 \longrightarrow Mo(S_2CN(CH_2)_4)_4 + NaCl \tag{36}$$

These complexes can be electrochemically and chemically oxidized (257, 351) and electrochemically reduced, and the diethyl dithiocarbamate complexes show voltametry in acetone versus sce, leading to the electron transfer scheme (scheme 37):

$$Mo(Et_2dtc)_4^+ \xrightarrow{-0.55V} Mo(Et_2dtc)_4 \xrightarrow{-0.66V} Mo(Et_2dtc)_4^- \xrightarrow{-1.18V} Mo(Et_2dtc)_4^{-2} \tag{37}$$

Uhlemann and Eckelman report that the monothio-β-diketonate complex is prepared by Eq. 38:

$$Mo(CO)_6 + PhCOCHC(Ph)SSC(Ph)CHC(Ph)O \longrightarrow Mo\left(\begin{array}{c} Ph \\ S—C \\ C \\ O—C \\ Ph \end{array}\right)_4 \quad (38)$$

Piovesana and Sestili (473) report the use of the Mo (III) starting material $(NBu_4)_3[MoCl_6]$ to produce $Mo(S_2CPh)_4$, $Mo(S_2CC_6H_4p\text{-}CH_3)_4$, and Mo-$(S_2CCH_2Ph)_4$ complexes. The crystal structure of the first complex (474) reveals the Mo to be dodecahedrally coordinated in a D_2 structure with $Mo–S_A = 2.543$ (1) Å (elongated tetrahedron) and $Mo–S_B = 2.475$ (1) Å (flattened tetrahedron).

The magnetic properties of these complexes are in some doubt (510). Thus, a moment of 2.11 B.M. was reported for $Mo((CH_2)_4dtc)$ (517), but moments of 1.0 and 0.8 B.M. were given for $Mo(Me_2dtc)_4$ (515) and Mo $(Et_2dtc)_4$ (516), respectively. Piovesana and Sestili (473) reported moments as high as 0.5 to 0.6 B.M. for dithiocarboxylate complexes, but observed a field dependence and variability with crystallization which implied that an impurity was responsible for the magnetism.

The D_{2d} dodecahedron has a d-orbital splitting that leaves the d_{xy} level lowest (4, 530, 532). This low-lying d_{xy} level is thought to contain the two electrons in $Mo(CN)_8^{-4}$, and a similar situation may obtain for all the Mo $(LL)_4$ complexes should these prove to be diamagnetic.

$Mo(CN)_4(RCN)_4$ can be prepared by the reaction of $Ag_4Mo(CN)_8$ with RBr (471a). For the case where $R{=}CH_3$ structural studies (471, 471a) reveal a dodecahedral coordination with the more strongly π-accepting isonitrile ligands occupying the B(flattened tetrahedron) positions.

Steele has reported the existence of the diamagnetic compounds $(NH_4)_4$ $[Mo(C_2O_4)_4]\cdot 8H_2O$ (512) and $K_4Mo(CO_3)_4\cdot K_2CO_3\cdot 2H_2O$ (513). The first compound was spectroscopically investigated by Basu (514). The method of preparation involves electrolysis of Mo(VI) solutions, and while the structures of these compounds are unknown, they could be further examples of 8-coordinate Mo(IV).

E. Cyclopentadienyl Compounds

There is an extensive class of compounds which contains two cyclopentadienyl ligands and two additional donors in a formally Mo (IV) complex (534, 547). Complexes of the form $(C_5H_5)_2MoX_2$ are known for $X_2 = H_2$, $(CH_3)_2$, $(SH)_2$, $(SR)_2$, S_4, Cl_2, Br_2, $(N_3)_2$, and P_2H_2 (534, 537, 546), and

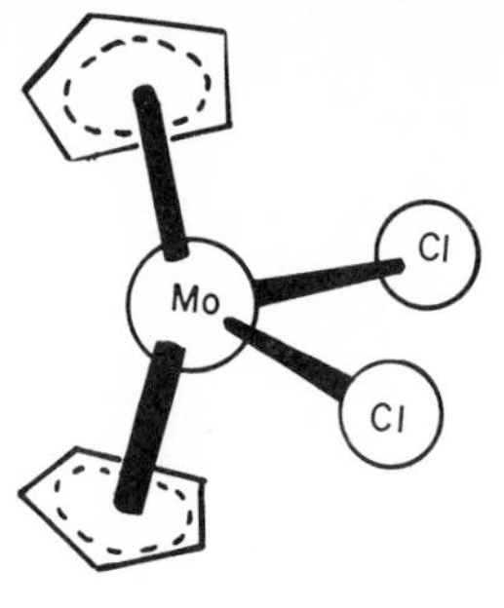

Fig. 27*a*. The structure of $(C_5H_5)_2MoX_2$ (see text for the nature of X).

$(C_5H_5)_2MoXL^+$ species are found for XL=$CH_3(PR_3)$, $Br(SMe_2)$, Br(CO), $O(NH)C_6H_4$, $S(NH_2)C_6H_4$, $SCH_2CH_2NH_2$, SCH_2CH_2SMe, $S_2C_6H_3CH_3$, cys, gly, and other amino acid ligands (534, 538, 545–547). Even dicationic compounds such as $(C_5H_5)_2Mo(en)^{+2}$ are known (534). Some of the compounds can be oxidized by one electron to formally Mo (V) species such as $(C_5H_5)_2MoX_2^+$(X = Cl, Br) (474), whereas the trihydride, $(C_5H_5)_2MoH_3^+$, formally d^0, Mo(VI) can be formed by protonation of $(C_5H_5)_2MoH_2$.

The compounds have been extensively probed chemically and structurally by Green, Prout, and others. Some of the solved structures are listed in Table XII, and Fig. 27*a* shows a representative complex $(C_5H_5)_2MoCl_2$ (479). Structural and bonding trends have been discussed, and some interesting chemical studies with derivatives of this series are discussed in Section IX. Generally, in the course of chemical reactions the two cyclopentadienyl ligands remain coordinated and essentially unchanged while the remaining two coordination sites are reactive.

Dias and Green (541–543) found that $(C_5H_5)_2Mo(SR)_2$ could serve as a bidentate ligand utilizing the two thiolate sulfurs to chelate a second metal. Complexes of the form $(C_5H_5)_2Mo(\mu\text{-}SR)_2MCl_2$ (M = Fe, Co, Pt) and $(C_5H_5)_2Mo(\mu\text{-}SR)_2M(\mu\text{-}SR)_2Mo(C_5H_5)_2$ (M = Ni, Pd, Pt) have been prepared by reaction of $(C_5H_5)_2Mo(SR)_2$ with MCl_2 in ethanol or thf. The Mo–Fe(R = Bu) complex was crystallographically investigated (483, 484), and the $Mo(S\text{-}n\text{-}Bu)_2Fe$ linkage was found to be bent in a manner analogous to di-μ-sulfido bridges found for Mo(V) (Section IV) with an MoS_2–FeS_2 dihedral angle of 148°. However, the long Mo–Fe distance of 3.66 Å rules out Mo–Fe bonding. The Mo–S and Fe–S distances average 2.464 and 2.386 Å, respectively. The complex ($(C_5H_5)_2Mo(\mu\text{-}SMe)_2)_2Ni$ was shown (544) to have square planar Ni^{+2} ion with Mo–Ni = 3.39 Å.

The amino acid complexes in this class (538) are of interest in that the α-C–H undergoes exchange with D_2O and racemization in weak base. Coordination of the amino group of the amino acid causes a labilization of an adjacent C–H position, and while the mechanism for this process is unknown, it might involve a reaction such as reaction 39:

$$(C_5H_5)_2Mo(-O-C(=O)-CH(R')-N(H)(R)-) \underset{+2H^+}{\overset{-2H^+}{\rightleftharpoons}} (C_5H_5)_2Mo(-O-C(=O)-C(R')=N(R)-) \quad (39)$$

F. Unusual Sulfur-Donor Complexes

An unusual Mo(IV) sulfur-donor complex was reported by Weiss and co-workers, who gave structural details for the reaction product of molybdenum(II) acetate (see Section VII) and ammonium *N,N*-di-*n*-propyldithiocarbamate in ethanol solution. The reaction product has the formula $Mo_2((n\text{-}Pr)_2dtc)_4$, but its X-ray crystal structure (475, 476) reveals that a C–S bond in two of the four dithiocarbamate ligands has been cleaved with the abstracted sulfur atoms, forming a bridge between the two Mo's. The remainder of the ligand, a thiocarboxamido group, is coordinated to the Mo through C and S in a roughly pi (side-on) fashion. The structure is shown in detail in Fig. 27*b*. The Mo_2S_2 group is not planar, and the dihedral angle between MoS_2 planes is 160.7°. The Mo–Mo distance of 2.71 Å is consistent with Mo–Mo bonding and is considerably shorter than the distances found in MoS_2Mo complexes of Mo(V).

It is likely that the great stability of the Mo_2S_2 four-membered ring provides much of the driving force for this process. In a sense this can be viewed as an oxidative addition of a C–S bond on Mo(II) or more precisely as an oxidative addition of two C–S bonds across an Mo(II)-Mo(II) linkage. In that case, the final product is formulated as a di-μ-sulfido-bridged Mo(IV) complex. Di-μ-sulfido bridging is not strange in Mo(IV) chemistry as witnessed by the extreme stability of MoS_2 (molybdenite) wherein six sulfurs surround the Mo atoms in a perfect trigonal prismatic array. It is significant that five of the six-donor atoms in the complex under discussion are sulfurs. Furthermore, inspection of the coordination geometry reveals a 6-coordinate polyhedron that appears to be distorted from an octahedron in the direction of a trigonal prism.

A second unusual grouping is found for the ion $Mo_3S_4(C_5H_5)_3^+$ crystallized as the $Sn(CH_3)_3Cl_2^-$ salt from the reaction of $(C_5H_5)Mo(CO)_3Cl$ and $(Sn(CH_3)_3)_2S$ in 1,2-dimethoxyethane (477). Here, a C_{3v} structure containing a triangular Mo_3 cluster is found with a single triply-bridging sulfur atom above the center of the triangle and three doubly bridging sulfurs below. The

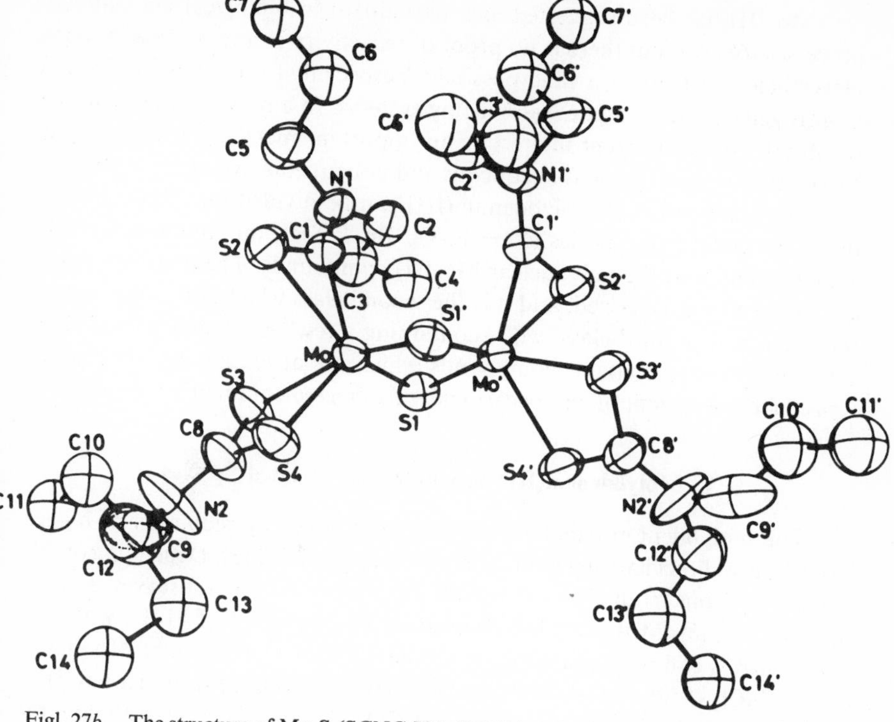

Figl. 27*b*. The structure of $Mo_2S_2(SCNC_3H_7)_2\ ((C_3H_7)_2dtc)_2$. (Reproduced with permission from Ref. 475.)

Mo–Mo distance of 2.81 Å is very close to that of Mo(V) complexes containing a di-μ-sulfido bridge.

A third class of unusual, formally Mo(IV), sulfur-donor complexes is that of the tris(1,2-dithiolenes) and these are discussed in Section VIII.

VI. THE CHEMISTRY OF MOLYBDENUM(III)

In comparison to the copiously studied Cr(III) complexes, the compounds of Mo(III) have been scantily investigated. The reason for this is undoubtedly that while Cr(III) complexes are inert to oxidation by all but the strongest oxidants, many Mo(III) complexes are readily oxidized by air and by a variety of mild oxidants. Despite this limitation, recent studies point to the existence of many complexes, and except for their reactivity toward oxidation, they generally resemble the corresponding Cr(III) compounds.

Mo(III) has been suggested as a participant in biological and catalytic processes (68–70), but there is no proof of its activity in any of these systems. Nevertheless, it remains a valid possibility, especially for the lowest Mo state in nitrogenase and catalytic N_2-reducing systems. In any event, the chemistry of Mo(III) holds inherent interest as an important member of the group of Mo oxidation states present in aqueous and nonaqueous systems.

The compounds of molybdenum(III) can be divided into three groups: the polymeric MoX_3 species (X = F, Cl, Br, I, SR), the mononuclear Mo complex ions, and the polynuclear Mo(III) compounds. The first two classes are very well characterized, and it is the second class which forms the bulk of this section. The third class, while containing a few well-characterized dinuclear species, also contains compounds which are at present poorly characterized from a structural and occasionally from a stoichiometric viewpoint.

A. Molybdenum(III) Trihalides and Related Species

The molybdenum trihalides are prepared either by reduction of higher halides or by direct halogen oxidation of the metal (8, 12, 13). Depending on the halogen, various methods have proved superior and a general survey is given by Larson (548). MoF_3 can be obtained by the reduction of MoF_5 with Mo (549). $MoCl_3$ can be prepared by reduction of $MoCl_5$ with either Mo metal, H_2, or $SnCl_2$ (550) and is slowly hydrolyzed and oxidized in moist air. It is insoluble in H_2O, dilute HCl, CCl_4, CH_3COCH_3, and C_6H_6, but apparently shows slight solubility in the donor solvents $C_2H_5OC_2H_5$, C_3H_8OH, and C_6H_5N (probably because of adduct formation). $MoBr_3$ (551) and MoI_3 (552) are prepared by direct combination of the elements under carefully controlled conditions. In the reaction between Mo and I_2 excess I_2 is required, apparently to suppress the formation of MoI_2 (552). MoI_3 is also formed in the reaction of $Mo(CO)_6$ with I_2 (575).

Each of the molybdenum trihalides has an extended lattice structure. $MoCl_3$ crystallizes in two polymorphs consisting of layered structures with cubic (α) and hexagonal (β) close-packed chlorides. The Mo's occupy adjacent octahedral holes, forming Mo_2 pairs with an Mo–Mo distance of 2.76 Å. $MoBr_3$ and MoI_3 are isostructural to $MoCl_3$ and, with respect to the Mo atoms, have a structure that can be viewed as two octahedra sharing a face (552).

The magnetic properties displayed in Table XV show the effective magnetic moment to decrease as the halogen changes from I to Br to Cl. This is probably indicative of antiferromagnetic coupling which is greatest for the Cl-bridged compound due to the closer approach of the Mo atoms. The analogous chromium compounds $CrCl_3$, $CrBr_3$, and CrI_3 all show full spin-only magnetic moments ($\sim$ 3.8 B.M.), indicating the tendency of Mo compounds to display stronger metal-metal interactions. The structural and magnetic features for these extended lattices carry through to some of the coordination compounds of Mo(III) discussed in Section VI. C.

Compounds of the form $Mo(SR)_3$ have been prepared by Jowitt and Mitchell (515) and by Brown et al. (553). These presumably polymeric materials show low magnetic moments and spectroscopic properties consistent with a (bridged) octahedral structure about the Mo similar to that found for $MoCl_3$ (586). The compounds are made by using organometallic, zerovalent, Mo starting materials which are oxidized by either the thiol or its disulfide. Equations 40 and 41,

Table XV
Complexes of Mo(III)

Compound	Color	μ_{eff} in B.M. (T°K)	Refs.
$MoCl_3$	Dark red	0.7(R.T.)	548, 550, 552
$MoBr_3$	Black	1.2(R.T.)	548, 552
MoI_3	Black	1.4(R.T.)	548, 552
$Mo(SC_6H_5)_3$	Black	0.77(293)	515, 553
$Mo(SCH_3)_3$	Black	0.81(293)	553
$Mo(SC_2H_5)_3$	Black	0.91(293)	553
$Mo(SC_5H_{11})_3$	Black	1.17(293)	553
K_3MoCl_6	Red	3.83(297)	554
		3.79(300)	555
		3.74(293) in 4*N* HCl	555
		3.70(R.T.)	555
$(NH_4)_3MoCl_6$	Red	3.83	556
$Li_3MoCl_6 \cdot 9H_2O$	Red	3.82	557
$K_3MoCl_6 \cdot 4KCl \cdot 6H_2O$	Red	3.2	557
$(NH_4)_3MoCl_6 \cdot NH_4Cl \cdot H_2O$	Red	3.70	557
K_3MoBr_6	Light red	—	558
$K_3Mo(NCS)_6 \cdot 4H_2O$	Yellow	3.79(300)	555, 559
$K_3Mo(NCS)_6$	Buff	—	559
$(NH_4)_3[Mo(NCS)_6] \cdot 4H_2O$	Yellow	3.70(295)	559
$(NH_4)_3[Mo(NCS)_6] \cdot H_2O \cdot C_2H_5OH$	Yellow	3.80(295)	559
$[Mo(OC(NH_2)_2)_6]Br_3$	Pale greenish yellow	—	560

Table XV (continued)

Compound	Color	μ_{eff} in B.M. (T°K)	Refs.
$Mo(H_2O)_6^{+3}$	Light yellow	3.69(in 2. 0*M* $CH_3C_6H_4SO_3H$ by NMR method)	561, 562, 563, 564
$K_4Mo(CN)_7 \cdot 2H_2O$	Green	1.75(295)	559, 565
$(NH_4)_2[MoCl_5H_2O]$	Orange	3.78	560, 556, 566, 557
$(NH_3CH_3)_2[MoCl_5(NCCH_3)]$	Orange	3.80	567
(R_3PH) $[Mo(NCS)_4(PR_3)]$ R = C_2H_5, C_4H_9	Red	—	568
$(N(C_2H_5)_4)[MoCl_4(py)_2]$	Golden yellow	—	569
$MoCl_3(py)_3$	Yellow	—	570, 571
$MoCl_3$(2,6-lutidine)$_3$	Yellow	3.20(299)	572
$MoCl_3(NCCH_3)_3$	Yellow	3.71	567, 570
$MoCl_3(OC(NH_2)_2)_3$	Yellow	3.72(293)	560, 573
$MoCl_3(SC(NH_2)_2)_3$	Orange yellow	3.71(293)	560, 573
$MoCl_3(SC(NHC_6H_5)_2)_3$	—	3.76(295)	573
$MoCl_3(OCHN(CH_3)_2)_3$	Yellow	—	560
$MoBr_3$(4-pic)$_3$	—	3.86	508, 574
$MoBr_3(NCCH_3)_3$	Yellow brown	3.76	508
$MoBr_3(NCC_2H_5)_3$	Yellow brown	3.51	508
$MoBr_3(NC\text{-}n\text{-}C_3H_8)_3$	Yellow brown	3.61	508
$MoBr_3(P(C_6H_5)_3)_2(CH_3CN)$	Yellow	3.89	508
$MoBr_3(As(C_6H_5)_3)_2(n\text{-}C_3H_8CN)$	Orange yellow	3.83	508
$MoI_3(py)_3$	Yellow	3.6(293)	575
$[MoBr_2(OC(NH_2)_2)_4]Br$	—	3.78(293)	573
[Mo(phen)$_3$] [Cl]$_3$	Red	3.83	576
[Mo(phen)$_3$][I]$_3$	Chocolate	3.84	576
[Mo(phen)$_3$] [Br]$_3$	Orange	3.84	576
[Mo(bipy)$_3$] [Cl]$_3$	Red	3.66	576
[Mo(bipy)$_3$] [I]$_3$	Chocolate	3.84 3.93(294)	576 577
[Mo(bipy)$_3$] [Br]$_3$	Orange yellow	3.84	576
[Mo(2-picolylamine)$_3$] $(ClO_4)_3 \cdot 3H_2O$	Yellow	3.70(293)	578

Table XV (continued)

Compound	Color	μ_{eff} in B.M. (T°K)	Refs.
$[Mo(en)_3](NCS)_3$	Yellow	3.65(300)	579
$[Mo((NH_2)_2C_6H_4)_3](NCS)_3$	Snuff	3.64(300)	579
$[Mo(bigH)_3](NCS)_3$	Yellow	3.72(300)	579
$[Mo(C_6H_5bigH)_3](NCS)_3$	Yellow	3.74(300)	579
$[Mo(pbz)_3](NCS)_3$	Dark violet	3.64(300)	579
$Mo(S_2PF_2)_3$	Red	3.53	494
$Mo(N\text{-}MeSal)_3$	Red brown	3.71(295)	580
$Mo(acac)_3$	Red brown	3.82(302)	581, 583
$Mo(CH_3COCHCOCF_3)_3$	Grey green	3.76(302)	582, 583
$[Mo(phen)_2Br_2]Br$	Brown	3.60(297)	572
$[Mo(phen)_2I_2]I$	Violet	3.69(293)	572
$[Mo(bipy)_2I_2]I \cdot CH_3CN$	Dark red	3.76(293)	572
$[Mo(bipy)_2Cl_2][MoCl_4(bipy)]$	Crimson	3.62(297)	577, 569
$[Mo(bipy)_2Br_2][Mo(bipy)Cl_4]$	Light brown	3.62(298)	577
$[Mo(en)_2(NCS)_2][NCS]$	Orange yellow	3.68	579
$Cs_3Mo_2Cl_9$	Red brown	0.6(R.T.)	610, 611
$Cs_3Mo_2Br_9$	Red	—	610, 611
$Mo_2Cl_6(SC(NH_2)_2)_3$	—	0.59(293)	573
$Mo_2Cl_6(SCNH_2NH(C_6H_5))_3$	—	0.57(293)	573
$MoX_2(NH_2CH_2CH_2NH)(en)$ [X = Cl, Br]	Brown	0.6–0.7(R.T.)	585

$$Mo(CO)_6 + 3/2\ RSSR \longrightarrow Mo(SR)_3 + 6CO \tag{40}$$

$$Mo(CO)_6 + 3RSH \longrightarrow Mo(SR)_3 + 3/2\ H_2 + 6CO \tag{41}$$

are possible formulations for the net observed reactions. Alternatively (arene)$Mo(CO)_3$ can be used as the Mo(0) starting material. The compounds are stable to water and dilute acid even at 300°C, indicating a degree of hydrolytic stability consistent with the proposed polymeric formulation.

B. Mononuclear Complexes

Monomeric Mo(III) compounds have a $4d^3$ electronic configuration, and the expected octahedral 6 coordination has been confirmed in most cases.

Several claims of 7 coordination have been made, but only for $Mo(CN)_7^{-4}$ is the evidence strong for a heptacoordinate structure.

1. Structural Studies

Only a few x-ray structural studies have been performed on Mo(III) complexes, and these have uniformly revealed octahedral coordination. Spectroscopic and magnetic properties tend to substantiate this result for other compounds. Crystal structures solved by x-ray techniques are for K_3MoF_6 (587), K_3MoCl_6 (588), $K_3Mo(NCS)_6 \cdot H_2O \cdot CH_3COOH$ (589), and $MoCl_3(py)_3$ (590).

The K_3MoF_6 structure shows a perfectly octahedral Mo coordination with Mo–F = 2.00(2) Å (587), while K_3MoCl_6 has an octahedral structure with Mo–Cl averaging 2.44 Å (588).

In the case of $K_3Mo(NCS)_6 \cdot H_2O \cdot CH_3COOH$, the structure determination not only confirms the octahedral coordination, but also establishes the N-bound nature of the ambidentate thiocyanate ligand [Mo–N = 2.09 Å (589)]. Previously, Lewis, Nyholm, and Smith (559) had assigned an N-bonded structure on the basis of infrared studies [ν(C–S) ~ 800 cm^{-1} compared to ν(C–S) ~ 700 cm^{-1} for S-bonded forms] and the isomorphism of $K_3[Mo(NCS)_6]$ to $K_3[Cr(NCS)_6]$ but not to $K_3[Rh(SCN_6]$. The Mo–N–C linkage is very nearly linear with the small deviations from linearity attributable to crystal-packing forces. Since Mo(III) prefers the N donor of SCN, it may be considered a hard acid (524) or class A (523) metal, and this would make it the only trivalent ion in the second transition series to have this character. In general, however, the softness of an ion toward a given ligand depends critically on the nature of the other ligands in the coordination sphere (524), and thus Mo(III) could possess "soft" character in other situations.

The crystal sturcture of $MoCl_3(py)_3$ reveals a meridional arrangement of Cl's and py's in a C_{2v} complex (structure **15**), with Mo–N averaging 2.19 Å and Mo–Cl averaging 2.43 Å (590).

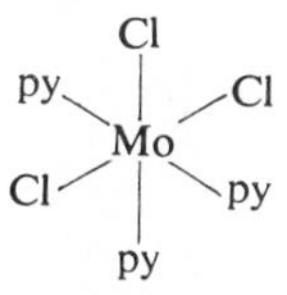

Cl
py Cl
Mo
Cl py
py

15

2. Preparation and Properties

The first and most thoroughly studied Mo(III) complex ion is $MoCl_6^{-3}$, which has been isolated as Li^+, K^+, and NH_4^+ salts. As seen in Table XV, a number of these are isolated as double salts, which led some early investigators to postulate 7-coordinate Mo. However, crystallographic studies (588) rule out this possibility. The standard method (591) for the preparation of $MoCl_6^{-3}$ involves the electrolytic reduction of "molybdate" in hydrochloric acid. Electrolysis usually yields a mixture of $MoCl_6^{-3}$ and $Mo(H_2O)Cl_5^{-3}$, with the hexachloro potassium salt precipitating out of 12*N* hydrochloric acid and alcohol. This electrolytic procedure was originated by Chilessoti and developed in detail by Wardlaw and co-workers (566). More recently, Brencic and Cotton (557) described a somewhat simpler (but more expensive!) synthesis of $MoCl_5H_2O^{-2}$ and $MoCl_6^{-3}$, employing the compound $Mo_2(OOCCH_3)_4$ [which must be prepared from $Mo(CO)_6$ (see Section VII)] according to Eq. 42:

$$Mo_2(O_2CCH_3)_4 + HCl(aq.) + O_2 \longrightarrow MoCl_5(H_2O)^{-2} + MoCl_6^{-3} \tag{42}$$

Hydrolysis of $MoCl_6^{-3}$ in 6*M* HCl gives pure $[MoCl_5(H_2O)]^{-2}$, whereas chloride substitution for water in saturated aqueous HCl is used to isolate the hexachloride. The procedures given are for the NH_4^+ salts of these species (557).

The analogous hexabromo and pentabromo complexes are also prepared by the electrochemical procedure of Wardlaw (558). It has been reported (574) that the hexaiodo complex is formed when K_3MoBr_6 is placed in concentrated HI, but no MoI_6^{-3} salt has been isolated to date.

Other complexes of the MoL_6 type are $Mo(NCS)_6^{-3}$, $Mo(OC(NH_2)_2)_6^{+3}$, and $Mo(H_2O)_6^{+3}$. Due to the special importance of the latter species it is discussed separately in Section VI. B. 5. The $Mo(NCS)_6^{-3}$ ion can be prepared by electrolytic procedures (592) similar to those used for the hexachloro and hexabromo salts. Interestingly, the salts $K_3Mo(NCS)_6$ or $(NH_4)_3Mo(NCS)_6$ precipitate either as tetrahydrates or, in the presence of an appropriate acid, as $(K,NH_4)_3Mo(NCS)_6 \cdot H_2O \cdot HA$, where HA = HCl, KHC_2O_4, C_2H_5OH, C_3H_7OH, or CH_3COOH. This series of isolated compounds has also led some to suggest 7 coordination for the molybdenum, but crystallographic evidence (589) shows the presence of 6-coordinate $Mo(NCS)_6^{-3}$ in the crystals. The hexaurea ion is prepared by the reaction of excess urea with $MoBr_3(urea)_3$ and is isolated as the bromide salt (589).

Monosubstituted and disubstituted complexes of the form MoX_5L^{-2} and $MoX_4L_2^-$ are rare, and those few which are known are prepared by direct ligand substitution from the MoX_6^{-3} species. The ions $MoCl_5H_2O^{-2}$ (508, 558) and $MoCl_5CH_3CN^{-2}$ (587) have been investigated. Andruchow and DiLiddo (593) studied the kinetics of hydrolysis of $MoCl_6^{-3}$ as shown in reactions 43 and 44:

$$MoCl_6^{-3} + H_2O \longrightarrow MoCl_5(H_2O)^{-2} + Cl^- \quad (43)$$

$$MoCl_5(H_2O)^{-2} + H_2O \longrightarrow MoCl_4(H_2O)_2^{-1} \quad (44)$$

At 0°C, over a pH range 3.2 to 4.5 (phthalate buffer), the pseudo first-order rate constant for the first equation was found to be $1.33 \pm 0.09 \times 10^{-4}$ sec^{-1}. The second reaction was very slow at that temperature (although it apparently takes place more rapidly at pH > 5.5). Direct comparison with Cr(III) chemistry is not reasonable since $CrCl_6^{-3}$, unlike other Cr(III) species, is manifestly unstable in aqueous solution. Nevertheless, Andruchow and DiLiddo stated that their results on $MoCl_6^{-3}$ indicated that Mo complexes may react more rapidly than their chromium counterparts. This is more clearly demonstrated in the reactions of $Mo(H_2O)_6^{+3}$ which are discussed below. Complexes of the form $MoX_4L_2^-$ have been reported but not studied in any detail (Table XV).

Complexes of the form MoX_3L_3 have been prepared in greater number. Apparently the substitution of three X's by neutral ligands is quite facile, whereas further substitution (except in the aquo or urea case) is more difficult. These same MoX_3L_3 complexes can also be prepared by the direct combination of MoX_3 (X = Cl, Br, I) and L (py, CH_3CN, etc.) (571, 594). In general, these complexes show some solubility and are indicated to be monomeric in those solvents in which they dissolve without decomposition. For example, $MoCl_3(py)_3$ (571), while insoluble in H_2O, is slightly soluble in 95% ethanol, benzene, and acetone, and soluble in methanol, glacial acetic acid, and pyridine. The MoX_3L_3 complexes can also be formed from $Mo(CO)_4X_2$ complexes by an anaerobic disproportionation reaction (570) in dry donor solvents such as CH_3CN, C_2H_5CN, $C_6H_5CH_2CN$, C_6H_5CN, $CH_3CH_2CH_2CN$, $CH_3(CH_2)_3CN$, $(CH_3)_2CHCN$, py, thf, and C_4H_8S (tetrahydrothiophene). The product is in most cases MoX_3L_3, where X = Cl or Br and L is one of the aforementioned solvents. [The other products of the reaction are not identified conclusively, but they may be of the form $Mo(CO)_n L_{6-n}$ and thus are Mo(O) species.] Many of these are thought to have meridional configurations with approximate C_{2v} symmetry as found for $MoCl_3(py)_3$ (590).

Species of stoichiometric composition MoX_3L_4 (567, 570) can be isolated, where X = Cl, Br; L = CH_3CN and X = Cl; L = py. In these cases the MoX_3L_4 species lose one donor molecule upon heating in vacuo at 30 to 40°C. It seems possible that the fourth L is a lattice solvent molecule, but the presence of a 7-coordinate Mo(III) in the crystal has not been completely eliminated.

One fairly well-authenticated (565) 7-coordinate Mo(III) complex is

$K_4Mo(CN)_7 \cdot 2H_2O$, which can be prepared from K_3MoCl_6 and excess KCN (556, 565, 595). Here, the electronic, infrared, Raman and EPR spectra and the magnetic susceptibility are consistent with a monocapped trigonal prismatic structure (C_{2v}) in the solid state for $K_4Mo(CN)_7 \cdot 2H_2O$, whereas a pentagonal bipyramidal (D_{5h}) structure may exist in solution and in the anhydrous crystal of $K_4Mo(CN)_7$ (565).

Another possibly 7-coordinate complex is $Mo(CO)_2(diars)I_3$, formed by reaction of I_2 with $Mo(CO)_4$ diars in boiling $CHCl_3$ (596). It has a μ_{eff} of 1.40 B.M. and CO stretching frequencies at 1905 and 1960 cm^{-1}. Shilov (69) reported the isolation of a Mo(III) carbonyl complex formed upon CO inhibition of a dinitrogen reducing system, but few details were given.

Mononuclear Mo(III) complexes with multidentate ligands have not been extensively explored. Tris complexes of bidentate ligands have been reasonably characterized for acac and related β-diketonates, phen, 2-(aminomethyl)pyridine, and more recently en, bigH, and 2-(2-pyridylbenzamidazole) (579). $Mo(acac)_3$ is prepared by substitution in $MoCl_6^{-3}$ (583) or by reaction of acacH with $Mo(CO)_6$ (581, 582). The dark-red compound is quite air sensitive and can be handled only in inert atmosphere. Other tris (β-diketonates) are prepared analogously. Tris phenanthroline and bipyridyl complexes were originally reported by Steele (516), but later studies by Walton (577), Kleinberg (569), and co-workers suggested that some of these can be formulated as $[MoCl_2(phen)_2][MoCl_4phen]$ or $[phenH^+][MoCl_4(phen)]$ with analogous formulations for the bipyridyl complexes. It is difficult to tell which formulation is correct or in fact whether all parties are working with the same species. However, for $[Mo(bipy)_3]I_3$ there seems to be general agreement as to the tris formulation. Recently, Ghosh and Prasad (519) prepared a variety of bis and tris (bidentate) complexes by using $Mo(NCS)_6^{-3}$ as a starting material. The tris (bidentate) complexes are summarized in Table XV.

The use of starting materials of lower oxidation state, particularly $Mo(CO)_6$, has been quite fruitful in the synthesis of Mo(II) and Mo(III) species. Protonic ligands apparently react with the hexacarbonyl by a redox process. For example, the reaction to produce $Mo(acac)_3$ (581, 582) is probably of the form shown in reaction 45,

$$Mo(CO)_6 + 3CH_3COCH_2COCH_3(acacH) \longrightarrow Mo(acac)_3 + 3/2\ H_2 + 6CO \qquad (45)$$

and Calderazzo et al. (580) have shown that reactions of this sort are general for other protonic ligands and have used it to prepare $Mo(N\text{-Mesal})_3$.

The only monomeric MoS_6 complex of Mo(III) that has been prepared is $Mo(S_2PF_2)_3$ (494), which is produced by reaction 46.

$$MoCl_n + 2nF_2PS_2H \longrightarrow 2Mo(S_2PF_2)_3 + n\text{-}3(F_2PS_2)_2 (n = 4 \text{ or } 5) \tag{46}$$

where $MoCl_4$ or $MoCl_5$ is reduced by excess ligand.

In addition to the tris complexes, a more limited degree of substitution on MoX_6^{-3} by neutral bidentate ligands produces MoX_4L^{n-} and MoX_2L^{n+}, with a number of examples shown in Table XV.

Dubois et al. (569, 597, 598) studied the electrochemical properties of complexes containing both chloride and pyridine-type ligands (e.g., py, phen, terpy). These complexes display at least one reversible polarographic reduction [presumably to a mononuclear Mo(II) species]. Significantly, the ease of reduction is a function of the number of phen, bipy, or terpy ligands in the complex. For the complexes $MoCl_4(phen)^-$, $MoCl_3(py)_3$, $MoCl_3py(phen)$, $MoCl_2(phen)_2^+$, and $MoCl_3(terpy)$, the first polarographic reduction waves occur at -1.20, -1.47, -1.01, -0.64, and -0.75 V, respectively (vs. sce). The presence of phen, bipy, or terpy ligands lowers the $E_{1/2}$ and makes reduction easier. This is easily rationalized in terms of the electronic structures of the complexes. The Mo(III) complex has a $(t_{2g})^3$ configuration in assumed octahedral symmetry. The one-electron reduction process probably leads to a $(t_{2g})^4$ configuration due to the substantial $10D_q$ values and the lowered values of interelectronic repulsion in these complexes and in accord with the tendency of second- and third-transition-row complexes to possess low-spin configurations. Any ligand possessing empty orbitals which can overlap with the t_{2g} metal orbitals ($4d_{xz}$, $4d_{yz}$, and $4d_{xy}$) will delocalize and stabilize them and cause a less negative reduction potential. Pyridine has this pi-acceptor ability, but the more rigidly held bipyridine and phenanthroline complexes possess more extensively delocalized pi systems and are better equipped to stabilize the t_{2g} orbitals. The terpyridyl system has a further extended pi system with which to interact with the metal t_{2g} levels.

When terpy or two phens are present, a second reversible wave, less negative than the reduction potential of the ligand, also appears. Further reductions at more negative potentials undoubtedly involve predominantly ligand orbitals. It should be noted that Herzog and Schneider (599) have prepared $Mo(bipy)_3$ by reduction of $Mo(bipy)_3^{+3}$ by $Li_2(bipy)$. Here, as in other extensive work of Herzog and co-workers (599), the occupancy of predominantly ligand orbitals must play a key role in stabilizing the formally very low metal oxidation states.

The possible relevance of these observations is that the reduction potential of a particular oxidation state depends strongly on the nature of the ligand system. For example, in biological systems the formally Fe^{+2}/Fe^{+3} couple may be involved in proteins which have $E^{\circ\prime}$ values anywhere from -0.44 V (ferredoxins) to $+0.35$ V (cytochrome *c*). In Mo enzymes the values for the Mo couples are not definitely known. Thus the value of -0.64 V versus sce for $Mo(phen)_2Cl_2^+$ found here becomes $E^{\circ\prime} = -0.40$ V versus a normal hydrogen electrode, and is well in the range of possible biological values, especially for nitrogenase. Thus Mo(III) and even Mo(II) cannot yet be ruled out as either an electron-transfer agent or as a site for binding of substrate molecules in Mo enzymes.

3. Magnetic and EPR Studies

All the mononuclear Mo(III) complexes show magnetic moments in the range 3.6 to 3.9 B.M. Westland and Murüthi have stated (570) (but not given explicit data) that moments as low as 3.4 were found for MoX_3L_3 complexes, where L is an oxygen or sulfur donor. For an octahedral d^3 complex, the effective magnetic moment is given by the expression (555)

$$\mu_{\text{eff}} = \left(1 - \frac{4K\lambda}{10D_q}\right)2[S(S + 1)]^{1/2}$$

where $2(S(S+1))^{1/2}$ is the spin-only moment of 3.87 B.M., $10D_q$ is the splitting between the $^4A_{2g}$ ground state and the $^4T_{2g}$ excited state (obtainable from the first spin-allowed band in the electronic spectrum) (see below), λ is the free-ion spin-orbit coupling constant, and K is a reduction factor that recognizes the effects of covalency. For a fully ionic complex, K is 1.0, but its value is lower than this in real situations. The parameter absorbs both the effect of covalent bonding on the nature of the interacting orbitals and the effect of delocalization in reducing the effective spin-orbit coupling constant. If K is set equal to 1.0, λ to 267 cm^{-1} (the free-ion value), and $10D_q$ to 20,000 cm^{-1} then μ_{eff} is calculated to be 3.67 B.M. The covalent bonding ($K < 1.0$) would tend to increase μ_{eff} as would distortions from O_h symmetry which disallow the type of orbital mixing required to lower the moment. In addition, $10D_q$ ranges between 15,000 and 30,000 cm^{-1} for the complexes studied. Thus theory predicts that the moments be clustered around 3.7 to 3.8 B.M., and this has indeed been found (Table XV).

Electron paramagnetic resonance studies have not been substantially explored in the investigation of Mo(III) complexes. Jarrett (600) investigated $Mo(acac)_3$ doped in the diamagnetic host $Al(acac)_3$. This same system was reinvestigated by Schoffman (49), who found principle g values of 4.30, 3.46, and 1.93 and no molybdenum hyperfine splitting. The absorption is considered due to the $(-1/2 \rightarrow 1/2)$ transition of the spin-quartet ground state. This signal cannot be confused with one from Mo(V) and, in fact, exposure of the mixed crystal to air leads to the appearance of a typical Mo(V) signal. This latter signal may be from an intermediate formed in the oxidation of $Mo(acac)_3$ to the diamagnetic final product, $Mo_2O_3(acac)_4$ (49). Other studies of Mo(III) EPR give little detailed information (601–603). In all cases, however, absorptions near $g = 1.96$ are seen for mononuclear complexes, whereas $g = 1.944$ is found for $((C_2H_5)_3NH)_3Mo_2Cl_9$ and $g = 2.014$ for $Cs_3Mo_2Br_9$. Possible interpretations are considered by Boyd and co-workers (603).

4. *Electronic Spectral Studies*

The absorption spectra for d^3 (530, 604, 605) ions have been successfully interpreted using a ligand field approach. The ground-state term for a rigorously octahedral $4d^3$ complex is $^4A_{2g}$, corresponding to a $(t_{2g})^3$ configuration containing three unpaired electrons. Spin-allowed excitations are possible to $^4T_{2g}[(t_{2g})^2(e_g)^1]$, $a^4T_{1g}[(t_{2g})^2(e_g)^1]$, and $b^4T_{1g}[(t_{2g})(e_g)^2]$, where the configurations in brackets represent the strong field limits. Three spin-allowed transitions should in principle be seen, and these have been well studied for V(II) and Cr(III) species (530). In addition, the possibility arises of intraconfigurational spin-forbidden bands within the $(t_{2g})^3$ manifold. These are expected to be more intense for Mo(III) than for the $3d^3$ configurations of first-transition-row ions due to the increased value of the spin-orbit coupling constant for Mo. For the most part, the experimental results confirm the predictions of the ligand field model, and these results and assignments are summarized in Table XVI along with some values from chromium(III) complexes for comparison.

The spin-forbidden bands appear more distinctly in the Mo spectrum as compared to the chromium spectrum, in part due to their greater separation from the spin-allowed bands and in part due to their greater extinction coefficients. The greater separation is probably due to smaller values of interelectronic repulsion parameters (making the spin-forbidden bands lower in energy) and greater values of $10D_q$ (making the spin-allowed bands higher in energy), as expected for a second-transition-row ion. The first two spin-allowed bands are observed for a number of compounds, but the third band has not been observed, which is expected from its predicted high energy. The electronic spectrum of Mo(III) offers useful information to those seeking to synthesize new Mo(III) complexes. This is very well illustrated in the studies on the $Mo(H_2O)_6^{+3}$ ion. For example, the latest numbers reported for $Mo(H_2O)_6^{+3}$ (564) are very similar to those reported for $Mo(OC(NH_2)_2)_6^{+3}$, and this lends support to the formulation of both these cationic species as containing MoO_6 coordination spheres.

The power of the crystal-field approach for Mo(III) complexes was illustrated by Jørgenson (600) who, in 1958, predicted that MoF_6^{-3} would show an absorption spectrum (at energies below 35,000 cm^{-1}) of four bands. The predicted energies calculated from empirical ligand-field parameters are 11,000 cm^{-1}($^4A_{2g} \longrightarrow {^2T_{1g}}, {^2E_g}$), 16,000 cm^{-1} ($^4A_{2g} \longrightarrow {^2T_{2g}}$), 24,500 cm^{-1}

($^4A_{2g} \longrightarrow {}^4T_{2g}$), and 30,000 cm^{-1} ($^4A_{2g} \longrightarrow {}^4T_{1g}$) (606). In 1969, the spectrum was first measured (587) (but only above 17,000 cm^{-1}) by reflectance, and maxima were found at 23,500 and 29,700 cm^{-1}. The agreement is evident and lends further support to the validity of crystal- or ligand-field treatments of Mo(III) ions.

For trisubstituted species of the form MoX_3L_3 there are two possible isomers: meridional (C_{2v}) and facial (C_{3v}). Westland and Murüthi (570) have used electronic and NMR spectroscopy to distinguish between them.

In the electronic spectra of MoX_3L_3 species in solution of L, bands near 8,000, 14,000, 20,000, and 24,000 cm^{-1} are seen (570). These have been reasonably assigned to the $^4A_{2g} \longrightarrow (^2E_g, {}^2T_{1g})$, $^4A_{2g} \longrightarrow {}^2T_{2g}$, $^4A_{2g} \longrightarrow {}^4T_{2g}$, and $^4A_{2g} \longrightarrow {}^4T_{1g}$ transitions using the notation of the octahedral group. However, the symmetry is less than octahedral, and in C_{3v} the new states become A_2 (from A_{2g}), E (from E_g), A_2 and E (from T_{2g}), and A_1 and E (from T_{1g}). The prediction is therefore that under higher resolution in C_{3v}, the various transitions will be split. This should be particularly noticeable in the spin-forbidden bands which, due to their essentially intraconfigurational character [they are within the $(t_{2g})^3$ manifold in the octahedral notation], are expected to be sharper than the spin-allowed bands. Westland and Murüthi (570) succeeded in resolving the splitting in the quartet-doublet transitions and found a larger number of bands than can be accommodated by the C_{3v} model (when they discounted the possible effect of spin-orbit coupling which should be smaller than the observed splitting). A C_{2v} meridional structure is thus considered. Here, the states are A_2 (from A_{2g}), A_1 and A_2 (from E_g), A_2, B_1, and B_2 (from T_{1g}), and A_1, B_1 and B_2 (from T_{2g}). With this number of states there is no problem in assigning all of the bands, and while the detailed assignment is equivocal, the C_{2v} structure seems likely to be correct. Further support for the assignment comes from proton NMR spectroscopy. In $MoCl_3(NC(CH_2)_2CH_3)_3$ two signals assignable to CH_3 protons can be discerned (despite the broadness of the lines in the paramagnetic compound). Since these signals are in a 1:2 intensity ratio, the weaker peak is assigned to the CH_3 of the ligand on the C_2 axis, whereas the stronger peak is due to the two equivalent CH_3's on the ligands related by that axis. (Note that the presence of any symmetry axis in this molecule requires that the ligands be assumed to be freely rotating.) The crystallographic evidence for $MoCl_3$-$(py)_3$ combined with the spectroscopy indicate that, unlike the tricarbonyls and trioxo species which are invariably facial when part of a 6-coordination sphere, the meridional isomer is found for the MoX_3L_3 species. Perhaps the occurrence of the meridional isomer can be viewed as sterically favored in the present case, whereas the oxo and carbonyl ligands prefer the mutually cis arrangement of donor atoms in order to maximize the pi bonding with the metal.

TABLE XVI
Electronic Absorption Spectra of Some Mo(III) Complexes,[a] cm^{-1}
(molar extinction coefficients in parenthesis)

Complex (medium)	$^4A_{2g} \longrightarrow {}^2E_g, {}^2T_{1g}$	$^4A_{2g} \longrightarrow {}^2T_{2g}$	$^4A_{2g} \longrightarrow {}^4T_{2g}$	$^4A_{2g} \longrightarrow a\ {}^4T_{1g}$	Charge transfer	Refs.
MoF_6^{-3} (reflectance of K_3MoF_6)			23,500	29,700	38,200	587
$MoCl_6^{-3}$ (in 9*M* HCl)	9,500(1.4)	14,800(2.0)	19,400(23)	24,200(38)		574
$MoBr_6^{-3}$ (in 8.8*M* HBr)	9,500(3.9)	14,500(5.5)	18,300(33)	23,200(ref.)		574
MoI_6^{-3} (in 57% HI)	9,200	13,800	16,650	20,050		574
$Mo(H_2O)_6^{+3}$ (in 1.0*M* $CH_3C_6H_4SO_3H$)			26,300(17)	31,250(25)		564
(in 1.0*M* CF_3SO_3H)			27,000(16)	32,500(28)		562
$Mo(OC(NH_2)_3)_6^{+3}$						
$Mo(OC(NH_2)_2)_6\ Br_3$ (reflectance)			25,900	31,000		560
$MoCl_5(H_2O)^{-2}$ (in 1*M* HCl)	9,800(2.3)	15,200(2.1)	21,000(25)	25,800(55)		574
$MoCl_5(NCCH_3)^{-2}$ (reflectance)	9,100	14,400	21,600	25,800		567
$MoBr_5(H_2O)^{-2}$ (in 1*M* Br)	8,500(4.0)	15,300(3.1)	19,700(29)	25,500(51)		574
$Mo(py)_3Cl_3$(reflectance)	8,800(7.0)	14,500(7.0)			27,200(2,950) 33,600(2,570)	574
$Mo(py)_3Br_3$(reflectance)	8,800(6.5)	14,700(7.6)			27,200(3,800) 32,700(7,080)	574
$Mo(NCCH_3)_3Cl_3$(reflectance)	8,400 8,750	13,500	25,200	28,900		567
$Mo(S_2PF_2)_3$			20,020(205)	24,310(99)	30,720(4,990) 35,110(5,280) 42,420(16,760)	494

TABLE XVI (continued)

Complex (mediam)	$^4A_{2g} \longrightarrow {}^2E_g, {}^2T_{1g}$	$^4A_{2g} \longrightarrow {}^2T_g$	$^4A_{2g} \longrightarrow {}^4T_{2g}$	$^4A_{2g} \longrightarrow a^4T_{1g}$	Charge transfer	Res.
$Mo(acac)_3$ (in C_6H_6 solution)			23,300		27,000(5,000)	582
CrF_6^{-3}	15,700		14,900	22,700		530
$CrCl_6^{-3}$			13,180	18,700		530
$Cr(H_2O)_6^{+3}$	15,000(2)		17,400(13)	24,600(15)		530

[a]In addition to the data noted in the table, Komorita et al. (560), Carmichael et al. (577), and Westland and Murüthi (570) each gave additional spectroscopic information. Komorita and Westland considered MoX_3L_3 complexes, whereas Carmichael et al. studied various bipyridyl derivatives of Mo(III).

The phosphorescence of Mo(III) complexes has been briefly studied (607). Of 15 Mo(III) complexes investigated, only three exhibited detectable phosphorescence. The phosphorescence was assigned to the $^2E_g \longrightarrow {}^4A_{2g}$ transition of the octahedral Mo(III) ion. The emission was detected between 9,000 and 10,000 cm^{-1} for $Mo(urea)_3Cl_3$, $Mo(urea)_3Br_3$, and Mo $(thiourea)_3Cl_3$. The fact that these emissions are in the infrared make their detection difficult, and it seems likely that further instances of Mo(III) emission will be recorded as the equipment used in its detection improves.

5. *The Hexaaquomolybdenum(III) Ion*

The first claim for the hexaaquomolybdenum(III) ion was made by Hartmann and Schmidt (608), who observed similar visible spectra in the 15,000 to 18,000 cm^{-1} region in 4*M* HCl, 4*M* HBr, and 4*M* H_2SO_4 for a green ion. They attributed these absorptions to the presence of $Mo(H_2O)_6^{+3}$ in all cases. As noted by Komorita et al. (560) and Bowen and Taube (561), the low energy of this band is not reasonable for the hexaaquo ion based upon ligand-field consideration. The $Mo(H_2O)_6^{+3}$ ion should show its first absorption band well above 20,000 cm^{-1}, and the presence of this ion in the work of Hartmann and Schmidt is unlikely. Rather, a dinuclear species is probably responsible for the absorption at ~ 16,000 cm^{-1} (see below).

Bowen and Taube (561) produced the first evidence for the hexaaquo ion. Using 1*M* *p*-toluenesulfonic acid ($CH_3C_6H_4SO_3H$) or 1*M* trifluoromethylsulfonic acid (CF_3SO_3H) as media, they hydrolyzed the $MoCl_6^{-3}$ ion. The ions of these acids are poorly coordinating, and coordination solely by aquo ions is expected upon hydrolysis in these solutions. The absorption bands of $MoCl_6^{-3}$ disappear over a period of two days, and cation-exchange chromatography was used to purify and help identify the aquo ion. The hexaaquo ion, as eluted from the column, has a faint yellow color, and its lowest energy absorption band was reported by Taube to be at 34,100 cm^{-1}($\varepsilon = 6.0 \times 10^2$) with a second band at 39,500 cm^{-1}(7.2×10^2) in 1.5*M* $CH_3C_6H_4SO_3H$.

Kustin and Toppen (563) repeated this work, and although they substantially reported the same results, they used additional precautions and did not find absorption maxima at 34,100 and 39,500 cm^{-1}. They reported that exposure to air did, however, increase the absorbance at 34,100 cm^{-1} thereby implying that in the original work much of the absorption may have been formed by reaction of air with $Mo(H_2O)_6^{+3}$. The Evans NMR method (utilizing the *p*-CH_3 protons of *p*-toluenesulfonic acid) was used to estimate a magnetic moment of 3.69 B.M. for the hexaaquo ion.

Sasaki and Sykes (564) also repeated the experiments and found a spectrum which was most reasonable for the hexaaquo ion of Mo(III). Recent work of Bowen and Taube (552) confirms these results. The spectrum clearly

TABLE XVII
Reaction Parameters for Formation of $Mo(H_2O)_5X^{-2}$ from $Mo(H_2O)_6^{+3}$[a]

X	$k_f(25°)M^{-1}\ s^{-1}$[b]	ΔH† kcal mole^{-1}	ΔS† cal mole^{-1} °k^{-1}	$K_{eq}(M^{-1})$
Cl	$(4.60 \pm 0.23) \times 10^{-3}$	23.5 ± 0.6	9.6 ± 2.1	10.8
NCS	$(2.68 \pm 0.13) \times 10^{-1}$	16.3 ± 0.4	-6.4 ± 1.3	10^5

[a]Data from Ref. 564 in 1.0M $CH_3C_6H_4SO_3H$
[b]Pseudo first-order constants k are expressed as $k = k_f[Cl^-] + k_b$

shows the presence of *d-d*-type absorptions (Table XVI) and decreased absorbance in the ultraviolet region from that seen previously (561, 562). Sykes attributed the ultraviolet absorption peaks to the presence of the di-μ-oxo-molybdenum(V) dimer. Clearly, the closest approach to the pure hexaaquo ion has been made in this study (564), and the early studies (561, 563) serve to illustrate the great air sensitivity in the ion.

The equilibrium and kinetics for the reaction of $Mo(H_2O)_6^{+3}$ with chloride and thiocyanate were studied (563, 564), and the data for the formation of $Mo(H_2O)_5X^{+2}$ are summarized in Table XVII. The rate constants are greater than those for the corresponding chromium reactions. An associative pathway is considered likely in view of the H^+ independence of the rate, the ratio of NCS^- and Cl^- rate constants, and the small value of ΔH† compared to the corresponding Cr(III) reactions. The associative (S_{N2}) pathway is more reasonable for Mo than for Cr in view of the existence of at least one *bona fide* 7-coordinate complex of Mo(III) and the absence of the same in Cr(III) chemistry. Thus (at this very early stage in its investigation) the $4d^3$ Mo(III) species do not appear to be as kinetically inert as their Cr(III) counterparts due to the greater availability of the associative mechanism. This may have important consequences in the chemistry of Mo(III) .In particular, the expected chemical inertness of the $4d^3$ ion cannot be viewed as a sufficient criterion to dismiss the participation of Mo(III) in enzymes or other catalytic systems.

C. Dinuclear and Polynuclear Complexes

Dinuclear complexes containing chloro or bromo bridges and non-bridged dimers containing strong Mo–Mo bonds are well characterized. Complexes reported to have oxo or hydroxo bridges are not as well characterized from a structural point of view.

Compounds of the form $M_3[Mo_2X_9]$ have been isolated by a variety of procedures. Lewis et al. (610) isolated $Cs_3[Mo_2Cl_9]$ by addition of CsCl to concentrated HCl solutions of Mo(III), whereas Saillant and Wentworth prepared $Cs_3Mo_2X_9$ (X = Cl, Br) by combination of MoX_3 and CsX at high temperature (611). Bennett et al. (612) prepared the binuclear chloride by electrolytic oxidation of $Mo_2Cl_8^{-4}$ in concentrated HCl. Recently, reaction 47,

$$Mo(CO)_4Cl_3^- + MoCl_6^{-2} \longrightarrow Mo_2Cl_9^{-3} + 4CO \quad (47)$$

was used by Delphin et al. (613, 614) in the preparation of the propylammonium salt of $Mo_2Cl_9^{-3}$. If $MoCl_6^-$ is used in place of $MoCl_6^{-2}$, reaction 48,

$$Mo(CO)_4Cl_3^- + MoCl_6^- \longrightarrow Mo_2Cl_9^{-2} + 4CO \quad (48)$$

leads to the $Mo_2Cl_9^{-2}$ ion [a mixed-valent Mo(III)-Mo(IV) complex] (613, 614). This CO elimination reaction seems quite general, and the trinuclear ion $Mo_3Cl_{12}^{-3}$ (614) can be synthesized according to Eq. 49:

$$Mo(CO)_4Cl_3^- + Mo_2Cl_9^{-2} \longrightarrow Mo_3Cl_{12}^{-3} + 4CO \quad (49)$$

The mixed metal complex $CrMoCl_9^{-3}$ has recently been prepared (614a) as the NBu_4^+ salt according to Eq. 50

$$CrCl_2 + NBu_4Cl + (NBu_4)_2[MoCl_6] \xrightarrow{CH_2Cl_2} (NBu_4)_3[CrMoCl_9] \quad (50)$$

Structurally, the $Mo_2X_9^{-3}$ species consist of two MoX_6 octahedra sharing a face, as shown in Fig. 28. In $Mo_2Cl_9^{-3}$ shown in Fig. 28 the Mo–Mo distance is 2.67 Å, whereas in $Mo_2Br_9^{-3}$ the distance is 2.78 Å (615). The magnetic moments of these complexes are 0.6 to 0.7 B.M. and 1.26 B.M., respectively, indicating a strong degree of spin-spin coupling between the Mo atoms which is reminiscent of the pairwise interactions found in the polymeric $MoCl_3$ and $MoBr_3$. Spectroscopic properties (610, 611) in some ways resemble MoX_6^{-3} ions and in other ways resemble $W_2Cl_9^{-3}$, but the detailed assignments remain enigmatic (611).

Substitution of three halides in the binuclear ion by neutral ligands is

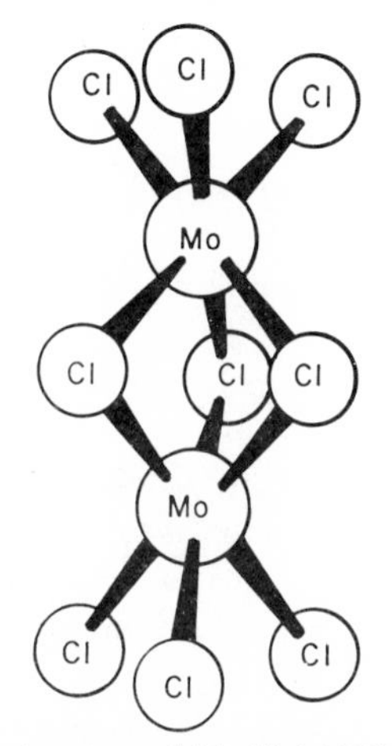

Fig. 28. The structure of $Mo_2X_9^{-3}$ (615) (X = Cl, Br).

possible which leads to the formation of neutral complexes of the form $Mo_2X_6L_3$, with individual examples given in Table XV. None of these species are structurally characterized, although the persistence of the trichloride bridge seems likely.

Lindoy et al. (492) reacted K_3MoCl_6 with a wide variety of thiol ligands including 8-mercaptoquinoline, 2-aminobenzenethiol, 2-(2-mercaptoethylpyridine), 3-ethylthiopropane-1-thiol, and ethylthiobenzoyl acetate. In all cases, brown (colors are described as from dark greenish brown to brownish black) complexes were produced which were quite insoluble (no molecular weight possible) and which possessed magnetic moments from 0.3 to1.56 B.M. per Mo. These results indicate that at least binuclear and possibly polymeric complexes are present. Typically, stoichiometries such as $Mo_2L_4Cl_2(H_2O)_3$, $Mo_2L_2Cl_4(H_2O)_5$, or $Mo_4L_5Cl_7(H_2O)_7$ are considered to be consistent with the analytical data. None of these complexes is in any way structurally characterized, and even the stoichiometric formulations must be considered as tentative.

The brown hygroscopic solids formulated as $MoCl_3 \cdot 3H_2O$ and $MoBr_3 \cdot 3H_2O$ are prepared by the electrolytic reduction of molybdenum trioxide (487, 616) in the absence of monovalent cations. While these compounds remain poorly characterized, they are probably polymeric and are useful as starting materials for the preparation of the complexes is lower oxidation states (487).

The green species observed in acid solutions of Mo(III) appears to be a dimeric oxo- or hydroxo-bridged complex. Few discrete complexes of this genre have been isolated (617, 618). One of these is the EDTA complex formulated as $LiH[Mo_2O_2(H_2O)_2(edta)] \cdot 5H_2O$ (617). The complex anion was suggested to have the form shown in structure 16:

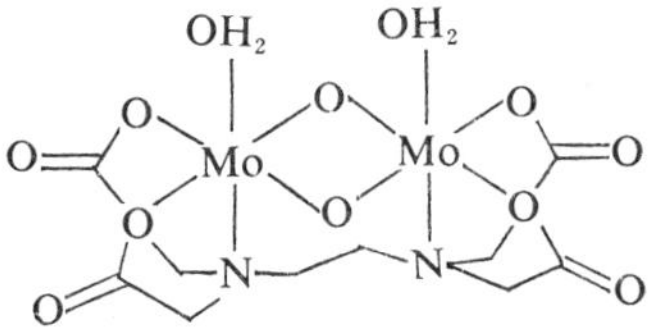

16

Significantly, although there is little doubt that the complex contains Mo(III), it is distinctly diamagnetic. In contrast, the electronic absorption spectrum contains bands which can be assigned to transitions within an $S = 3/2$, $4d^3$ spin-manifold. This dichotomy can be resolved by invoking strong antiferromagnetic coupling of two Mo(III) centers, similar to the situation that occurs in oxygen- and sulfur-bridged Fe systems (15, 619). The electronic spectrum consists of bands at 13,000 cm^{-1} ($\varepsilon = 25$), 16,200 cm^{-1} ($\varepsilon = 40$), 23,000 cm^{-1} ($\varepsilon = 48$), 27,900 cm^{-1} ($\varepsilon = 880$), and 34,800

cm^{-1} (ε = 1120). The first three bands have been assigned to spin-forbidden transitions $^4A_{2g} \rightarrow {}^2E_g$, $^4A_{2g} \rightarrow {}^2T_{1g}$, and $^4A_{2g} \rightarrow {}^2T_{2g}$, respectively. These are very intense for spin-forbidden transitions, and the intensity enhancement may be attributed to the presence of the coupled center. This complex is oxidizable in a reversible four-electron, four-proton process to Mo_2O_4-(edta)$^{-2}$, as discussed in Section IV (428a).

Mitchell and Scarle (618) investigated the probable dinuclear complexes $Mo_2O_2Cl_2(H_2O)_6$, $Mo_2O_2Cl_2(H_2O)_2(bipy)_2$, $Mo_2Cl_2(H_2O)_4(pyrrole)_2$, and others. These diamagnetic species are also formulated as containing di-μ-oxo bridges, and a strong broad infrared band from 670 to 690 cm^{-1} is assigned to a vibration in the Mo_2O_2 bridge (618). The electronic spectra show low energy transitions similar to the edta complex and the dinuclear aquo ions (620) discussed below. This is supportive of a common structural unit in these species.

Ardon and Pernick (620) investigated the acidic Mo(III) solutions prepared using Jones or cadmium reductors. Ion-exchange techniques allow the separation of a blue-green ion as the major product. The blue-green ion has an Mo oxidation state of (III), a probable overall charge of +4, and a probable charge per molybdenum of +2. Working in *p*-toluenesulfonic acid solutions the ion is shown to be diamagnetic by the Gouy method and has absorption maxima at 16,000 cm^{-1} (ε = 43), 17,500 cm^{-1} (ε = 39), and 27,800 cm^{-1} (ε = 306). Ardon and Pernick prefer to formulate the ion as a dihydroxy-bridged species (structure 17),

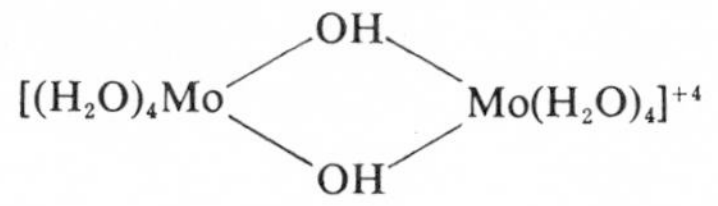

$$[(H_2O)_4Mo(\mu\text{-}OH)_2Mo(H_2O)_4]^{+4}$$

17

but they cannot rule out a structure containing a monooxo bridge.

Very recently (380b) the X-ray structure of the complex anion $Mo_2(OH)_2$-(edta) $(CH_3CO_2)^-$ was determined in its potassium salt. The anion has two hydroxy bridges, an acetate bridge as well as an ethylenediamine bridge supplied by the edta ligand (structure 17a).

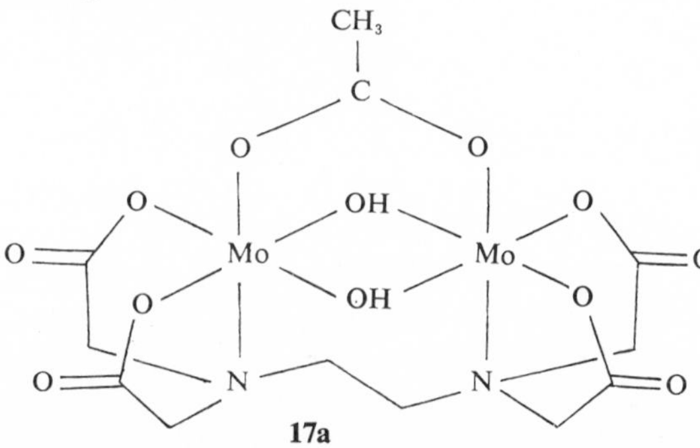

CH₃, C, O, O, O, OH, O, O, Mo, Mo, O, O, OH, O, O, N, N, O

17a

The short Mo–Mo distance of 2.43Å is consistent with considerable metal-metal bonding. gnSiificantly, this complex was prepared by the same procedure (electrolysis of $Mo_2O_4(edta)^{-2}$ in acetic acid-acetate solution) as had been used to synthesize the complexes which were formulated as having structure 16. As additional structures in this class become known it may be that the dihydroxy bridge will be recognized as the common structural unit for aqueous dimeric Mo(III) systems.

Both Brown et al. (620a) and Mitchell and Scarle (620b) have claimed the synthesis of complexes of the form $Mo_2(R_2dtc)_6$ (R = Ph (620a), Et (620b)). Recently, Nieuwpoort et al. (620c) using the same procedure as Brown found $Mo(Ph_2dtc)_4$ as the product. It is of further interest to note that the $Mo_2(Et_2dtc)_6$ species claimed by Mitchell and Scarle (620b) shows identical epr parameters to $Mo(Et_2dtc)_4^+$. Clearly these interesting species warrant further investigation.

There are several known Mo(III) compounds in which a direct Mo–Mo bond is present. These include $Mo_2(CH_2Si(CH_3)_3)_6$ (621) and $Mo_2(N(CH_3)_2)_6$ (622), where the Mo–Mo bond has been crystallographically found at 2.167 and 2.214 Å, respectively. Both compounds display approximate D_{3d} symmetry with a staggered arrangement of ligands. The Mo–Mo bond is considered to be a triple bond using metal d_{z^2}, d_{xz}, and d_{yz} orbitals on each Mo to accommodate the six metal electrons in bonding molecular orbitals. The triple bond allows free rotation about Mo–Mo, which is consistent with the staggered D_{3d} configuration being assumed for steric reasons. Analogous compounds have been prepared with neopentyl and benzyl ligands wherein the ligands [as is $CH_2Si(CH_3)_3$] are stabilized toward alkene elimination reactions by lacking β-hydrogens (623).

The complex $[Mo(C_5H_5)(SCH_3)_2]_2$ was structurally characterized (624), and it was shown to have four bridging methylsulfide groups with a short Mo–Mo distance of 2.603(a) Å. The sulfurs form a square (S–S = 2.96 Å) and the Mo's lie above and below the plane of that square. The one-electron oxidation product [Mo(III), Mo(IV)] maintains the same structure with Mo–Mo = 2.617(4) Å and S–S = 2.90 Å (624).

VII. THE CHEMISTRY OF MOLYBDENUM(II)

The chemistry of Mo(II) can be divided into three distinct subclasses. The first class contains those (largely organometallic) complexes with carbonyl, isonitrile, nitrosyl, phosphine, arsine, and/or hydride ligands in which monomeric Mo(II) predominates and coordination numbers 6 and 7 are common. The second class consists of dimeric complexes, many of which contain the tightly bound Mo_2^{+4} unit characterized by a quadruple Mo–Mo bond. Included in this class are a few examples of dimeric intervalent species where the average oxidation state is 2.5. The third class consists of cluster compounds having octahedral Mo_6 cores and bridging anionic ligands. In general, the members of each of these classes are not directly convertible to the members of any other class, thus the chemistry within each class is to date self-contained. Only the first two classes are considered here.

Except for nitrogenase, the presence of Mo(II) in enzymes would seem virtually ruled out by the available data. For nitrogenase, on the other hand, the reduction potential is sufficiently low and sufficiently little is known of the role of Mo that Mo(II) must still be considered as a dark-horse candidate for the lower redox state of Mo. This is especially the case in view of the known inhibiting effect of CO on nitrogenase and the high frequency of occurrence of CO in Mo(II) complexes. However, to add caution to these considerations, we note that it seems equally, if not more, likely that CO inhibition takes place at one of the Fe sites of nitrogenase, and this would render the above consideration irrelevant.

A. Monomeric Mo(II) Complexes

The complexes in this class are usually associated with organometallic chemistry and the majority have carbonyls in their coordination spheres. However, the existence of, for example, $Mo(diars)_2Cl_2$ shows that carbonyls are not absolutely necessary to stabilize the divalent state. NO^+ and RNC complexes are also known, but these ligands bear strong resemblance to CO. Both 6- and 7-coordinate species are found for this class and the interconversion of these is of some interest. The complexes are generally synthesized from $Mo(CO)_6$ by oxidation and substitution reactions. There is to date no aqueous chemistry of monomeric Mo(II).

Complexes in this class were reported by Lewis and Nyholm (625–627), Stiddard (628), Wilkinson (629), and their co-workers. Here, substituted carbonyls are oxidized by halogens to give apparently 7-coordinate species. Representative reactions include reactions 51 to 53:

$$Mo(CO)_4(bipy) + Br_2 \longrightarrow Mo(CO)_3(bipy)Br_2 + CO \tag{51}$$

$$Mo(CO)_4(dth) + I_2 \longrightarrow Mo(CO)_3(dth)I_2 + CO \tag{52}$$

$$Mo(CO)_2(diars)_2 + 2Br_2 \longrightarrow [Mo(CO)_2(diars)_2Br]Br_3 \tag{53}$$

Subsequently, many substituted carbonyl complexes have been oxidized by Cl_2, Br_2 or I_2 to give mixed-ligand Mo(II) complexes containing halide carbonyl and neutral ligands. The neutral ligands used include mono-, bi-, tri-, and tetradentate phosphines, as well as mono- and bidentate arsines Representative complexes prepared by this route include $Mo(PEt_3)_2$–

$(CO)_3Cl_2$ (522), $Mo(AsMe_2Ph)_2(CO)_3Cl_2$ (522), $[Mo(CO)_2(phen)_2I]I_3$ (630), $[Mo(CO)_3(triars)Br]Br$ (626), $Mo(CO)_2(triars)I_2$ (626), $[Mo(CO)_3(ttas)Br]Br$ (627), $Mo(CO)_2(triphos)I_2$ (631), and $Mo(CO)_3(dppe)Br_2$ (632).

The alternate technique of first reacting $Mo(CO)_6$ with halogens and then adding the neutral ligands was used by Colton and co-workers (633–640). Here, the initial reaction shown by equation 54 (633, 634, 640),

$$Mo(CO)_6 + X_2 \longrightarrow Mo(CO)_4X_2 + 2CO \; [X = Cl, Br, I] \quad (54)$$

is followed by ligand replacement reactions to produce a large variety of complexes. Either CO or Cl or both can be replaced by mono- or bidentate ligands. Some representative preparations include Eqs. 55, 56, and 57:

$$Mo(CO)_4Cl_2 + 2PPh_3 \longrightarrow Mo(PPh_3)_2(CO)_3Cl_2 + CO \quad (55)$$

$$Mo(CO)_4Cl_2 + 2NaR_2dtc \longrightarrow Mo(CO)_3(R_2dtc)_2 + 2NaCl + CO \quad (56)$$

$$Mo(CO)_4Br_2 + 2dpam \longrightarrow MoBr_2(CO)_2(dpam)_2 + 2\,CO \quad (57)$$

Most of the complexes are formulated as having 7-coordinate geometries. When the stoichiometry is such as to suggest 8-coordination, such as in $MoBr_2(CO)_2(dpam)_2$, it is invariably found that one of the bidentate ligands (dpam in this case) is acting in a monodentate mode. Other complexes prepared using this method include $Mo(CO)_3(AsPh_3)_2Cl_2$ (633), $Mo(CO)_3(SbPh_3)_2Br_2$ (634), $Mo(CO)_2(dppm)_2Cl_2$ (641), and $Mo(CO)_3(dppm)X_2$ (642).

Each of the above complexes has two or three carbonyls, and Drew and collaborators (643–645) have established 7-coordination for three representative complexes by x-ray crystallographic studies. $[Mo(CO)_2(diars)_2I]I_3$ (645) has a monocapped trigonal-prismatic (1:4:2) structure with I capping a square face consisting of the two diars ligands. The two CO's occupy the remaining edge to complete a near C_{2v} coordination sphere. $Mo(CO)_2(dpam)_2Br_2$ (643) has a distorted monocapped octahedral (1:3:3) structure with one monodentate and one bidentate dpam ligand. One CO caps a face consisting of the other CO, As from the monodentate dpam, and one As from the bidentate dpam. The remaining triangular face consists of the other As from the bidentate ligand and the two Br's. $Mo(CO)_3(dppe)Br_2$ (644) also has a distorted capped octahedral structure with CO in the cap, 2 CO's and P in the capped face, and P and two Br's in the uncapped face. It seems probable that the coordination spheres are fluxional in solution, but detailed solution studies have not been reported.

Recently (645a), the structures of the complexes $[rac\text{-}o\text{-}C_6H_4(AsMePh)_2Mo(CO)_3I_2] \cdot CHCl_3$ and $[meso\text{-}o\text{-}C_6H_4(AsMePh)_2Mo(CO)_3I_2]$ have been crystallographically determined. In each case a 7-coordinate Mo is found wherein the geometry is describable as a monocapped trigonal prism with one of the iodo ligands capping a square face. A discussion of 7-coordinate Mo(II) structures is given (645a) using the considerations of Kepert (645b).

Many of the yellow 7-coordinate complexes of the type $Mo(PR_3)_2(CO)_3X_2$ (X = Cl, Br) readily lose one mole of CO to form diamagnetic blue complexes, $Mo(PR_3)_2(CO)_2X_2$ (635, 637, 522). The complex $Mo(PPh_3)_2(CO)_2Br_2$ has been structurally investigated (646), and has been shown to contain a 6-coordinate Mo(II) ion. The geometry is described as an "unusual non-octahedral 6-coordinate arrangement" (646). The CO dissociation reaction is reversible, and the blue compounds are described as reversible CO carriers. A number of other complexes, including $Mo(CO)_3(R_2dtc)_2$, also have this interesting property (635, 638) (reaction 58).

$$Mo(CO)_3(R_2dtc)_2 \rightleftarrows Mo(CO)_2(R_2dtc)_2 + CO \tag{58}$$

Recent work (647) showed that these R_2dtc^- and related $R_2PS_2^-$ complexes react with *unsubstituted* acetylene according to reaction 59,

$$Mo(CO)_2(dtc)_2 + HC{\equiv}CH \longrightarrow Mo(CO)(dtc)_2(HC_2H) + CO \tag{59}$$

but do not react with ethylene. Remarkably, the chemical shift of the acetylenic protons occurs at 12.3 ppm downfield of TMS, and no C≡C stretch is observed in the infrared, indicating an unusual mode of bonding. It is postulated (647) that both pi levels of acetylene act as donors in order to satisfy the $18e^-$ requirement of Mo. The complex is thus formulated as containing a metallocyclopropenium ring. Treatment of the complex with acid leads to production of ethylene. The CO-carrying and acetylene-binding abilities of this compound indicate that 6-coordinate Mo(II) complexes are coordinately unsaturated, and it has not, of course, gone unnoticed that CO is a potent inhibitor and C_2H_2 is a substrate of nitrogenase, whereas ethylene does not bind.

In this respect it is interesting that Creedy and coworkers (647a), find that $Mo(CO)_2(Et_2dtc)_2$ reacts with EtO_2CNNCO_2Et to give $Mo(EtO_2CNNCO_2Et)(Et_2dtc)_2$ and with excess of the diazene to give $Mo(EtO_2CNNCO_2Et)_2(Et_2dtc)_2$. Hydrolysis of these two complexes leads to the hydrazine, $EtO_2CNHNHCO_2Et$, with the Mo products being $MoO(R_2dtc)_2$ and $MoO_2(R_2dtc)_2$, respectively. This reaction serves as a model for one of the postulated steps in the reduction of N_2 to NH_3 and coupled with earlier results (401) illustrates that this step can be affected by using either coordinately unsaturated Mo(II) or Mo(IV) systems.

The complex $(C_5H_5)_2Mo(CO)$ upon irradiation in the presence of acetylene also forms an acetylene complex $(C_5H_5)_2Mo(C_2H_2)$ (648). However, here the acetylenic protons appear at 7.68 ppm downfield from TMS and a C=C stretch is found at 1616 cm^{-1}. In this case, the $Mo(C_5H_5)_2$ fragment has a $16e^-$ core and the acetylene uses only one set of pi orbitals for bonding. The contrast between the binding of acetylene on 14 and 16-electron Mo cores is notable.

Halocarbonylanions of the form $Mo(CO)_4X_3^-$ are prepared by halide addition at the dihalotetracarbonyls (639), or by halogen oxidation of $Mo(CO)_5X^-$ (649). These chloro compounds are of use in the synthesis of $Mo_2Cl_9^{-2}$, $Mo_2Cl_9^{-3}$, and $Mo_3Cl_{12}^{-3}$ (613, 614) (Section VI). The nature of these reaction products and the known structure of $W(CO)_4Br_3^-$ (650) combine to suggest a C_{3v} capped octahedral structure for $Mo(CO)_4X_3^-$, with one CO capping a face consisting of the other three CO's, while the halides make up the uncapped face.

The compounds $[Mo(CNR)_7]X_2$ [R = Me, X = I; R = t-C_4H_9, X = PF_6] have been prepared either from $Mo(CO)_6$ and RCN or by reaction of $Ag_4[Mo(CN)_8]$ and t-C_4H_9I in $CHCl_3$ containing AgCN (651). They show only one CN stretch at ~2140 cm^{-1} and have visible absorption near 25,000 cm^{-1} (ε = 1000). The complexes $Mo(CNR)_5X_2$ are prepared from $Mo(CO)_4X_2$ and CNR [R = p-tolyl, cyclohexyl] (652), whereas the complex $[Mo(CNR)_6I]I$ is formed by the reaction of $Ag_4[Mo(CN)_8]$ and t-C_4H_9I (653). The complexes $[Mo(CN{-}t{-}Bu)_7](PF_6)_2$ (653a) and $[Mo(CN{-}t{-}Bu_6)I]I$ (653) have been crystallographically investigated and in each case a near C_{2v}, 7-coordinate monocapped trigonal-prism is found. For the iodo complex the I occupies the capping position in a square face of the prism.

The formally Mo(II) complexes $Mo(NO)(R_2dtc)_3$ are also 7-coordinate and are likewise prepared from $Mo(CO)_6$ as the ultimate starting material (645) (reaction 60).

$$Mo(CO)_6 \xrightarrow[-78°C]{Cl_2} Mo(CO)_4Cl_2 \xrightarrow{NOCl} Mo(NO)Cl_3 \xrightarrow{Nadtc} MoNO(dtc)_3 \qquad (60)$$

This complex has the 7-coordinate structure of a pentagonal-bipyramid with NO in the axial position (655, 656).

The complexes $MoX_2(diars)_2$ (X = Cl, Br, I) were prepared by Lewis, Nyholm, and Smith (657) from the anaerobic reaction of acidic Mo(III) halide solutions with diars in water and/or alcohol. The compounds are paramagnetic showing near spin-only room temperature magnetic moments of 2.8 to 2.9 B.M. They are presumed to be 6-coordinate and a ground-state configuration derived from octahedral $(t_{2g})^4$ seems likely. The $MoCl_2(diars)_2$ complex shows a possible d-d band at 11,000 cm^{-1} and a probable charge transfer at 17,000 cm.$^{-1}$ The 6-coordinate complexes $Mo(diars)(CO)_2I_2$ (596) and $Mo(py)_2(CO)_2X_2$ (X = Cl, Br) are also found to be paramagnetic (658). These paramagnetic low-spin $4d^4$ complexes have not been studied in great detail.

In certain cases (522), dimerization occurs to give $[MoX_2(CO)_2(PMe_2Ph)_2]_2$ (522), but only in a few instances have dimeric complexes been found.

The compounds discussed above are predominately monomeric. The 6-coordinate complexes can be either paramagnetic or diamagnetic, whereas the 7-coordinate species are diamagnetic. The 6- and 7-coordinate complexes seem readily interconvertible in a number of cases, and the range of geometries found for the 7-coordinate structures would be consistent with a small energy difference between the various 7-coordinate geometries (659). If monomeric Mo(II) is present in nitrogenase, it would be reasonable to assume that some of these properties could be involved in the explanation for that presence.

B. Dimeric Mo(II) Complexes

The complex $Mo_2(O_2CCH_3)_4$ was first prepared by Wilkinson and coworkers (660, 661) by refluxing molybdenum hexacarbonyl and CH_3COOH in the presence of acetic anhydride. Other carboxylates are made in an analogous manner. The $Mo_2(O_2CCH_3)_4$ molecule was shown by Lawton and Mason (662) to have a quadruply- bridged structure analogous to copper acetate dihydrate (663), but containing an extremely short Mo–Mo bond length of 2.11 Å. The D_{4h} structures illustrated in Fig. 29 show the acetates to be symmetrically bridging and equivalent.

Some additional carboxylates can be prepared by ligand-replacement

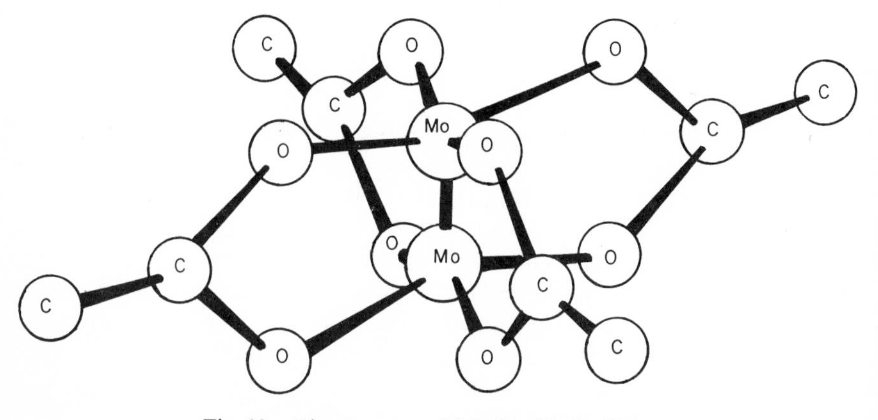

Fig. 29. The structure of $Mo_2(O_2CCH_3)_4$ (662).

reactions using $Mo_2(O_2CCH_3)_4$ as a starting material (664). $Mo_2(O_2CCF_3)_4$ prepared by this procedure was shown by Cotton and Norman to have a structure totally analogous to the acetate analog, with Mo–Mo = 2.090(4) Å, and its volatility allowed a mass spectral analysis to be performed (664).

The dinuclear carboxylates form adducts with donor molecules such as py, PPh_3, and MeOH, with $Mo_2(O_2CCF_3)_4$ forming more stable adducts than does $Mo_2(O_2CCH_3)_4$ (661, 665, 666). In $Mo_2(O_2CCF_3)_4 \cdot 2py$ the Mo–Mo distance was found (665) to have increased from 2.090(4) to 2.129(2) Å [while the ν(Mo–Mo) in the Raman decreased from 397 to 367 cm^{-1}] (664, 665, 667). The nitrogen of the pyridine ligand is found along the Mo–Mo axis at a distance 2.548 Å. This extreme length is considered a consequence of the strong Mo–Mo bond in the trans position (665). The importance of steric factors in stabilizing the complexes toward aerial oxidation was noted by Hochberg et al. (668). Complexes which are prevented from achieving axial coordination appear to undergo aerial oxidation at generally slower rates than those not so prevented. For example, Mo_2 (anthracene-9-carboxylate)$_4$ and Mo_2(1-napthoate)$_4$ are stable in air (as ground solids) for one year, Mo_2(benzoate)$_4$ is stable for two months, whereas Mo_2(2-napthoate)$_4$ is stable (to color change) for only two days. The aryl carboxylate complexes give mass spectral patterns containing parent molecular ion peaks and fragments containing Mo in average oxidation states 2, 2.5, and 3 (668).

Brencic and Cotton (669) demonstrated the utility of $Mo_2(O_2CCH_3)_4$ as a starting material for the preparation of a number of Mo(II), Mo(III), and mixed-valent [Mo(II), Mo(III)] chloro complexes. The reaction products of $Mo_2(OCCH_3)_4$ with HCl can be controlled by varying the concentration of acid and the temperature, and sometimes by using alcohol in the reaction solution. Using these techniques, the complexes $K_4(Mo_2Cl_8)$ (669), $K_4[Mo_2Cl_8] \cdot 2H_2O$ (669), $K_3Mo_2Cl_7 \cdot 2H_2O$ (669), $Rb_3Mo_2Cl_7 \cdot 2H_2O$ (669), $(enH_2)_2Mo_2Cl_8 \cdot 2H_2O$ (670), $(NH_4)_5Mo_2Cl_9 \cdot H_2O$ (671), $Rb_3Mo_2Cl_8$ (612), and $Cs_3Mo_2Cl_8$ (612) were prepared.

The structures of $K_4[Mo_2Cl_8] \cdot 2H_2O$ (672), $(NH_4)_5[Mo_2Cl_8]Cl \cdot H_2O$ (671), and (enH_2) $(Mo_2Cl_8) \cdot 2H_2O$ (670) were solved crystallographically, and contrary to prior claims, each contains the $Mo_2Cl_8^{-4}$ ion with virtually identical dimensions, that is, Mo–Mo = 2.14 Å, Mo–Cl = 2.47 Å, and Mo–Mo–Cl = 105°. The Cl–Cl distances across the Mo–Mo bond and on one Mo atom are both around 3.4 ± 0.1 Å. Thus the structure is grossly described as a cube of chlorides with the short Mo–Mo bond embedded within it. As illustrated in Figs. 29 and 30, the Mo–Mo bond and the basic $Mo_2X_8^{-2}$ D_{4h} structure are maintained through the ligand substitution reaction from $Mo_2(O_2CCH_2)_4$ to $Mo_2Cl_8^{-4}$.

The bonding in all the Mo_2^{+4} dimers can be described by a common scheme, originally developed by Cotton (673, 674), for the (valence electron)

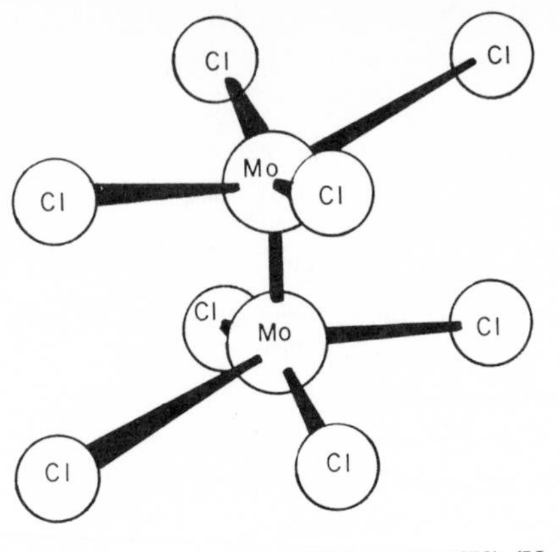

Fig. 30. The structure of $Mo_2X_8^{-4}$ (666, 667, 670) (X = Cl, Br).

isoelectronic $Re_2Cl_8^{-2}$ ion. This scheme is based on qualitative and semi-empirical MO treatments. In the coordinate system chosen, the Z axis is taken along the Mo–Mo bond, whereas the X and Y axis are such as to allow the ligands to fall in the XZ and YZ planes. In this system the s, p_x, p_y, and $d_{x^2-y^2}$ on each Mo are used in sigma bonding with the eight ligand donor atoms. This leaves d_{z^2}, (d_{xz}, d_{yz}), and d_{xy} orbitals for Mo–Mo bonding. The d_{z^2} orbitals (with some s and p_z contribution) form a metal-metal sigma bond, the (d_{xz}, d_{yz}) orbitals of the two metals combine to form two pi bonds, while the remaining d_{xy} orbitals can form a delta bond. Since there are eight electrons to be used in Mo–Mo bonding, these exactly fill the sigma, two pi, and delta bonding levels, producing a quadruple bond and a closed-shell structure. The bond order of 4 is in excellent accord with the very short bond lengths and assumed high bond strengths for these complexes. Furthermore, the presence of the delta bond receives strong support from the eclipsed nature of the ligands in these D_{4h} complexes. The eclipsed conformation, upon moving to the sterically favored staggered structure would loose $d_{xy} - d_{xy}$ overlap and destroy the delta bond. Apparently, the delta component is sufficiently strong to prevent this.

Recently, theoretical (675, 676) and spectroscopic studies (677–678) have given strong support to this basic bonding scheme. $K_4Mo_2Cl_8$ has an absorption band with origin at 17,897 cm^{-1} which exhibits a vibrational spacing of 351 cm^{-1} (677, 678). Notably, 350 cm^{-1} is the Raman band which is resonance enhanced by irradiation at 514.5 nm (678a). The electronic absorption is assigned to the $\delta \rightarrow \delta^*$ transition within the Mo–Mo quadruple bond (678). Transitions in this area of the spectrum would thus seem to be diagnostic of the presence of an Mo–Mo bond. Some spectra of representative complexes are shown in Table XVIII.

TABLE XVIII
Properties of Dinuclear Mo(II) Compounds

Complex	Mo–Mo (Å)	ν(Mo–Mo) Raman (cm^{-1})	ν max cm^{-1}	(ε)	Refs.
$Mo_2(O_2CCH_3)_4$	2.11(1)	406	22,700	(60)	661, 662 667, 677 680, 691
$Mo_2(O_2CCF_3)_4$	2.090(4)	397	23,000	(100)	667, 664
$Mo_2(O_2CC_2H_5)_4$		400			661, 666
$Mo_2(O_2C(n\text{-}C_3H_7))_4$		402			661, 666
$Mo_2(O_2C(n\text{-}C_3F_7))_4$			23,000	(100)	677
$Mo_2(O_2CPh)_4$		404			666
$Mo_2(O_2CC_6H_{11})_4$		397			661, 666
$[Mo_2(CH_3CO_2C_2H_5)_4](CF_3SO_3)_4$					686
$MoCr(O_2CCH_3)_4$					692
$Mo_2(OCCH_3)_4 \cdot 2py$		363			661, 665
$Mo_2(OCCF_3)_4 \cdot 2py$	2.129(2)	367 (368) 343 (in pyridine)	19,700		665, 667, 680
$Mo_2(O_2CCF_3)_4 \cdot 2MeOH$		386			666
$Mo_2(O_2CCF_3)_4 \cdot 2PPh_3$		377			666
Mo_2^{+4}(aq.)			19,800 27,000	(37) (40)	562
$[Mo_2(en)_4]Cl_4$			20,700 27,800	(483) (36)	562
$K_4Mo_2(SO_4)_4$		370			667, 562
$K_4Mo_2(SO_4)_4 \cdot 2H_2O$	2.110(3)				667
$K_3Mo_2(SO_4)_4 \cdot 2H_2O$	2.164(3)				682
$Mo_2(O_2CCF_3)_4$					686
$Mo_2(C_3H_5)_4$	2.183				693
$Mo_2(Etxan)_4 \cdot 2thf$	2.125		17,900 23,400 27,500	(700) (6,500) (11,000)	687, 688
$Mo_2(Etxan)_4$			17,700 23,500 27,000	(600) (5,000) (17,000)	688
$Mo_2(SOCPh)_4$					688

TABLE XVIII (continued)

ComplexÅ)	Mo–Mo	ν(Mo–Mo) Raman (cm^{-1})	νmax (cm^{-1})	(ε)	Refs.
$K_4[Mo_2Cl_8]$		345, 347			669, 679
$Cs_4[Mo_2Br_4]$		335			666
$(NH_4)_5[Mo_2Cl_8]\cdot Cl\cdot H_2O$	2.150	350 (339)			667, 680
$(enH_2)_2Mo_2Cl_8\cdot 2H_2O$	2.134	350			667, 670
$K_4[Mo_2Cl_8]\cdot 2H_2O$	2.139(4)	345			667, 670, 672, 669
$Mo_2Cl_8^{-4}$(6*M* HCl)			19,300	(340)	685
$Rb_3Mo_2Cl_8$	2.38(1)				612
$Cs_3Mo_2Cl_8$					612
$Cs_3Mo_2Br_8$	2.439(7)				683
$Li_4[Mo_2(CH_3)_8]\cdot$4thf	2.148(3)				694
$Mo_2Cl_4(py)_4$		348	17,500 13,750	(130) (850)	684, 685
$Mo_2Cl_4(bipy)_2$		338	18,850	(2,200)	684, 785
$Mo_2Cl_4(CH_3CN)_4$		347	16,600	(1,800)	685
$Mo_2Cl_4(PhCN)_4$		352	15,360	(3,000)	685
$Mo_2Cl_4(S(CH_3)_2)_4$		358	17,100	(1,100)	685
$Mo_2Cl_4(S(C_2H_5)_2)_4$		348	16,950	(1,800)	685
$Mo_2Cl_4(S_2C_4H_8)_2$		357	16,700	(1,500)	685
$Mo_2Cl_4(dth)_2$		359	15,900		685
$Mo_2Cl_4(dtd)_2$		345	15,770	(1,400)	685
$Mo_2Cl_4(dtdd)_2$		425	15,770	(1,400)	685
$Mo_2Cl_4(P(n\text{-}C_4H_9)_3)_4$		350	17,000	(1,300)	680
$Mo_2Cl_4(P(OCH_3)_3)_4$		347	16,800	(1,200)	680
$Mo_2Cl_4(dmpe)_2$		349	15,000	(2,500)	685
$Mo_2Br_4(py)_4$		335	15,200 21,650 24,000	(930) (620) (1,600)	685
$Mo_2Br_4(bipy)_4$		330	17,400		685
$Mo_2Br_4[S(CH_3)_2]_4$		350	16,750		685
$Mo_2Br_4(dtd)_2$		345	15,450		685
$Mo_2Br_4(P(n\text{-}C_4H_9)_3)_4$		342			685

Two other noncrystallographic techniques have found use in probing the presence and nature of the strong Mo-Mo bond. The Raman effect has proved exceedingly useful (666, 667, 679, 680). Due to the symmetrical nature of most of these complexes, the ν(Mo–Mo) vibration is forbidden in the infrared, but it is fully allowed in the Raman. Furthermore, it has been found to be very intense and easily observed, corresponding to a vibrational transition between 340 and 410 cm^{-1}, and serves as a powerful tool for detecting the presence of the Mo–Mo bond. In addition to the high polarizability changes associated with the Mo–Mo vibration (which cause the high Raman intensity), a significant diamagnetic anisotropy is associated with this bond. This has been found to have a distinct effect on the proton magnetic resonance of coordinated ligands (681) causing a substantial deshielding and downfield shift. While this technique has some potentially useful applications, its limitations (681) make it a secondary or supplementary technique to Raman, electronic spectral, and, where available, crystallographic studies.

$Mo_2Cl_8^{-4}$ and $Mo_2(O_2CCH_3)_4$ each serve as starting materials for the preparation of other complexes in this class. The work of Bowen and Taube (561, 562) provided an additional stimulation in this respect. By dissolving $Mo_2Cl_8^{-4}$ in CF_3SO_3H, they adduced strong evidence for the existence of the aquo ion Mo_2^{+4} (aq.). Its spectroscopic properties and the reisolation of the dinuclear core as sulfato and ethylenediamine complexes substantiate its integrity in solution. The aquo ion displays characteristic electronic absorption at 19,800 cm^{-1} (ε = 37) (561).

The crystal structure (667) of $K_4Mo_2(SO_4)_4 \cdot 2H_2O$ again displays the strong Mo–Mo bond [2.110(3) Å]. The sulfates symmetrically bridge the two Mo's, but there is a 20° dihedral angle between MoMoOO and OOS planes giving the molecule overall C_{4h} symmetry. In the axial positions, a long Mo–O contact (to O from another dimer) of 2.593 Å is found compared to the Mo–O bond of 2.136 Å in the bridging sulfates.

The one-electron oxidation product of $Mo_2(SO_4)_4^{-4}$ can be crystallized as $K_3Mo_2(SO_4)_4 \cdot 2H_2O$. The structures of the two compounds [containing $Mo_2(SO_4)_4^{-4}$ and $Mo_2(SO_4)_4^{-3}$ units] are remarkably similar (682), with the latter having a somewhat longer Mo–Mo distance of 2.164(3) Å, but a short Mo–O distance of 2.064 Å. Cyclic voltametry has indicated a half wave potential of +0.22V vs. sce between the $Mo_2(SO_4)_4^{-4}$ and $Mo_2(SO_4)_4^{-3}$ ions (682a). The $Mo_2(SO_4)_4^{-3}$ ion generated in an EPR cavity shows $g_{\parallel} = 1.891$ and $g_{\perp} = 1.909$ with $A_{\parallel} = 4.5 \times 10^{-4}$ and $A_{\perp} = 22.9 \times 10^{-4}$ cm^{-1} (coupling to ^{95}Mo). In agreement with these values the magnetic moment of $K_3Mo_2(SO_4)_4 \cdot 3.5H_2O$ is 1.65 B.M. The 2.5 oxidation state would appear to be at least moderately stable in H_2SO_4 solutions and reducible to the dimeric Mo(II) species by a Jones reductor (682b). Recently (682b), the one electron oxidation product of $Mo_2(O_2CC_3H_7)_4$ was observed in the epr and found to have $g_{\parallel} = g_{\perp} = 1.941$, $A_{\parallel} = 35.6 \times 10^{-4}$ cm^{-1} and $A_{\perp} = 17.8 \times 10^{-4}$ (for ^{95}Mo).

The cationic complex is unstable in CH_3CN and is related to $Mo_2(O_2CC_3H_7)_4$ by a quasi-reversible wave at + 0.39V vs. sce (682c).

One-electron oxidation products of chloro and bromo dimolybdenum(II) species have also been isolated, but in these cases, a marked structural change occurs. As shown in Fig. 31, the complex $Rb_3Mo_2Cl_8$ (612) can be viewed as a bioctahedron sharing a face in which one of the atoms in the shared face is missing. Thus it seems derived from the $Mo_2Cl_9^{-3}$ structure (Fig. 28) by removing one bridging chlorine. The Mo–Mo distance of 2.38 Å indicates strong Mo-Mo bonding; not as strong as that in $Mo_2Cl_8^{-4}$ (Mo–Mo = 2.14 Å), but stronger than in $Mo_2Cl_9^{-3}$ (Mo–Mo = 2.67 Å). The ion $Mo_2Br_8^{-3}$ has the same structure as $Mo_2Cl_8^{-3}$, with an Mo–Mo distance of 2.439(7) Å compared to 2.82 Å in $Mo_2Br_9^{-3}$ (683). Remarkably, the Mo–Br–Mo angle for the bridging bromide is only 54.3(2)°, indicating the Mo–Mo bond can assume its required distance even at the expense of drastically decreasing the bond angles of the bridging bromides from their nominal values. These compounds, in which the metal has an average oxidation state of 2.5, are thus found to have Mo–Mo bond distances (and presumably bond strengths) between the values for Mo(II)–Mo(II) and Mo(III)–Mo(III) dimers.

A number of neutral complexes in which four of the halo ligands are replaced by neutral ligands have been prepared (680, 681, 684, 685). Although detailed structural information is not yet present, the stoichiometry, electronic spectra, Raman scattering, and analogies with structurally characterized Re complexes leave no doubt that these maintain the quadruply bound Mo_2^{+4} core. Complexes are listed in Table XVIII along with their spectral characteristics.

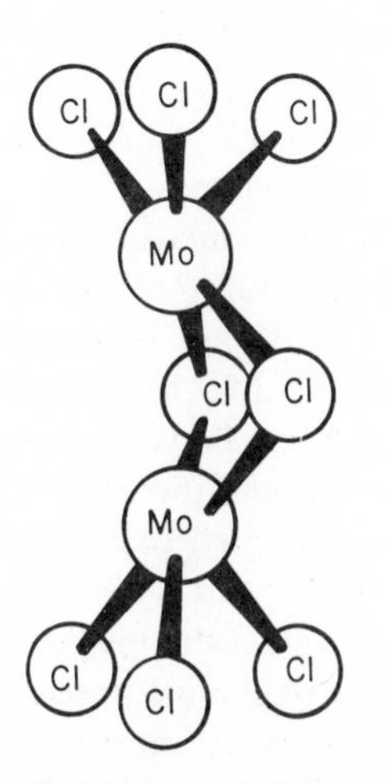

Fig. 31. The structure of $Mo_2X_8^{-3}$ (612, 683) (X = Cl, Br). Note added in proof: A recent report [F. A Cotton and B. J. Kalbacher, *Inorg. Chem.*, *15*, 552 (1976)] indicates bridging hydride in this molecule.

The bidentate ligands $CF_3SO_3^-$, $EtOC(O)CH_3$ (686), $Etxan^-$ (687), and ^-OSCPh (688) can completely replace the chlorides. The complex $Mo_2(Etxan)_4 \cdot 2thf$ was structurally characterized and found to contain the D_{4h}-type quadruply bridged, quadruply bonded structure with Mo–Mo = 2.125(2) Å. This contrasts sharply with the reaction product of R_2dtc^- with $Mo(O_2CCH_3)_4$ where the di-μ-sulfido structure discussed in Section V is found (475, 476).

Recently, other ligands such as N,N′-diphenylbenzamidinato (PhNC(Ph)NPh$^-$) and dimethyl phosphonium dimethylylid ($(CH_2)_2P(CH_3)_2^-$) have been shown to form Mo_2L_4 compounds (688a, 688b). The amidinato complex (688a) has a structure similar to the carboxylates with Mo–Mo = 2.090 Å and a strong Raman band at 410 cm^{-1}. Several new adducts of $Mo_2(O_2CCF_3)_4$ have also been recently isolated and the interesting (but extremely air sensitive) complex $(NEt_4)_4[Mo_2(CN)_8]$ has been prepared by treatment of $Mo_2(O_2CCF_3)_4$ with $(NEt_4)CN$ (688c).

Mixed metal complexes have been prepared which involve strong (bond order 3.5 or 4.0) heteronuclear metal-metal bonding. $CrMo(O_2CCH_3)_4$ (692) can be prepared by adding $Mo(CO)_6$ in $CH_3COOH:CH_2Cl_2$ solution (5:1) to refluxing $Cr_2(O_2CCH_3)_4 \cdot 2H_2O$ in $CH_3COOH:(CH_3CO)_2O$. (1:1). $MoW(O_2CC(CH_3)_3)_4$ is prepared (692a) along with $Mo_2(O_2CC(CH_3)_3)_4$ by refluxing a 3:1 mixture of $W(CO)_6$ and $Mo(CO)_6$ with pivalic acid in o-dichlorobenzene. The heterobinuclear Mo–W complex is selectively oxidized by I_2 and removed from the reaction mixture as $[MoW(O_2CC(CH_3)_3)_4]I$. The crystal structure of $MoW(O_2CC(CH_3)_3)_4I(CH_3CN)$ has been solved (692a) and reveals a structure analogous to the homodinuclear tetracarboxylates with an Mo–W distance of 2.194 Å. The iodo and CH_3CN ligands are coordinated in the axial positions to W and Mo, respectively. The crystals are paramagnetic and show an epr signal with $g = 1.873$. The facile preparation of the oxidized complex may mean that the Mo–W complex is more stable in the oxidized form than the dimolybdo species. This would be in accord with the general tendency of W complexes to be more stable in higher oxidation states than their Mo analogs.

The subject of quadruple and other multiple metal-metal bonds has recently been reviewed (692b)

The complex $[(C_6H_6)(C_3H_5)MoCl]_2$ prepared by Green and Silverthorn (689) is an example of a dinuclear Mo(II) complex which bears little resemblance to the complexes that form the subject of this section. The diamagnetic complex has two bridging chlorides (690), one pi-bound benzene and one pi-bound allyl per Mo, an Mo–Mo contact of 3.94 Å, and bond angles which indicate no Mo–Mo bonding to be present.

Class scrutiny of the preparative procedures for Mo(II) complexes reveals $Mo(CO)_6$ as the ultimate starting material in almost all preparations. Exceptions (488, 625, 651, 653) are rare and involve use of $Mo(CN)_8^{-4}$, Mo(III) species or RCN ligands. Thus for the vast majority of these Mo(II) complexes, there have been, to date, no simple and/or general methods presented for their preparation from higher oxidation states. In a few cases, Mo(II) compounds have been oxidized irreversibly to Mo(III), Mo(V) or Mo(VI) compounds, but the potentials required for these reactions are not known. There are reports of reduction of Mo(III) by Cr^{+2}-edta solutions (561) and by diarsines (625), but the details of the reactions remain obscure. We truly do not yet know enough about of the potentials at which Mo(III) or higher states are reducible to (either dimeric or monomeric) Mo(II) to be able to assess the possibility of Mo(II) being present in nitrogenase.

VIII. MOLYBDENUM TRIS(DITHIOLENE) AND RELATED COMPLEXES

The 1,2-ethylenedithiolate and 1,2-benzenedithiolate ligand systems have been called dithiolenes to distinguish them from saturated 1,2-dithiolate and 1,1- and 1,3-dithiolate ligands whose complexes display vastly different structural and chemical properties. In particular, dithiolene complexes readily undergo reversible one-electron transfer reactions, leading to the existence of complexes in a range of oxidation states. However, unlike complexes of most other ligands (with bipy being a notable exception), the change in oxidation state cannot necessarily be delegated to the metal alone. In fact, the oxidation state of the metal, in many cases, is not readily assignable in an unambiguous manner. It is for this reason (our unwillingness to assign a definite oxidation state in some cases) that we discuss the dithiolenelike complexes of Mo in this separate section (thereby avoiding any ambiguity).

Chemical and structural aspects of dithiolene complexes have been reviewed (695–697). Tris(dithiolene) type complexes of Mo are known for the sulfur ligands $S_2C_2R_2^{-2}$ [R = CN(mnt) (497, 698–700), Ph (701, 702), *p*-$CH_3C_6H_4$ (702), CH_3 (702), CF_3 (703, 704), H (705, 706)], $S_2C_6H_3CH_3^{-2}$ (tdt) (707–709), $S_2C_6H_2(CH_3)_2^{-2}$ (710), $S_2C_6F_4$ (711), $S_2C_6Cl_4$ (712), and $S_2C_6H_4^{-2}$ (709), and the selenium ligand $Se_2C_2(CF_3)_2^{-2}$ (713, 714).

Initial studies led to the synthesis of neutral 6-coordinate tris chelates. Early reactions allowed formally analogous complexes to be prepared from $Mo(CO)_6$ (703), $MoCl_5$ (709), or MoO_4^{-2} (707). For example, reactions 61 and 62 give high yield preparations.

$$Mo(CO)_6 + S_2C_2(CF_3)_2 \longrightarrow Mo[S_2C_2(CF_3)_2]_3 + 6\,CO \quad (61)$$

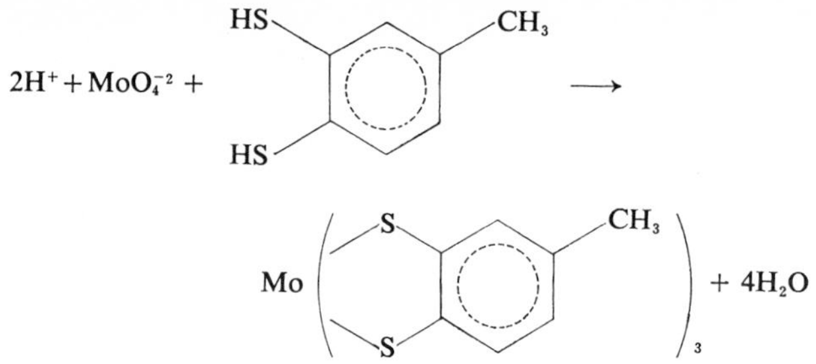

$$2H^+ + MoO_4^{-2} + \text{(HS)}_2\text{-benzene-}CH_3 \longrightarrow Mo\left(S_2\text{-benzene-}CH_3\right)_3 + 4H_2O \qquad (62)$$

Neutral complexes of $S_2C_2Ph_2$ are prepared by adding various Mo compounds (702, 709) to P_4S_{10}-benzoin reaction mixtures.

All of the neutral chelates undergo two reversible one-electron reductions, as demonstrated polarographically and voltammetrically. In the case of the $S_2C_2(CF_3)_2$ ligand, all three members of the reversible electron transfer series, namely, $Mo(S_2C_2CF_3)_3$, $[Ph_4As][Mo(S_2C_2(CF_3)_2)_3]$, and $[Ph_4As]_2[Mo(S_2C_2(CF_3)_2)_3]$, have been isolated (704) by using reactions 63 and 64:

$$Mo(S_2C_2(CF_3)_2)_3 \xrightarrow{N_2H_4} Mo\,(S_2C_2(CF_3)_2)_3^{-2} \qquad (63)$$

$$Mo(S_2C_2(CF_3)_2)_3 + Mo(S_2C_2(CF_3)_2)_3^{-2} \longrightarrow 2Mo(S_2C_2(CF_3)_2)_3^{-1} \qquad (64)$$

The same reactions occur for $Mo(tdt)_3$ (712). The dianionic complex $Mo(mnt)_3^{-2}$ can be prepared by the direct reaction of $MoCl_5$ and mnt^{-2}, using EtOH as the reaction medium in air (497). In general, the highly reduced (anionic) forms of the complexes are stabilized with respect to aerial oxidation when the ligand has electron-withdrawing substituents (e.g., CN^- or CF_3^-).

Structurally, these compounds are of great interest. The neutral tris (dithiolene) complexes were the first examples of trigonal-prismatic coordination in molecular 6-coordinate complexes. The trigonal-prismatic coordination was initially assigned to neutral $Mo(dithiolene)_3$ complexes by

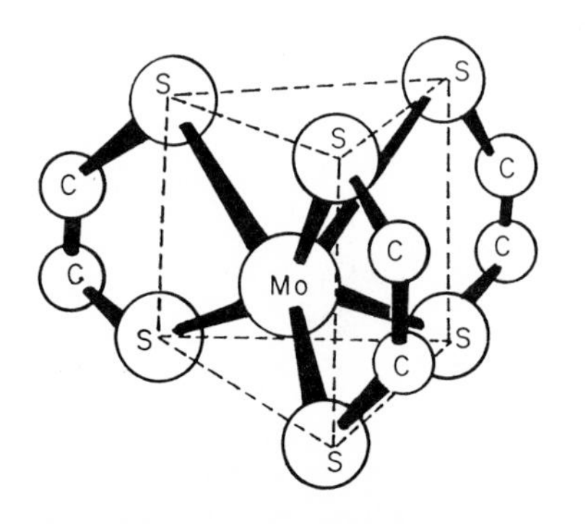

Fig. 32. The structure of $Mo(S_2C_2H_2)_3$ (706).

comparison of powder diffraction patterns and electronic spectra (708) with the structurally characterized $Re(S_2C_2Ph_2)_3$ (715). This has been fully confirmed by x-ray studies on $Mo(S_2C_2H_2)_3$ (706) and $Mo(Se_2C_2(CF_3)_2)_3$ (714).

The structure of $Mo(S_2C_2H_2)_3$, shown in Fig. 32, has C_{3h} symmetry. The deviation from D_{3h} symmetry is due to a bending back of the chelate rings which produces a dihedral angle of 18° between the S_2C_2 and MoS_2 planes of each chelate ring. A similar structural feature is found in $Mo(Se_2C_2(CF_3)_2)_3$ (714). The "diselenolene" structure (714) has the additional interesting feature of a distinctly short interligand Se–Se contact of 3.22 Å. This is taken as supportive of those electronic structural models which entertain weak interdonor atom bonding forces as a stabilizing feature of trigonal-prismatic coordination for the oxidized complexes (709, 714).

The reduced (anionic) complexes, as typified by $Mo(mnt)_3^{-2}$, do not possess trigonal-prismatic coordination, but neither do they have near-octahedral structures. The geometry of $Mo(mnt)_3^{-2}$ (716, 717), as shown in Fig. 33, is by all criteria (718) midway between octahedral and trigonal-prismatic limits. The gross structure of the dianion appears to be dictated by the bite and electronic structural features of the ligand combined with the nominal *d*-electron configuration of the metal. Simply viewed, the change in geometry from the oxidized (trigonal prismatic) to the reduced (in between prismatic and octahedral) forms reflects an increased electron density on the ligands and therefore an increased interligand repulsion. This may be another way of saying that much of the interligand bonding which may contribute to stabilizing the trigonal prism (in the oxidized form) is eliminated by the addition of two extra electrons. Twisting and elongation of the MoS_6 coordination

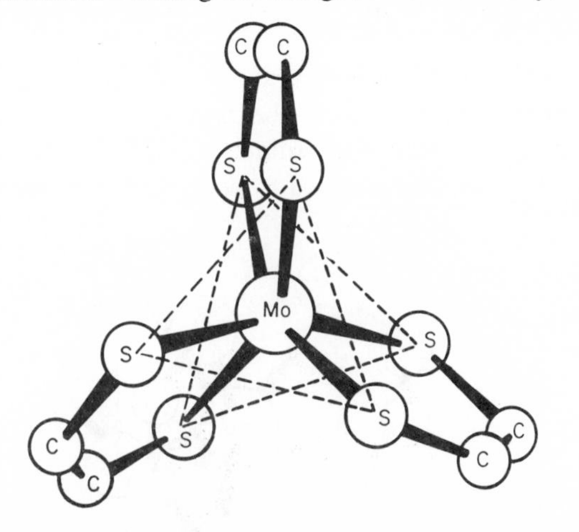

Fig. 33. The structure of $Mo(mnt)_3^{-2}$ in $(Ph_4As)_2$ $[Mo(mnt)_3]$ (716, 717). (The cyano groups are omitted for clarity.)

sphere in the direction of the octahedral complex allow relaxation of the ligand repulsions in the dianions (717).

Both the neutral and dianionic tris complexes are diamagnetic, but the intermediate monoanionic forms are paramagnetic and show EPR signals consistent with doublet ground states (701, 708, 709). The g and A values are shown in Table VIII.

As discussed briefly above, the designation of oxidation state in this class of compounds has been a bone of contention since the earliest days of their study (695). The problem can be focused by reference to the complex $V(S_2C_2Ph_2)_3$ (701). If the ligands here are designated as 1,2-dithiolates, then this is formally a V(VI) complex, and the required $(4p)^5$-metal configuration is clearly not reasonable. However, these ligands are not innocent of electron-transfer capability. Thus they can be formulated by either of two limiting structures (structures **18** and **19**),

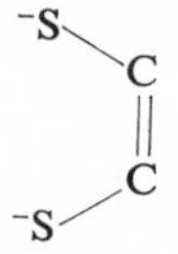

18 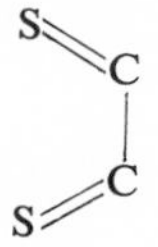 **19**

which differ by two electrons. Then, depending on how one designates the ligands in complexes such as $V(S_2C_2Ph_2)_3$ or $Mo(S_2C_2R_2)_3$, one can formulate the complex as containing M(VI), M(IV), M(II), or M(0). Clearly, this must not be done arbitrarily.

Molecular-orbital calculations (702, 709) have been performed, in part, to address the problem of oxidation-state formulation of the complexes. The most flexible scheme (709) allows the limiting formulations of the set of complexes MoL_3^0, MoL_3^{-1}, and MoL_3^{-2} as either Mo(VI), Mo(V), and Mo(IV) or Mo(II), Mo(I), and Mo(0). The designation depends on whether the highest filled e level (in D_{3h} symmetry) is assigned as a ligand or metal level. When assigned as a ligand level the higher Mo oxidation states are called for, but when assigned to the metal the lower set of oxidation states prevails. In fact, this e level is highly delocalized and cannot unequivocally be assigned to metal or ligand. Thus physical methods rather than calculations should be called upon if one wishes to designate the oxidation state (although a case can be made to forego the effort entirely).

The application of photoelectron spectroscopy to determine the binding energy of inner Mo electrons may hold promise in this respect. Recent studies (710) [using species such as $(NH_4)_2MoS_4$ and MoS_2 as standards] support Mo(VI) formulations for $Mo(S_2C_2H_2)_3$ and $Mo(S_2C_6H_2(CH_3)_2)_3$. Oddly, $(NBu_4)_2[Mo(mnt)_3]$ would also be formulated as Mo(VI) by this technique.

However, the mnt ligands in $Mo(mnt)_3^{-2}$ have dimensions consistent with their formulation as dianions, which would make the complex one of Mo(IV) (717).

Although it is clear that the arguments are not straightforward, the higher set of oxidation states would seem to comprise a slightly better overall description of the electronic structures than do the lower oxidation states. This is especially true since the EPR properties (Table VIII) of the MoL_3^- species are consistent with an Mo(V) configuration (719).

Recently, complexes containing MoS_3N_3 coordination spheres have been prepared which appear closely related to the MoS_6 tris(1,2-dithiolene) complexes. These complexes are of two types, structures **20** and **21**,

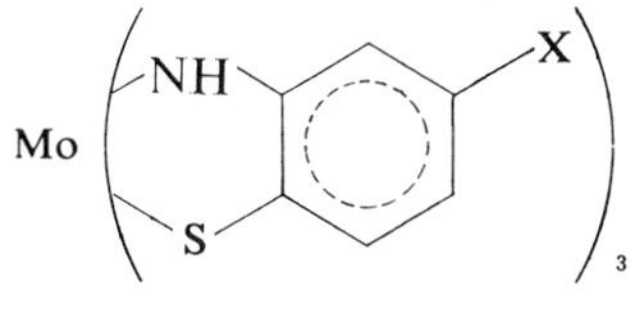

X = H, Cl
(78, 720)

20

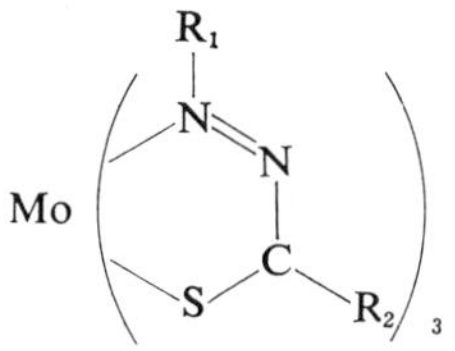

R_1 = H, R_2 = Ph, p–$MeOC_6H_4$
R_1 = Ph, R_2 = CH_2Ph, Ph
(721)

21

and show spectroscopic properties (78, 720) and reversible voltammetric behavior (720, 721), indicative of their strong resemblence to the dithiolenes. The voltammetric potentials, however, reveal these compounds to be far more difficult to reduce than the dithiolenes. The complexes are of particular interest as they are prepared from MoO_4^{-2} by formally nonredox reactions, for example, Eq. 65:

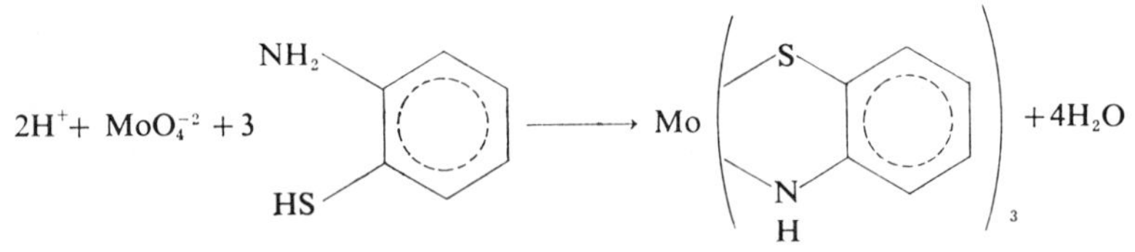

$$2H^+ + MoO_4^{-2} + 3\ (NH_2,\ HS\text{-benzene}) \longrightarrow Mo(\ldots)_3 + 4H_2O \qquad (65)$$

The ligand amino group thus becomes deprotonated, even though the reaction is carried out in acid solution. If we formulate this as an Mo(VI) complex, then this behavior is consistent with the proclivity of high oxidation states to cause deprotonation of coordinated ligands. The thiohydrazide complex $Mo(PhCSNNH)_3$ (721) can apparently be further deprotonated to form $Mo(PhCSNNH)_2(PhCSNN)^-$ (721). The high formal oxidation state can be said to have enhanced the acidity of the coordinated ligand. These compounds are of interest because they possess the ability to transfer both protons and electrons, a feature which seems to be characteristic of at least one

Mo enzyme (i.e., xanthine oxidase) and has been suggested as common to all Mo enzymes (74, 78).

The dithiolenes and these NS ligands are among those few ligands which are able to fully strip Mo of its oxo ligands while maintaining the formal high oxidation states of IV, V, and VI. The reason for this must lie in the ability of these ligands to act as strong sigma and pi donors while maintaining sufficient pi-acceptor quality to prevent complete charge transfer from ligand to metal.

Attempts have been made to explore the possibility of dithiolenelike behavior in O-donor ligands. Pierpont (163) allowed $Mo(CO)_6$ to react with tetrachloro-*o*-benzoquinone to obtain a dimeric complex of the form $[Mo(O_2C_6Cl_4)_3]_2$. This complex, despite its stoichiometric resemblance to the tris dithiolenes, does not resemble them in chemical or structural properties. Its dimeric structure contains ligand bridges with each Mo having a distorted octahedral coordination. The complex does not display reversible electron-transfer behavior and clearly does not exhibit the dithiolene character of its sulfur homolog. However, recently (163a) the complex $Mo(PQ)_3$(PQ = 9, 10-phenanthrenequinone) has been isolated and structurally characterized. It shows a near trigonal-prismatic structure with the unusual feature of one of the chelate rings severely bent back from the MoO_2-plane causing a dihedral angle of 60°. This bending seems required to establish a stacking interaction in which two phenanthroquinone ligands on two adjacent complexes lie parallel at an average distance of 3.2–3.3Å. It is not clear whether this interaction is in someway responsible for the coordination structure.

A few mixed ligand complexes have been prepared which contain dithiolene ligands. The complexes $Mo(S_2C_2Ph_2)_2(S_2C_2H_2)$, $Mo(S_2C_2Ph_2)_2(mnt)$, and $Mo(S_2C_2Ph_2)_2(dppe)$ are formed from the reaction of $Mo(S_2C_2Ph_2)_2$-$(CO)_2$ with $S_2C_2H_2^{-2}$, mnt^{-2}, or dppe, respectively (722). Reaction of Mo-$(S_2C_2Ph_2)_2(CO)_2$ with sulfide ion produces the interesting binuclear complex $Mo_2S_2(S_2C_2Ph_2)_4$, which can also be prepared from the P_4S_{10}-benzoin reaction mixture and $MoCl_5$ (722). The P_4S_{10}-benzoin reaction product with ammonium paramolybdate (486) gives a complex formulated as Mo_2S_4-$(S_2C_2Ph_2)_2$. These compounds presumably contain bridging sulfide, but their detailed structures are not yet known from crystallographic study. The same comment applies to the species formulated as $Mo_2(tdt)_5$ (723). The mixed ligand complexes $Mo(Et_2dtc)(SNHC_6H_4)_2$, $Mo(Et_2dtc)(S_2C_6H_4)_2$, and $Mo(S_2P(\text{–i–}C_3H_7)_2)(SNHC_6H_4)_2$ have recently been isolated and their interesting EPR properties have been studied (237a, 258b) as discussed in section IV.

Oxo dithiolene complexes of Mo(IV) (486, 497) have been discussed in Section VI, along with Mo(IV) chemistry. Nitrosyl dithiolene complexes are discussed in Section IX.

IX. MOLYBDENUM NITROGEN CHEMISTRY

The chemistry of Mo with simple nitrogen ligands is of great interest in view of the key role played by Mo enzymes in the metabolism of nitrogenous material. In this section, we review recent results on the chemistry of Mo, predominantly in lower oxidation states, with simple N-containing ligands. A brief discussion of related hydride and zerovalent complexes is included for completeness.

A. Complexes of Dinitrogen and Their Reactions

Molecular N_2 complexes have been reviewed many times in recent years (36, 726–731). Except in a limited number of cases, these complexes of dinitrogen have resisted attempts at reduction, and in that sense they do not model the nitrogenase enzyme. This is not to say that these complexes are irrelevant to our understanding of N_2 fixation. On the contrary, $M–N_2$ binding may be a required precursor to the enzymatic reduction, and the identification of trends in dinitrogen binding and activation may help focus on the nature of the enzyme active site.

While dinitrogen complexes of all metals may be of value in this respect, in this review, we summarize only the work on dinitrogen complexes of Mo. We must emphasize that there is no direct evidence that the Mo in nitrogenase forms a dinitrogen complex in the catalytic cycle. In fact, it has been suggested that N_2 may form a complex with Fe prior to being brought to the Mo site for reduction (28). Thus Mo complexes of N_2, especially in the low valence states discussed here, are not certain to be relevant to Mo action in N_2ase.

With that qualification in mind, it is nonetheless notable that in complexes of Mo and W some success has been achieved in reducing bound dinitrogen in structurally well-characterized compounds. The fact that group VI complexes undergo the reaction is worthy of continued scrutiny.

We can conveniently classify dinitrogen complexes of Mo into four groups. The first and most intensely studied group is that which contains *trans*-$Mo(N_2)_2(dppe)_2$ and related complexes and derivatives. The second group consists of $(CH_3C_6H_5)Mo(PPh_3)_2N_2$ and other mononuclear and binuclear complexes which contain a pi-bound (hexahapto) aryl ligand. The complexes in these two groups contain zero- or monovalent Mo. The third group consists of complexes wherein a high-valent Mo complex serves as a Lewis acid for the exposed end of a dinitrogen complex of another metal [e.g., $(PMe_2Ph)_4ClReN_2MoCl_4(OMe)$]. The three crystal-structure determinations completed (306, 732, 733) for Mo dinitrogen complexes span these three classes. The fourth class is a limited one consisting of dinitrogen derivatives of molybdocene and its permethyl derivative. The fifth part of this section discusses some less well-defined Mo-containing systems in which successful reduction of dinitrogen has been achieved.

1. *Trans-Mo(N_2)$_2$(dppe)$_2$ and Related Species*

a. Synthesis and Structure. The species trans-Mo(N_2)$_2$(dppe)$_2$ is by far the most thoroughly studied Mo-dinitrogen complex. It was first reported by Hidai et al. (734–736). The original preparation (736) involved reaction of Mo(acac)$_3$, dppe, $AlEt_3$ and N_2 in toluene at $-40°C$ with a five-day waiting period prior to isolation of the product in 13% yield. This complex has since been prepared by a number of other, fortunately briefer, procedures. Atkinson, Mawby, and Smith (737) reported the reduction of $MoCl_2$(dppe)$_3$ or $MoCl_3$(dppe)$_2$ with 2% Na(Hg) in thf under N_2 as giving a pure product, but reaction times and yield were not given. George and Seibold (738) also use Na(Hg) in the reduction of $MoCl_4$(dppe) in thf under Na with excess dppe present. Here the reaction time is 6 hr and the yield is 36%. Chatt and co-workers (739, 740) reported 88% yield for the preparation of *trans*-Mo(N_2)$_2$(dppe)$_2$ from $MoCl_3$(thf)$_3$ and Na(Hg) in the presence of dppe and N_2 with a 3 hr reaction time. The overall pathway for this preparation is shown in Eq. 66.

$$MoCl_5 \xrightarrow[\text{reflux}]{\text{EtCN}} MoCl_4(EtCN)_2 \xrightarrow[CH_2Cl_2/\text{thf}]{\text{Zn}} MoCl_3(\text{thf})_3 \xrightarrow[\text{dppe},\ N_2]{\text{Na(Hg)}} \textit{trans}\text{-}Mo(N_2)_2(\text{dppe})_2 \tag{66}$$

This appears to be the reaction of choice for synthetic purposes.

Complexes analogous to *trans*-Mo(N_2)$_2$(dppe)$_2$ have been formed with other bidentate phosphine and arsine ligands, as well as with monodentate phosphines. In the latter case, a *cis*-dinitrogen arrangement has been reported for $Mo(N_2)_2(PMe_2Ph)_4$ (738, 739, 741). All complexes are listed in Table XIX along with their NN stretching vibrations. Each of these compounds has Mo in a zero oxidation state, and they are generally stored under N_2, Ar, or vacuum and kept free of moisture to avoid decomposition. Details of the syntheses appear in the appropriate references (734–747).

The complex *trans*-Mo(N_2)$_2$(dppe)$_2$ was shown crystallographically by Uchida et al. (733) to have the centrosymmetric structure of Fig. 34. The Mo–N length of 2.01(1) Å is comparable to Mo–N single bond lengths, whereas the N–N distance at 1.10 Å does not differ significantly from that in free N_2. The Mo–N–N angle of 172.8(2)° shows a small nonlinearity.

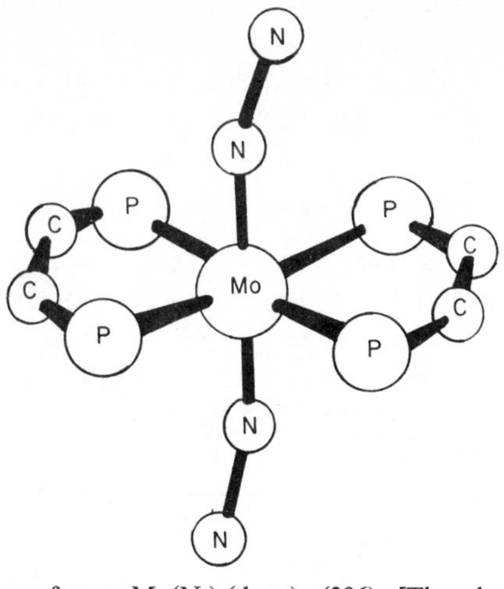

Fig. 34. The structure of *trans*-$Mo(N_2)_2(dppe)_2$ (306). [The phenyl groups of the dppe ligands are omitted for clarity (733).]

The perturbed NN stretching vibration has proved a valuable indicator for the presence of coordinated dinitrogen and for the structure of the complex. Most of the bis(dinitrogen) complexes show two bands in the N_2 stretching region. For example (748), *trans*-$Mo(N_2)_2(dppe)_2$ in CS_2 shows a strong band at 1973.5 cm^{-1} assignable to the antisymmetric stretch (A_{2u} in assumed D_{4h} symmetry) and a weak band at 2038.6 cm^{-1} assignable to the symmetric stretch (A_{1g} in D_{4h}). Presumably, the low intensity of the latter is due to its forbidden character in D_{4h}. The fact that it has any intensity at all is attributed (748) to either the slightly bent nature of the N_2MoN_2 linkage or to local environmental conditions which considerably lower the effective symmetry. We note, however, that for the bend to cause the vibration to become active it must remove the center of symmetry found in the crystal.

Darensbourg (748) noted a linear relationship between ν(N–N) and the absolute intensity of the vibrational absorption band. Furthermore, the infrared intensity of the N_2 stretch is said to be largely determined by the extent of pi electronic charge transferred from the transition metal to dinitrogen during the N_2 stretching motion. A similar process, metal to ligand pi-electron donation, has been suggested as responsible for the frequency decrease. In agreement with this notion, in related dinitrogen complexes the intensity generally increases and frequency decreases as the nuclear charge on the metal decreases. Additional recent work substantiates the idea that significant charge is transferred to N_2 on complex formation, with the terminal nitrogen actually being somewhat more negative than in N_2 and displaying substantial Lewis basicity (743, 749).

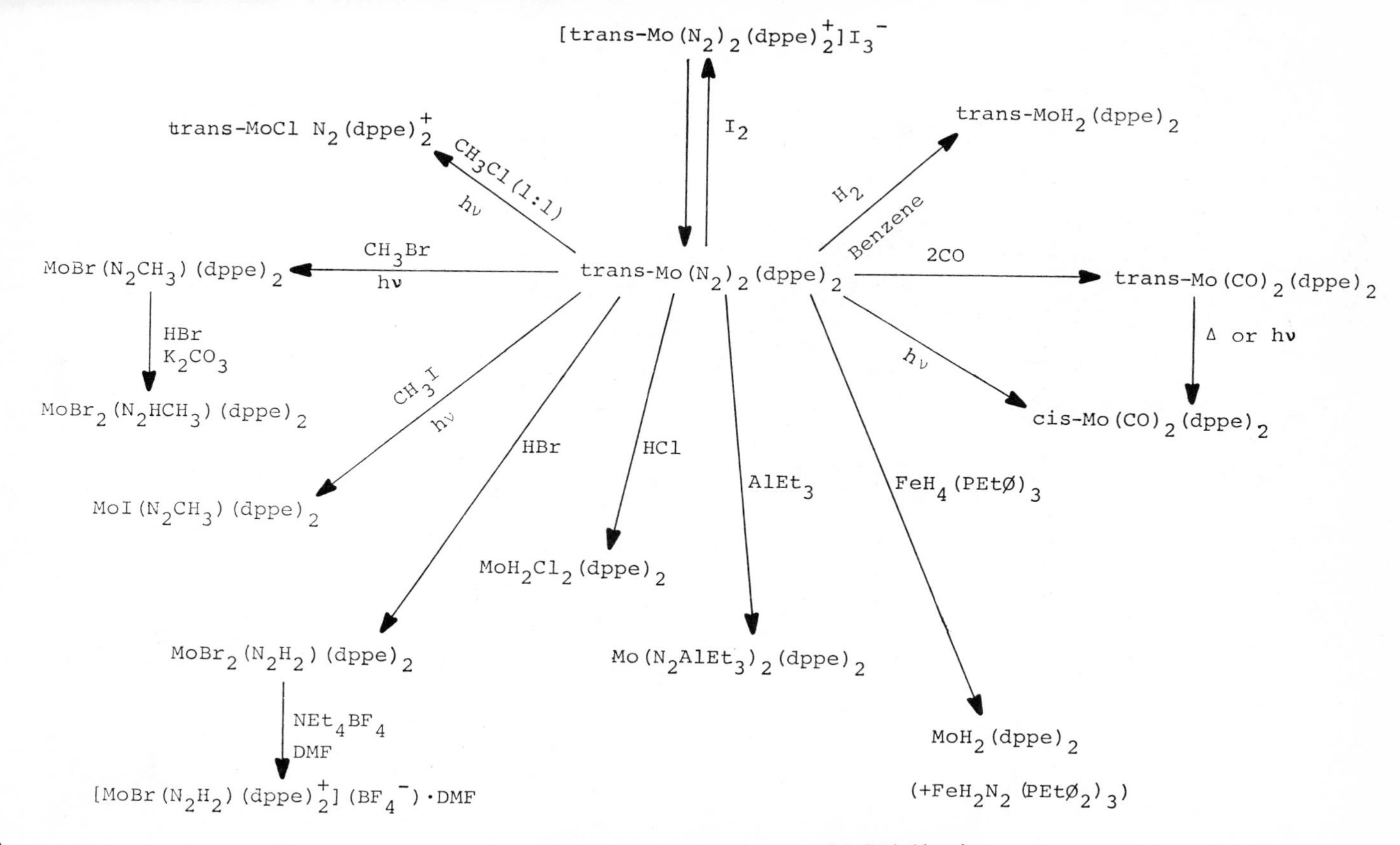

Fig. 35. Reaction scheme for *trans*-$Mo(N_2)_2(dppe)_2$.

The other ligands in the coordination sphere also effect ν(NN). In the *trans*-bis(N_2) complexes, the frequency of the antisymmetric (A_{2u}) $\nu(N_2)$ mode decreases as the basicity of the other ligands increases (735). Again, the obvious interpretation lies in the increased $\pi(M \rightarrow N_2)$ donation in the complexes permitted by the more strongly basic ligands.

In the one well-characterized cis complex, *cis*-$Mo(N_2)_2(PMe_2Ph)_4$, both NN stretches are very intense, as expected for the C_{2v} symmetry of the coordination sphere ($Mo(N_2)_2P_4$) of this complex (741).

b. Reactions. The reactivity of *trans*-$Mo(N_2)_2$(dppe)$_2$ has been investigated in a number of laboratories. While many reagents cause irreversible decomposition of the complex with liberation of N_2, a surprisingly large number of well-defined reactions have been executed. A reaction diagram is shown in Fig. 35.

(1) Adduct Formation. The simplest reaction is that of adduct formation, where the lone pair on the unbound end of the coordinated dinitrogen serves as a Lewis base. Both 1:1 and 2:1 adducts have been reported with trialkylaluminum as the Lewis acid, and those which appear to be well characterized are reported in Table XIX. The reduction in $\nu(N_2)$ upon adduct formation is noted as evidence for weakening of the NN bond. The possible significance of this observation is discussed below.

(2) Formation of Mo(I) Complexes. George and Seibold (750) have shown by cyclic voltammetry that $Mo(N_2)_2$(dppe)$_2$ undergoes a reversible one-electron oxidation (E_p = 0.20 V (anodic), E_p (cathodic) = 0.27 V versus sce, on Pt in dmf with $(C_2H_5)_4N^+ClO_4^-$ as electrolyte). Using iodine as the oxidant in CH_3OH, they were able to isolate the paramagnetic salt $[Mo(N_2)_2(dppe)_2^+]I_3^-$. The magnetic moment of 1.97 B.M., while nominally consistent with the presence of one unpaired electron in this formally $4d^5$, Mo(I) complex, is unusual insofar as it is greater than the spin-only moment (1.73 B.M.) and stands in contrast to the 1.66 B.M. found for the $[Mo(CO)_2$-$(dppe)_2^+]I_3^-$.

The oxidation of *trans*-$Mo(N_2)_2$(dppe)$_2$ by I_2, in the hands of Miniscloux et al. (751), is claimed to yield either $[(C_6H_6)(dppe)MoI]_2N_2$ or Mo_2I_3(dppe)N_2, depending on the solvent (benzene vs. toluene, respectively). Furthermore, these workers use *n*-heptyliodide in toluene to synthesize a complex formulated as $Mo_2I_2(dppe)_4N_2$. Molecular weight and magnetic measurements would help in the proper formulation of these species, and these interesting reactions deserve further study.

Related Mo(I) complexes have also been identified. *Trans*-$MoCl(N_2)$-(dppe)$_2$ is prepared [along with *trans*-$Mo(N_2)_2$(dppe)$_2$] from the reaction of $MoOCl_2$(dppe) (thf) with Zn dust under N_2 in the presence of excess dppe

TABLE XIX
Dinitrogen Complexes of Molybdenum

Complex	ν(N–N), cm^{-1} (medium)	Refs.
trans-$Mo(N_2)_2(PEt_2Ph)_4 \cdot C_7H_8$	1965 (toluene)	741
trans-$Mo(N_2)_2(PPh_2Me)_4$	1925 vs (benzene)	738
cis-$Mo(N_2)_2(PMe_2Ph)_4$	2018 s, 1939 (benzene)	738
	2014, 1951 (thf)	739
	2022 vs, 1940 vs (mull)	741
trans-$Mo(N_2)_2(PMe_2Ph)_2(PhSCH_2CH_2SPh)$	1925 (toluene)	740
trans-$Mo(N_2)_2(Ph_2PCH{=}CHPPh_2)_2$	1995 vs (KBr)	739
trans-$Mo(N_2)_2(dppm)_2$	2020 vw, 1970 vs (KBr)	736
trans-$Mo(N_2)_2(dppe)_2$	2040 vw, 1971 vs (CH_2Cl_2)	738, 739, 734–736
trans-$Mo(N_2)_2(dppp)_2$	2010 vw, 1925 vs (KBr)	736
trans-$Mo(N_2)_2(diars)_2$	2044 vw, 1950 vs (CH_2Cl_2)	738
trans-$Mo(N_2)_2(dpae)_2$	2041 vw, 1970 vs (CH_2Cl_2)	738
trans-$Mo(N_2)_2(arphos)_2$	2044 vw, 1971 vs (CH_2Cl_2)	738
[*trans*-$Mo(N_2)_2(dppe)_2](I_3)$	2047 s (mull)	738, 747
[*trans*-$Mo(N_2)_2(arphos)_2]$ (I_3)	2043 s (mull)	738, 750
trans-MoCl (N_2) $(dppe)_2$	1970	737, 751 752
trans-$MoBrN_2(dppe)_2$	1966 (CsI)	752
trans-$MoClN_2(Ph_2PCH{=}CHPPh_2)_2$		754
trans-$Mo(N_2AlEt_3)_2(dppe)_2$	1975 vs, 2100 w (toluene)	749
trans-$Mo(N_2AlEt_3)_2(PEt_2Ph)_4$	1875 vs (toluene)	749
cis-$Mo(N_2AlEt_3)_2(PMe_2Ph)_4$	1938 vs, 1855 vs, 2040 (w) (toluene)	749
trans-$Mo(N_2)(N_2AlMe_3)$ $(dppe)_2$	1979	775
$((CH_3)_3C_6H_3)Mo(PPh_3)_2N_2$	1983 (toluene)	689
$(CH_3C_6H_5)Mo(Ph_2MeP)_2N_2$	1970 vs	689
$(C_6H_6)Mo(P(C_6H_{11})_2Me)_2N_2$	1975 vs (nujol)	764
$(CH_3C_6H_5)Mo(PPh_3)_2N_2$	1988 vs (toluene)	689, 736
$[(CH_3)_3C_6H_3]$ $(dmpe)MoN_2$	1973 (benzene)	732, 764
$[[(CH_3)_3C_6H_3]$ $(dmpe)MoH]_2N_2(PF_6)_2$	1937 vw (nujol)	764
$[[(CH_3)_3C_6H_3]$ $(dmpe)MoH]_2N_2(BF_4)_4$	1937 vw (nujol)	764

TABLE XIX (continued)

Complex	ν(N–N), cm^{-1} (medium)	Refs.
$[(C_6H_6)(PPh_3)_2Mo]_2N_2$	1919 (Raman)	698
$[(CH_3C_6H_5)(PPh_3)_2MoN_2Fe(dmpe)(C_5H_5)](BF_4)$	1930 vs	689
$(PMe_2Ph)_4ClReN_2MoCl_4(OMe)$	1660	306, 767
$(PMe_2Ph)_4ClReN_2Mo_2OCl_5(Et_2O)^a$	1680	743, 776
$(PMe_2Ph)_4ClReN_2MoOCl_3(Et_2O)^a$	1810	743, 776

[a] A number of adducts of $ReCl(N_2)(PMe_2Ph)_4$ have been isolated (743), and while they definitely contain ReNNMo linkages, their detailed nature remains in some doubt.

(737). This compound is also prepared from *trans*-$Mo(N_2)_2(dppe)_2$ upon irradiation with an equimolar amount of CH_3Cl (752). The bromo analog, *trans*-$MoBrN_2(dppe)_2$, has also been reported (752) as synthesized by photochemical procedure.

(3) Ligand Substitution. *Trans*-$Mo(N_2)_2(dppe)_2$ reacts readily with CO in the presence of light in various solvents to give exclusively *cis*-$Mo(CO)_2$-$(dppe)_2$ (736, 750, 753) as the isolated product. In agreement with general trends in metal carbonyl chemistry, the *cis*-dicarbonyl complex appears to be the thermodynamically more stable form. However, the *trans*-dicarbonyl complex may be formed at a significantly greater rate from the *trans*-dinitrogen complex. Early work using infrared and NMR evidence supported the existence of the *trans*-dicarbonyl species as an intermediate in the reaction to form the *cis*-dicarbonyl product (750). More recently (741, 754), the *trans*-dicarbonyl was isolated by excluding light from the reaction mixture. The isolated complex rapidly and irreversibly isomerizes in light to the more stable *cis*-dicarbonyl. It thus appears that N_2 and CO have different stereochemical preferences in the compounds $MoL_2(dppe)_2$ (L = CO, N_2). There must be a fine balance between steric and electronic factors for these complexes, with steric factors favoring trans placement of the bulky dppe ligands and electronic factors favoring *cis*-dicarbonyl or *cis*-bis(dinitrogen) structures for these pi-acceptor ligands (754). For CO the balance favors the electronic effect whereas for N_2 the steric factor dominates presumably because of its weaker pi-acceptor nature. It is of interest that $Mo(CO)_2(dppe)_2^+$ apparently has a trans structure. The increased positive charge may again switch the balance as it both reduces $M \rightarrow L(\pi)$ back donation while increasing steric effects by decreasing the effective size of Mo.

Trans-$Mo(N_2)_2(dppe)_2$ reacts with aryl and alkyl nitriles to give the monosubstitution product trans-$Mo(N_2)(RCN)(dppe)_2$. For the case R = CH_3 the reaction is reversible and the bis(dinitrogen) complex can be regenerated by bubbling N_2 through a toluene solution of the nitrile complex (754b). Reaction of trans-$Mo(N_2)_2(dppe)_2$ with isonitriles leads (754c) to the formation of the complexes trans-$Mo(CNR)_2(dppe)_2$. Structural studies on the $R=CH_3$ complex confirm the trans arrangement but reveal a bent C–N–C (methyl) linkage (156°). It is found that these isonitrile complexes readily protonate at their nitrogen atoms to give a carbyne-like ligand in the complex ion $[Mo(CNHMe)_2(dppe)_2^{+2}]$. Monoprotonation and monoalkylation reactions have also been found to occur (754c).

Trans-$Mo(N_2)_2(dppe)_2$ also reacts reversibly with ethylene (754a) at high temperature according to Eq. 67:

$$Mo(N_2)_2(dppe)_2 + C_2H_4 \xrightarrow[\text{toluene}]{\text{refluxing}} Mo(C_2H_4)(dppe)_2 + 2N_2 \qquad (67)$$

This reaction is unusual insofar as it involves an $18e^-$, $4d^6$ octahedral complex reacting with ethylene to produce an apparently 5-coordinate, 16-electron species which retains the $4d^6$ Mo configuration. While one might wonder if the phosphine ligands remain intact in this reaction, the reversibility of the reaction indicates that if the phosphines do indeed rearrange, the process is probably intramolecular and fully reversible. The structure of the ethylene complex would be of considerable interest.

Trans-$Mo(N_2)_2(dppe)_2$ reacts with H_2 (741) to give *cis* or *trans*-$MoH_2(dppe)_2$, with the stereochemistry determined by the solvent. The complexes *cis*-$Mo(N_2)_2(PMe_2Ph)_4$ and *trans*-$Mo(N_2)_2(PEt_2Ph)_4$ react with H_2 to produce MoH_4L_4 (L = PMe_2Ph, PEt_2Ph) (741).

The cis complex $Mo(N_2)_2(PMe_2Ph)_4$ undergoes substitution of two phosphines by $PhSCH_2CH_2SPh$ to yield *trans*-$Mo(N_2)_2(PMe_2Ph)_2(PhSCH_2CH_2SPh)$. This compound is unstable and decomposes at room temperature with the evolution of gas (749).

The complex cis-$Mo(N_2)(PMe_2Ph)_4$ reacts with CO_2 in an unusual reaction (754d) to form the binuclear complex $(PMe_2Ph)_3COMo(\mu\text{–}CO_3)_2MoCO(PMe_2Ph)_3$. Crystallographic studies reveal two bridging carbonates which are bidentate on one Mo and monodentate on the second Mo in such a way as to give 7-coordination spheres to each Mo.

(4) Intermolecular Ligand-Exchange Reactions. Since there is a strong possibility of more than one metal being at the active site of N_2ase, there is incentive to look at N_2 ligand-transfer reactions as a possible step in the catalytic cycle for N_2 reduction.

Hidai et al. (735) looked at the reaction of *trans*-$Mo(N_2)_2(dppe)_2$ with $CoH_3(PPh_3)_3$ in benzene and toluene under argon. In benzene, reaction 68 occurs.

$$trans\text{-}Mo(N_2)_2(dppe)_2 + CoH_3(PPh_3)_3 \longrightarrow trans\text{-}MoH_2(dppe)_2 + CoH(N_2)PPh_3 \qquad (68)$$

N_2 ligand transfer from Mo to Co would seem to be implicated, although as yet there is no mechanistic information on this reaction. A similar reaction with $FeH_4(PEtPh_2)_3$ was studied by Aresta and Sacco (741), and reactions 69 and 70 are suggested to explain the results in benzene solution for 1:1 and 2:1 Fe to Mo ratios, respectively ($P = PEtPh_2$).

$$trans\text{-}Mo(N_2)_2(dppe)_2 + FeH_4P_3 \longrightarrow MoH_2(dppe)_2 + FeH_2N_2P_3 + N_2 \quad (69)$$

$$trans\text{-}Mo(N_2)_2(dppe)_2 + 2FeH_4P_3 \longrightarrow MoH_2(dppe) + 2FeH_2N_2P_3 + H_2 \quad (70)$$

These reactions are of interest in view of the presence of both Fe and Mo in nitrogenase, and they demonstrate some of the complexities in the relationship between potential N_2 and H_2 binding sites in the enzyme.

(5) Reduction of Coordinated Dinitrogen. Chatt and co-workers (740, 742) discovered that treatment of these Mo (and W) bis(dinitrogen) complexes with strong acid causes the reduction of bound dinitrogen to a (still bound) diimide or hydrazine level ligand and, when monodentate phosphines are present, to free ammonia.

Trans-$Mo(N_2)_2(dppe)_2$ reacts with HCl or HBr, but the final products differ for the two acids (740). HBr gives reactivity analogous to the tungsten complex according to reaction 71,

$$trans\text{-}Mo(N_2)_2(dppe)_2 + 2HBr \longrightarrow MoBr_2(N_2H_2)(dppe)_2 + N_2 \quad (71)$$

whereas HCl produces the dihydride dichloride according to reaction 72,

$$trans\text{-}Mo(N_2)_2(dppe)_2 + 2HCl \longrightarrow MoH_2Cl_2(dppe)_2 + 2N_2 \quad (72)$$

In the complex $MoBr_2(N_2H_2)(dppe)_2$ one of the halides is labile, and the salt $[MoBr(N_2H_2)(dppe)_2]BF_4 \cdot HCOMe_2$ can be isolated by treatment with $(NEt_4)BF_4$ in dmf. This latter complex is apparently analogous to the cation $WCl(NNH_2)(dppe)_2^+$, shown by Heath et al. (755) to contain a near-linear (Mo–N–N = 171°) hydrazido linkage. The presumed structure of the cationic Mo complex is then structure **22**:

H H
N
||
N
P P
Mo
P P
Br

22

The structure of the neutral $MoBr_2(N_2H_2)(dppe)_2$ complex remains in doubt, and Chatt et al. (740) favor a monohaptodiimide linkage (structure **23**) based largely on a 350-cm^{-1} separation of the two NH bands in

Mo—N(H)=N—H

23

the infrared. This contrasts with the 200-cm^{-1} separation for $[MoX(N_2H_2)(dppe)_2]X$ compounds. Furthermore, the tungsten analogs $WX_2(N_2H_2)(dppe)_2$ show an asymmetric multiplet in the N–H region of their NMR spectra below $-70\,°C$. These data support the unsymmetrical monohapto *N*-diimine structure, in contrast to either the known hydrazido or the (as yet) unknown symmetrical dihapto-*N*,*N*-diimide structure. Sellman (726, 756), however, has argued, based mostly on chemical shift data, that the hydrazido structure may be present in the 7-coordinated complexes as well.

One would be remiss if one did not trace down the ultimate source of the reducing power in these systems. In all cases, the complexes used are of zero-valent Mo, and strong reductants [typically Na(Hg)] are required for their formation. Thus while the conditions of the actual reduction of N_2 are relatively mild, the conditions used to prepare the reactants are not and the claim that these reactions represent the reduction of dinitrogen under mild conditions must be viewed in that light.

The initiation of the redox process by strong acid is a significant feature of this work. Proton transfer to dinitrogen is a necessary companion to electron transfer in order to produce the final product. In this case, electron transfer from Mo will not occur unless proton concentration is quite high. The results (739, 740) point to the potential significance of proton transfer in N_2 reduction and adds substance to mechanistic speculations (74) which contain distinct proton-transfer steps as opposed to hydrolysis following reduction.

Furthermore, these reactions apparently occur at monomeric Mo sites, and this observation shows that reduction in nitrogenase does not necessarily require more than one metal at the active site.

Other reactions which *trans*-$Mo(N_2)_2(dppe)_2$ undergoes involve either acylation or alkylation of the terminal nitrogen. Treatment of either trans-$Mo(N_2)_2(dppe)_2$ (754b, 757) or trans-$Mo(N_2)(RCN)(dppe)_2$ (754b) with an acyl chloride gives the reaction products shown in equation 73. Preliminary structural data (quoted in ref. 754b) on $MoCl(dppe)_2N_2COR$ reveal an acylazo complex having Mo–N–N = 171.7, N–N–C = 116.9°, Mo–N = 1.81 and N–N = 1.24Å.

$$\textit{trans}\text{-}Mo(N_2)_2(dppe)_2 + RCOCl \longrightarrow MoCl(dppe)_2N_2COR \underset{NEt_3}{\overset{HCl}{\rightleftarrows}} MoCl_2(dppe)_2N_2HCOR \quad (73)$$

The final product resembles the diimide complex formed on treating *trans*-$Mo(N_2)_2(dppe)_2$ with HBr (740).

George and Iske (752) reported photochemical reaction of *trans*-Mo-$(N_2)_2(dppe)_2$ with alkylhalides. With CH_3I the new complex $MoI(N_2CH_3)$-$(dppe)_2$ was formed. Diamantis et al. (758, 759) reported some of the same photochemical reactions and some additional unusual reactivity with thf as the reaction solvent. Upon irradiation in the presence of CH_3Br, complexes containing a tetrahydropyridizine ligand are thought to be formed (759), with the hydrocarbon chain of the ligand probably originating from the thf.

The *trans*-$Mo(N_2)_2(dppe)_2$ complex was reported (760) to be reduced by $Fe_4S_4(SEt)_4^{n-}$ cluster compounds, and it was further claimed that NH_3 could be formed upon acidification with HCl. This report (760) was not substantiated (761), and it seems clear that the small amount of NH_3 formed must come from an impurity present in the *trans*-$Mo(N_2)_2(dppe)_2$ complex used in the initial studies (760).

In a recent note, Chatt et al. (762) found that reaction of *cis*-$Mo(N_2)_2$-$(PMe_2Ph)_4$ with H_2SO_4 in CH_3OH or thf liberated ammonia. The yields are reasonable (20–36%) based on the assumption of only one dinitrogen undergoing reduction and are even better (approaching 100% reduction of one dinitrogen) for the W analogs. The difference between these results and those for *trans*-$Mo(N_2)_2(dppe)_2$ is startling. The complexes differ in two important ways. First, the stereochemistry is different (cis vs. trans), and second, the complex whose dinitrogen is fully reducible to NH_3 has monodentate ligands. This second factor may be the key. Phosphines stabilize the lower oxidation states, and the chelating dppe will not readily dissociate. Under these circumstances the Mo is not able to lose its remaining electrons and go all the way to the Mo(VI) state. Therefore, Mo cannot deliver all six of its electrons to dinitrogen, and the reduction stops at the diimide or hydrazide level. In the $Mo(N_2)_2(PPh_2Me)_4$ system, the monodentate phosphine can readily dissociate and be replaced by a harder ligand (CH_3OH or H_2SO_4). The harder ligand stabilizes the higher oxidation state and, along with protonation of the nitrogens, facilitates the electron transfer from Mo to N.

Chatt et al. (762) suggest a mechanism for the nitrogenase enzyme wherein a harder ligand (such as N or O) substituted for a softer ligand presumably S in the enzyme) to facilitate reduction of substrate. The idea is intriguing.

Recently, Brulet and van Tamelen (762a) reported that treatment of *trans*-$Mo(N_2)_2(dppe)_2$ with HCl or HBr in *N*-methylpyrrolidone solution also produced good yields of NH_3 (0.286 moles of NH_3 per mole of Mo). Other solvents, such as propylene carbonate and triethylphosphate, were effective, but to a much lower degree. The complex $MoBr_2(N_2H_2)(dppe)_2$ will also yield ammonia by this treatment and is thus considered a likely intermediate in the reduction of *trans*-$Mo(N_2)_2(dppe)_2$. It has been suggested (762a) that a second Mo may act as a reducing agent and that an Mo nitride is formed as an intermediate. It is also possible that ligand substitution in the monomeric complex (i.e., replacement of dppe by *N*-methylpyrrolidone) may facilitate reduction of bound N_2, and thus the basic reaction may be similar to that described by Chatt et al. (762). An interesting observation (762a) is that $MoBr_2(^{15}N_2H_2)(dppe)_2$ and *trans*-$Mo(N_2)_2(dppe)_2$ undergo complete isotopic exchange during 1 hr in thf at 25°C.

Recently, Chatt et al. (762b) were able to isolate the complex $MoCl_2(NNH_2)(PMe_2Ph)_3$ by using HCl in place of H_2SO_4 in a reaction which occurs quite rapidly. This complex, when treated with H_2SO_4 in CH_3OH gives NH_3 in yields comparable to those achieved with $Mo(N_2)_2(PMe_2Ph)_4$. By appropriate treatment of the analogous W complex, $W(N_2)_2(PMePh_2)_4$, a complex containing bound N_2H_3 is also prepared (and also yields NH_3 when treated with H_2SO_4 in CH_3OH). These observations, coupled with the preparation of $MoCl(N_2H)(dppe)_2$, leads to the reasonable suggestion that bound N_2H, N_2H_2 and N_2H_3 species are intermediates in the reduction of N_2 to NH_3. A mechanism is suggested in which an $Mo{-}N{-}NH_3^+$ linkage upon protonation yields NH_4^+ and an Mo—N grouping, the latter rapidly hydrolyzing to form NH_4^+ (762b)

The results of Chatt, van Tamelen, and co-workers demonstrate the feasibility of reducing N_2 to NH_3 using a system based on Mo (or W). While these results are for stoichiometric (and not catalytic) processes, they represent a step forward in the development of systems which will reduce N_2 under mild conditions (762c).

2. *(Ar)Mo(PR$_3$)$_2$N$_2$ and Related Complexes*

A second class of well-defined $Mo{-}N_2$ compounds is found in the complexes $(Arene)Mo(phosphine)_2N_2$ and those related to them. In this class, the dinitrogen is found as either a near-linear bridging ligand in dinuclear complexes or as a linear terminal group in mononuclear species. Although the first complex in this class was probably reported by Hidai (735), most of the work in this area has been done by Green and his collaborators (689, 732, 763, 764).

The scheme for the preparation and interconversion of some of the key

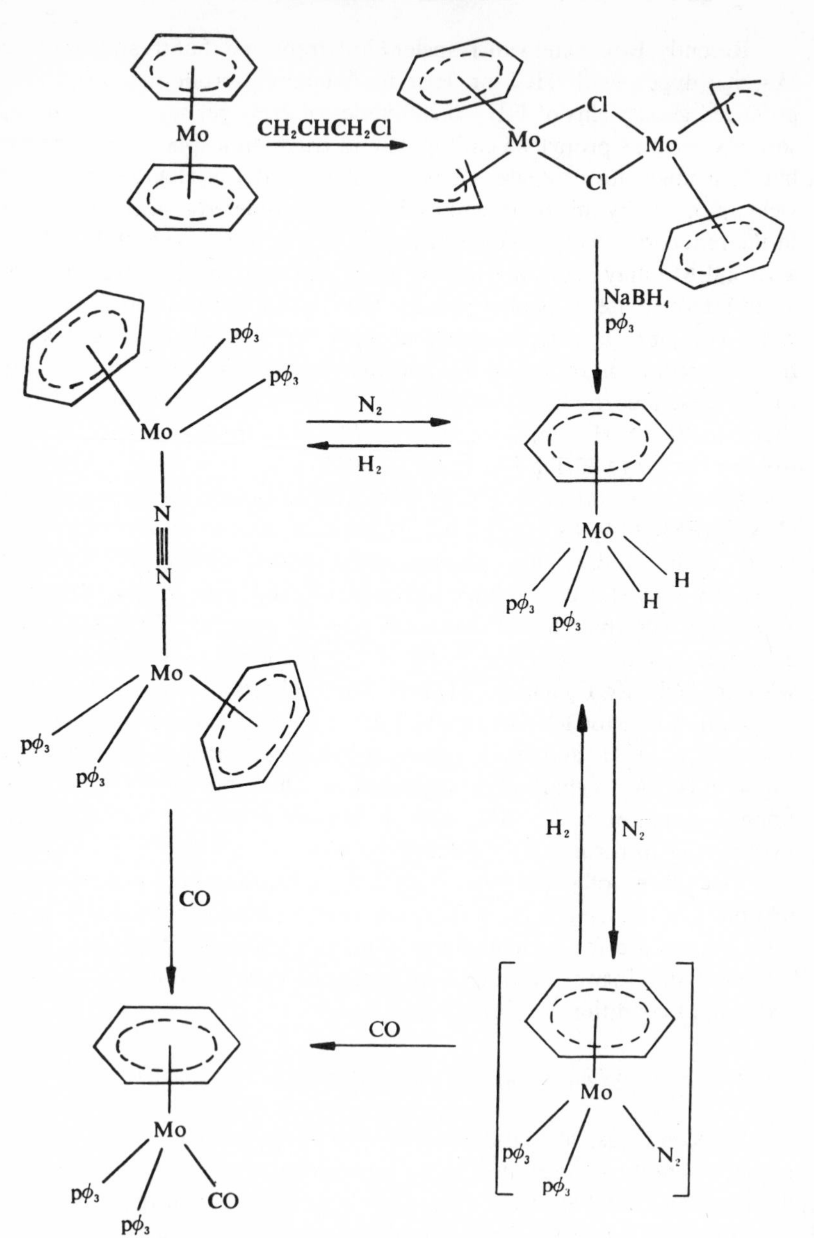

Fig. 36. Reaction scheme for $(C_6H_6)Mo(PPh_3)_2N_2$ (689, 732, 763, 764).

complexes for the benzene-triphenylphosphine series is shown in Fig. 36. With mesitylene or toluene as the arene, the mononuclear dinitrogen complexes (of the type shown in the brackets in the figure) are formed. The related binuclear complex $[(C_6H_3(CH_3)_3Mo(dmpe)]_2N_2$ prepared by Green and Silverthorn (689, 763) was structurally characterized by Forder and Prout (732). The centrosymmetric complex shown in Fig. 37 has a near-linear bridging N–N group and an N–N distance of 1.145(7) Å. Reactions 74 and 75, which form the dinitrogen complexes,

$$2ArMo(PR_3)_2H_2 + N_2 \longrightarrow 2H_2 + (ArMo(PR_3)_2)_2N_2 \quad (74)$$

$$ArMo(PR_3)_2H_2 + N_2 \longrightarrow H_2 + ArMo(PR_3)_2N_2 \quad (75)$$

clearly demonstrate the ability of N_2 to replace H_2 in the coordination sphere of Mo. This recalls that the N_2ase enzyme evolves H_2, even in the presence of excess N_2 as a substrate. It is possible that the N_2 binding site is also a dihydride site, with N_2 freeing the dihydrogen molecule upon binding. However, since this type of reaction has been chemically demonstrated for Fe (and Co), as well as for Mo, it is not clear which metal in nitrogenase (if any) can be said to display this reaction.

Treatment of the binuclear N_2 complexes with acid (HBF_4) leads to protonation of both Mo atoms, forming the binuclear N_2 bridged species (764) (reaction 76):

$$[((CH_3)_3C_6H_3)Mo(dmpe)]_2N_2 + 2H^+ \longrightarrow ((CH_3)_3(C_6H_3)MoH(dmpe)\text{-}N_2\text{-}MoH(dmpe)((CH_3)_3C_6H_3)_3{}^{+2} \quad (76)$$

The reaction of $(CH_3C_6H_5)Mo(PPh_3)_2N_2$ with $[(C_5H_5)Fe(dmpe)(Me_2CO)^+]BF_4^-$ in acetone gives a brown solid formulated as $[(C_5H_5)Fe(dmpe)N_2Mo(PPh_3)_2(CH_3C_6H_5)^+]BF_4^-$. This compound shows a dinitrogen molecule

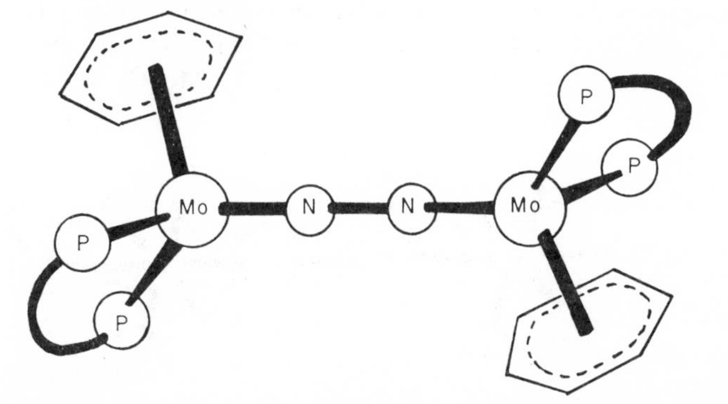

Fig. 37. The structure of $((CH_3)_3C_6H_3)Mo(dppe)(\mu\text{-}N_2)Mo(dppe)((CH_3)_3C_6H_3)$ (732). (The phenyl groups of the dppe ligands are omitted for clarity.)

bridging Fe and Mo in a complex which probably contains a linear Fe–NN–Mo linkage (689). While this is the only reportedc omplex containing the Mo-N_2Fe grouping, its significance with respect to nitrogenase is not thereby guaranteed.

In none of these complexes has the bound N_2 been caused to undergo any chemical reaction (besides, of course, complex formation and dissociation).

3. *M′–N≡N–Mo Complexes*

One potential strategy for the activation of dinitrogen which has been considered is the end-on coordination of dinitrogen at both N's. This mode of coordination would greatly reduce the strength of the N–N bond and could presumably make the coordinated N–N more susceptible to reduction. We note that while much evidence points to successful weakening of the N–N bond, there is no evidence for subsequent facile reduction of the resulting bound N_2. Chatt's group (743, 765, 766) used this approach extensively, and recently, Mercer (306, 767) determined the structure of an Mo(V) complex containing a dinitrogen bridge connecting the Mo(V) to a Re(I) group. The complex is prepared (306) according to reaction 77,

$$ReCl(N_2)(PMe_2Ph)_4 + MoCl_4(thf)_2 \longrightarrow [(PMe_2Ph)_4ClReN_2MoCl_4(OCH_3)]\cdot HCl \quad (77)$$

in CH_2Cl_2-methanol. Although the starting materials contain Re(I) and Mo-(IV), the product presumably contains Mo(V) as evidenced by its magnetic moment of 1.85 B.M. [One wonders if this complex might contain Mo(IV) coordinated by CH_3OH since EPR spectra, which would definitively characterize the Mo oxidation state, are not reported, and although it is stated that hydrogen atoms are clearly revealed in the difference map, the absence of hydrogen on coordinated oxygen was not explicitly noted.] The coordination spheres of Mo and Re are shown in Fig. 38. The key distances are Mo–N = 1.90 Å, Re–N = 1.81 Å, N–N = 1.18 Å, Mo–Cl = 2.40 Å (mean),

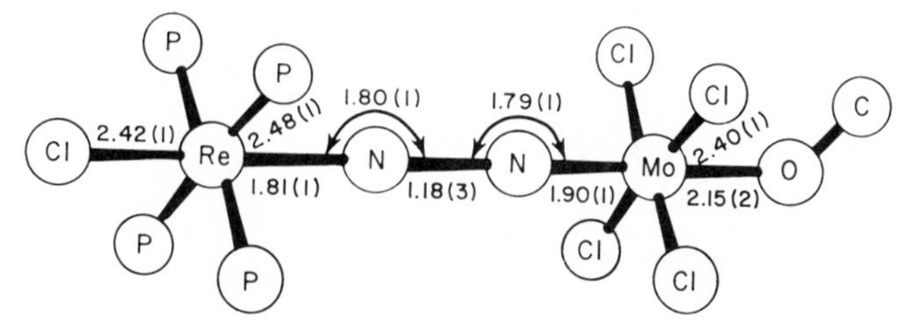

Fig. 38. The structure of $Cl(PMe_2Ph)_4ReN_2MoCl_4(OCH_3)$ in $[Cl(PMe_2Ph)_4ReN_2MoCl_4(OCH_3)]\cdot CH_3OH\cdot HCl$ (306).

Mo–O = 2.15 Å, Re–P = 2.48 Å, and Re–Cl = 2.42 Å. The bridge is distinctly linear with Re–N–N = 179.6 Å and Mo–N–N = 178.7 Å, and the Cl and P atoms are eclipsed. The bonding in this compound can be described as including substantial $M \rightarrow N_2$ donation via the pi orbitals (767). This is consistent with the lengthening of the N–N distance to 1.18 Å from, for example, the distance of 1.12 Å in $Re(PMe_2Ph)_4N_2Cl$ (768). Furthermore, the Mo–N and Re–N distances both appear to be considerably shorter than single-bond distances. The Re–N distance of 1.82 Å is substantially shorter than that in $Re(PMe_2Ph)_4N_2Cl$ (1.97 Å), indicating stronger bonding of Re to N_2 when the opposite end is coordinated by Mo.

Recently (768a), the trinuclear complex trans-$(MoCl_4(N_2ReCl(PMe_2Ph)_4)_2]$ has been prepared and structurally characterized. A near linear ClReNNMoNNReCl grouping is found and the N–N distance averages 1.28 with Mo–N = 1.99 and Re–N = 1.75Å.

These complexes provide strong confirmation that the strategy evolved is imminently successful in weakening the NN bond. It remains to be seen if chemists can take advantage of this weakness to reduce the bound dinitrogen.

4. Molybdocene Derivatives

Thomas and Brintzinger (769, 770) synthesized the molybdocene compound $[Mo(C_5H_5)_2]_x$. Despite its apparently polymeric nature, it serves as a useful starting material to produce such species as $(C_5H_5)_2MoH_2$, $(C_5H_5)_2MoCl_2$, $(C_5H_5)_2MoCO$, and $(C_5H_5)_2Mo(RCCR)$ (R = CF_3 Me, Ph) (777). The intermediate molybdocene can presumably be generated from $(C_5H_5)_2MoCl_2$ and Na(Hg), and a dinitrogen complex can be formed under pressure (Eq. 78).

$$(C_5H_5)_2MoCl_2 + Na(Hg) + N_2 \xrightarrow{350\ atm} [(C_5H_5)_2MoN_2]_x \qquad (78)$$

Acid hydrolysis of this compound does not yield ammonia, although the analogous titanocene complex does. An important difference between the molybdocene and titanocene is that while the former is a $16e^-$ species, the latter is a $14e^-$ species and thus has two potential coordination sites. This is thought to allow edge-on coordination by the dinitrogen.

Thomas also synthesized $[Mo(C_5(CH_3)_5)_2]_2$ by the pathway, Eqs. 79 to 81:

$$MoCl_5 + 3\ NaC_5(CH_3)_5 + 2NaBH_4 \longrightarrow [C_5(CH_3)_5]_2MoH_2 + 5NaCl + C_5(CH_3)_5 + B_2H_6 \qquad (79)$$

$$(C_5(CH_3)_5)_2MoH_2 + 2CHCl_3 \longrightarrow (C_5(CH_3)_5)_2MoCl_2 + 2CH_2Cl_2 \qquad (80)$$

$$2(C_5(CH_3)_5)_2MoCl_2 + 4Na(Hg) \longrightarrow [(C_5(CH_3)_5)_2Mo]_2 + 4NaCl \qquad (81)$$

This compound does not have any hydridic character and is straightforwardly assigned the dimeric molybdocene structure. It undergoes reactions totally analogous to the polymeric $[Mo(C_5H_5)_2]_x$ and reversibly reacts with molecular nitrogen as shown in Eq. 82:

$$[(C_5(CH_3)_5)_2Mo]_2 + 2N_2 \xrightarrow{250\ atm} (C_5(CH_3)_5)_2MoN_2 \tag{82}$$

Otsuka et al. (771) and Thomas (770) also prepared the related acetylene complexes in a manner analogous to the preparation of the nitrogen complex (Eq. 83),

$$(C_5H_5)_2MoCl_2 + Na(Hg) + RC{\equiv}CR \longrightarrow (C_5H_5)_2Mo\langle \overset{C-R}{\underset{C-R}{\|}} \tag{83}$$

where R = C_6H_5, CF_3, and CH_3. The unsubstituted acetylene complex was prepared by irradiation of $(C_5H_5)_2MoCO$ in the presence of acetylene (648).

Of interest to the mechanism of nitrogenase activity is the reactivity of the R = CH_3 complex, which is susceptible to attack by HCl, but not by pure water, to yield pure *cis*-2-butene according to Eq. 84:

$$(C_5H_5)_2Mo\langle \overset{C-CH_3}{\underset{C-CH_3}{\|}} + 2HCl \longrightarrow \overset{CH_3-C-H}{\underset{CH_3-C-H}{\|}} + (C_5H_5)_2MoCl_2 \tag{84}$$

Analogous ethylene complexes can be prepared and reduced to ethane upon treatment with HCl. Here, the mimicry of the enzyme is incomplete as nitrogenase will reduce acetylene, but will not bind or reduce ethylene. Thomas (770a) recently presented evidence that CF_3CN binds in a dihapto fashion to the molybdecene core. Protonation of $(C_5H_5)_2Mo(NCCF_3)$ by HCl in H_2O leads to $(C_5H_5)_2MoCl(CF_3CNH_2)^+$ which is reducible by $NaBH_4$ in basic solution to yield NH_3. If the dihapto formulation for the nitrile complex is confirmed, then this would constitute evidence that the side-on bound triple bond may be more susceptible to protonation and/or reduction than is its end-on counterpart.

5. Dinitrogen Reduction in Other Mo Systems

There have been several studies of dinitrogen reduction where the chemical systems involved are less well understood. These studies usually consist of bubbling N_2 into aqueous solutions of a "catalyst," although true catalytic activity has seldom been achieved. Haight and Scott (56), and subsequently Yatsimirskii and Pavlova (57), bubbled N_2 into 3 to 4*M* acid solutions of molybdate and found small amounts of ammonia produced, with reduction affected by Zn (56, 57), or electrolytically (56).

Schrauzer and co-workers (59–65) evolved a system utilizing BH_4^-, $Na_2Mo_2O_4(cys)_2$, and various so-called cocatalysts (such as Fe^{+3}, Fe–S clusters, ATP) in pH 9.6 borate buffer, which appears to be capable of reducing small (far less than catalytic) amounts of nitrogen. Strong evidence is reported that diimide is formed as the primary reduction product (65). The small amount of ammonia produced in this system presumably arises from the disproportion of N_2H_2 to N_2 and N_2H_4, with the N_2H_4 subsequently being reduced to NH_3. The extremely small yields in these experiments coupled with the uncertain nature of some of the solution species (see Section IV and Ref. 7) make it difficult to be certain of the reactive species. The data are, however, consistent with reduction at a monomeric Mo site, and although Mo(IV) has been claimed (65, 66) as the reduction-active state, recent work (65d) indicates that Mo(III) may be present in some of these systems. Although only low yields of NH_3 are obtained in noncatalytic reactions, these systems do efficiently catalyze the reduction of C_2H_2 to C_2H_4.

Recently, evidence from electrochemical studies has been presented (65d) which strongly implicates a monomeric Mo(III) state in the reduction of C_2H_2 by a molybdocysteine catalyst system. On the other hand (65c), recent chemical studies using a "molybdocyano" model system have been taken as supporting a monomeric Mo(IV) state in the reduction of nitrogenase substrates. Additional work will hopefully establish the detailed nature of the reactive species in these systems.

Shilov and co-workers (58, 69) reported a system active in reducing N_2 at pressures above 50 atm. This system works in aqueous or water-alcohol solutions and uses molybdenum species as "true catalysts" and Cr^{+2}, Ti^{+3}, or V^{+2} as reductants. The systems work best at high pH ($\geq$ 10.5), and although the reason for the preference for basic solutions is not yet clear, it may in part be due to enhanced reducing power of the reductants in basic media. The Ti^{+3} system has been discussed in most detail. Either hydrazine or both ammonia and hydrazine are formed, depending on the reaction conditions. In contrast to the Schrauzer model, the reaction is effectively inhibited by CO. Interestingly, a golden-yellow Mo compound [with absorption maximum reported at 24,700 cm^{-1} (ε = 2300)] is isolated from the Ti^{+3}–Mo–CO reaction mixture. The oxidation state of Mo in the complex (determined by titration with MnO_4^- after Fe^{+3} oxidation) is said to be III. The catalyst-reductant system evolves H_2 in the absence of substrates and reduces acetylene (although these two reactions may not require Mo). Finally, the presence of divalent ions such as Mg^{+2}, Ca^{+2}, and Ba^{+2} seems to enhance the yield of hydrazine and decrease H_2 production. Mechanistically, scheme 85 has been proposed:

Mo(III) — N ≡ N — Mo(III), H–O–Ti(III), H–O–Ti(III), Ti(III)–O—M—O–Ti(III) ⟶ Mo(IV)–N(H)—N(H)–Mo(IV), O–Ti, O–Ti, Ti–O—M—O–Ti (85)

This is a four-electron process, with each of the four metals transferring one electron to N_2 to give the hydrazine level product. The transfer of two protons is implied to accompany electron transfer. The divalent ion is suggested to serve the role of a spacer, keeping the Ti-OH groups from approaching each other and liberating H_2. This system, lacking any ligands other than water, hydroxo, or oxo, nevertheless shows some distinct analogies to the activity of nitrogenase. Significantly, it is found that only V can substitute for Mo in these systems, and a system containing only V^{+2} is an effective reductant of dinitrogen to hydrazine (772, 773).

B. Arylazo and Alkylazo Complexes

Studies of arylazo complexes are of potential significance as an aid in further understanding the chemistry of Mo with multiply bound NN species. As a ligand, arylazo resembles nitrosyl and has limiting formulations, RN_2^+ and RN_2^-, corresponding to linear and bent Mo–N–N linkages, respectively (777). The known complexes of Mo with this ligand (as with NO, discussed in Section IX. E) seem closer to the RN_2^+ formulation.

The arylazo-molybdenum linkage is well known in a number of carbonyl and cyclopentadienyl complexes. King and Bisnette (788, 799) prepared the red compounds $(C_5H_5)Mo(CO)_2(N{=}NAr)$ (Ar = p-$CH_3C_6H_4$, p-$CH_3OC_6H_4$, p-$NO_2C_6H_4$,1-napthalide) by reacting $(ArN_2^+)(BF_4^-)$ with $(C_5H_5)Mo(CO)_3^-$. The compound with Ar = p-tolyl reacts with triphenylphosphine or CH_3SSCH_3 according to reactions 86 and 87, respectively:

$$(C_5H_5)Mo(CO)_2(N_2Ar) + PPh_3 \longrightarrow (C_5H_5)Mo(CO)(PPh_3)(N_2Ar) \quad (86)$$

$$(C_5H_5)Mo(CO)_2(N_2Ar) + CH_3SSCH_3 \longrightarrow [(C_5H_5)Mo(SCH_3)(N_2Ar)]_2 \quad (87)$$

The Mo–N bond would seem quite strong in these compounds as it remains unbroken in these reactions. In fact, the compounds $(C_5H_5)Mo(CO)_2N_2Ar$ display thermal stability, and the Ar = *p*-$CH_3C_6H_4$ derivative sublimes at 110°C at 0.1-mm pressure (779). The complex $(C_5H_5)Mo(CO)_2(N_2C_6H_5)$ was also prepared by Green et al. (780) from reaction of phenylhydrazine ($PhNHNH_2$) with $(C_5H_5)Mo(CO)_3H$. A related alkylazo compound is prepared from the Mo hydride and trimethylsilydiazomethane according to Eq. 88:

$$(C_5H_5)Mo(CO)_3H + Me_3SiC\overset{-}{H}N{=}\overset{+}{N} \xrightarrow{-CO} (C_5H_5)(CO)_2MoN{=}NCH_2SiMe_3 \qquad (88)$$

The trimethyl silyl group may play some role in stabilizing this complex (781).

Trofimenko (782) prepared complexes of the form $RB(pz)_3Mo(CO)_2$-N_2Ar [R = H, pz (pz = 1-pyrazolyl), Ar = C_6H_5, *p*-FC_6H_4, *m*-FC_6H_4, *p*-$NO_2C_6H_4$, *o*-$CH_3C_6H_4$, etc.]. The overall pathway from $Mo(CO)_6$ is expressed by Eq. 89:

$$Mo(CO)_6 + RB(pz)_3^- \longrightarrow RB(pz)_3Mo(CO)_3^- \xrightarrow{N_2R^+} RB(pz)_3Mo(CO)_2(N{=}NAr) \qquad (89)$$

These complexes are clearly analogous to those of cyclopentadienyl and illustrate the similarity of $RB(pz)_3^-$ and $C_5H_5^-$ as ligands. The $RB(pz)_3$-complexes are remarkably stable. They do not react with PPh_3 or CH_3SSCH_3, as do their cyclopentadienyl analogs, and $(HB(pz)_3)Mo(CO)_2(N_2C_6H_5)$ withstands hot 70% H_2SO_4, 15-hr irradiation, and H_2(Pt) hydrogenation reactions. This complex does, however, react with NOCl to yield $(HB(pz)_3)$-$Mo(NO)(N_2Ph)Cl$ (783) and with halogens at −70° to form the intensely colored blue complexes $[(HB(pz)_3)Mo(N_2Ph)X]_2$ (X = Cl, Br, I) (784). These latter complexes in analogy to the nitrosyl compound $[(C_5H_5)MoNOI]_2$ (785) are formulated as containing bridging arylazo ligands, bridging halides, and an Mo–Mo bond (structure **24**).

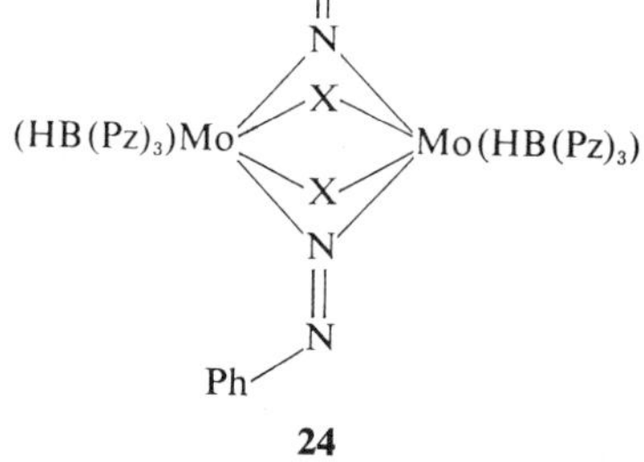

Ph–N=N, X, $(HB(Pz)_3)Mo$, $Mo(HB(Pz)_3)$, X, N=N–Ph

24

The crystal structure of $(HB(pz)_3)Mo(CO)_2(N_2Ph)$ was solved by Avitabile et al. (786) and shows the MoN_2Ph grouping to have a near-linear Mo–N–N linkage, but a bent N–N–C linkage (structure **25**).

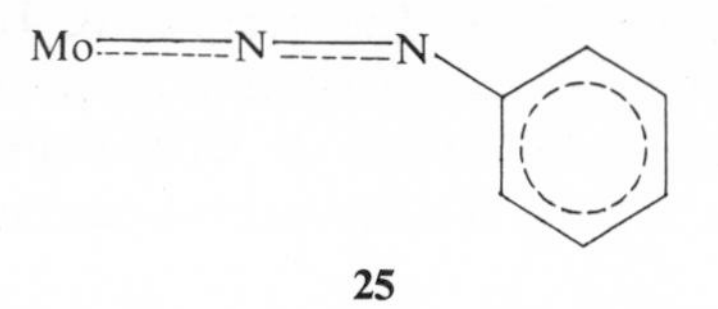

25

The key distances and angles are Mo–N = 1.83 Å, N–N = 1.21 Å, Mo-N-N = 1.74°, and N–N–C = 121°.

The infrared spectra of arylazo complexes in the NN region are complicated by coupling of ν(N–N) with vibrations in the aryl ring. Detailed studies by Sutton (787) on $(RB(pz)_3)Mo(CO)_2(N_2Ar)$ complexes led to assignment of the band to ν(NN) in the region of 1530 to 1580 cm^{-1}, whereas a band in the 1600 to 1630 cm^{-1} region is considered to have a contribution from the asymmetric –N=N–C(=C)(–H) stretch. Studies using F^{19} NMR (788) have probed the pi-acceptor nature of the PhN_2^+ ligand in a number of Mo derivatives and it is concluded that PhN_2^+ is a weaker pi acceptor than NO^+.

Arylazo and alkylazo complexes have also been prepared by reacting higher oxidation state, oxo molybdenum species with hydrazines. Bishop et al. (789) found that $MoO_2(R_2dtc)_2$ (R = Me, Et, Ph) in CH_3OH reacted with substituted hydrazines ($R'NHNH_2$) in the presence of excess NaR_2dtc to give the azo complexes $(R'NN)Mo(R_2dtc)_3$ ($R' = CH_3$, Ph, $\overset{O}{\overset{\|}{C}}Ph$). NMR spectra suggest that these complexes are structurally similar to the $MoNO(R_2dtc)_3$ complex (655, 656), and a pentagonal-bipyramidal structure with $-N{=}NR'$ at one apex is suggested. The Mo–N–N linkage is expected to be linear. The compound $Mo(NNCOPh)(R_2dtc)_3$ reacts with HCl, with the liberation of N_2 and the formation of $MoCl(R_2dtc)_3$ (reaction 90):

$$MoO_2(R_2dtc)_2 \xrightarrow[R_2dtc^-]{Ph\overset{O}{\overset{\|}{C}}-NH-NH_2} Mo(NN\overset{O}{\overset{\|}{C}}Ph)(R_2dtc)_3 \longrightarrow MoCl(R_2dtc)_3 \qquad (90)$$

Alklyazo complexes have been prepared by the photochemical alkylation of *trans*-$Mo(N_2)_2(dppe)_2$ (752, 758), and these have been discussed in the section on dinitrogen complexes.

Recently (789a) the photochemical alkylation has been accomplished with cyclohexyl iodide as the alkyl halide. The structure of the resulting alkyldiazenido derivative, trans-MoI($N_2C_6H_{11}$)(dppe)$_2$ was determined crystallographically and a near linear Mo–N–N linkage was found. The key angles are Mo–N–N = 176°, NNC = 142° and a length of 1.95Å for the Mo–N bond is reported.

The protonation of alkylazo complexes potentially leads to diimide- or hydrazide-level complexes, and these are the subject of Section IX. C.

C. Complexes of Hydrazines, Diimines, and Related N Donors

Intermediates at the diimine (also called diimide, diazene) and hydrazine levels of oxidation have been suggested many times during mechanistic speculations on nitrogen fixation (28, 29, 65, 74). Recent results strongly implicate the production of these species (in their bound forms) in key steps during the course of nitrogenase activity (37). While only a few studies on Mo complexes have been completed, the results point to some interesting mechanistic possibilities. Various techniques have found utility in the synthesis of complexes of these ligands. In some cases, diimide- or hydrazine-level complexes have been prepared by a nonredox reaction between the ligand and a suitable metal starting material. However, more often product formation involves a redox step. Hydrazine complexes have been prepared by reaction of diimide-level species with an Mo hydride, and some hydrazido complexes have been prepared by the alkylation or protonation of arylazo compounds. Likewise, some diimide-level complexes have been prepared by reduction (actually protonation and internal electron transfer) of bound dinitrogen.

Otsuka and co-workers (790) studied the reaction of $(C_5H_5)_2MoH_2$ with PhNNPh, MeOC(=O)NNC(=O)OMe, EtOC(=O)NNC(=O)OEt, and PhC(=O)NNC(=O)Ph. With azobenzene the reaction proceeds through the formation of the hydridohydrazine complex to the formation of the *N,N′*-dihaptodiimine complex. The transformation is shown in reaction 91

$$(C_5H_5)_2MoH_2 + PhN{=}NPh \longrightarrow [(C_5H_5)_2Mo(H)\text{–}N(Ph)\text{–}N(H)Ph] + RNNR \longrightarrow (C_5H_5)_2Mo(\eta^2\text{-}PhN{=}NPh) + RNHNHR \quad (91)$$

and the reaction is analogous to that of $(C_5H_5)_2MoH_2$ with olefins (791).

The conditions involve heating the reactants to 70°C in thf, but at 10 to 30°, where no appreciable net reactions occur, $(C_5H_5)_2MoH_2$ catalyzes the isomerization of *cis*-PhNNPh to *trans*-PhNNPh. This catalysis points to an interaction prior to the insertion of PhNNPh into Mo–H (790). In the case of the azodicarboxylates, the hydrido-hydrazino complex can be isolated, but the final product appears to contain an MoNNCO metallocycle (structure **26**) similar to that found for the Pt complex $(PPh_3)_2Pt(PhCONNCOPh)$ (792).

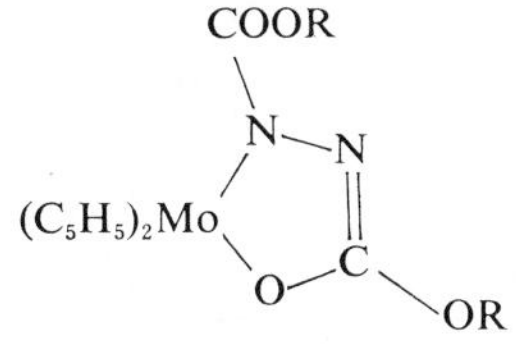

COOR

N—N

$(C_5H_5)_2Mo$

O—C

OR

26

Mechanistically, the formation of the final product can proceed in both of the above reactions via a $(C_5H_5)_2Mo$ intermediate. Azodibenzoyl reacts analogously to the azodicarboxylate. Treatment with HCl leads to $(C_5H_5)_2$-$MoCl_2$ and dibenzoylhydrazine. Scheme 92,

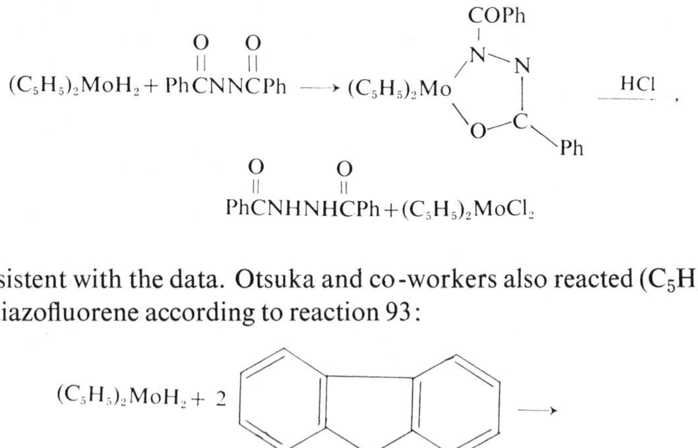

$$(C_5H_5)_2MoH_2 + PhC(=O)N{=}NC(=O)Ph \longrightarrow (C_5H_5)_2Mo[N(COPh)NC(Ph)O] \xrightarrow{HCl} PhC(=O)NHNHC(=O)Ph + (C_5H_5)_2MoCl_2 \quad (92)$$

is consistent with the data. Otsuka and co-workers also reacted $(C_5H_5)_2MoH_2$ with diazofluorene according to reaction 93:

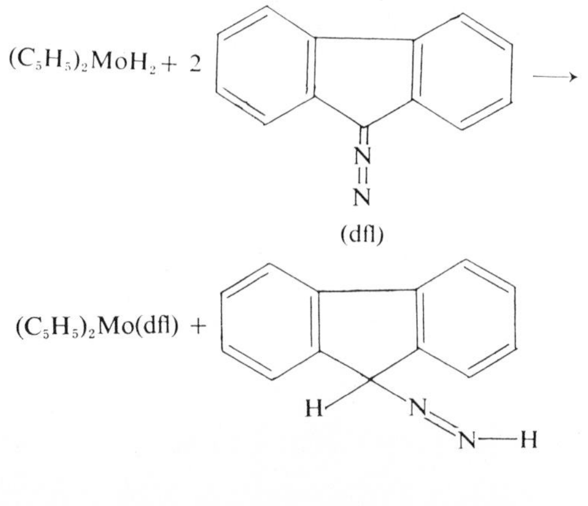

$$(C_5H_5)_2MoH_2 + 2\ (\text{dfl}) \longrightarrow (C_5H_5)_2Mo(\text{dfl}) + C_{13}H_8(H)N{=}N{-}H \quad (93)$$

(dfl)

This interesting compound has been suggested to contain a side-on bound C=N=N leakage, but its detailed formulation must await crystallographic study said to be in progress (790).

McCleverty and co-workers (793) found that complexes of the form $[(C_5H_5)MoNOX_2]$ (X = Cl, Br, I) reacted with unsymmetrically disubstituted hydrazines, $R_1R_2NNH_2$, to produce compounds of the form $((C_5H_5)MoNOX)_2(NNR_1R_2)$. A representative complex was shown to have the structure in Fig. 39. The hydrazido ligand bridges the two Mo's in an unsymmetrical fashion. The most likely bonding description for this species involves the bridging N donating four electrons (two sigma and two pi) to Mo(1) and two electrons to Mo(2), thereby satisfying the $18e^-$ requirements for both Mo's.

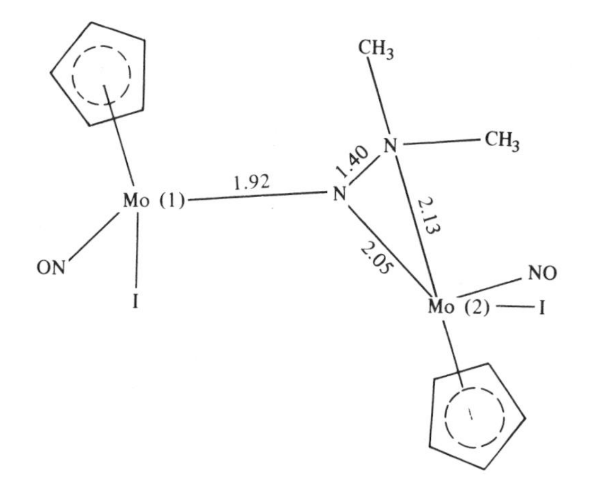

Fig. 39. The structure of $((C_5H_5)Mo(NO)X)_2(NNR_1R_2)$ (793).

Recently (793a), the complex $[(C_5H_5)Mo(NO)I(NH_2NHPh)][BF_4]$ was prepared and subjected to X-ray crystallographic analysis. The monomeric complex displays a clearly *bidentate phenylhydrazine group.* The Mo–N bond lengths to the hydrazine are 2.184 and 2.134 while the N–N distance is 1.430Å. The location of the hydrogens shows that both nitrogens are reasonably formulated as sp^3 hybridized atoms. This is the first example of a hydrazine bound in a dihapto fashion to a single metal.

Mitchell (794) investigated the reaction of hydrazine with binuclear Mo(III) and Mo(V) complexes. Polymeric structures are suggested for the reaction products, but their detailed nature is not yet clear.

The azo complexes $Mo(N_2R')(R_2dtc)_3$ (789) can be alkylated by MeI or $[Et_3O^+][BF_4^-]$ to give cationic complexes $[Mo(N_2R'R'')(R_2dtc)_3]^+$ (R″ = Me, Et). The alkyl group is found on the terminal nitrogen (i.e., the one *not* bonded to Mo). The crystal structure (794a) of $Mo(N\text{–}NEt(Ph))(S_2CN(CH_2)_5)_3$ confirms a pentagonal-bipyramidal coordination on Mo with angles Mo–N–N = 170°, N–N–C(Et) = 118°, N–N–C(Ph) = 116°, C(Et)–N–C(Ph) = 123° and distances Mo–N = 1.72 and N–N = 1.37Å. The complexes $MoCl(N_2HR')(R_2dtc)_3$ and $[Mo(N_2HR')(R_2dtc)_3^+](BF_4^-)$ are isolated from the reaction of $Mo(N_2R')(R_2dtc)_3$(R′ = CH_3, C_6H_5) with HCl and HBF_4, respectively. These perhaps resemble the acidification products of *trans*-$Mo(N_2)_2(dppe)_2$, namely, $MoBr_2(N_2H_2)(dppe)_2$ and $MoBr(N_2H_2)(dppe)_2^+$ discussed previously (740). Remarkably, the N-bound proton has not been observed in the NMR spectrum. MoO_2L_2 (L = oxine or R_2dtc) reacts with dimethylhydrazine to give $MoO(NNMe_2)L_2$, and these Mo(VI) complexes have low ν(Mo–O) at around 890 cm^{-1} [(776).

Green and Sanders reacted $(C_5H_5)_2Mo(CO_3)^-$ with $N_2CH_2CO_2Et$ and found a complex of composition $(C_5H_5)Mo(CO)_3(N_2CH_2CO_2Et)$ (795). Combined structural (796, 797) and spectroscopic (798) data show that the CH_2 group of diazoacetic ester ligand adds to one of the CO ligands, generating a metal carbene linkage which along with the nitrogen of the ligand completes a five-membered chelate ring. The structure is shown in Fig. 40.

Remarkably, this complex can be protonated or deprotonated, the deprotonated form can be methylated and the methylated form can be protonated, while the five-membered ring and carbene coordination remain

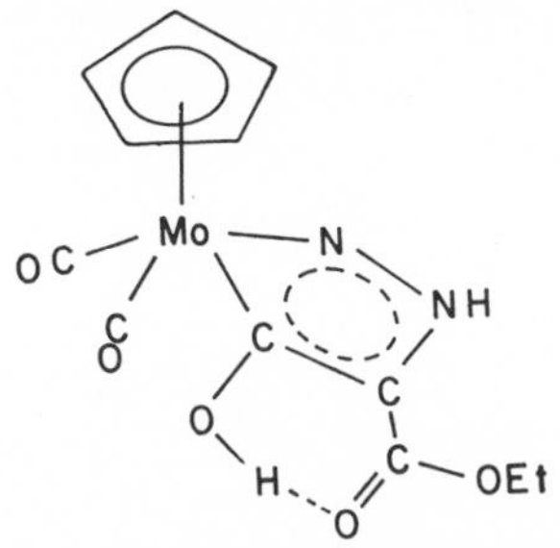

Fig. 40. The structure of $(C_5H_5)Mo(CO)_2(C(OH)C(COOEt)NHN)$ (796,797).

intact. The scheme for these interconversions is shown in Fig. 41. The structure of the protonated, methylated complex (796) has been solved crystallographically, and while there are still some grounds for uncertainty as to the position of the proton in some of the intermediates, the formulations shown are strongly suggested by the data.

The oxidation level of the NN bond in these compounds is open to question. The N–N distance of 1.35 Å in the protonated methyl derivative

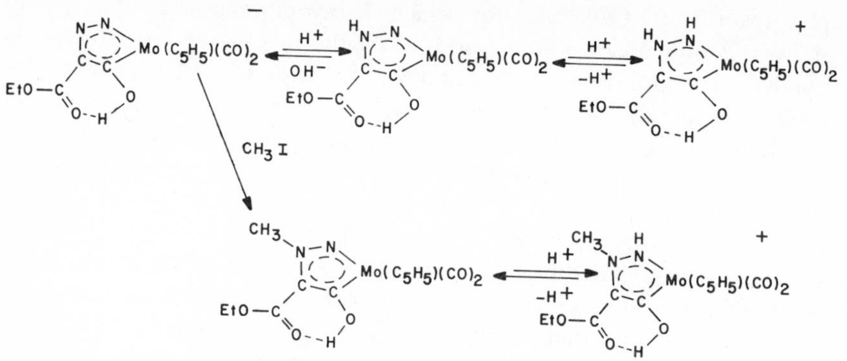

Fig. 41. The reactions of $(C_5H_5)Mo(CO)_2(C(OH)C(COOEt)NHN)$ (795, 798).

indicates some multiple bond character in this system but the delocalized binding in the five-membered ring makes further dissection of the electronic structure unproductive at this point.

The oxo Mo(IV) complex $MoOCl_2(PMe_2Ph)_3$ is found to react with PhCONHNHR (R = Ph, 1-napthyl, p-$CH_3OC_6H_4$, p-$CH_3C_6H_4$, or p-ClC_6H_4) to give complexes which are formulated as $MoCl_2(NR)(PhCON_2R)(PMe_2Ph)$ (799). The $PhCON_2R$ serves as a bidentate ligand forming a five-membered chelate ring. This complex can be described (799) by a number of resonance forms which individually define the oxidation state as Mo(II), Mo(IV), or Mo(VI). The true electronic structure embodies a linear combination of these possible forms, and the predominant contributor would be determinable from the crystal structure, which although apparently solved (799) has not been published at this time. The complex presumably contains an arylimido ligand trans to the phosphine, with chlorides cis to each other and trans to the bidentate chelate (structure **27**).

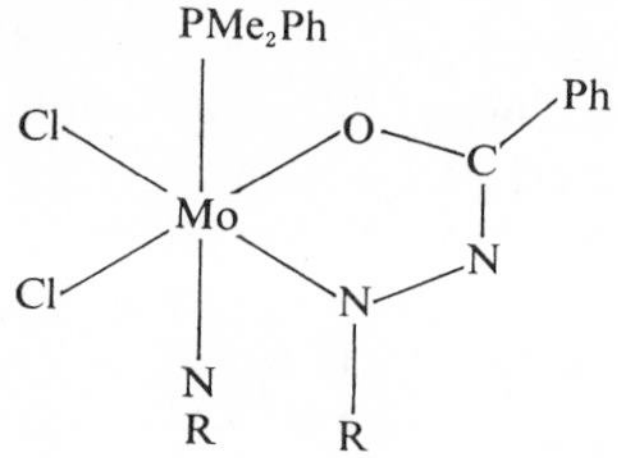

27

The diimide and hydrazido complexes $MoBr_2(N_2H_2)(dppe)_2$ and $MoBr(N_2H_2)(dppe)_2^+$ formed by treatment of $Mo(N_2)_2(dppe)_2$ (740) with HBr have been discussed previously in Section IX.A.

Treatment of $Mo(Et_2dtc)_2(CO)_2$ with N_2H_4 results in the dinuclear complex $(Mo(Et_2dtc)_2CO)_2N_2H_4$ for which X-ray diffraction analysis (799b) shows 7-coordinate Mo atoms bridged by the hydrazine. The structure is not symmetrical having Mo–N distances of 2.36 and 2.44Å while the N–N distance is found at 1.44Å.

Recently, Sellman et al. (799a) have isolated the interesting compounds μ-$N_2H_2[Mo(CO)_5]_2$, μ-$N_2H_4(Mo(CO)_5)_2$, $Mo(CO)_5N_2H_4$, and $Mo(CO)_5$-NH_3. The diimine complex μ-$N_2H_2[Mo(CO)_5]$ is prepared by H_2O_2 oxidation of $Mo(CO)_5N_2H_5$ using Cu^{+2} as a catalyst. The hydrazine and NH_3 complexes are prepared by ligand substitution from $Mo(CO)_6$ or $Mo(CO)_5$-(thf). The structures and mechanisms of interconversion are considered (799a) and it is noted that a comparative study of analogous Cr, Mo and W complexes of this type reveals the Mo species to be least stable and most reactive.

D. Mo Nitrido Complexes

Nitrido complexes are the nitrogen analogs of the ubiquitous oxo complexes. Their study holds particular appeal as they are potential models for intermediates in the reduction of dinitrogen, and the (formally) N^{3-} ligand is the deprotonated form of NH_3, the coveted product of N_2 fixation. Unfortunately, in most of the cases discussed below, the source of the coordinated N is not N_2 but rather azides, which upon two-electron reduction by a metal can liberate N_2 and leave the terminal Mo–N bond. The Mo–N bond strongly resembles the Mo–O_t bond, with N having both sigma- and pi-donor electrons available, and the high values found for ν(Mo–N) attest to its multiple-bond character. The general subject of nitride chemistry has been reviewed by Griffith (800).

Mo(VI) nitrido complexes have been claimed since 1906 when Rosenheim (801) reported the preparation of $K_3[MoO_3N]$ from MoO_3, and KNH_2 in liquid ammonia. Dehnicke and co-workers (802, 803) prepared $MoNCl_3$ by the reaction of (explosive) chloroazide with $MoCl_5$.

The crystal structure (804, 805) of $MoNCl_3$ reveals tetrameric $(MoNCl_3)_4$ units with unsymmetrical Mo–N–Mo linkages, as shown in Fig. 42. The N is multiply bound to one Mo at the shorter, average Mo–N distance of 1.66 Å, but maintains sufficient basicity to bind to another Mo at an average distance of 2.16 Å forming a near-linear Mo–N–Mo linkage (167 and 178°). The position trans to N, although shown as empty, is, in fact, occupied by a Cl from another tetramer at a Mo–Cl distance considerably longer than the Mo–Cl bonds. This structure demonstrates a strong and short Mo–N bond and the trans effect expected for this group.

The adduct of $MoNCl_3$ with triphenylphosphine, $MoNCl_3[PPh_3]_2$, and the salts $[N(C_2H_5)_4]_2[MoNCl_5]$ have also been reported (803). Bereman

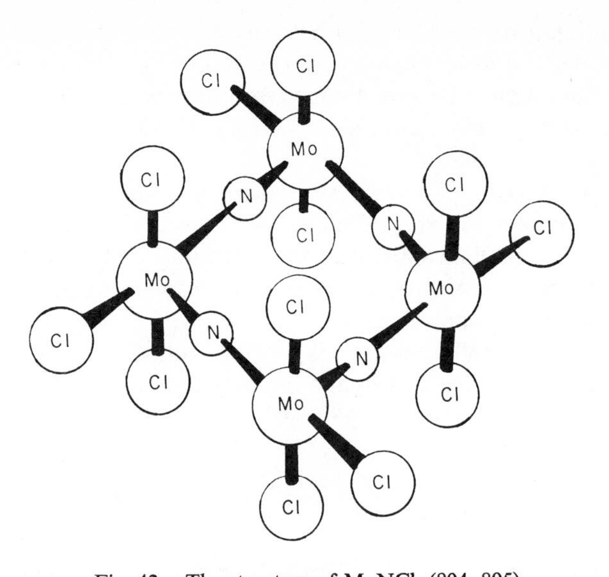

Fig. 42. The structure of $MoNCl_3$ (804, 805).

reacted (806) $MoCl_5$ or $MoCl_4$ and $NEt_4^+N_3^-$ in anhydrous CH_2Cl_2 to obtain $[NEt_4][MoNCl_4]$, although Chatt and Dilworth (807) claimed difficulty in reproducing this experiment, believing only oxo products to be obtained.

Mo(VI)-N compounds can be prepared (807) by reacting orange solutions of $MoCl_4(thf)_2$ (or $MoCl_4(CH_3CN)_2$) with $(CH_3)_3SiN_3$ and the appropriate ligand. In this manner the complexes $MoCl_3N(bipy)$, $MoCl_3N(OPPh_3)_2$, $MoN(Me_2dtc)_3$, and $MoN(Et_2dtc)_3$ are prepared. The last two compounds, each presumably 7 coordinate, can also be prepared from $MoCl(R_2dtc)_3$ (R = Me, Et) and $Si(CH_3)_3N_3$, or NaN_3 in CH_3CN.

The nitrido complexes of the form $MoN(R_2dtc)_3$ [$R_2 = (CH_3)_2$ and $(CH_2)_5$] react with sulfur or propylene sulfide in refluxing CH_3CN to give the thionitrosyl complex $MoNS(S_2CNR_2)_3$ (808). The NMR of the complex is consistent with the pentagonal-bipyramidal structure found for the nitrosyl analog, $MoNO(R_2dtc)_3$ (654–656), and the NS bond shows a strong band at 1100 cm^{-1}. The sulfur is extractable from the complexes by tributylphosphine (808).

Trimethylsilylazide can also be used in the preparation of Mo(V)-nitrido complexes. Addition of PPh_3 to the orange reaction product of $MoCl_4(thf)_2$ and trimethylsilylazide in CH_2Cl_2 yields the complex $MoCl_2N(PPh_3)_2$, which can also be prepared from $MoCl_3(thf)_3$, $Si(CH_3)_3N_3$, and PPh_3. The complex displays ν(Mo–N) at 1049 cm^{-1} and is paramagnetic, showing a magnetic moment of 1.52 B.M. and an EPR spectrum that displays both Mo and P hyperfine splitting [with $A(^{95,97}Mo) = 49$ G and $a(^{31}P) = 24.5$ G].

The ^{31}P splitting gives rise to a 1:2:1 pattern, indicating equivalent P's on the ESR time scale. The structure presumably contains 5-coordinate Mo(V), and structural elaboration of these preliminary results is awaited. The corresponding complex $MoCl_2N(bipy)$ was also reported, with $\nu(Mo–N)$ at the low value of 948 cm^{-1}.

Trans-$Mo(N_2)_2(dppe)_2$ reacts (808a) with trimethylsilylazide to give the azidonitrido complex $MoN(N_3)(dppe)_2$ which when treated with HX (X = Cl, Br) yields the imido complex $MoX_2(NH)(dppe)_2$. This latter complex reacts with Et_3N to give $MoXN(dppe)_2$ which can be treated with HX to regenerate the imido complex.

Scott and Wedd (809) reacted alkyl azides with $MoCl_4(PR_3)_2$ in an attempt to make organo-imido (organo-nitrene) compounds. However, reaction 94 is believed to obtain in dry CH_2Cl_2.

$$2\,MoCl_4(P(n\text{-}C_3H_7)Ph_2)_2 + 2p\text{-}CH_3C_6H_4SO_2N_3 \longrightarrow$$
$$2\,MoCl_4(P(n\text{-}C_3H_7)Ph_2)_2NO + 2N_2 + p\text{-}CH_3C_6H_4SSO_2C_6H_4CH_3\text{-}p \qquad (94)$$

The Mo product is monomeric in CH_2Cl_2, shows a magnetic moment of 1.78 B.M. and an EPR spectrum with $g = 1.94$ and $A(^{95,97}Mo) = 50$ G, and is formulated as containing both phosphine oxide and phosphine imine ligands, $MoCl_4(NP(n\text{-}C_3H_7)Ph)_2)(OP(n\text{-}C_3H_7)(Ph)_2)$. The infrared absorptions at 1128 and 1093 cm^{-1} are assigned to P–O and P–N stretching vibrations, respectively (809). In view of the unique formulation, this complex deserves further structural characterization.

E. Nitrosyl Complexes

NO^+ is isoelectronic with N_2, and the study of nitrosyl complexes may offer significant insight into the mode of N_2 binding and activation (see Refs. 810 and 811 for pertinent examples). Furthermore, NO is a potent inhibitor of nitrogenase, and the comparative chemistry of NO and N_2 complexes may be of use in understanding this effect. Finally, nitrosyl complexes of Mo have recently been studied as precursors to olefin disproportionation catalysts. General aspects of the chemistry of metal nitrosyls have been reviewed (811, 812).

In early work the complex $K_4[Mo(CN)_5NO]\cdot 2H_2O$ was prepared (813) by the reaction of MoO_4^{-2}, KCN, and NH_2OH. This compound was later (814) formulated as $K_4Mo(CN)_5(OH)_2(NO)$ and was presumed to contain 8-coordinate Mo(II). However, x-ray studies (815) of $K_4[Mo(CN)_5NO]$ confirmed the presence of the octahedral pentacyanonitrosyl ion with approximate C_{4v} symmetry. The short Mo–N bond [1.95(3) Å], the near-linear Mo–N–O linkage [175(3)°], and the N–O distance [1.23(4) Å] are consistent with coordinated NO^+. The CN trans to NO shows Mo–C = 2.20(4) Å compared to the 2.12 Å average for the four equatorial Mo–C's. The ion $Mo(CN)_5NO^{-4}$ is thus isoelectronic with $Mo(CO)_6$ and of similar structure. Its chemistry has not been studied in great detail, but its oxidation product, $Mo(CN)_5NO^{-3}$, has been studied by EPR (375), and the data are consistent with retention of the C_{4v} monosubstituted octahedral structure upon one-electron oxidation.

Prior to 1964 the other known nitrosyl complexes also contained cyclopentadienyl ligands (540, 816). While some very interesting studies have been done on cyclopentadienyl-nitrosyl complexes (see, e.g. Refs. 793, 817, and 818), they are not detailed in this review. Additionally, a number of complexes containing both NO and CO are known (819), but they too are not discussed.

In 1964, two papers appeared (820, 824) which provided much impetus and ground work for further studies on Mo nitrosyls. Cotton and Johnson (820) reported the synthesis of $Mo(NO)_2Cl_2$ by the action of NOCl on $Mo(CO)_6$. The soluble complexes $Mo(NO)_2L_2Cl_2$ could be prepared using $Mo(NO)_2Cl_2$ as a starting material. Also in 1964, Canziani et al. (821) reported the compound $Mo(NO)_2Cl_2(C_2H_5OH)_2$ as being formed by simultaneous bubbling of NO and HCl through a solution of $Mo(CO)_6$ in ethanol, and this complex also proved useful for subsequent syntheses.

1. $Mo(NO)_2X_2$ Compounds

The compound $Mo(NO)_2Cl_2$ was initially (820) prepared by direct reaction of NOCl and $Mo(CO)_6$ at room temperature in CH_2Cl_2. The green solid is insoluble in nondonor solvents such as C_6H_6 or CCl_4 (but will dissolve in the presence of appropriate ligands). Infrared evidence [ν(N–O)'s at 1805 and 1609 cm^{-1}] is taken to indicate a *cis*-dinitrosyl arrangement, and a polymeric structure containing dichloride bridges (with a kinked chain forced by the *cis* nitrosyls) has been suggested. With nitrosyl formulated as NO^+, the complex contains $4d^6$, Mo(0), and the presumed octahedral structure about Mo is consistent with the observed diamagnetism. The analogous $Mo(NO)_2Br_2$ and $Mo(NO)_2I_2$ compounds were prepared by Johnson (827). $Mo(NO)_2Br_2$ is prepared from $Mo(CO)_6$ and NOBr, whereas $Mo(NO)_2I_2$ is prepared from $Mo(NO)_2Br_2$ and KI in acetone. $Mo(NO)_2X_2$ [X = Cl, Br] has also been prepared (823) by the reaction of $Mo(CO)_4X_2$ (633) with NO. The above two preparative procedures both require $Mo(CO)_6$ as the starting material, as shown in Eqs. 95 and 96:

$$Mo(CO)_6 + 2NOCl \longrightarrow Mo(NO)_2Cl_2 + 6CO \quad (95)$$

$$Mo(CO)_6 + Cl_2 \xrightarrow{-2CO} Mo(CO)_4Cl_2 \xrightarrow{2NO} Mo(NO)_2Cl_2 + 4CO \quad (96)$$

An alternate preparative procedure, termed reductive nitrosation, allows the preparation of $Mo(NO)_2Cl_2$ by reaction of $MoCl_5$ and NO in CH_2Cl_2 (824), C_6H_5Cl, $CHCl_3$, or CCl_4 (825). $MoOCl_3$ can also serve as a starting material for this reaction (825).

2. *Neutral Complexes of $Mo(NO)_2L_2X_2$ and $Mo(NO)_2(LL)X_2$*

Cotton and Johnson (820, 822) demonstrated that $Mo(NO)_2X_2$ readily reacts with neutral monodentate ligands to form 6-coordinate complexes, $Mo(NO)_2L_2X_2$ (L = $CH_3C_6H_4NH_2$, $C_6H_{11}NH_2$, PPh_3, $AsPh_3$, py), which show two distinct ν(N–O) bands in the infrared. If again the nitrosyl is taken as NO^+, the complexes are formally Mo(0), $4d^6$ and are expected to have near-octahedral structures.

The complex $MoCl_2(NO)_2(PPh_3)_2$ was crystallographically investigated by Visscher and Caulton (826), as shown in Fig. 43. The *cis*-dinitrosyl arrangement suggested by infrared is confirmed, and the average NO distance of 1.19 Å would seem closer to the NO^+ formulation. The phosphines are in trans positions to minimize steric repulsions which are known to be substantial

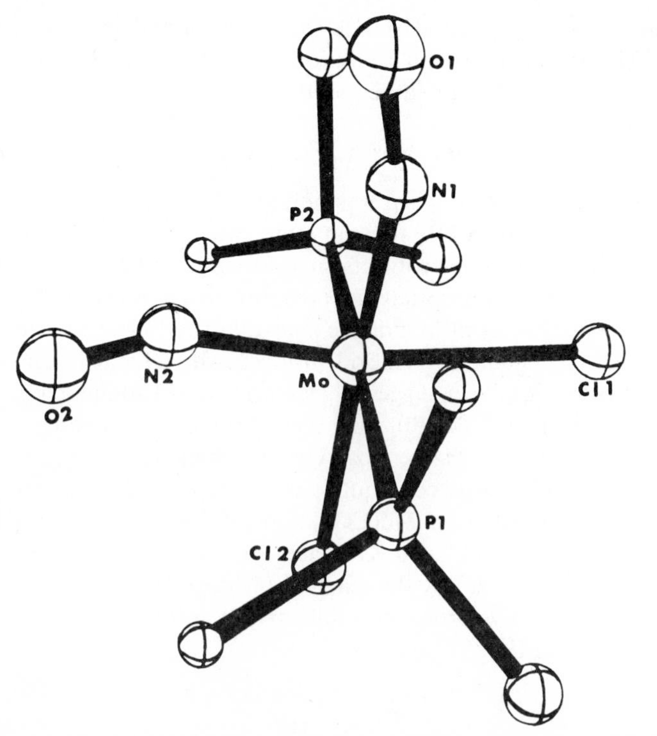

Fig. 43. The structure of $Mo(NO)_2Cl_2(PPh_3)_2$. (Reproduced with permission from ref. 826.)

for *cis* phosphines (827). The cis arrangement of the NO groups is attributed to the electronic preference for this arrangement, demonstrated by pi-interacting ligands. Significantly, the Mo–N–O linkage is substantially bent, averaging 162°. The bending of the NO ligands is correlated with the low symmetry of the complex which removes the degeneracy in the π^*(NO) levels [although more sophisticated interpretations are possible (see Ref. 811)]. It is significant that each NO is bent in the direction of one of the phosphine ligands.

ESCA studies (828) of $MoCl_2(NO)_2(PPh_3)_2$ show only one N(1s) peak, indicating equivalence of the NO ligands. The Mo binding energies and N(1s) binding energy are consistent with a small net negative charge on NO, and nitrosyl is thereby implicated as a strong pi acceptor—stronger than CO in analogous complexes. The result is consistent with a mode of binding of NO, intermediate between linear NO^+ and bent NO^- (120°), and is not inconsistent with the crystal structure analysis.

The complexes $Mo(NO)_2(EPh_3)_2X_2$ (E = P, As; X = Cl, Br) are also prepared by the action of NO on the tricarbonyl $Mo(CO)_3(EPh_3)_2X_2$ (829). Equation 97 describes the synthesis starting from $Mo(CO)_6$ (829, 830):

$$Mo(CO)_6 \xrightarrow{X_2} [Mo(CO)_4X_2] \xrightarrow{EPh_3} Mo(CO)_3X_2(EPh_3)_2 \xrightarrow{NO} Mo(NO)_2X_2(EPh_3)_2 \qquad (97)$$

While the formation of these products from $Mo(NO)_2Cl_2$ gives a stereospecific product containing *cis* NO's, *trans* L's, and *cis* Cl's, the reaction of NO with the tricarbonyl presumably gives varying amounts of the three possible isomers which contain *cis* NO, depending on the particular L and X (829, 830).

The reductive nitrosation process (824) can also be used to prepare $Mo(NO)_2Cl_2(OPPh_3)_2$ by reacting $MoCl_4(PPh_3)_2$ in hexane with NO and $C_2H_5AlCl_2$. In reacting $MoCl_5$ with NO in CH_2Cl_2 in the presence of PPh_3, the compound $Mo(NO)Cl_3(OPPh_3)$ is formed as an intermediate, with $Mo(NO)_2Cl_2(OPPh_3)_2$ as the final product (831). In general, complex mixtures are found during the reductive nitrosations (825).

Canziani et al. (821) found that ethanol in $Mo(NO)_2(EtOH)_2Cl_2$ was easily replaced by other monodentate ligands (821, 832) forming $Mo(NO)_2L_2Cl_2$ complexes (L = PPh_3, $OPPh_3$, $PhMe_2P$, py, pyO, C_6H_5NC, $C_6H_5CH_2CN$) or by neutral bidentate ligands forming $Mo(NO)_2(LL)Cl_2$ complexes [LL = bipy, phen, 1,2-bis(diphenylphosphine oxide)ethane]. With monoanionic ligands (L = acac, oxine, or R_2dtc) both C_2H_5OH and Cl are replaced by L to yield $Mo(NO)_2L_2$ (832).

$Mo(NO)_2X_2$ reacts with dpam to give the complex $Mo(NO)_2(dpam)_2X_2$ (823). NMR spectra (utilizing the methylene protons of the ligand) indicate that both dpam ligands are monodentate and are probably trans to each other. This method of preparation gives no evidence for a bidentate dpam ligand. The bidentate dpam complexes $Mo(NO)_2(dpam)X_2$ can be prepared [along with $Mo(NO)_2(dpam)_2X_2$] from the reaction of NO with $Mo(CO)_2$-$(dpam)_2X_2$. The reactivity of $Mo(CO)_2(dpam)_2X_2$ decreases in the order Cl > Br > I which is attributed to steric effects (823). With dppm the predominant product of the reaction with $Mo(NO)_2X_2$ is $Mo(NO)_2(dppm)X_2$, but low solubility has hampered detailed investigation (823).

Feltham et al. (833) studied the reaction of $Mo(NO)_2Cl_2$ with diars and en to give $MoCl_2(NO)_2(diars)$ and $MoCl_2(NO)_2(en)$. When left anaerobically in CH_3OH solutions these complexes change their color from the characteristic green to yellow. A yellow complex was isolated from the diars preparation and found to have the same composition, $Mo(NO)_2Cl_2(diars)$, as the starting compound. The presence of a single NO band at 1650 cm^{-1} and bands at 1160, 1030, and 970 cm^{-1} led Feltham to formulate this as a hyponitrite-bridged complex (structure **28**):

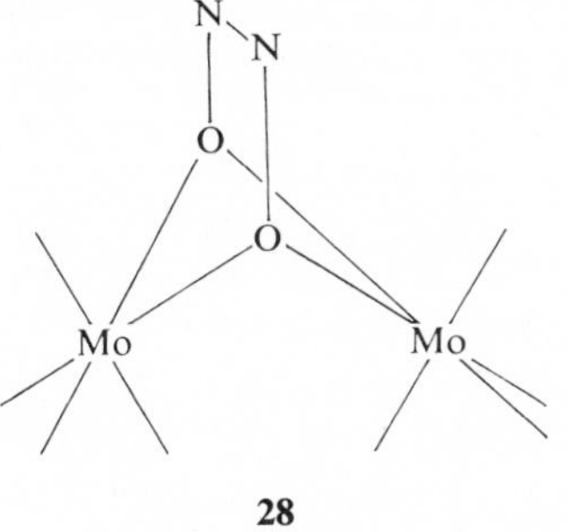

28

Trofimenko (782) prepared complexes of the form $(RB(pz)_3)MoCl(NO)_2$ and $(RB(pz)_3)MoCl_2NO$ by the reaction of $(RB(pz)_3)Mo(CO)_3^-$ with NOCl. These are analogous to complexes of C_5H_5, but seem substantially more stable. The complex $HB(Pz)_3Mo(NO)(N_2Ph)Cl$ is prepared by the reaction of $HB(Pz)_3(Mo(CO)_2N_2Ph)$ with NOCl (783).

3. *Mo(NO)Cl₃ and Derivatives*

The compound $Mo(NO)Cl_3$ can be prepared (834) (but apparently not in pure form) from NOCl and $(Mo(CO)_4Cl_2)$, which in turn is formed by the reaction of $Mo(CO)_6$ with Cl_2 at $-78°C$ (830). Like $Mo(NO)_2Cl_2$ (the probable contaminant), the species $Mo(NO)Cl_3$ is presumed to be polymeric. The green complex [after discounting the $Mo(NO)_2Cl_2$ impurity] shows a single ν(N–O) at 1590 cm^{-1}. It reacts in benzene with bipy to give $Mo(NO)Cl_3$-(bipy) and with Ph_3PO to give $MoNOCl_3(OPPh_3)_2$. The latter complex is also prepared by the reaction with Ph_3P in benzene or by reaction of $Mo(CO)_3$-

$(PPh_3)Cl_2$, $Mo(CO)_4(PPh_3)_2$, or $Mo(CO)_3(PPh_3)_3$ with NOCl in CH_2Cl_2. It is also believed to be formed as an intermediate in the reductive nitrosation of $MoCl_5$ by NO in CH_2Cl_2 in the presence of PPh_3 (831).

4. *$Mo(NO)_2(R_2dtc)_2$ and $MoNO(R_2dtc)_3$*

$Mo(NO)_2(R_2dtc)_2$ ($R = CH_3$, C_2H_5) is prepared from $Mo(NO)_2Cl_2$ and NaR_2dtc (833). The NMR and infrared data are consistent with a *cis*-octahedral structure (C_{2v}). Hindered rotation about C═N causes nonequivalence of the R groups on each R_2dtc^- ligand, and at 25° the proton NMR shows $Mo(NO)_2((CH_3)_2dtc)_2$ to have two methyl signals of equal intensity which at higher temperatures coalesce to a single line. The process is reversible and has been attributed to stereochemical lability of the 6-coordination sphere.

The complex $MoNO(R_2dtc)_3$ can be prepared by the reaction of $Mo(NO)Cl_3$ with NaR_2dtc (654, 835) and has been structurally characterized by Bernal and Brennan (655, 656). The Mo is 7-coordinate, showing a pentagonal-bipyramidal structure with NO occupying one of the apical sites. Two of the dtc ligands lie in the axial plane, while the third ligand spans axial and equatorial positions. At room temperature, the NMR of $MoNO((CH_3)_2dtc)_3$ in the CH_3 region shows a 2:2:1:1: pattern, consistent with the pentagonal-bipyramidal geometry. From 60 to 128°, interconversion of the N-methyl groups and of the two types of $(CH_3)_2dtc$ ligands occurs such that a single CH_3 resonance is observed. A mechanism is suggested in which one of the dtc ligands becomes monodentate, forming a 6-coordinate intermediate; however, the authors (834) readily admit that nonbond-breaking processes involving polytopal rearrangement of the 7-coordinate complex cannot be eliminated.

5. *Anionic Molybdenum Nitrosyls*

$Mo(NO)_2X_2$ can react with $R^+ X^-$ to produce compounds of the form $(R^+)_2(Mo(NO)_2X_4^{-2})$ [$R = Ph_4As$, X = Cl (820), $R = NEt_4$, X = Br (822); $R = N(C_4H_9)_4$, X = I (822)] and with $R^+ Y^-$ to give $(R^+)_2Mo(NO)_2X_2Y_2^{-2}$ [$R = Ph_4P$, X = Cl, Y = Br (725)]. These complexes are green and display the two characteristic infrared bands of *cis*-dinitrosyl compounds. They are presumed to have C_{2v} octahedral structures and do not differ in any substantive way (except charge) from the compounds of the previous sections.

Reaction of CH_2Cl_2 solutions of $(Ph_4P)_2[Mo(NO)_2Cl_2Br_2]$ with Na_2mnt, Na_2i-mnt, or $K_2S_2C_6Cl_4$ afforded the nitrosyl dithiolate complexes of form $[Ph_4P]_2$ $[Mo(NO)_2L_2]$ (725). The previously reported (724) $Mo(NO)_2(mnt)_2^{-2}$ preparation from $Mo(NO)_2Cl_2$ and Na_2mnt apparently gives a mixture containing $Mo(NO)_2(mnt)_2^{-2}$, $MoO(mnt)_2^{-2}$, and $Mo(mnt)_3^{-2}$, all of which are green dianions. The complexes $Mo(NO)_2(S_2C_6Cl_4)_2^{-2}$ and $Mo(NO)(mnt)_2^{-2}$ are of the dithiolene variety (695), and each displays two reversible one-electron oxidations by cyclic voltammetry. The 1,1-dithiolate complex $Mo(NO)_2$-$(i\text{-mnt})_2^{-2}$ displays no such behavior. Reaction of a solution containing Mo-$(NO)_2(S_6C_6Cl_4)_2^{-2}$ with I_2 causes a color change from red to green, presumably coincident with the formation of the monoanion $Mo(NO)_2(S_2C_6Cl_4)_2^{-1}$. This species diplays an EPR spectrum with $\langle g \rangle = 2.0031$ and $\langle a \rangle_{Mo} = 26$ G. The infrared spectra of the dianions each display two strong peaks in the 1500 to 1800 cm^{-1} region, consistent with *cis* NO groups. $\nu(NO)$ is found to decrease in the order dtc > mnt > $S_2C_6Cl_4$, and this has been rationalized in terms of the effects of charge and the differing pi-acceptor ability of the ligands (725.)

6. *Cationic Molybdenum Nitrosyls*

The useful intermediate $[Mo(NO)_2(CH_3CN)_4]$ $(PF_6)_2$ is prepared by reaction of $Mo(CO)_3(CH_3CN)_3$ and $NO^+PF_6^-$ in CH_3CN (836). The green dication has a probable *cis*-dinitrosyl structure, as evidenced by $\nu(NO)$ bands at 1863 and 1780 cm^{-1}, and is reactive toward monoanionic bidentate ligands forming the known compounds $Mo(NO)_2((C_2H_5)_2dtc)_2$ and $Mo(NO)_2(acac)_2$ which retain the *cis*-dinitrosyl unit. The complex *cis*-$Mo(NO)_2(CH_3CN)_4^{+2}$ can also be prepared by reaction of $Mo(NO)_2Cl_2$ with $AgBF_4$ in CH_3CN (837).

The NMR spectrum of $Mo(NO)_2(CH_3CN)_4^{+2}$ shows two signals of equal intensity in the methyl region (837). Exchange reactions of the nitrile ligands were studied by observing the disappearance of the NMR signal when CD_3CN is added to CH_3NO_2 solutions of the complex. Within 1 hour at 25° *one* of the two NMR signals disappears completely, while the other remains unchanged. The rate is unaffected by CD_3CN concentration, implicating a dissociative pathway. It is assumed that the labile CH_3CN ligands are trans to the two *cis* NO groups, and this is strongly supported by the fact that complex **29**,

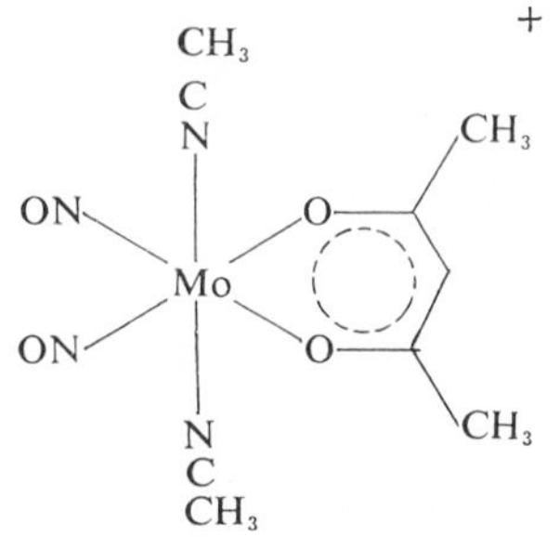

+
CH₃
C
N
CH₃
ON
O
Mo
ON
O
N
CH₃
C
CH₃

29

does not undergo exchange at 25°, while complex **30**,

```
            CH3      +
            C
            N
            |
   ON\      |      /Cl
       \    |    /
          Mo
       /    |    \
   ON/      |      \NCCH3
            |
            N
            C
            CH3
```

30

only exchanges one CH_3CN ligand. The stereospecificity of the exchange is explained by postulating stereochemical rigidity in the 16-electron, 5-coordinate intermediate $Mo(NO)_2(CH_3CN)_3^{+2}$ (837) and selective lability of the group trans to NO.

McCleverty et al. (838) prepared the complexes $Mo(NO)(CNR)_5^+$ and $Mo(NO)(CNR)_4I$ by treatment of $[(C_5H_5)Mo(NO)X_2]_2$ (X = Br, I) with the alkyl isocyanide. At room temperature the complex $[Mo(NO)(CNR)_5]I$ is reactive toward primary amines $R'NH_2$ by addition to CNR according to reaction 98:

$$MoNO(CNR)_5^+ + R'NH_2 \longrightarrow Mo(NO)(CNR)_4C(NHR)(NHR') \qquad (98)$$

wherein a formal diaminocarbene is coordinated to Mo (presumably trans to NO). At higher temperature only the substitution reaction occurs (reaction 99).

$$MoNO(CNR)_5^+ + R'NH_2 \longrightarrow MoNO(CNR)_4(NH_2R')^+ + CNR \qquad (99)$$

7. *General Considerations and the Catalyzed Olefin Disproportionation Reaction*

In general, the isolated molybdenum NO complexes seem to rigorously obey the 18-electron rule. A large number of complexes are known with 6-coordinate structures containing *cis* dinitrosyls and a variety of other ligands. If the nitrosyls are formulated as NO^+, these are all $4d^6$, Mo(O) compounds analogous to $Mo(CO)_6$ and $Mo(PF)_3)_6$ and $Mo(CN)_5NO^{-4}$. When the stoichiometry is such as to disallow 6-coordination in a monomeric structure, then 6-coordination is presumably achieved by polymerization [as in $Mo(NO)_2X_2$ and $Mo(NO)X_3$]. Known complexes containing one nitrosyl are either 6-coordinate Mo(O) species or Mo(II) species having

7-coordinate geometries, consistent with the results on other monomeric Mo(II) compounds (Section VII). On the whole, synthetic manipulation of the various types of compounds is quite straightforward, and the recent finding of catalytic activity for derivatives of these compounds is therefore cause for excitement.

Olefin disproportionation (also called dismutation or metathesis) reactions have been known for some time to be catalyzed by heterogeneous Mo or W catalysts. More recently, homogenous systems have been found by Zuech (839, 840) and others (841) to be active catalysts (841a). Some of the most active systems utilized $MoCl_2(NO)_2L_2$ (where L = PPh_3 or py) and organoaluminum compounds such as $(CH_3)_3Al_2Cl_3$ or $C_2H_5AlCl_2$. As an example of the reaction catalyzed, if 2-heptene is left in contact with the catalyst for 1 h, the resulting olefin fraction contains 2-butene (12%), 5-decene (27%), and 2-heptene (61%). The alkyl substituents on the double bonds appear to migrate to form all possible combinations, and the equilibrium expressed in Eq. 100 represents the process.

$$CH_3CH = CHCH_2CH_2CH_2CH_3 \rightleftharpoons CH_3CH = CHCH_3 + CH_3CH_2CH_2CH_2CH = CHCH_2CH_2CH_2CH_3 \quad (100)$$

Other Mo nitrosyl derivatives also appear to be catalytically active (840, 841), but non-nitrosyl Mo compounds are not nearly as effective. The reaction is reversible and attains thermodynamic equilibrium (i.e., the same final distribution of olefins is obtained independent of the proportions of starting materials, and the observed proportions agree well with those calculated from free energy data).

An incubation (or preformation) period is required after addition of the organoaluminum species to the Mo nitrosyl. Hughes (842) reported that during this period the color changed from green to brown and most significantly, at least in the case of $Mo(py)_2(NO)_2Cl_2$, the two strong ν(N–O) bands disappeared. Kinetic studies (824) are complicated by poisoning reactions which are not yet understood, but comparative reaction studies are possible. The reaction is first order with respect to catalyst, but of variable order (n = 0.7–1.7) with respect to olefin. The process is stereospecific (843) with, for example, *cis*-2-pentene, giving *cis*-2-butene and *cis*-3-hexane. An activation energy for 4-nonene equilibration is estimated at 7.0 kcal mole^{-1}, whereas activation entropy is estimated at -42 eu.

Recently (843a, 843b) a mechanism for the olefin disproportionation has been proposed involving transition metal carbenes in a chainlike reaction. These suggestions seem to fit the experimental observations while avoiding some of the problems (843a) inherent in earlier proposals.

F. Mo Hydrides and Zerovalent Phosphines

Several review articles and books on transition metal hydrides have been published in recent years (844–849). The hydride complexes of Mo which have been found to date contain either (C_5H_5), CO, or tertiery phosphines in their coordination spheres in addition to the hydrido ligand(s). The hydrido carbonyls (847) will not be discussed.

Hydrides have been suggested as present in biological systems, but there is no firm evidence, and the point has been considered for xanthine oxidase in Section II.

Hydrido cyclopentadienyl complexes of Mo have been known since the early work of Fischer (850) and Wilkinson (539) and their collaborators on $(C_5H_5)_2MoH_2$ and $(C_5H_5)_2MoH_3^+$. The dihydride is prepared from NaC_5H_5 and $MoCl_5$, but in good yields only if $Na^+BH_4^-$ is also present. The dihydride has a structure of the general $(C_5H_5)_2MoXY$ type discussed in Section V (478). A chemically accessible lone pair remains in this $4d^2$ system and has distinct chemical implications for the reactivity of the species, expecially toward protonation.

Green used $(C_6H_6)_2Mo$ as starting materials to prepare $(C_6H_6)Mo(PR_3)_3$ (R = Me, Et, etc.) complexes and found that these can be singly or doubly protonated to yield mono or dihydride derivatives (851).

The hydrido phosphine complexes of Mo (which do not contain CO, cyclopentadienyl, or arene groups) are of the type MoH_2L_4, $MoH_2Cl_2L_4$, and MoH_4L_4, where L is a tertiery phosphine (or L_2 is a ditertiery phosphine). The MoH_4L_4 complexes, where L = $MePh_2P$, PMe_2Ph, and $MoH_4(dppe)_2$, were reported by Penella (852) and are characterized by vibrational spectra [ν(Mo–H) = 1800, 1714 cm^{-1}; δ(Mo–H) = 558, 460 cm^{-1} for $(PPh_2Me)_4MoH_4$] and proton NMR spectra (1:4:6:4:1 quintet for four equivalent protons coupling to four equivalent ^{31}P, with J_{P-H} = 34 Hz, centered at τ 12.20).

Complexes of the type $MoH_2(dppe)_2$ were prepared by Frigo et al. (853) from the reaction of $Mo(acac)_3$, $Al(i\text{-}C_4H_9)_3$, dppe, and H_2 gas in benzene. The complex shows a single ν(Mo–H) at 1730 cm^{-1} in CH_2Cl_2, consistent with a *trans*-dihydrido structure. The dihydrido complex reacts slowly with N_2 at 20°C in benzene to yield *trans*-$Mo(N_2)_2(dppe)_2$. The isomeric *cis*-$MoH_2(dppe)_2$ has been claimed by Aresta and Sacco (741). The complex $MoH_2Cl_2(dppe)_2$ was prepared from $Mo(N_2)_2(dppe)_2$ by treatment with HCl (740), whereas $MoH(acac)(dppe)_2$ was prepared by reaction of $Mo(acac)_3$ and dppe with $AlEt_3$ in toluene under argon (854).

The solid-state structure of $MoH_4(PMePh_2)_4$ was determined by Guggenberger (472) and found to have a dodecahedral 8-coordination about

Mo. The hydride ligands occupy the *A* positions of the dodecahedron (elongated tetrahedron), whereas the bulkier phosphines occupy the *B* positions (flattened dodecahedron). The nonrigidity of the coordination sphere of this and related complexes has been noted (855–857) by NMR studies. Intriguing intramolecular rearrangement pathways based on the NMR and structural data have been suggested for this and other H_xML_4 molecules.

A few examples of zerovalent Mo complexes which do not contain CO, $C_5H_5^-$, or arene ligands have been reported. $Mo(dppm)_3$ was isolated as a by-product in the preparation of $Mo(N_2)_2(dppm)_2$ (736) from $Mo(acac)_3$, dppm, and $AlEt_3$ under N_2. $Mo(PH_2Ph)_6$ was prepared by reduction of $MoCl_3(thf)_3$ by Na(Hg) or Mg in the presence of excess PH_2Ph under Ar or N_2. The complex $Mo(PF_3)_6$ had been known for some time (858). It thus appears that phosphine ligands can fully replace carbonyls and form 6-coordinate MoP_6, octahedral complexes. The complex $Mo(bipy)_3$ was prepared by reduction of $Mo(bipy)_3^{+3}$ with Li_2bipy (599). Here, although the formal oxidation state is zero, the additional electrons probably reside in orbitals which have a great deal of ligand character.

G. Catalyzed Amine Oxidation

Green and co-workers (859, 860) used dicyclopentadienyl molybdenum compounds for a catalytic process in which amines were oxidized to aldehydes and ketones. The mechanism proposed is shown in Fig. 44, and the reaction is notable in that it involves a proposed hydride transfer to Mo from a carbon adjacent to a coordinated nitrogen. While the reactions involved are vastly different, this same process has been suggested by some (109) to explain the $2e^-$, 1-proton transfer known to occur in xanthine oxidase. In xanthine oxidase, however, the proposed (109) hydride transfer is accompanied by nucleophilic attack on C and the C–N bond is not broken, while in the present case, attack by H_2O (or OH^-) leads to fission of the CN bond (i.e., hydrolysis). It would be interesting to see if nitrogen heterocyclics can be hydroxylated by this system as models for the hydroxylation of purines and pyrimidines by xanthine and aldehyde oxidase.

Acknowledgment

Parts of this review were initiated while the author was on the Faculty of the State University of New York at Stony Brook where he held a Camille and Henry Dreyfus Teacher-Scholar Grant. Thanks are due to Professors Zvi Dori, John Enemark, Paul A. Ketchum, Gilbert P. Haight, Jr., Stephen J. Lippard, W. Robert Scheidt, Jack T. Spence, Henry Taube, and Raymond Weiss for informing me of results prior to publication. Special thanks are

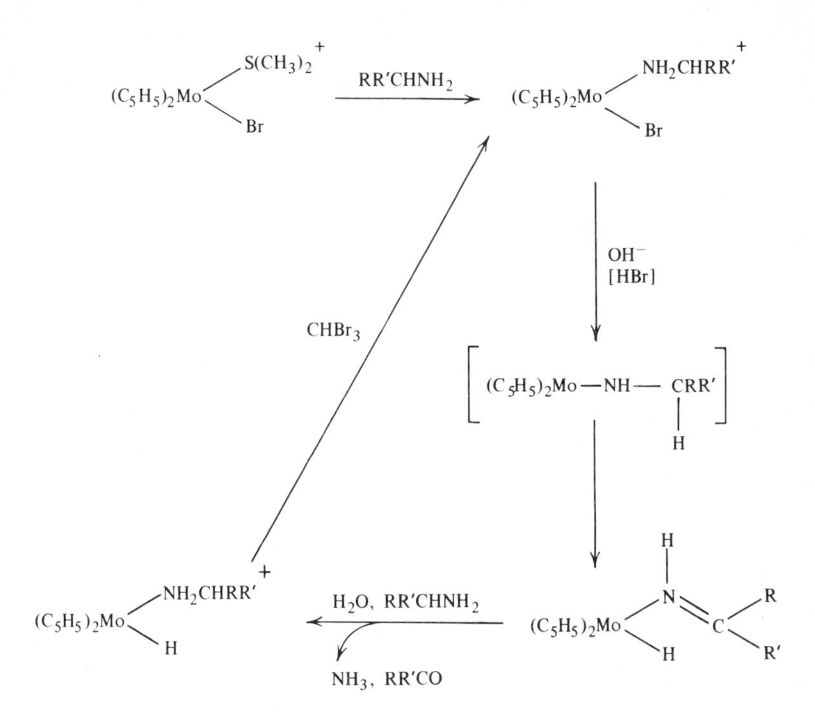

Fig 44. Proposed scheme for the catalysis of amine oxidation by Mo complexes (859, 860).

given to my colleagues James, L. Corbin, William E. Newton, John McDonald, Narayanankutty Pariyadath, and Gary Watt foı critically reading all or parts of this manuscript and for many helpful discussions. I am greatful to Steve Dunbar and Donna Current for assistance in figure preparation, Dorothy McNeil for superb typing, and Jeannette Stiefel for critical proofreading.

References

1. F. A. Cotton and G. Wilkinson, *Advanced Inorganic Chemistry*, Wiley New York, 1972.
2. C. K. Jørgensen, *Structure and Bonding*, *1*, 234 (1966).
3. D. H. Killeffer and A. Linz, *Molybdenum Compounds, Their Chemistry and Technology*, Interscience, New York, 1952.

4. P. C. H. Mitchell, *Coord. Chem. Rev.*, *1*, 315 (1966).
5. P. C. H. Mitchell, *Quart. Rev.*, 103 (1966).
6. J. T. Spence, *Coord. Chem. Rev*, *4*, 475 (1969).
7. J. T. Spence, *Metal Ions in Biological Systems*, Vol. 5, H. Sigel, Ed., Dekker, in press.
7a. F. L. Bowden in Techniques and Topics in Bioinorganic Chemistry, C. A. McAuliffe, Ed., Halsted Press, J. Wiley and Sons, N.Y. (1975) p. 205.
7b. B. Spivack and Z. Dori, *Coord. Chem. Rev.*, *17*, 99 (1975).
7c. F. A. Schroeder, *Acta Cryst.*, *B31*, 2294 (1975).
8. C. L. Rollinson, *Comprehensive Inorganic Chemistry*, Vol. III, J. C. Bailar and A. F. Trotman-Dickenson, Ed., Pergamon, Oxford, England, 1973.
8a. D. L. Kepert, *The Early Transition Metals*, Academic Press, New York, 1972.
9. J. Selbin, *Angew. Chem.* (*Intl. ed.*), *5*, 712 (1966).
10. W. P. Griffith, *Coord, Chem. Rev.*, *5*, 459 (1970).
11. R. A. Walton, *Prog Inorg. Chem.*, *16*, 1 (1972).
12. J. H. Centerford and R. Colton, *Halides of the Second and Third Row Transition Metals*, Interscience, New York, 1968.
13. J. E. Fergusson, *Halogen Chemistry*, Vol. 3, V. Gutmann, Ed., Academic, New York, 1967, p. 227.
14. K. W. Barnett and D. W. Slocum, *J. Organometal. Chem.*, *44*, 1 (1972).
15. R. H. Holm, *Endeavour*, *34*, 38 (1975).
16. J. Chatt, *Proceedings of the International Conference on Chemistry and Uses of Molybdenum*, P. C. H. Mitchell, Ed., Climax Molybdenum Co., 1974.
17. J. Levy, J. J. R. Campbell, and T. H. Blackburn, *Introductory Microbiology*, Wiley, New York, 1973.
18. E. J. Hewitt, *Biol. Rev.*, *34*, 333 (1959).
19. E. J. Underwood, *Trace Elements in Human and Animal Nutrition*, 3rd ed., Academic Press, New York, 1971.
20. S. H. Mudd, F. Irreverre, and L. Laster, *Science*, *156*, 1599 (1967).
21. H. J. Cohen, R. T. Drew, J. L. Johnson, and K. V. Rajagopalan, *Proc. Nat. Acad. Sci. U.S.A.*, *70*, 3655 (1973).
22. E. N. Mishustin and V. K. Shilnikova, *Biological Fixation of Atmospheric Nitrogen*, Macmillan, London, 1971.
23. M. J. Dilworth, *Ann. Rev. Plant Physiol.*, *25*, 81 (1974).
24. J. R. Benemann, *Science*, *181*, 164 (1973).
25. E. J. Carpenter and J. L. Culliney, *Science*, *187*, 551 (1974).
26. F. J. Bergersen and E. H. Hipsley, *J. Gen. Microbiol.*, *60*, 61 (1970).
27. J. R. Postgate, Ed., *The Chemistry and Biochemistry of Nitrogen Fixation*, Plenum Press, London, 1971.
28. R. W. F. Hardy, R. C. Burns, and G. W. Parshall, *Inorganic Biochemistry*, Vol. 2, G. L. Eichhorn, Ed., Elsevier, New York, 1973, p. 787.
29. R. W. F. Hardy, R. C. Burns, and G. W. Parshall, *Bioinorganic Chemistry, Adv. in Chem.* #100, American Chemical Society Publications, Washington D.C., 1971, p. 219.
30. *Proceedings of the International Conference on Nitrogen Fixation*, W. E. Newton and C. J. Nyman, Eds., Washington State University Press, in press.
31. J. R. Benemann and R. C. Valentine, *Adv. Microbiol. Physiol.*, *8*, 59 (1972).
32. F. J. Bergersen, *Ann. Rev. Plant. Physiol.*, *22*, 121 (1971).
33. S. L. Streicher and R. C. Valentine, *Ann. Rev. Biochem.*, *42*, 279 (1973).
34. J. R. Postgate, *Nature*, *226*, 25 (1970).

35. R. R. Eady and J. R. Postgate, *Nature, 249*, 805 (1974).
36. J. Chatt and G. J. Leigh, *Chem. Soc. Rev., 1*, 121 (1972).
37. W. A. Bulen, *Proceedings of the International Conference on Nitrogen Fixation*, June 1974, W. E. Newton and C. J. Nyman, Eds., Washington State University Press, in press.
38. M. N. Walker and L. E. Mortenson, *J. Biol. Chem., 249*, 6356 (1974).
39. W. G. Zumft, L. E. Mortenson, and G. Palmer, *Eur. J. Biochem., 46*, 525 (1974).
40. W. H. Orme-Johnson, W. H. Hamilton, T. L. Jones, M.-Y. W. Tso, and R. H. Burris, *Proc. Nat. Acad. Sci. U.S.A., 69*, 3142 (1972).
41. B. E. Smith and G. Lang, *Biochem. J., 137*, 169 (1974).
42. B. E. Smith, D. J. Lowe, and R. C. Bray, *Biochem. J., 135*, 331 (1973).
43. W. G. Zumft, W. C. Cretney, T. C. Huang, L. E. Mortenson, and G. Palmer, *Biochem. Biophys. Res. Commun., 48*, 1525 (1972).
44. G. Palmer, J. S. Multani, W. C. Cretney, W. G. Zumft, and L. E. Mortenson, *Arch. Biochem. Biophys., 153*, 325 (1972).
45. M. C. W. Evans and S. L. Albrecht, *Biochem. Biophys. Res. Commun., 61*, 1187 (1974).
46. S. L. Albrecht and M. C. W. Evans, *Biochem. Biophys. Res. Commun., 55*, 1009 (1973).
47. L. C. Davis, V. K. Shah, W. J. Brill, and W. H. Orme-Johnson, *Biochim. Biophys. Acta, 256*, 512 (1972).
48. R. H. Burris and W. H. Orme-Johnson, *Proceedings of the International Symposium on Nitrogen Fixation*, W. E. Newton and C. J. Nyman, Eds., Washington State University Press, in press.
49. A. J. Schoffman, Ph.D. Thesis, Polytechnic Institute of Brooklyn, Dissertation Abstract, *31*, 3311B (1970).
50. H. H. Nagatani and W. J. Brill, *Biochem. Biophys. Acta, 362*, 160 (1974).
51. J. R. Benemann, C. E. McKenna, R. F. Lie, T. G. Traylor, and M. D. Kamen, *Biochem. Biophys. Acta, 264*, 25 (1972).
52. J. R. Benemann, G. M. Smith, P. T. Kostel, and C. E. McKenna, *FEBS Lett., 29*, 219 (1973).
53. R. C. Burns, W. H. Fuchsman, and R. W. F. Hardy, *Biochem. Biophys. Res. Commun., 42*, 353 (1971).
54. R. C. Burns, J. T. Stasny, and R. W. F. Hardy, *Proceedings of the International Conference on Nitrogen Fixation*, W. E. Newton and C. J. Nyman, Eds., Washington State University Press, in press.
55. C. E. McKenna, *Proceedings of the International Conference on Nitrogen Fixation*, W. E. Newton and C. J. Nyman, Eds., Washington State University Press, in press.
56. G. P. Haight and R. Scott, *J. Am. Chem. Soc., 86*, 743 (1965).
57. K. B. Yatsimirskii and Y. K. Pavlova, *Dokl. Akad. Nauk. SSSR, 165*, 130 (1965).
58. A. Shilov, N. Denisov, O. Efimov, N. Shuvalov, N. Shuvalova, and A. Shilova, *Nature, 231*, 461 (1974).
59. G. N. Schrauzer and G. Schlesinger, *J. Am. Chem. Soc., 92*, 1809 (1970).
60. G. N. Schrauzer and P. A. Doemeny, *J. Am. Chem. Soc., 93*, 1608 (1971).
61. G. N. Schrauzer, G. Schlesinger and P. A. Doemeny, *J. Am. Chem. Soc., 93*, 1803 (1971).
62. G. N. Schrauzer, P. A. Doemeny, G. W. Kiefer, and R. H. Frazier, *J. Am. Chem. Soc., 94*, 3604 (1972).
63. G. N. Schrauzer, P. A. Doemeny, R. H. Frazier, and G. W. Kiefer, *J. Am. Chem. Soc., 94*, 7378 (1972).
64. G. N. Schrauzer, G. W. Kiefer, P. A. Doemeny, and H. Kisch, *J. Am. Chem. Soc., 95*, 5582 (1973).

65. G. N. Schrauzer, G. W. Kiefer, K. Tano, and P. A. Doemeny, *J. Am. Chem. Soc., 93*, 641 (1974).
65a. G. N. Schrauzer, G. W. Kiefer, K. Tano and P. R. Robinson, *J. Am. Chem. Soc., 97*, 6088 (1975).
65b. G. N. Schrauzer, *Angew. Chem. Int. Edit., 14*, 514 (1975).
65c. G. N. Schrauzer, P. R. Robinson, E. L. Moorehead and T. M. Vickrey, *J. Am. Chem. Soc., 97*, 7069 (1975).
65d. D. A. Ledwith and F. A. Schultz, *J. Am. Chem. Soc., 97*, 6591 (1975).
66. M. Ichikawa and S. Meshitsuka, *J. Am.Chem. Soc., 95*, 3411 (1973).
67. R. E. E. Hill and R. L. Richards, *Nature, 233*, 114 (1971).
68. D. Werner, S. A. Russell, and H. J. Evans, *Proc. Nat. Acad. Sci. U.S.A., 70*, 339 (1973)
69. N. T. Denisov, V. F. Shuvalov, N. I. Shuvalova, A. K. Shilova, and A. E. Shilov, *Dokl. Akad. Nauk. SSSR, 195*, 879 (1970).
70. N. T. Denisov, G. G. Terekhina, N. I. Shuvalova, and A. E. Shilov, *Kinet. and Catal., 14*, 819 (1973).
71. H. J. Sherrill, J. Hunter Nibert, and J. Selbin, *Inorg. Nucl. Chem. Lett., 10*, 845 (1974).
72. A. P. Khrushch, A. E. Shilov, and T. A. Vorontsova, *J. Am. Chem. Soc., 96*, 4987 (1974).
73. W. E. Newton, J. L. Corbin, P. W. Schneider, and W. A. Bulen, *J. Am. Chem. Soc., 93*, 268 (1971).
74. E. I. Stiefel, *Proc. Nat. Acad. Sci. U.S.A., 70*, 988 (1973).
75. G. A. Walker and L. E. Mortenson, *Biochem., 13*, 2382 (1974).
76. M. G. Yates, *Eur. J. Biochem., 29*, 386 (1972).
77. M. Walker and L. E. Mortenson, *Biochem. Biophys. Res. Commun,. 54*, 669 (1973).
78. E. I. Stiefel and J. K. Gardner, *Proceedings of the International Conference on Chemistry and Uses of Molybdenum*, P. C. H. Mitchell, Ed., Climax Molybdenum Co., 1974; and *J. Less-Common Metals., 36*, 521 (1974).
79. T. A. Verontsova and A. E. Shilov, *Kinet. and Catal., 14*, 1166 (1973).
80. E. J. Hewitt, *Plant Biochemistry*, D. H. Northcote, Ed., University Park Press, Baltimore, Maryland 1974.
81. A. Nason, *Bacteriol. Rev., 26*, 16 (1962).
82. L. E. Schrader, G. L. Ritenour, G. L. Eilrich, and R. H. Hageman, *Plant Physiol., 43*, 930 (1968).
83. A. Nason, *The Enzymes*, Vol. 7, P. D. Boyer, H. Lardy, and K. Myrbach, Eds., Academic, New York, 1963, p. 587.
84. P. Forget, *Eur. J. Biochem., 18*, 442 (1971).
85. C. H. MacGregor, C. A. Schnaitman, D. E. Normansell, and M. G. Hodgins, *J. Biol. Chem., 249*, 5321 (1974).
86. R. H. Garrett and A. Nason, *J. Biol. Chem., 244*, 2870 (1969).
87. P. Forget and D. V. Dervartanian, *Biochim. Biophys. Acta, 256*, 600 (1972).
88. I. Fridovich and P. Handler, *J. Biol. Chem., 237*, 916 (1962).
89. B. A. Notton and E. J. Hewitt, *Biochim. Biophys. Res. Commun., 44*, 702 (1971).
90. B. A. Notton and E. J. Hewitt, *Biochim. Biophys. Acta, 275*, 355 (1972).
91. G. P. Haight, Jr., *Acta Chem. Scand., 15*, 2012 (1961).
92. G. P. Haight, Jr., P. Mohilner, and A. Katz, *Acta Chem. Scand., 16*, 221 (1962).
93. G. P. Haight, Jr., and A. Katz, *Acta Chem. Scand., 16*, 569 (1962).
94. J. M. Kolthoff and I. Hodara, *J. Electroanal. Chem., 5*, 2 (1963).
95. J. T. Spence, *Arch. Biochem. Biophys., 137*, 288 (1970).
96. E. P. Guymon and J. T. Spence, *J. Phys. Chem., 70*, 1964 (1966).
96a. C. D. Garner, M. R. Hyde, F. E. Mabbs, and V. I. Routledge, *Nature 252*, 579 (1974);

J. Chem. Soc. Dalton, 1180 (1975); M. R. Hyde and C. D. Garner, *J. Chem. Soc. Dalton,* 1186 (1975).
96b. R. D. Taylor and J. T. Spence, *Inorg. Chem., 14,* 2815 (1975).
96c. P. A. Ketchum, R. C. Taylor, and D. C. Young, *Nature, 259,* 202 (1976).
97. P. Kroneck and J. T. Spence, *Biochem., 12,* 5020 (1973).
98. P. C H. Mitchell and R. J. P. Williams, *Biochim. Biophys. Acta, 86,* 39 (1964).
99. J. T. Spence, M. Heydanek, and P. Hemmerich, *Magnetic Resonance in Biological Systems,* A. Ehrenberg, B. G. Malmstrom, and T. Vänngard, Eds., Pergamon, Oxford, England, 1967, p. 269.
100. G. Colovos and J. T. Spence, *Biochem., 11,* 2542 (1972).
101. P. Hemmerich and J. T. Spence, *Flavins and Flavoproteins,* E. C. Slater, Ed., Elsevier, Amsterdam, Netherlands, 1966, p. 82.
102. J. Selbin, J. Sherrill, and C. Bigger, *Inorg. Chem., 13,* 2544 (1974).
103. M. M. Jezewska, *Eur. J. Biochem., 36,* 385 (1973).
104. V. Massey, *Iron-Sulfur Proteins,* Vol. I, W. Lovenberg, Ed., Academic, 1973, p. 301.
105. R. C. Bray and T. C. Swann, *Struct. Bonding. 11,* 107 (1972).
106. K. N. Murray, J. G. Watson, and S. Chaykin, *J. Biol. Chem., 241,* 4798 (1966).
107. L. I. Hart, M. A. McGartoll, H. R. Chapman, and R. C. Bray, *Biochem. J., 212,* 523 (1970).
108. J. S. Olson, D. P. Ballou, G. Palmer, and V. Massey, *J. Biol. Chem., 249,* 4350, 4363 (1974).
109. D. Edmondson, D. Ballou, A. Van Heuvelen, G. Palmer, and V. Massey, *J. Biol. Chem., 248,* 6135 (1973).
110. J. C. Swann and R. C. Bray, *Eur. J. Biochem., 26,* 407 (1972).
111. D. Edmondson, V. Massey, G. Palmer, L. M. Beachman III, and G. B. Elion, *J. Biol. Chem., 247,* 1597 (1972).
112. R. C. Bray, *Proceedings of the International Conference on Chemistry and Uses of Molybdenum,* P. C. H. Mitchell, Ed., Climax Molybdenum Co., 1974.
113. V. Massey, H. Komai, G. Palmer, and G. B. Elion, *J. Biol. Chem., 245,* 2837 (1970).
114. R. C. Bray and P. F. Knowles, *Proc. Roy. Soc. A, 302,* 351 (1968).
115. R. C. Bray and T. Vänngard, *Biochem. J., 114,* 725 (1969).
116. R. C. Bray, P. F. Knowles, F. M. Pick, and T. Vänngard, *Biochem. J., 107,* 601 (1968).
117. W. N. Lipscomb, *Quart. Rev. Chem. Soc., 1,* 319 (1972).
118. K. V. Rajagopalan, I. Fridovich, and P. Handler, *J. Biol. Chem., 237,* 922 (1962).
119. R. L. Felsted, A. E. Y. Chu, and S. Chaykin, *J. Biol. Chem., 248,* 2580 (1973).
120. T. A. Krenitsky, S. M. Niel, G. B. Elion, and G. H. Hitchings, *Arch. Biochem. Biophys., 150,* 585 (1972).
121. U. Branzoli and V. Massey, *J. Biol. Chem., 249,* 4339 (1974).
122. K. V. Rajagopalan and P. Handler, *J. Biol. Chem., 242,* 4097 (1967).
123. K. V. Rajagopalan and P. Handler, *J. Biol. Chem., 239,* 2022 (1964).
124. K. V. Rajagopalan and P. Handler, *J. Biol. Chem., 239,* 2027 (1964).
125. U. Branzoli and V. Massey, *J. Biol. Chem., 249,* 4347 (1974).
126. I. Fridovich, *J. Biol. Chem., 241,* 3162 (1966).
127. H. J. Cohen and I. Fridovich, *J. Biol. Chem., 246,* 359, 367 (1971).
128. H. J. Cohen, I. Fridovich, and K. V. Rajagopalan, *J. Biol. Chem., 246,* 374 (1971).
129. D. L. Kessler and K. V. Rajagopalan, *J. Biol. Chem., 247,* (1972).
130. J. L. Johnson, H. J. Cohen, and K. V. Rajagoplan, *J. Biol. Chem., 249,* 5046 (1974).
131. R. R. Eady, B. E. Smith, K. A. Cook, and J. R. Postgate, *Biochem. J., 128,* 655 (1972).
132. T. C. Huang, W. G. Zumft, and L. E. Mortenson, *J. Bacteriol., 113,* 884 (1973).
133. G. Nakos and L. E. Mortenson, *Biochem., 10,* 455 (1971).

134. D. W. Israel, R. L. Howard, H. J. Evans, and S. A. Russell, *J. Biol. Chem., 249*, 500 (1974).
135. P. Forget, *Eur. J. Biochem.*, *42*, 325 (1974).
136. C. A. Nelson and P. Handler, *J. Biol. Chem., 243*, 5368 (1968).
137. S. T. Smith, K. V. Rajagopalan, and P. Handler, *J. Biol. Chem.*, *242*, 4108 (1967).
138. D. V. Dervartanian and R. Bramlett, *Biochim. Biophys. Acta*, *220*, 443 (1970).
139. R. K. Thauer, G. Fuchs, U. Schnitker, and K. Jungermann, *FEBS Lett.*, *38*, 45 (1973).
140. J. R. Andreesen and L. G. Ljungdahl, *J. Bacteriol.*, *116*, 867 (1973).
141. D. Kleiner and R. H. Burris, *Biochim. Biophys. Acta*, *212*, 417 (1970).
142. G. Nakos and L. E. Mortenson, *Biochim. Biophys. Acta*, *227*, 576 (1971).
143. S. P. J. Albracht and E. C. Slater, *Biochim. Biophys. Acta*, *223*, 457 (1970).
144. R. J. Downey, *Biochem. Biophys. Res. Commun.*, *50*, 920 (1973).
145. J. A. Pateman, D. J. Cove, B. M. Rever, and D. B. Roberts, *Nature*, *201*, 58 (1964).
146. C. Scazzochio, *Proceedings of the International Conference on Chemistry and Uses of Molybdenum* P. C. H. Mitchell, Ed., Climax Molybdenum Co., 1974.
147. A. Nason, K.-Y. Lee, S.-S. Pan, P. A. Ketchum, A. Lamberti, and J. DeVries, *Proc. Nat. Acad. Sci. U.S.A.*, *68*, 3242 (1971).
148. K.-Y. Lee, S.-S Pan, R. Erikson, and A. Nason, *J. Biol. Chem.*, *49*, 3941 (1974).
149. K.-Y. Lee, R. Erikson, S.-S. Pan, G. Jones, F. May, and A. Nason, *J. Biol. Chem.*, *249*, 3953 (1974).
150. S.-S. Pan, R. H. Erickson, K.-Y. Lee, and A. Nason, *Proceedings of the International Conference on Nitrogen Fixation*, W. E. Newton and C. J. Nyman, Ed., Washington State University Press, in press.
151. V. L. Ganelin, N. P. L'vov, N. S. Sergeev, G. L. Shaposhnikov, and V. L. Kretovich, *Dokl. Akad. Nauk. SSSR*, *26*, 1236 (1972).
152. C. E. McKenna, N. P. L'vov, V. L. Ganelin, N. S. Sergeev, and V. L. Kretovich, *Dokl. Akad. Nauk. SSSR*, *217*, 228 (1974).
153. C. A. Adams, G. M. Warnes, and D. J. D. Nicholas, *Biochim. Biophys. Acta*, *235*, 398 (1974).
154. R. J. P. Williams and R. A. D. Wentworth, *Proceeding of the International Conference on Chemistry and Uses of Molybdenum*, P. C. H. Mitchell, Ed. Climax Molybdenum Co., 1974.
155. R. J. P. Williams, *Biochem. Soc. Trans.*, *1*, 1 (1973).
156. F. M. Pick, M. A. McGartoll, and R. C. Bray, *Eur. J. Biochem.*, *18*, 65 (1971).
157. R. C. Bray and L. S. Meriwether, *Nature, 212*, 467 (1966).
158. J.-C. Hwang and R. H. Burris, *Biochem. Biophys. Acta*, *283*, 339 (1972).
160. H. J. M. Bowen, *Trace Elements in Biochemistry*, Academic, New York 1966.
161. E. Diemann and A. Müller, *Coord. Chem. Rev.*, *10*, 79 (1973).
162. R. Stomberg, *Acta Chem. Scand.*, *24*, 2024 (1970).
163. C. G. Pierpont, H. H. Downs, and T. G. Rukavina, *J. Am. Chem. Soc.*, *96*, 5574 (1974). C. G. Pierpont and H. H. Downs, *J. Am. Chem. Soc., 88,* 2123 (1975).
163a. C. G. Pierpont and R. M. Buchanan, *J. Am. Chem. Soc., 97,* 4912 (1975).
164. G. B. Hargreaves and R. D. Peacock, *J. Chem. Soc.*, 4390 (1958).
165. M. Mercer, *Chem. Commun.*, 119 (1967).
166. B. M. Gatehouse and P. Leverett, *J. Chem. Soc. (A)*, 849 (1969).
167. H. T. Evans, Jr., *Perspect. Struct. Chem.*, *4*, 1 (1971).
168. F. A. Cotton and R. C. Elder, *Inorg. Chem.*, *3*, 397 (1964).
168a. R. J. Butcher, B. R. Penfold, H. K. J. Powell and C. J. Wilkins, Abstracts of the

Spring Meeting of the American Crystallographic Association, March 24–28, 1974 Abstract 59.

169. J. J. Park, M. D. Glick, and J. L. Hoard, *J. Am. Chem. Soc.*, *91*, 301 (1968).
170. F. A. Cotton, S. M. Morehouse, and J. S. Wood, *Inorg. Chem.*, *3*, 1603 (1964).
171. L. O. Atovmyan, Y. A. Sokolova, and V. V. Tkachey, *Dokl. Akad. Nauk. SSSR*, *195*, 1355 (1970).
172. L. O. Atovmyan, V. V. Tkachey, and T. G. Shishova, *Dokl. Akad. Nauk. SSSR*, *205*, 609 (1972).
172a. C. G. Pierpont and R. M. Buchanan, *J. Amer. Chem. Soc.*, *97*, 6451 (1975).
172b. J. E. Godfrey and J. M. Waters, *Cryst. Struct. Comm.* *4*, 5 (1975).
173. B. Kojić-Prodić, Z. Ruzić-Toroś, D. Grdenic, and L. Golic, *Acta Cryst.*, *B30*, 300 (1974).
174. B. Kamenar and M. Penavic, *Cryst. Struct. Commun.*, *2*, 41 (1973).
175. L. O. Atovmyan and Y. A. Sokolova, *Chem. Commun.*, 649 (1969).
176. A. Kopwillem, *Acta Chem. Scand.*, *26*, 2941 (1972).
177. L. Ricard, J. Estienne, P. Karagiannidis, P. Toledano, A. Mitchell, and R. Weiss, *J. Coord. Chem.*, *3*, 277 (1974).
177a. R. Weiss, personal communication.
177b. W. E. Newton, D. C. Bravard, and J. W. McDonald, *Inorg. Nucl. Chem. Lett.*, *11*, 219 (1975).
178. V. S. Sergienko, M. A. Porai-Koshits, and T. S. Khodashova, *Zh. Strukt. Khim.*, *13*, 461 (1972).
179. D. Grandjean and R. Weiss, *Bull. Soc. Chim. France*, *34*, 3049 (1967).
180. L. R. Florian and E. R. Corey, *Inorg. Chem.*, *7*, 722 (1968).
181. L. O. Atovmyan and O. N. Krasochka, *Chem. Commun.*, 1670 (1970).
182. R. H. Fenn, *J. Chem. Soc. (A)*, 1764 (1969).
183. D. Grandjean and R. Weiss, *Bull. Soc. Chim. France*, *34*, 3044 (1967).
184. I. Larking and R. Stomberg, *Acta Chem. Scand.*, *24*, 2043 (1970).
185. R. Stomberg, *Acta Chem. Scand.*, *24*, 2024 (1970).
186. J. M. LeCarpentier, R. Schlupp, and R. Weiss, *Acta Cryst.*, *B28*, 1278 (1972).
187. R. Stomberg, *Acta Chem. Scand.*, *22*, 1076 (1968).
188. J. M. LeCarpentier, A. Mitschler, and R. Weiss, *Acta Cryst.*, *B28*, 1288 (1972).
189. A. Mitschler, J. M. LeCarpentier, and R. Weiss, *Chem Commun.*, 1260 (1968).
189a. J. Dirand, L. Ricard, and R. Weiss, *Inorg. Nucl. Chem. Letters*, *11*, 661 (1975).
190. R. Stomberg, L. Trysberg, and I. Larking, *Acta Chem. Scand.*, *24*, 2678 (1970).
191. J. A. Connor and E. A. Ebsworth, *Adv. Inorg. Chem. Radiochem.*, *6*, 279 (1964).
192. N. A. Johnson and E. S. Gould, *J. Org. Chem.*, *39*, 407 (1974).
193. C.-C. Su, J. W. Reed, and E. S. Gould, *Inorg. Chem.*, *12*, 337 (1973).
194. Molybdenum Catalyst Bibliography, Climax Molybdenum Company, New York, 1950–1964, 1964–1967, 1968–1969.
195. D. S. Honig and K. Kustin, *Inorg. Chem.*, *11*, 65 (1972).
195a. J. J. Cruywagen and E. F. C. H. Rohwer, *Inorg. Chem.*, *14*, 3136 (1975).
196. L. Pauling, *The Nature of the Chemical Bond*, Cornell University Press, Ithaca, New York, 1960.
197. J. Aveston, E. W. Anacker, and J. S. Johnson, *Inorg. Chem.*, *3*, 735 (1974).
198. J. Burclová, J. Prášilova, and P. Benes, *J. Inorg Nucl. Chem.*, *35*, 909 (1973).
199. P. M. Harrison and T. G. Hoy, *Inorganic Biochemistry*, G. L. Eichhorn, Ed., Elsevier, New York, 1973, p. 253.
200. A. Müller, E. Diemann, and C. K. Jørgensen, *Structure and Bonding*, *14*, 23 (1973).

201. A. Müller, H. Schultze, W. Sievert, and N. Weinstock, *Z. Anorg. Allg. Chem.*, *403*, 310 (1974).
202. G. Gattow and W. Flindt, *Naturwissenschaften*, *57*, 245 (1970).
203. P. J. Aymonino, A. C. Ranade, E. Diemann, and A. Müller, *Z. Anorg. Allg. Chem.*, *371*, 300 (1969).
204. P. J. Aymonino, A. C. Ranade, and A. Muller, *Z. Anorg. Allg. Chem.*, *371*, 295 (1969).
205. A. Müller and W. Sievert, *Z. Anorg. Allg. Chem.*, *403*, 267 (1974).
206. A. Müller, H. Schulze, W. Sievert, and N. Weinstock, *Z. Anorg. Allg. Chem.*, *403*, 310 (1974).
207. A. Müller, E. Diemann, and H. H. Heinsen, *Chem. Ber.*, *104*, 975 (1971).
208. A. Müller, E. Ahlborn, and H. Heinson, *Z. Anorg. Allg. Chem.*, *386*, 102 (1971).
209. E. Königer-Ahlborn and A. Müller, *Angew. Chem. (Intl. Ed).*, *13*, 672 (1974).
210. D. Coucouvanis, *Prog. Inorg. Chem.*, *11*, 233 (1969).
211. B. Spivack and Z. Dori, *Chem. Commun.*, 1716 (1970).
212. A. Bartecki and D. Dembicka, *Inorg. Chim. Acta*, *7*, 610 (1973).
213. G. M. Clark and W. P. Doyle, *J. Inorg. Nucl. Chem.*, *28*, 381 (1966).
214. A. Müller, E. Diemann, F. Neumann, and R. Menge, *Chem. Phys. Lett.*, *16*, 521 (1972).
215. N. Weinstock, H. Schulze, and A. Müller, *J. Chem. Phys.*, *59*, 9 (1973).
216. F. A. Cotton and R. M. Wing, *Inorg. Chem.*, *4*, 867 (1965).
217. H. Siebert, *Z. Anorg. Allg. Chem.*, *275*, 225 (1954).
218. R. H. Busey and O. L. Keller, Jr., *J. Chem. Phys.*, *44*, 215 (1964).
219. F. Mathey and J. Bensoam, *Tetrahedron*, *27*, 3965 (1971).
220. B. G. Ward and F. E. Stafford, *Inorg. Chem.*, *7*, 2569 (1968).
221. T. V. Iorns and F. E. Stafford, *J. Am. Chem. Soc.*, *88*, 4819 (1966).
222. C. G. Barraclough and D. J. Kew, *Austral. J. Chem.*, *23*, 2387 (1970).
223. M. L. Larson and F. W. Moore, *Inorg. Chem.*, *5*, 801 (1966).
224. M. G. B. Drew, G. W. A. Fowles, D. A. Rice, and K. J. Shanton, *J. Chem. Soc. Chem. Commun.*, 614 (1974).
225. A. J. Edwards and B. R. Stevenson, *J. Chem., Soc. (A)*, 2503 (1968).
226. A. Beuter and W. Sawodny, *Angew. Chem (Intl. Ed.)*, *11*, 1020 (1972).
227. Y. A. Buslaev, Y. V. Kokunov, V. A. Bochkareva, and E. M. Shustorovich, *Zh. Strukt. Khim*, *13*, 526 (1972) through *Chem. Abst.*, *77*, 95042*S*.
228. C. G. Barraclough and J. Stals, *Austral. J. Chem.*, *19*, 741 (1966).
229. H. L. Krauss and W. Huber, *Chem. Ber.*, *94*, 2864 (1961).
230. S. M. Horner and S. Y. Tyree, Jr., *Inorg. Chem.*, *1*, 122 (1962).
231. W. M. Carmichael and D. A. Edwards, *J. Inorg. Nucl. Chem.*, *30*, 2641 (1968).
232. H. M. Newmann and N. C. Cook, *J. Am. Chem. Soc.*, *79*, 3026 (1957).
233. E. Wendling, *Bull. Soc. Chim. France*, *427* (1965).
234. Y. A. Buslaev and R. L. Davidovich, *Russ. J. Inorg. Chem.*, *10*, 1014 (1965).
235. W. P. Griffith and T. D. Wickins, *J. Chem. Soc. (A)*, *675* (1967).
236. B. J. Brisdon and D. A. Edwards, *Inorg. Nucl. Chem. Lett.*, *10*, 301 (1974).
237. N. Pariyadath and E. I. Stiefel, to be published.
237a. N. Pariyadath, W. E. Newton, and E. I. Stiefel, Submitted for publication.
238. J. F. Allen and H. M. Newmann, *Inorg. Chem.*, *3*, 1612 (1964).
239. F. W. Moore and R. E. Rice, *Inorg. Chem.*, *7*, 2511 (1968).
240. M. M. Jones, *J. Am. Chem. Soc.*, *81*, 3188 (1959).
241. H. Gehrke, Jr., and J. Veal. *Inorg. Chim. Acta*, *4*, 623 (1969).

242. K. Yamanouchi and S. Yamada, *Inorg. Chim. Acta*, *9*, 161 (1974).
243. T. J. Pinnavaia and W. R. Clements, *Inorg. Nucl. Chem. Lett.*, *7*, 1127 (1971).
244. B. M. Craven, K. Ramey, and W. B. Wise, *Inorg. Chem. 10*, 2626 (1971).
245. A. F. Isbell and D. T. Sawyer, *Inorg. Chem.*, *10*, 2449 (1971).
246. T. Dupuis and C. Duvall, *Anal. Chim. Acta*, *4*, 173 (1950).
247. A. I. Vogel, *A Textbook of Quantitative Inorganic Analysis*, 3rd ed., Wiley, New York, 1961.
248. L. W. Amos and D. T. Sawyer, *Inorg. Chem*, *13*, 78 (1974).
249. A. F. Isbell, Jr., and D. T. Sawyer, *Inorg. Chem.*, *10*, 2449 (1972).
250. L. Malatesta, *Gazz. Chim. Ital.*, *69*, 752 (1939).
251. L. Malatesta, *Gazzetta*, *69*, 408 (1939).
252. F. W. Moore and M. L. Larson, *Inorg. Chem.*, *6*, 998 (1967).
253. R. N. Jowitt and P. C. H. Mitchell, *J. Chem. Soc.* (A), 1702 (1970).
254. W. E. Newton, J. L. Corbin, D. C. Bravard, J. E. Searles, and J. W. McDonald, *Inorg. Chem.*, *13*, 1100 (1974).
255. R. Colton and G. G. Rose, *Austral. J. Chem.*, *23*, 1111 (1970).
256. R. Barral, C. Bocard, I. Seree de Roch and L. Sajus, *Kinet. and Catal.*, *14*, 130 (1973).
257. N. Nieuwpoort, *Proceedings of the International Conference on Chemistry and Uses of Molybdenum*, P. C. H. Mitchell, Ed., Climax Molybdenum Co., 1974.
258. W. E. Newton, J. W. McDonald, and J. L. Corbin, to be published. W. E. Newton, private communication.
258a. W. E. Newton, G. J.-J. Chen., and J. W. McDonald, Submitted for publication.
259. H. Diebler and R. E. Timms. *J. Chem. Soc. (A)*, 273 (1971).
260. P. F. Knowles and H. Diebler, *Trans. Farad. Soc.*, *64*, 977 (1968).
261. K. Kustin and S. Liu, *J. Am. Chem. Soc.*, *95*, 2487 (1973).
262. W. F. Marzluff, *Inorg. Chem.*, *3*, 395 (1964).
263. R. L. Pecsok and D. T. Sawyer, *J. Am. Chem. Soc.*, *78*, 5496 (1956).
264. W. P. Griffith and T. W. Wickins, *J. Chem. Soc.*, 400 (1968).
265. R. S. Taylor, P. Gans, P. F. Knowles, and A. G. Sykes, *J. Chem. Soc. Dalton*, 24 (1972).
266. R. J. Kula, D. T. Sawyer, S. I. Chan, and C. M. Finley, *J. Am. Chem. Soc.*, *85*, 2930 (1963).
267. S. I. Chan, R. J. Kula, and D. T. Sawyer, *J. Am. Chem. Soc.*, *86*, 377 (1964).
268. D. S. Honig and K. Kustin, *J. Am. Chem. Soc.*, *95*, 6525 (1973).
269. N. A. Kostromina, *Russ. Chem. Rev.*, *42*, 261 (1973).
270. R. H. Petit, B. Briat, A. Müller, and E. Diemann, *Mol. Phys.*, *27*, 1373 (1974).
271. G. Kruss, *Ann. Chem.*, *225*, 29 (884).
272. A. Müller, E. Diemann, A. C. Ranade, and P. J. Aymonino, *Z. Naturforsch.*, *24b*, 1247 (1969).
273. E. Diemann and A. Müller, *Spectrochim. Acta*, *26A*, 215 (1970).
274. A. Müller, B. Krebs, O. Glemser, and E. Diemann, *Z. Naturforsch.*, *22b*, 1235 (1967).
275. A. Müller, W. Rittner, and G. Nagarajan, *Z. Physik. Chem.*, *54*, 229 (1967).
276. M. Cousins and M. L. H. Green, *J. Chem. Soc.*, 1567 (1964).
277. A. Müller, V. V. K. Rao, and E. Diemann, *Chem. Ber.*, *104*, 461 (1971).
278. Y. Yamanouchi and S. Yamada, *Inorg. Chim. Acta*, *9*, 83 (1974).
279. A. Kay and P. C. H. Mitchell, *J. Chem. Soc. (A)*, 2421 (1970).
280. M. Naarova, J. Podlahava, and J. Podlaha, *Coll. Czech. Chem. Commun.*, *33*, 1991 (1968).
281. W. P. Griffith, *J. Chem. Soc. (A)*, 211 (1969).

282. R. Kergoat and J. E. Guerchais, *Rev. Chim. Minerale*, *10*, 583 (1973).
283. J. I. Gelder, J. H. Enemark, G. Wolterman, D. A. Boston, and G. P. Haight, *J. Am. Chem. Soc.*, *97*, 1616 (1975).
284. D. Grandjean and R. Weiss, *Bull. Soc. Chim. France*, 3054 (1967).
285. L. O. Atovmyan, O. A. Dyachenko and E. B. Lobkovskii, *Zh. Struct. Khim.*, *11*, 469 (1970).
286. J. G. Scane, *Acta. Cryst.*, *23*, 85 (1967).
287. P. M. Boorman, C. D. Garner, F. E. Mabbs, and T. J. King, *J. Chem. Soc. Chem. Commun.*, 663 (1974).
288. B. J. Corden, J. A. Cunningham, and R. Eisenberg, *Inorg. Chem.*, *9*, 356 (1970).
289. M. G. B. Drew, G. M. Egginton, and J. D. Wilkins, *Acta Cryst.*, *B30*, 1895 (1974).
290. D. E. Sands and A. Zalkin, *Acta Cryst.*, *12*, 723 (1959).
291. A. B. Blake, F. A. Cotton, and J. S. Wood, *J. Am. Chem. Soc.*, *86*, 3024 (1964).
292. J. R. Knox and C. K. Prout, *Acta Cryst.*, *B25*, 2281 (1969).
293. F. A. Cotton and S. M. Morehouse, *Inorg. Chem.*, *4*, 1377 (1965).
294. J. R. Knox and C. K. Prout, *Acta Cryst.*, *B25*, 1857 (1969).
295. L. T. J. Delbaere and C. K. Prout, *Chem. Commun.*, 162 (1971).
296. M. G. B. Drew and A. Kay, *J. Chem. Soc. (A)*, 1846 (1971).
297. D. L. Stevenson and L. F. Dahl, *J. Am. Chem. Soc.*, *89*, 3721 (1967).
298. B. Spivack, A. P. Gaughan, and Z. Dori, *J. Am. Chem. Soc.*, *93*, 5266 (1971).
299. B. Spivack and Z. Dori, *J. Chem. Soc. Dalton*, 1173 (1973).
300. D. H. Brown and J. A. D. Jeffreys, *J. Chem. Soc. Dalton*, 732 (1973).
301. M. G. B. Drew and A. Kay, *J. Chem. Soc. Dalton*, 1851 (1971).
302. J. I. Gelder and J. H. Enemark, private communication from J. H. Enemark.
303. J. F. Dahl, P. D. Frisch, and G. R. Gust, *Proceedings of the First International Conference on the Chemistry and Uses of Molybdenum,* P. C. H. Mitchell, Ed., Climax Molybdenum Co., 1974, p. 134.
304. B. Spivack, Z. Dori, and E. I. Stiefel, *Inorg. Nucl. Chem. Lett.*, *11,* 501 (1975).
305. L. D. C. Bok, J. G. Leipoldt, and S. S. Basson, *Acta Cryst.*, *B26*, 684 (1970).
306. M. Mercer, *J. Chem. Soc. Dalton.*, 1637 (1974).
307. R. V. G. Ewens and M. W. Lister, *Trans. Farad. Soc.*, *34*, 1358 (1938).
308. J. Donohue, *Inorg. Chem.*, *4*, 921 (1965).
309. F. A. Cotton, S. M. Morehouse, and J. S. Wood, *Inorg. Chem.*, *4*, 922 (1965).
310. R. M. Wing and K. P. Callahan, *Inorg. Chem.*, *8*, 2303 (1969).
310a. L. Ricard, C. Martin, R. Wiest, and R. Weiss, *Inorg. Chem.*, *14*, 2301 (1975).
310b. B. M. Gatehouse, E. K. Nunn, J. E. Guerchais and R. Kergoat, *Inorg. Nucl. Chem. Lett.*, *12*, 23 (1976).
311. B. Spivack and Z. Dori, *Proceedings of the First International Conference on Chemistry and Uses of Molybdenum*, P. C. H. *Molybdenum,* Ed., Climax Molybdenum Co. 1974.
312. J. A. Beaver and M. G. B. Drew, *J. Chem. Soc. Dalton*, 1376 (1973).
312a. L. J. DeHayes, H. C. Faulkner, W. H. Doub, Jr., and D. T. Sawyer, *Inorg. Chem.*, 14, 2110 (1975).
313. C. K. Jørgensen, *Acta Chem. Scand.*, *11*, 73 (1957).
314. H. B. Gray and C. R. Hare, *Inorg. Chem.*, *1*, 363 (1962).
315. C. J. Ballhausen and H. B. Gray, *Inorg. Chem.*, *1*, 111 (1962).
316. R. A. D. Wentworth and T. S. Piper, *J. Chem. phys.*, *41*, 3885 (1964).
316a. C. D. Garner, I. H. Hillier, J. Kendrick, and F. E. Mabbs, *Nature*, *258*, 134 (1975).
317. H. So and M. T. Pope, *Inorg. Chem.*, *11*, 1441 (1974).
318. S. M. Horner and S. Y. Tyree, *Inorg. Chem.*, *2*, 568 (1963).

319. B. J. Brisden, D. A. Edwards, D. H. Machin, K. S. Murray, and R. A. Walton, *J. Chem. Soc (A)*, 1825 (1967).
320. T. M. Brown, D. K. Pings, L. R. Lieto, and S. J. DeLong, *Inorg. Chem.*, *5*, 1695 (1966).
321. C. K. Jørgensen, *Mol. phys.*, *2*, 309 (1959).
322. J. G. Scane and R. M. Stephens, *Proc. Phys. Soc.*, *92*, 833 (1967).
323. O. Piovesana and C. Furlani, *Inorg. Nucl. Chem. Lett.*, *3*, 535 (1967).
324. E. A. Allen, B. J. Brisdon, D. A. Edwards, G. W. A. Fowles, and R. G. Williams, *J. Chem. Soc.*, 4649 (1963).
324a. P. M. Boorman, C. D. Garner and F. E. Mabbs, *J. Chem. Soc. Dalton,* 1299 (1975).
324b. C. D. Garner, M. R. Hyde, F. E. Mabbs, and V. I. Routledge, *J. Chem. Soc. Dalton,* 1175 (1975).
325. G. P. Haight, Jr., *J. Inorg. Nucl. Chem.*, *24*, 663 (1962).
326. M. G. B. Drew and I. B. Tomkins, *J. Chem. Soc. (D)*, 22 (1970).
327. D. P. Rillema and C. H. Brubaker, Jr., *Inorg. Chem.*, *9*, 397 (1970).
328. K. Feenan and G. W. A. Fowles, *Inorg. Chem.*, *4*, 310 (1965).
329. D. A. Edwards, *J. Inorg. Nucl. Chem.*, *27*, 303 (1965).
330. S. M. Horner and S. Y. Tyree, Jr., *Inorg. Chem.*, *1*, 122 (1962).
331. H. E. Pence and J. Selbin, *Inorg. Chem.*, *8*, 353 (1969).
332. A. V. Butcher and J. Chatt, *J. Chem. Soc. (A)*, 2356 (1971).
333. D. L. Kepert and R. Mandyczewsky, *J. Chem. Soc. (A)*, 530 (1968).
334. J. Lewis and R. Whyman, *J.Chem. Soc.*, 6027 (1965).
335. S. F. A. Kettle and R. V. Parish, *Spectrochim. Acta*, *21*, 1087 (1965).
336. W. M. Carmichael and D. A. Edwards, *J. Inorg. Nucl. Chem.*, *32*, 1199 (1970).
337. Y. M. Byr'ko, M. B. Polinskaya ,and A. I. Busev, *Russ. J. Inorg. Chem.*, *18*, 1479 (1973).
338. G. M. Larin, P. M. Solozhenkin, and E. V. Semenov, *Dokl. Akad. Nauk. SSSR*, *214*, 1343 (1974).
339. A. Felty and E. Sennewald, *Z. Anorg. Allg. Chem.*, 358 (1968).
340. P. C. H. Mitchell and R. J. P. Williams, *J. Chem. Soc.*, 4570 (1962).
341. E. B. Fleischer and T. S. Srivastava, *Inorg. Chim. Acta*, *5*, 151 (1971).
342. J. Fuhrhop, K. M. Kadish, and D. G. Davis, *J. Am. Chem. Soc.*, *95*, 5140 (1973).
343. I. R. Beattie and G. A. Ozin, *J. Chem. Soc.* (*A*), 1692 (1969).
344. R. A. Walton, P. C. Crouch, and B. J. Brisdon, *Spectrochim. Acta (A)*, *24*, 601 (1968).
345. H. H. Patterson and J. L. Nims, *Inorg. Chem.*, *11*, 520 (1972).
346. R. D. Dowsing and J. F. Gibson, *J. Chem. Soc.* (*A*), 655 (1967).
347. H. Funk, F. Schmeil, and H. Scholz, *Z. Anorg. Allg. Chem.*, *310*, 86 (1961).
348. P. D. Rillema and C. H. Brubaker, Jr., *Inorg. Chem.*, *8*, 1645 (1969).
349. D. A. McClung, L. R. Dalton, and C. H. Brubaker, *Inorg. Chem.*, *5*, 1985 (1966).
350. B. R. McGarvey, *Inorg. Chem.*, *5*, 476 (1966).
351. J. F. Rowbottom and G. Wilkinson, *Inorg. Nucl. Chem. Lett.*, *9*, 675 (1973).
352. D. G. Blight, D. L. Kepert, R. Mandyczewsky, and K. R. Trigwell, *J. Chem. Soc. Dalton*, 313 (1972).
353. H. Funk and H. Böhland, *Z. Anorg. Allg. Chem.*, *324*, 168 (1963).
354. L. S. Meriwether, W. F. Marzluff, and W. G. Hodgson, *Nature*, *212*, 465 (1966).
355. T. Huang and G. P. Haight, Jr., *J. Am. Chem. Soc.*, *92*, 2336 (1970).
356. T. J. Huang and G. P. Haight, *Chem, Commun.*, 985 (1969).
357. T. Huang and G. P. Haight Jr., *J. Am. Chem. Soc.*, *93*, 611 (1971).
358. P. T. Manoharan and M. T. Rogers, *J. Chem. Phys.*, *49*, 5510 (1968).

359. C. R. Hare, I. Bernal, and H. B. Gray, *Inorg. Chem.*, *1*, 831 (1962).
360. H. Kon and N. E. Sharpless, *J. Phys. Chem.*, *70*, 105 (1966).
361. K. DeArmond, B. B. Garrett, and H. S. Gutowsky, *J. Chem. Phys.*, *42*, 1019 (1965).
362. I N. Marov, Y. N. Dubrov, V. K. Belyaeva, and A. N. Ermakov, *Russ. J. Inorg. Chem.*, *17*, 396 (1972).
363. R. G. Hayes, *J. Chem. Phys.*, *44*, 2210 (1966).
364. E. L. Muetterties, *Inorg. Chem.*, *4*, 769 (1965).
365. E. L. Muetterties, *Inorg. Chem.*, *12*, 1963 (1973).
366. N. S. Garif'yanov, V. N. Fedotov, and N. S. Kucheryavenko, *Izv. Akad. Nauk. SSSR, Ser. Khim.*, 743 (1964).
367. I. N. Marov, V. K. Belyaeva, Y. N. Dubrov, A. N. Ermakov, and P. A. Korovaikov, *Russ. J. Inorg. Chem.*, *15*, 1701 (1970).
368. I. N. Marov, V. K. Belyaeva, Y. N. Dubrov, and A. N. Ermakov, *Russ. J. Inorg. Chem.*, *17*, 515 (1972).
369. N. S. Garif'yanov, *Dokl. Akad. Nauk. SSSR*, *190*, 1368 (1970).
370. I. N. Marov, Y. N. Dubrov, V. K. Belyaeva, and A. N. Ermakov, *Dokl. Akad. Nauk. SSSR*, *177*, 1166 (1967).
371. I. N. Marov, V. K. Belyaeva, A. N. Ermakov, and Y. N. Dubrov, *Russ. J. Inorg. Chem.*, *14*, 1391 (1969).
372. I. F. Gainulin, N. S. Garif'yanov, and V. V. Trachevskii, *Iszv. Akad. Nauk. SSSR*, *10*, 2176 (1969).
373. N. S. Garif'yanov and S. E. Kamenev, *Russ. J. Phys. Chem.*, *43*, 609 (1969).
374. D. I. Ryabchikov, I. N. Marov, Y. N. Dubrov, V. K. Belyaeva, and A. N. Ermakov, *Dokl. Akad. Nauk. SSSR*, *167*, 629 (1966).
375. R. G. Hayes, *J. Chem. Phys.*, *47*, 1692 (1967).
376. P. C. H. Mitchell and R. D. Scarle, *Proceedings of the First International Conference on Chemistry and Uses of Molybdenum*, P. C. H. Mitchell, Ed. Climax Molybdenum Co. 1974.
377. G. R. Lee and J. T. Spence, *Inorg. Chem.*, *11*, 2354 (1972).
378. H. K. Saha and M. C. Halder, *J. Inorg. Nucl. Chem.*, *33*, 3719 (1971).
379. R. G. James and W. Wardlaw, *J. Chem. Soc.*, 2145 (1927).
380. B. Jezowska-Trzebiatowska, M. F. Rudolf, L. Natkaniec, and H. Sabat, *Inorg. Chem.*, *13*, 617 (1974).
380a. T. Glowiak, M. Sabat, H. Sabat, and M. F. Rudolf, *J. Chem. Soc. Chem. Commun.*, 712 (1975).
380b. G. K. Kneale, A. J. Geddes, Y. Sasaki, T. Shibahara, and A. G. Sykes, *J. Chem. Soc. Chem. Commun.*, 356 (1975).
381. H. K. Saha and A. K. Banerjee, *J. Inorg. Nucl. Chem.*, *34*, 697 (1972).
382. A. Sabatini and I. Bertini, *Inorg, Chem.*, *5*, 204 (1966).
383. J. P. Brunett and M. J. F. Leroy, *J. Inorg. Nucl. Chem.*, *36*, 289 (1974).
384. H. Sabat, M. F. Rudolf, and B. Jezowska-Trzebiatowska, *Inorg. Nucl. Chim. Acta*, *7*, 365 (1973).
385. A. M. Golub, V. A. Grechikhina, V. V. Trachevskii, and N. V. Ul'ko, *Zh. Neorg. Chem.*, *18*, 1120 (1973).
386. W. Andruchow, Jr., and R. D. Archer, *J. Inorg. Nucl. Chem.*, *34*, 3185 (1972).
387. J. W. Buchler and K. Rohbock, *Inorg. Nucl. Chem. Lett.*, *8*, 1073 (1972).
387a. W. R. Scheidt, personal communication.
388. D. H. Brown, D. R. Russell, and D. W. A. Sharp, *J. Chem. Soc.* (*A*), 18 (1966).
389. D. A. Edwards and G. W. A. Fowles, *J. Chem. Soc.*, 24 (1961).
390. L. A. Dalton, R. D. Bereman, and C. H. Brubaker, Jr., *Inorg. Chem.*, *8*, 2477 (1969).

391. D. I. Ryabchikov, I. N. Marov, Y. N. Dubrov, U. K. Belyaeva, and A. N. Ermakov, *Dokl. Akad. Nauk., SSSR, 167*, 629 (1966).
392. J. Selbin and J. Sherrill, *J. Chem. Soc. Chem. Comm.*, 120 (1973).
393. B. Blues, D. H. Brown, R. G. Perkins, and J. J. P. Stewart, *Inorg. Chim. Acta, 8*, 67 (1974).
394. L. Natkaniec, M. Rudolf, and B. Jezowska-Trzebiatowska, *Theoret. Chim. Acta, 28*, 193 (1973).
395. A. T. Casey, D. J. Mackey, R. L. Martin, and A. H. White, *Aust. J. Chem., 25*, 477 (1972).
396. F. A. Cotton, D. L. Hunter, L. Ricard, and R. Weiss, *J. Coord. Chem., 3*, 259 (1974)
397. R. Colton and G. G. Rose, *Austral. J. Chem., 21*, 883 (1968).
398. R. Colton and G. R. Scollary, *Austral. J. Chem., 21*, 1427 (1968).
399. R. N. Jowitt and P. C. H. Mitchell, *J. Chem. Soc. (A)*, 2631 (1969).
400. W. E. Newton, J. L. Corbin, and J. W. McDonald, *J. Chem. Soc. Dalton*, 1044 (1974).
401. P. W. Schneider, D. C. Bravard, J. W. McDonald, and W. E. Newton, *J. Am. Chem. Soc., 94*, 8640 (1972).
402. R. Kirmse, *Z. Chem., 13*, 187 (1973).
403. P. C. H. Mitchell, *J. Inorg. Nucl. Chem., 26,* 1967 (1964).
404. M. L. Larson and F. W. Moore, *Inorg. Chem., 2*, 881 (1973).
405. C. B. Riolo and F. Guerrieri, *Ann. Chim. (Rome), 57*, 873 (1967) through *Chem. Abst., 67*, 104708M (1967).
406. H. Sabat, M. F. Rudolf, and B. Jezowska-Trzebiatowska, *Inorg. Chim. Acta, 7*, 366 (1973).
407. V. I. Spitsyn, I. D. Kolli, and T. Wen-Hsia, *Russ. J. Inorg. Chem., 9*, 54 (1964); *Zh. Neorg. Khim., 9*, 99 (1964).
408. I. Bernal, *Chem. Ind.*, 1343 (1966).
409. L. R. Melby, *Inorg. Chem., 8*, 349 (1969).
410. P. C. H. Mitchell, *J. Chem. Soc. (A), 146* (1969).
411. P. C. H. Mitchell, *J. Inorg. Nucl. Chem., 25*, 963 (1963).
412. H. K. Saha and A. K. Banerjee, *J. Inorg. Nucl. Chem., 34*, 1861 (1972).
413. S. Wajda and A. Zarzecznv, *Nukleonika, 19*, 33 (1974) through *Chem. Abst., 81*, 69217r (1974).
414. D. H. Brown, P. G. Perkins, and J. J. Stewart, *J. Chem. Soc. Dalton*, 1105 (1972).
415. C. M. French and J. H. Garside, *J. Chem. Soc.*, 2006 (1962).
416. J. H. Garside, *J. Chem. Soc.*, 6634 (1965).
417. R. M. Wing and K. P. Callahan, *Inorg. Chem., 8*, 871 (1969).
418. M. Ardon and A. Pernick, *Inorg. Chem., 12*, 2484 (1973).
419. L. R. Melby, *Inorg. Chem., 8*, 1539 (1969).
420. J. T. Spence and J. Y. Lee, *Inorg. Chem., 4*, 385 (1965).
421. D. H. Brown and J. MacPherson, *J. Inorg. Nucl. Chem., 34*, 1705 (1972).
422. D. H. Brown and J. MacPherson, *J. Inorg. Nucl. Chem., 33*, 4203 (1971).
423. D. H. Brown and J. MacPherson, *J. Inorg. Nucl. Chem., 32*, 3309 (1970).
424. M. Cousins and M. L. H. Green, *J. Chem. Soc. (A)*, 16 (1969).
425. L. V. Haynes and D. T. Sawyer, *Inorg. Chem., 6*, 2147 (1967).
426. D. Hruskova, J. Podlahova, and J. Podlaha, *Coll. Czech. Chem. Commun., 35*, 2738 (1970).
427. Y. Sasaki and A. G. Sykes, *J. Chem. Soc. Dalton*, 1468 (1974).
428. J. Kloubek and J. Podlaha, *J. Inorg. Nucl. Chem., 33*, 2981 (1971).
428a. V. R. Ott and F. A. Schultz, *J. Electroanal. Chem., 59*, 47 (1975), *61*, 81 (1975).
429. R. L. Dutta and B. Chatterjee, *J. Ind. Chem. Soc., 47*, 657 (1970).

430. B. Jezowska-Trzebiatowska and M. Rudolf, *Roczniki. Chem.*, *44*, 745 (1970).
431. R. Mattes and G. Lux, *Angew. Chem.*, *86*, 598 (1974).
432. H. K. Saha and M. G. Halder, *J. Inorg. Nucl. Chem.*, *34*, 3097 (1972).
433. R. G. James and W. Wardlaw, *J. Chem. Soc.*, 2726 (1928).
434. R. N. Jowitt and P. C. H. Mitchell, *Chem. Commun.*, 605 (1966).
435. P. M. Treichel and G. R. Wilkes, *Inorg. Chem.*, *7*, 1182 (1966).
436. B. Spivack and Z. Dori, *J. Chem. Soc. Chem. Commun.*, 909 (1973).
437. T. Sakurai, H. Okabe, and H. I. Soyama, *Bull. Japan Petrol. Inst.*, *13*, 243 (1971).
438. J. P. Fackler and J. A. Fetchin, *J. Am. Chem. Soc.*, *92*, 7648 (1970).
439. W. Beck, W. Danzer, and G. Thiel, *Angew. Chem. (Intl. ed.)*, *12*, 582 (1973).
440. L. Sacconi and R. Cini, *J. Am. Chem. Soc.*, *76*, 4239 (1954).
441. Y. Yoshino, I. Taminaga, and S. Uchida, *Bull. Chem. Soc. Japan*, *44*, 1435 (1971).
442. J. T. Spence and H. Y. Y. Chang, *Inorg. Chem.*, *2*, 319 (1963).
443. A. Kay and P. C. H. Mitchell, *Nature*, *219*, 267 (1968).
444. A. Kay and P. C. H. Mitchell, *J. Chem. Soc. Dalton*, 1388 (1972).
445. T. M. Tam and J. H. Swinehart, Abstracts, *1974 Pacific Conference on Chemistry and Spectroscopy*, San Francisco, Ca., October 16–18, 1974, Abst. #101.
446. E. Bayer and P. Krauss, *Z. Naturforsch.*, *24b*, 776 (1969).
447. P. Kroneck and J. T. Spence, *J. Inorg. Nucl. Chem.*, *35*, 3391 (1973).
448. P. Kroneck and J. T. Spence, *Inorg. Nucl. Chem. Lett.*, *9*, 177 (1973).
449. R. F. Stephenson and F. A. Schultz, *Inorg. Chem.*, *12*, 1762 (1973).
450. R. E. Benesch and R. Benesch, *J. Am. Chem. Soc.*, *77*, 5877 (1955).
451. G. E. Clement and T. P. Hartz, *J. Chem. Educ.*, *48*, 395 (1971).
452. J. T. Spence and M. Heydanek, *Inorg. Chem.* *6*, 1489 (1967).
453. E. M. Kosower and N. S. Kosower, *Nature*, *224*, 117 (1969).
454. H. C. Freeman, *Inorganic Biochemistry*, G. L. Eichhorn, Ed., Elsevier, New York 1973.
455. M. Ardon and A. Pernick, *J. Am. Chem. Soc.*, *95*, 6871 (1973).
455a. T. Ramasami, R. S. Taylor, and A. G. Sykes, *J. Am. Chem. Soc.*, *97*, 5918 (1975).
456. M. LaMache-Duhameaux, *Rev. Chim. Minerale*, *5*, 459 (1968).
457. M. LaMache, *J. Less-Common Metals*, *39*, 179 (1975).
458. V. W. Day and J. L. Hoard, *J. Am. Chem. Soc.*, *90*, 3374 (1968).
459. K. Stadnicka, *Roczniki. Chem.*, *47*, 2021 (1973).
460. V. C. Adam, U. A. Gregory, and B. T. Kilbourn, *Chem. Commun.*, 1400 (1970).
461. J. Chatt, L. Manojlović-Muir, and K. W. Muir, *Chem. Commun.*, 655 (1971).
462. L. Manojlović-Muir, *J. Chem. Soc. (A)*, 2796 (1971).
463. L. Manojlović-Muir and K. W. Muir, *J. Chem. Soc. Dalton*, 686 (1972).
464. L. Richard and R. Weiss, *Inorg. Nucl. Chem. Lett.*, *10*, 217 (1974).
465. A. J. Edwards, A. J. Peacock, and A. Said, *J. Chem. Soc.*, 4643 (1962).
466. J. E. Davies and B. M. Gatehouse, *J. Chem. Soc. Dalton*, 184 (1974).
467. L. Manojlović-Muir, *Inorg. Nucl. Chem. Lett.*, *9*, 59 (1973).
468. M. G. B. Drew, J. D. Wilkins, and A. P. Wolters, *J. Chem. Soc. Chem. Commun.*, 1278 (1972).
469. J. L. Hoard and H. H. Nordsieck, *J. Am. Chem. Soc.*, *61*, 2853 (1939).
470. J. L. Hoard, T. A. Hamor, and M. D. Glick, *J. Am. Chem. Soc.*, *90*, 3177 (1968).
471. F. H. Cano and D. W. J. Cruickshank, *J. Chem. Soc. D*, 1617 (1971).
471a. M. Novotny, D. F. Lewis, and S. J. Lippard, *J. Am. Chem. Soc.*, *94*, 6961 (1972).
471b. R. V. Parish and P. G. Simms, *J. Chem. Soc. Dalton.*, 2389 (1972).
472. L. J. Guggenberger, *Inorg. Chem.*, *12*, 2295 (1973).
473. O. Piovesana and L. Sestili, *Inorg. Chem.*, *13*, 2745 (1974).

474. M. Bonamico, G. Dessy, U. Fares, and L. Scaramuzza, *J. Chem. Soc. Dalton*, 2079 (1975).
475. L. Ricard, J. Estienne, and R. Weiss, *J. Chem. Soc. Chem. Commun.*, 906 (1972).
476. L. Ricard, J. Estienne, and R. Weiss. *Inorg. Chem.*, *12*, 2182 (1973).
477. P. J. Vergamini, H. Vahrenkamp, and L. F. Dahl, *J. Am. Chem. Soc.*, *93*, 6327 (1971).
478. M. Gerloch and R. Mason, *J. Chem. Soc.*, 296 (1965).
479. K. Prout, T. S. Cameron, R. A. Forder, S. R. Critchley, B. Denton, and G. V. Rees, *Acta Cryst.*, *B30*, 2290 (1974).
480. C. K. Prout, G. B. Allison, L. T. J. Dellaere, and E. Gare, *Acta Cryst.*, *B28*, 3043 (1972).
481. J. R. Knox and C. K. Prout, *Acta Cryst.*, *B25*, 2013 (1969).
482. J. R. Knox and C. K. Prout, *Acta Cryst.*, *B25*, 2482 (1969).
483. T. S. Cameron and C. K. Prout, *Chem. Commun.*, 161 (1971).
484. T. S. Cameron and C. K. Prout, *Acta Cryst.*, *B28*, 453 (1972).
485. A. V. Butcher and J. Chatt, *J. Chem. Soc. (A)*, 2652 (1970).
486. J. A. McCleverty, J. Locke, B. Ratcliff, and E. J. Wharton, *Inorg. Chim. Acta*, *3*, 283 (1969).
487. S. Midollini and M. Bacci, *J. Chem. Soc. (A)*, 2964 (1970).
488. M. Novotny and S. J. Lippard, *Inorg. Chem.*, *13*, 828 (1974).
489. H. A. O. Hill and M. M. Norgett, *J. Chem., Soc. (A)*, 1476 (1966).
490. S. J. Lippard and B. J. Russ, *Inorg. Chem.*, *6*, 1943 (1967).
491. J. van de Poel and H. M. Neumann, *Inorg. Chem.*, *7*, 2087 (1968).
492. L. F. Lindoy, S. E. Livingstone, and T. M. Lockyer, *Austral. J. Chem.*, *18*, 1549 (1965).
493. L. K. Atkinson, A. H. Mawby, and D. C. Smith, *Chem. Commun.*, 1399 (1970).
494. R. G. Cavell and A. R. Sanger, *Inorg. Chem.*, *11*, 2011 (1972).
495. R. Barral, C. Bocard, I. Seree de Roch, and L. Sajus, *Tetrahedron Lett.*, *1693* (1972).
496. M. Green, R. B. L. Osborn, and F. G. A. Stone, *J. Chem. Soc. (A)*, 3083 (1968).
497. E. I. Stiefel, L. E. Bennett, Z. Dori, T. H. Crawford, C. Simo, and H. B. Gray, *Inorg. Chem.*, *9*, 281 (1970).
498. E. G. Arumyunyan, A. S. Antsishkina, and E. Y. Balma, *Zh. Strukt. Khim.*, *11*, 2400 (1966).
499. S. J. Lippard, H. Nozaki, and B. J. Russ, *Chem. Commun.*, *3*, 119 (1967).
499a. P. R. Robinson, E. O. Schlemper and R. K. Murmann, *Inorg. Chem.*, *14*, 2035 (1975).
500. R. K. Murmann and P. R. Robinson, *Inorg. Chem.*, *14*, 203 (1975).
501. G. W. A. Fowles, *Prog. Inorg. Chem.*, *6*, 1 (1964).
502. M. L. Larson and F. W. Moore, *Inorg. Chem.*, *3*, 285 (1964).
503. P. J. H. Carnell, R. E. McCarley, and R. D. Hogue, *Inorg. Syn.*, *10*, 49 (1967).
504. S. M. Horner and S. Y. Tyree, *Inorg. Chem.*, *2*, 568 (1963).
505. E. A. Allen, B. J. Brisdon, D. A. Edwards, G. W. A. Fowles, and R. G. Williams, *J. Chem. Soc.*, 4649 (1963).
506. C. J. Horn and T. M. Brown, *Inorg. Chem.*, *11*, 1970 (1972).
507. E. A. Allen, B. J. Brisdon, and G. W. A. Fowles, *J. Chem. Soc.*, 4531 (1964).
508. E. A. Allen, K. Feenan, and G. W. A. Fowles, *J. Chem. Soc.*, 1636 (1965).
509. H. L. Nigam, R. S. Nyholm, and M. H. B. Stiddard, *J. Chem. Soc.*, 4531 (1964).
510. A. van den Bergen, K. S. Murray, and B. O. West, *Austral. J. Chem.*, *25*, 705 (1972).
511. G. Doyle, *Inorg. Chem.*, *10*, 2348 (1971).
512. M. C. Steele, *Austral. J. Chem.*, *10*, 368 (1957).
513. M. C. Steele, *Austral. J. Chem.*, *10*, 367 (1957).
514. M. Basu and S. Basu, *J. Inorg. Nucl. Chem.*, *31*, 3326 (1969).
515. R. N. Jowitt and P. C. H. Mitchell, *Inorg. Nucl. Chem. Lett.*, *4*, 39 (1968).

516. Z. B. Varadi and A. Nieuwpoort, *Inorg. Nucl. Chem. Lett.*, *10*, 801 (1974).
517. T. M. Brown and J. N. Smith, *J. Chem. Soc. (Dalton)*, 1614 (1972).
518. E. Uhlemann and U. Eckelmann, *Z. Chem.*, *12(8)*, 290 (1972).
519. G. B. Allison and J. C. Sheldon, *Inorg. Chem.*, *6*, 1493 (1967).
520. A. D. Westland and V. Uzelac, *Can. J. Chem.*, *48*, 2870 (1970).
521. E. L. Muetterties, *J. Am. Chem. Soc.*, *82*, 1082 (1960).
522. J. R. Moss and B. L. Shaw, *J. Chem. Soc. (A)*, 595 (1970).
523. S. Ahrland, J. Chatt, and N. R. Davies, *Quart. Rev.*, *12*, 265 (1958).
524. R. G. Pearson, *J. Am. Chem. Soc.*, *85*, 3533 (1963).
525. N. Serpone and R. C. Fay, *Inorg. Chem.*, *8*, 2379 (1969).
526. J. W. Faller and A. Davison, *Inorg. Chem.*, *6*, 182 (1967).
527. I. Douek, M. J. Frazer, Z. Goffer, M. Goldstein, B. Rimner, and H. A. Willis, *Spectrochim. Acta*, *A23*, 373 (1967).
527a. B. F. Studd and A. G. Swallow, *J. Chem. Soc. (A)*, 1961 (1968).
528. C. J. L. Lock and C. Wan *Chem. Commun.*, 1109 (1967).
529. G. A. Barbieri, *Atti. Accad. Lineei.*, *12*, 55 (1930).
530. A. B. P. Lever, *Inorganic Electronic Spectroscopy*, Elsevier, New York, 1970.
531. N. H. Furman and C. O. Miller, *Inorg. Syn.*, *3*, 160 (1950).
532. D. C. Bradley and M. H. Chisholm, *J. Chem. Soc.*, 2741 (1971).
533. S. J. Lippard, *Prog. Inorg. Chem.*, *8*, 120 (1967).
534. M. L. H. Green, *Pure Appl. Chem.*, *30*, 373 (1972).
535. M. L. H. Green and W. E. Lindsell, *J. Chem. Soc. (A)*, 1455 (1967).
536. J. C. Green, M. L. H. Green, and G. E. Morris, *J. Chem. Soc. Chem. Commun.*, 212 (1974).
537. R. L. Cooper and M. L. H. Green, *J. Chem. Soc. (A)*, 1155 (1967).
538. E. S. Gore and M. L. H. Green, *J. Chem. Soc. (A)*, 2315 (1970).
539. M. L. H. Green, J. A. McCleverty, L. Pratt, and G. Wilkinson, *J. Chem. Soc.*, 4854 (1961).
540. E. O. Fischer, O. Beckert, W. Hafner, and H. O. Stall, *Z. Naturforsch.*, *10b*, 598 (1955).
541. A. R. Dias and M. L. H. Green, *J. Chem. Soc. (A)*, 1951 (1971).
542. A. R. Dias and M. L. H. Green, *Chem. Commun.*, *962*, (1969).
543. A. R. Dias and M. L. H. Green, *J. Chem. Soc. (A)*, 2808 (1971).
544. K. Prout, S. R. Critchley, and G. V. Rees, *Acta Cryst.*, *B30*, 2305 (1974).
545. E. Gore, M. L. H. Green, M. G. Harriss, W. E. Lindsell, and H. Shaw, *J. Chem. Soc. (A)*, 1981 (1969).
546. M. G. Harris, M. L. H. Green, and W. E. Lindsell, *J. Chem. Soc. (A)*, 1453 (1969).
547. R. H. Crabtree, A. R. Dias, M. L. H. Green, and P. J. Knowles, *J. Chem. Soc. (A)*, 1350 (1971).
548. M. L. Larson, *Inorg. Syn.*, *12*, 165 (1970).
549. D. E. LaValle, R. M. Steele, M. K. Wilkinson, and H. L. Yokel, Jr., *J. Am. Chem. Soc.*, *82*, 2433 (1960).
550. A. K. Mallock, *Inorg. Syn.*, *12*, 179 (1970).
551. P. J. H. Carnell, R. E. McCarley, and R. D. Hogue, *Inorg Syn.*, *10*, (1967).
552. J. Lewis, D. Machin, R. S. Nyholm, P. J. Pauling, and P. W. Smith, *Chem. Ind.*, 259 (1960).
553. D. A. Brown, W. K. Glass, and C. O'Daly, *J. Chem. Soc. Dalton*, 1311 (1973).
554. C. Epstein and N. Elliot, *J. Chem. Phys.*, *22*, 634 (1954).
555. B. M. Figgis, J. Lewis, and F. E. Mabbs, *J. Chem. Soc.*, 3138 (1961).
556. R. J. Irving and M. C. Steele, *Austral. J. Chem.*, *10*, 490 (1957).

557. J. V. Brencic and F. A. Cotton, *Inorg. Syn.*, *13*, 170 (1970).
558. W. Wardlaw and A. J. I. Harding, *J. Chem. Soc.*, 1592 (1926).
559. J. Lewis, R. S. Nyholm, and P. W. Smith, *J. Chem. Soc.*, 4590 (1961).
560. T. Kumorita, S. Miki, and S. Yamada, *Bull. Chem. Soc. Japan*, *38*, 123 (1965).
561. A. R. Bowen and H. Taube, *J. Am. Chem. Soc.*, *93*, 3289 (1971).
562. A. R. Bowen and H. Taube, *Inorg. Chem.*, *13*, 2244 (1974).
563. K. Kustin and D. Toppen, *Inorg. Chem.*, *11*, 2851 (1972).
564. Y. Sasaki and A. G. Sykes, *Chem. Comm.*, *767* (1973).
565. G. R. Rossman, F. D. Tsay, and H. B. Gray, *Inorg. Chem.*, *12*, 825 (1973).
566. W. R. Bucknall, S. R. Carter, and W. Wardlaw, *J. Chem. Soc.*, *512* (1927).
567. P. W. Smith and A. G. Wedd, *J. Chem. Soc. (A)*, 231 (1966).
568. K. Issleib and B. Biermann, *Z. Anorg. Allg, Chem.*, *347*, 39 (1966).
569. D. W. DuBois, R. T. Iwamoto, and J. Kleinberg, *Inorg. Chem.*, *8*, 815 (1969).
570. A. D. Westland and N. Muriithi, *Inorg. Chem.*, *11*, 2971 (1972).
571. H. B. Jonassen and L. J. Bailin, *Inorg. Syn.*, *7*, 140 (1962).
572. A. D. Westland and N. Muriithi, *Inorg. Chem.*, *12*, 2356 (1973).
573. V. B. Evdokmov, V. V. Zelenstov, I. D. Kolli, T. Wen-hsia, and V. I. Spitsyn, *Dokl. Akad. Nauk. SSSR*, *145*, 1282 (1962).
574. C. Furlani and O. Piovesana, *Mol. Phys.*, *9*, 341 (1965).
575. C. Djordjevic, R. S. Nyholm, C. S. Pande, and M. H. B. Stiddard, *J. Chem. Soc. (A)*, *16*, (1966).
576. M. C. Steele, *Austral. J. Chem.*, *10*, 489 (1957).
577. W. M. Carmichael, D. A. Edwards, and R. A. Walton, *J. Chem. Soc. (A)*, *97* (1966).
578. G. J. Sutton, *Austral. J. Chem.*, *15*, 232 (1962).
579. S. P. Ghosh and K. M. Prasad, *Proceedings of the First International Conference on Chemistry and Uses of Molybdenum*, P. C. H. Mitchell, Ed., Climax Molybdenum Co., 1974.
580. F. Calderazzo, C. Floriani, R. Henzi, and F. L. Eplattenier, *J. Chem. Soc. (A)*, 1378 (1969).
581. M. L. Larson and F. W. Moore, *Inorg. Chem.*, *1*, 856 (1962).
582. T. G. Dunne and F. A. Cotton, *Inorg. Chem.*, *2*, 263 (1963).
583. K. Christ and H. L. Schlafer, *Angew. Chem. (Intl. ed.)*, *2*, 97 (1963).
584. R. Saillant and R. A. D. Wentworth, *Inorg. Chem.*, *8*, 1226 (1969).
585. D. A. Edwards, G. W. A. Fowles, and R. A. Walton, *J. Inorg. Nucl. Chem.*, *27*, 1999 (1965).
586. H. G. Schnering and H. Woehrle, *Naturwiss*,. *50*, 91 (1963).
587. L. M. Toth, G. D. Brunton, and G. P. Smith, *Inorg. Chem.*, *8*, 2694 (1969).
588. Z. Amilius, B. van Laar, and H. M. Rietveld, *Acta Cryst.*, *B25*, 400 (1969).
589. J. R. Knox and K. Eriks, *Inorg. Chem.*, *7*, 84 (1968).
590. J. V. Brencic, *Z. Anorg. Allg. Chem.*, *403*, 218 (1974).
591. K. H. Lohmann and R. C. Young, *Inorg. Syn.*, *4*, 97 (1953).
592. J. Mass and J. Sand, *Ber.*, *42*, 2295 (1909).
593. W. Andruchow, Jr., and J. Diliddo, *Inorg. Nucl. Chem. Lett.*, *8*, 689 (1972).
594. D. A. Edwards and G. W. A. Fowles, *J. Less-Common Metals*, *4*, 512 (1962).
595. R. C. Young, *J. Am. Chem. Soc.*, *54*, 1402 (1932).
596. J. Lewis, R. S. Nyholm, C. S. Pande, and M. H. B. Stiddard, *J. Chem. Soc.*, 3600 (1963).
597. D. W. DuBois, R. I. Iwamoto, and J. Kleinberg, *Inorg. Chem.*, *9*, 969 (1970).
598. D. W. DuBois, R. T. Iwamoto, and J. Kleinberg, *Inorg. Nucl. Chem. Lett.*, *6*, 53 (1970).

599. S. Herzog and I. Schneider, *Z. Chem.*, *2*, 24 (1962).
600. H. S. Jarrett, *J. Chem. Phys.*, *29*, 1298 (1957).
601. J. Owen and I. M. Ward in K. D. Bowers and J. Owen, *Rep. Prog. Phys.*, *18*, 304 (1955).
602. J. Owen and I. M. Ward, *Phys. Rev.*, *102*, 591 (1956).
603. P. D. W. Boyd, P. W. Smith, and A. G. Wedd, *Austral. J. Chem.*, *22*, 653 (1969).
604. C. K. Jørgenson, *Absorption Spectra and Chemical Bonding in Complexes*, Addison-Wesley, Reading, Mass. 1962.
605. C. J. Ballhausen, *Introduction to Ligand Field Theory*, McGraw-Hill, New York, 1962.
606. C. K. Jørgenson, *Acta Chem. Scand.*, *12*, 1539 (1958).
607. H. L. Schlafer, H. Gausmann, and H. Witzke, *J. Mol. Spec.*, *21*, 125 (1966).
608. H. Hartman and H. J. Schmidt, *Z. Physikal. Chem. Neue Folge*, *11*, 234 (1957).
609. C. Postmus and E. L. King, *J. Phys. Chem.*, *59*, 1216 (1955).
610. J. Lewis, R. S. Nyholm, and P. W. Smith, *J. Chem. Soc. (A)*, 57 (1969).
611. R. Saillant and R. A. D. Wentworth, *Inorg. Chem.*, *8*, 1226 (1969).
612. M. J. Bennett, J. V. Brencic, and F. A. Cotton, *Inorg. Chem.*, *8*, 1060 (1969).
613. W. H. Delphin and R. A. D. Wentworth, *J. Am. Chem. Soc.*, *95*, 7920 (1973).
614. W. H. Delphin, R. A. D. Wentworth, and M. S. Matson, *Inorg. Chem.*, *13*, 2552 (1974).
614a. M. S. Matson and R. A. D. Wentworth, *J. Amer. Chem. Soc.*, *96*, 7837 (1974).
615. R. Saillant, P. B. Jackson, W. E. Strieb, K. Folting, and R. A. D. Wentworth, *Inorg. Chem.*, *10*, 1453 (1971).
616. W. Wardlaw and R. L. Wormell, *J. Chem. Soc.*, 1087 (1927).
617. J. Kloubek and J. Podlaha, *Inorg. Nucl. Chem. Lett.*, *7*, 67 (1971).
618. P. C. H. Mitchell and R. D. Scarle, *J. Chem. Soc.*, 1809 (1972).
619. H. B. Gray and H. J. Schugar, *Inorganic Biochemistry*, G. L. Eichhorn, Ed., Elsevier, New York, 1970, p. 102.
620. M. Ardon and A. Pernick, *Inorg. Chem.*, *13*, 2275 (1974).
620a. D. A. Brown, B. J. Gordon, W. K. Glass, and C. J. O'Daley, *Proc. XVIth I.C.C.C.*, Toronto, 1972 p. 646.
620b. P. C. H. Mitchell and R. D. Scarle, *J. Chem. Soc., Dalton*, 110 (1975).
620c. A. Nieuwpoort, H. M. Cloessen, and J. G. M. van der Linden, *Inorg. Nucl. Chem. Lett.*, 11, 869 (1975).
621. F. Hug, W. Mowat, A. Shortland, and G. Wilkinson, *Chem. Commun.*, 1079 (1971).
622. M. Chisholm, F. A. Cotton, B. A. Frenz, and L. Shive, *J. Chem. Soc. Chem. Commun.*, 480 (1974).
623. W. Mowat and G. Wilkinson, *J. Chem. Soc. Dalton*, 1120 (1973).
624. N. G. Connelly and L. F. Dahl, *J. Am. Chem. Soc.*, *92*, 7470 (1970).
625. H. L. Nigam, R. S. Nyholm, and M. H. B. Stiddard, *J. Chem. Soc.*, 1806 (1960).
626. R. S. Nyholm, M. R. Snow, and M. H. B. Stiddard, *J. Chem. Soc.*, 6570 (1965).
627. C. D. Cook, R. S. Nyholm, and M. L. Tobe, *J. Chem. Soc.*, 4194 (1965).
628. M. H. B. Stiddard, *J. Chem. Soc.*, 4712 (1962).
629. H. C. E. Mannerskantz and G. Wilkinson, *J. Chem. Soc.*, 4454 (1962).
630. H. Behrens and J. Rosenfelder, *Z. Anorg. Allg. Chem.*, *352*, 61 (1967).
631. I. V. Howell and L. M. Venanzi, *Inorg. Chim. Acta*, *3*, 121 (1969).
632. J. Lewis and R. Whyman, *J. Chem. Soc.*, 5486 (1965).
633. R. Colton and I. B. Tomkins, *Austral. J. Chem.*, *19*, 1143 (1966).
634. R. Colton and I. B. Tomkins, *Austral. J. Chem.*, *19*, 1519 (1966).
635. R. Colton, G. R. Scollary, and I. B. Tomkins, *Austral. J. Chem.*, *21*, 15 (1968).
636. M. W. Anker, R. Colton, and I. B. Tomkins, *Rev. Pure Appl. Chem.*, *18*, 23 (1968).

637. M. W. Anker, R. Colton, and I. B. Tomkins, *Austral. J. Chem.*, *21*, 1159 (1968).
638. R. Colton and G. R. Scollary, *Austral. J. Chem.*, *21*, 1427 (1968).
639. J. A. Bowden and R. Colton, *Austral. J. Chem.*, *21*, 2657 (1968).
640. R. Colton and C. J. Rix, *Austral. J. Chem.*, *22*, 305 (1969).
641. M. W. Anker, R. Colton, and I. B. Tomkins, *Austral. J. Chem.*, *21*, 1143 (1968).
642. T. W. Beall and L. W. Houk, *Inorg. Chem.*, *13*, 2549 (1974).
643. M. G. B. Drew, A. W. Johans, A. P. Wolters, and I. B. Tomkins, *Chem. Commun.*, 819 (1971); M. G. B. Drew, *J. Chem. Soc. Dalton*, 626 (1972).
644. M. G. B. Drew, *J. Chem. Soc. Dalton*, 1329 (1972).
645. M. G. B. Drew and J. D. Wilkins, *J. Chem. Soc. Dalton*, 2664 (1973).
645a. J. C. Dewan, K. Hendrick, D. L. Kepert, K. R. Trigwell, A. H. White, and S. B. Wild, *J. Chem. Soc. Dalton*, 546 (1975).
645b. D. L. Kepert, *J. Chem. Soc. Dalton*, 617 (1973).
646. M. G. B. Drew, I. B. Tomkins, and R. Colton, *Austral. J. Chem.*, *23*, 2517 (1970).
647. J. W. McDonald, J. L. Corbin, and W. E. Newton, *J. Am. Chem. Soc.*, *97*, 1970 (1975).
647a. J. W. McDonald, W. E. Newton, C. T. C. Creedy and J. L. Corbin, *J. Organometal. Chem.*, *92*, C25 (1975).
648. K. L. Tang Wong, J. L. Thomas, and H. H. Brintzinger, *J. Am. Chem. Soc.*, *96*, 3694 (1974).
649. M. C. Ganorkar and M. H. B. Stiddard, *J. Chem. Soc.*, 3494 (1965).
650. M. G. B. Drew and A. P. Wolters, *Chem. Commun.*, 457 (1972).
651. M. Novotny and S. J. Lippard, *J. Chem. Soc. Chem. Commun.*, 202 (1973).
652. F. Bonati and G. Minghetti, *Inorg. Chem.*, *9*, 2642 (1970).
653. D. F. Lewis and S. J. Lippard, *Inorg. Chem.*, *11*, 621 (1972).
653a. D. L. Lewis and S. J. Lippard, *J. Am. Chem. Soc.* *97*, 2697 (1975).
654. B. F. G. Johnson and K. H. Al-Obaidi, *Chem. Commun.*, 876 (1968).
655. T. F. Brennan and I. Bernal, *Chem. Commun.*, 138 (1970).
656. T. F. Brennan and I. Bernal, *Inorg. Chim. Acta*, *7*, 283 (1973).
657. J. Lewis, R. S. Nyholm, and P. W. Smith, *J. Chem. Soc.*, 2592 (1962).
658. R. Colton and C. J. Rix, *Austral. J. Chem.*, *21*, 1155 (1968).
659. E. L. Muetterties and C. M. Wright, *Quart. Rev.*, *21*, 109 (1967).
660. E. Bannister and G. Wilkinson, *Chem. Ind.*, 319 (1960).
661. J. A. Stephenson, E. Bannister, and G. Wilkinson, *J. Chem. Soc.*, 2538 (1964).
662. D. Lawton and R. Mason, *J. Am. Chem. Soc.*, *87*, 921 (1965).
663. J. N. van Niekerk and F. R. L. Schoening, *Acta Cryst.*, *6*, 227 (1953).
664. F. A. Cotton and J. G. Norman, Jr., *J. Coord. Chem.*, *1*, 161 (1971).
665. F. A. Cotton and J. G. Norman, Jr. *J. Am. Chem. Soc.*, *94*, 5697 (1972).
666. A. P. Ketteringham and C. Oldham, *J. Chem. Soc. Dalton*, 1068 (1973).
667. C. L. Angell, F. A. Cotton, B. A. Frenz, and T. R. Webb, *J. Chem. Soc. Chem. Commun.*, 399 (1973).
668. E. Hochberg, P. Walks, and E. H. Abbott, *Inorg. Chem.*, *13*, 1824 (1974).
669. J. V. Brencic and F. A. Cotton, *Inorg. Chem.*, *9*, 351 (1970).
670. J. V. Brencic and F. A. Cotton, *Inorg. Chem.*, *8*, 2698 (1969).
671. J. V. Brencic and F. A. Cotton, *Inorg. Chem.*, *9*, 346 (1970).
672. J. V. Brencic and F. A. Cotton, *Inorg. Chem.*, *8*, 7 (1969).
673. F. A. Cotton, *Rev. Pure Appl. Chem.*, *17*, 25 (1967).
674. F. A. Cotton, *Inorg. Chem.*, *4*, 334 (1965).
675. J. G. Norman, Jr., and H. J. Kolari, *J. Chem. Soc. Chem. Commun.*, 303 (1974), 649 (1975).

676. J. G. Norman and H. J. Kolari, *J. Am. Chem. Soc.*, *97*, 33 (1975).
677. L. Dubicki and R. L. Martin, *Austral. J. Chem.*, *22*, 1571 (1969).
678. C. D. Cowman and H. B. Gray, *J. Am. Chem. Soc.*, *95*, 8177 (1973).
678a. R. J. H. Clark and M. L. Franks, *J. Am. Chem. Soc.*, *97*, 2691 (1975).
679. R. J. H. Clark and M. L. Franks, *J. Chem. Soc. Chem. Commun.*, 316 (1974).
680. J. San Filippo, Jr., and H. J. Sniadoch, *Inorg. Chem.*, *12*, 2326 (1973).
681. J. San Filippo, Jr., *Inorg. Chem.*, *11*, 3140 (1972).
682. F. A. Cotton, B. A. Frenz, and T. R. Webb, *J. Am. Chem. Soc.*, *95*, 4431 (1973).
682a. F. A. Cotton, B. A. Frenz, E. Pedersen, and T. R. Webb, *Inorg. Chem.*, *14*, 391 (1975).
682b. A. Pernick and M. Ardon, *J. Am. Chem. Soc.*, *97*, 1255 (1975).
682c. F. A. Cotton and E. Pedersen, *Inorg. Chem.*, *14*, 399 (1975).
683. F. A. Cotton, B. A. Frenz, and Z. C. Mester, *Acta Cryst.*, *B29*, 1515 (1973).
684. J. V. Brencic, D. Dobchik, and P. Segedin, *Monat. Chem.*, *105*, 142 (1974).
685. J. San Filippo, Jr., H. J. Sniadoch, and R. L. Grayson, *Inorg. Chem.*, *13*, 2121 (1974).
686. E. H. Abbott, F. Schoenewolf, Jr., and T. Backstrom, *J. Coord. Chem.*, *3*, 255 (1974), 649 (1975).
687. L. Richard, P. Karagiannidis, and R. Weiss, *Inorg. Chem.*, *12*, 2179 (1973).
688. D. F. Steele and T. A. Stephenson, *Inorg. Nucl. Chem. Lett.*, *9*, 777 (1973).
688a. F. A. Cotton, T. Inglis, M. Kilner, and T. R. Webb, *Inorg. Chem.*, *14*, 2023 (1975).
688b. E. Kurras, H. Mennenga, G. Oehome, U. Rosenthal, and G. Engelhardt, *J. Organometal. Chem.*, *84*, C13 (1975).
688c. C. D. Garner and R. G. Senior, *J. Chem. Soc. Dalton*, 1171 (1975).
689. M. L. H. Green and W. E. Silverthorn, *J. Chem. Soc. Dalton*, 301 (1973).
690. K. Prout and G. V. Rees, *Acta Cryst.*, *B30*, 2251 (1974).
691. W. K. Bratton, F. A. Cotton, M. Debeau, and R. A. Walton, *J. Coord. Chem.*, *1*, 121 (1971).
692. C. D. Garner and R. G. Senior, *Chem. Commun.*, 580 (1974).
692a. V. Katovic, J. L. Templeton, R. J. Hoxmeier, and R. E. McCarley, *J. Am. Chem. Soc.*, *97*, 5300 (1975).
692b. F. A. Cotton, *Chem. Soc. Rev.*, *4*, 27 (1975).
693. F. A. Cotton and J. R. Pipal, *J. Am. Chem. Soc.*, *93*, 544 (1971).
694. F. A. Cotton, J. M. Troup, T. R. Webb, D. H. Williamson, and G. Wilkinson, *J. Am. Chem. Soc.*, *96*, 3824 (1974).
695. J. A. McCleverty, *Prog. Inorg. Chem.*, *10*, 145 (1968).
696. G. N. Schrauzer, *Accts. Chem. Res.*, *2*, 72 (1969).
697. R. Eisenberg, *Prog. Inorg. Chem.*, *12*, 295 (1970).
698. M. Gerloch, S. F. A. Kettle, J. Locke, and J. A. McCleverty, *Chem. Commun.*, 30 (1966).
699. J. A. McCleverty, J. Locke, E. J. Wharton, and M. Gerloch, *J. Chem. Soc. (A)*, 816 (1968).
700. E. I. Stiefel, Z. Dori, and H. B. Gray, *J. Am. Chem. Soc.*, *89*, 3353 (1967).
701. J. H. Waters, R. Williams, H. B. Gray, G. N. Schrauzer, and H. W. Finck, *J. Am. Chem. Soc.*, *86*, 4198 (1964).
702. G. N. Schrauzer and V. P. Mayweg, *J. Am. Chem. Soc.*, *88*, 3235 (1966).
703. R. B. King, *Inorg. Chem.*, *2*, 641 (1963).
704. A. Davison, N. Edelstein, R. H. Holm, and A. H. Maki, *J. Am. Chem. Soc.*, *86*, 2799 (1964).
705. E. Hoyer and W. Schroth, *Chem. Ind. (London)*, 652 (1965).
706. A. E. Smith, G. N. Schrauzer, V. P. Mayweg, and W. Heinrich, *J. Am. Chem. Soc.*, 5798 (1965).

707. T. W. Gilbert, Jr., and E. B. Sandell, *J. Am. Chem. Soc.*, *82*, 1087 (1959).
708. E. I. Stiefel and H. B. Gray, *J. Am. Chem. Soc.*, *87*, 4012 (1965).
709. E. I. Stiefel, R. Eisenberg, R. C. Rosenberg, and H. B. Gray, *J. Am. Chem. Soc.*, *88*, 2956 (1966).
710. J. Finster, N. Meusel, P. Müller, W. Dietzsch, A. Meisel, and E. Hoyer, *Z. Chem.*, *13*, 146 (1973).
711. A. Callaghan, A. J. Layton, and R. S. Nyholm, *Chem. Commun.*, 399 (1969).
712. S. Koch, E. R. Purdy, and E. I. Stiefel, unpublished work.
713. A. Davison and E. T. Shawl, *Inorg. Chem.*, *9*, 1820 (1970).
714. C. G. Pierpont and R. Eisenberg, *J. Chem. Soc. (A)*, 2285 (1971).
715. R. Eisenberg and J. A. Ibers, *J. Am. Chem. Soc.*, *87*, 3776 (1965).
716. G. F. Brown and E. I. Stiefel, *Chem. Commun.*, 728 (1970).
717. G. F. Brown and E. I. Stiefel, *Inorg. Chem.*, *12*, 2140 (1973).
718. E. I. Stiefel and G. F. Brown, *Inorg. Chem.*, *11*, 434 (1972).
719. W. L. Kwik and E. I. Stiefel, *Inorg. Chem.*, *12*, 2337 (1973).
720. J. K. Gardner, N. Pariyadath, J.L.Corbin, and E. I. Stiefel, submitted for publication.
721. J. Dilworth, *Proceedings of the First International Conference on Chemistry and Uses of Molybdenum* P. C. H. Mitchell, Ed., Climax Molybdenum Co., 1974.
722. G. N. Schrauzer, V. P. Mayweg, and W. Heinrich, *J. Am. Chem. Soc.*, *88*, 5174 (1966).
723. A Butcher and P. C. H. Mitchell, *Chem. Commun.*, 176 (1967).
724. J. Locke and J. A. McCleverty, *Chem. Commun.*, 102 (1965).
725. N. G. Connelly, J. Locke, J. A. McCleverty, D. A. Phipps, and B. Ratcliff, *Inorg. Chem.*, *9*, 278 (1970).
726. D. Sellman, *Angew. Chem. (Intl. ed.)*, *13*, 640 (1974).
727. A. D. Allen, R. O. Harris, B. R. Loescher, J. R. Stevens, and R. N. Whiteley, *Chem. Rev.*, *73*, 11 (1973).
728. J. Chatt and R. L. Richards, *Chemistry and Biochemistry of Nitrogen Fixation*, J. R. Postgate, Ed., Plenum Press, New York, 1971.
729. Y. G. Borod'ko and A. E. Shilov, *Russ. Chem. Rev.*, *38*, 355 (1969).
730. A. D. Allen and F. Bottomley, *Accts. Chem. Res.*, *1*, 360 (1968).
731. R. Murray and D. C. Smith, *Coord. Chem. Rev.*, *3*, 429 (1968).
732. R. A. Forder and K. Prout, *Acta Cryst.*, *B30*, 2778 (1974).
733. T. Uchida, Y. Uchida, M. Hidai, and T. Kodama, *Bull. Chem. Soc. Japan*, *44*, 2883 (1971).
734. M. Hidai, K. Tominari, and Y. Uchida, *Chem. Commun.*, 1392 (1969).
735. M. Hidai, K. Tominari, Y. Uchida, and A. Misono, *Chem. Commun.*, 814 (1969).
736. M. Hidai, K. Tominari, and Y. Uchida, *J. Am. Chem. Soc.*, *94*, 110 (1972).
737. L. K. Atkinson, A. H. Mawby, and D. C. Smith, *Chem. Commun.*, 157 (1971).
738. T. A. George and C. D. Siebold, *Inorg. Chem.*, *12*, 2544 (1973).
739. J. Chatt and A. G. Wedd, *J. Organometal. Chem.*, *27*, C15 (1971).
740. J. Chatt, G. A. Heath, and R. L. Richards, *J. Chem. Soc. Dalton*, 2074 (1974).
741. M. Aresta and A. Sacao, *Gazz. Chim. Ital.*, *102*, 755 (1972).
742. J. Chatt, G. A. Heath, and R. L. Richards, *J. Chem. Soc. Chem. Commun.*, 1010 (1972).
743a. J. Chatt, J. R. Dilworth, G. J. Lĕigh, and R. L. Richards, *Chem. Commun.*, 955 (1970).
743b. J. Chatt, R. H. Crabtree, E. A. Jeffrey, and R. L. Richards, *J. Chem. Soc.*, 1167 (1973).
744. T. A. George and C. D. Seibold, *J. Organometal. Chem.*, *30*, C13 (1971).
745. T. A. George and C. D. Seibold, *Inorg. Nucl. Chem. Lett.*, *8*, 465 (1972).

746. M. L. H. Green and W. E. Silverthorn, *Chem. Commun.*, 557 (1971).
747. T. A. George and C. D. Seibold, *J. Am. Chem. Soc.*, *94*, 6859 (1972).
748. D. J. Darensbourg, *Inorg. Chem.*, *11*, 1436 (1972).
749. M. Aresta, *Gazz. Chim. Ital.*, *102*, 781 (1972).
750. T. A. George and C. D. Seibold, *Inorg. Chem.*, *12*, 2548 (1973).
751. C. Miniscloux, G. Martino, and L. Sajus, *Bull. Soc. Chim. France*, 2183 (1973).
752. T. A. George and S. A. D. Iske, Jr., *Proceedings of the International Conference on Nitrogen Fixation*, W. E. Newton and C. J. Nyman, Ed., Washington State University Press, in press.
753. D. J. Darensbourg, *Inorg. Nucl. Chem. Lett.*, *8*, 529 (1972).
754. L. K. Holden, A. H. Mawby, D. C. Smith, and R. Whyman, *J. Organometal .Chem.*, *55*, 343 (1973).
754a. T. Ito, T. Kokubo, T. Yamamoto, A. Tamamoto, and S. Ikeda, *J. Am. Chem. Soc.*, *96*, 1783 (1974).
754b. T. Tatsumi, M. Hidai, and Y. Uchida, *Inorg. Chem.*, *14*, 2531 (1975).
754c. J. Chatt, A. J. L. Pombeiro, R. L. Richards, G. H. D. Royston, K. W. Muir, and R. Walker, *J. Chem. Soc. Chem. Commun.*, 708 (1975).
754d. J. Chatt, M. Kubota, G. J. Leigh, F. C. March, R. Mason and D. J. Yarrow, *J. Chem. Commun.*, 1033 (1974).
755. G. A. Heath, R. Mason, and K. M. Thomas, *J. Am. Chem. Soc.*, *96*, 259 (1974).
756. D. Sellman, A. Brandl, and R. Endell, *J. Organometal. Chem.*, *49*, C22 (1973).
757. J. Chatt, G. A. Heath, and G. L. Leigh, *J. Chem. Soc. Chem. Commun.*, 444 (1972).
758. A. A. Diamantis, J. Chatt, G. J. Leigh, and G. A. Heath, *J. Organometal. Chem.*, *84*, C11 (1975).
759. A. A. Diamantis, J. Chatt, G. A. Heath, and G. J. Leigh, *J. Chem. Soc. Chem. Commun.*, 27 (1975).
760. E. E. Van Tamelen, J. A. Gladysz, and C. R. Brûlet, *J. Am. Chem. Soc.*, *96*, 3020 (1974).
761. J. Chatt, C. M. Elson, and R. L. Richards, *J. Chem. Soc. Chem. Commun.*, 189 (1974).
762. J. Chatt, A. J. Pearman, and R. L. Richards, *Nature*, *253*, 40 (1975).
762a. C. R. Brûlet and E. E. van Tamelen, *J. Am. Chem. Soc.*, *97*, 911 (1975),
762b. J. Chatt, A. J. Pearman and R. L. Richards, *J. Organometal. Chem.*, *101*, C45 (1975).
762c. J. Chatt, *J. Organometal. Chem.*, *100*, 17 (1975).
763. M. L. H. Green and W. E. Silverthorn, *Chem. Commun.*, 557 (1971).
764. M. L. H. Green and W. E. Silverthorn, *J. Chem. Soc. Dalton*, 2164 (1974).
765. J. Chatt, J. R. Dilworth, R. L. Richards, and J. R. Sanders, *Nature*, *224*, 1201 (1969).
766. J. Chatt, R. C. Fay, and R. L. Richards, *J. Chem. Soc. (A)*, 702 (1971).
767. M. Mercer, R. H. Crabtree, and R. L. Richards, *Chem. Commun.*, 808 (1973).
768. B. R. Davis and J. A Ibers, *Inorg. Chem.*, *10*, 578 (1971).
768a. P. D. Chadwick, J. Chatt, R. H. Crabtree, and R. L. Richards, *J. Chem. Soc. Chem. Commun.*, 351 (1975).
769. J. L. Thomas and H. H. Brintzinger, *J. Am. Chem. Soc.*, *94*, 1386 (1972).
770. J. L. Thomas, *J. Am. Chem. Soc.*, *95*, 1838 (1973).
770a. J. L. Thomas, *J. Am. Chem. Soc.*, *97*, 5943 (1975).
771. S. Otsuka, A. Nakamura, and H. Minamida, *Chem. Commun.*, 1148 (1969).
772. L. A. Nikonova, A. G. Ovcharenko, O. N. Efimov, V. A. Avilov, and A. E. Shilov, *Kinet. and Catal.*, *13*, 1427 (1972).
773. N. T. Denisov, E. I. Rudstein, N. I. Shuvalova, A. K. Shilova, and A. E. Shilov, *Dokl. Akad. Nauk. SSSR. 202*, 623 (1971).

774. L. J. Archer and T. A. George, *J. Organometal. Chem.*, *54*, C25 (1973).
775. J. Chatt, R. H. Crabtree, and R. L. Richards, *J. Chem. Soc. Chem. Commun.*, 534 (1972).
776. J. Chatt, J. R. Dilworth, R. L. Richards, and J. R. Sanders, *Nature*, *224*, 1201 (1969).
777. A. P. Gaughan, Jr., B. L. Haymore, J. A. Ibers, W. H. Myers, T. E. Nappier, Jr., and D. W. Meek, *J. Am. Chem. Soc.*, *95*, 6861 (1973).
778. R. B. King and M. B. Bisnette, *J. Am. Chem. Soc.*, *86*, 5694 (1964).
779. R. B. King and M. B. Bisnette, *Inorg. Chem.*, *5*, 300 (1966).
780. M. L. H. Green, T. R. Sanders, and R. N. Whiteley, *Z. Naturforsch.*, *23b*, 106 (1968).
781. M. F. Lappert and J. S. Poland, *Chem. Commun.*, 1961 (1969).
782. S. Trofimenko, *Inorg. Chem.*, *8*, 2675 (1969).
783. M. E. Deane and F. J. Lalor, *J. Organometal. Chem.*, *57*, C61 (1973).
784. M. E. Deane and F. J. Lalor, *J. Organometal. Chem.*, *67*, C19 (1974).
785. T. A. James and J. A. McCleverty, *J. Chem. Soc. (A)*, 1068 (1971).
786. G. Avitabile, P. Ganis, and M. Nemiroff, *Acta Cryst.*, *B27*, 725 (1971).
787. D. Sutton, *Can. J. Chem.*, *52*, 2634 (1974).
788. W. E. Carroll, M. E. Deane, and F. J. Lalor, *J. Chem. Soc. Dalton*, 1837 (1974).
789. M. W. Bishop, J. Chatt, and J. R. Dilworth, *J. Organomet. Chem.*, *73*, C59 (1974).
789a. V. W. Day, T. A. George, and S. O. A. Iske, *J. Am. Chem. Soc.*, *97*, 4127 (1975).
790. A. Nakamura, M. Aotake, and S. Otsuka, *J. Am. Chem. Soc.*, *96*, 3456 (1974).
791. A. Nakamura and S. Otsuka, *J. Am. Chem. Soc.*, *94*, 1886 (1972).
792. S. D. Ittel and J. A. Ibers, *Inorg. Chem.*, *12*, 2290 (1973).
793. W. G. Kita, J. A. McCleverty, B. E. Mann, D. Seddon, G. A. Sim, and D. I. Woodhouse, *J. Chem. Soc. Chem. Commun.*, 132 (1974).
793a. N. A. Bailey, P. D. Frisch, J. A. McCleverty, N. W. Walker, and J. Williams, *J. Chem. Soc. Chem. Commun.*, 350 (1975).
794. P. C. H. Mitchell and R. D. Scarle, *Nature*, *240*, 417 (1972).
794a. F. C. March, R. Mason, and K. M. Thomas, *J. Organometal. Chem. C43*, 96 (1975).
795. M. L. H. Green and J. R. Sanders, *Chem. Commun.*, 956 (1967).
796. C. K. Prout, T. S. Cameron, and A. R. Gent, *Acta Cryst.*, *B28*, 32 (1972).
797. J. R. Knox and C. K. Prout, *Acta Cryst.*, *B25*, 1952 (1969).
798. M. L. H. Green and J. R. Sanders, *J. Chem. Soc. (A)*, 1947 (1971).
799. J. Chatt and J. R. Dilworth, *J. Chem. Soc. Chem. Commun.*, 549 (1972).
799a. D. Sellmann, A. Brandt, and R. Endell, *J. Organometal. Chem.*, *97*, 229 (1975).
799b. J. A. Broomhead, J. Budge, J. H. Enemark, R. D. Feltham, J. I. Gelder, and P. L Johnson, Submitted for publication, J. H. Enemark, private communication.
800. W. P. Griffith, *Coord. Chem. Rev.*, *8*, 370 (1972).
801. A. Rosenheim and F. Jacobsohn, *Z. Anorg. Allg. Chem.*, *50*, 297 (1906).
802. K. Dehnicke and J. Strähle, *Z. Anorg. Allg. Chem.*, *339*, 171 (1970).
803. W. Kolitsch and K. Dehnicke, *Z. Naturforsch.*, *B25*, 1080 (1970).
804. J. Strähle, *Angew. Chem.* (*Intl. ed. England*), *8*, 925 (1969).
805. J. Strähle, *Z. Anorg. Allg. Chem.*, *375*, 239 (1970).
806. R. D. Bereman, *Inorg. Chem.*, *11*, 1149 (1972).
807. J. Chatt and J. R. Dilworth, *J. Chem. Soc. Chem. Commun.*, 517 (1974).
808. J. Chatt and J. R. Dilworth, *J. Chem. Soc. Chem. Commun.*, 508 (1974).
808a. J. Chatt and J. R. Dilworth, *J. Chem. Soc. Chem. Commun.*, 983 (1975).
809. D. Scott and A. G. Wedd, *J. Chem. Soc. Chem. Commun.*, 527 (1974).
810. J. H. Enemark and R. D. Feltham, *Proc. Nat. Acad. Sci. U.S.A.*, *69*, 3534 (1972).
811. J. H. Enemark and R. D. Feltham, *Coord. Chem. Rev.*, *13*, 339 (1974).

812. N. G. Connelly, *Inorg. Chim. Acta Rev.*, *6*, *48* (1972).
813. W. Hieber, R. Nast, and G. Gehring, *Z. Anorg. Chem.*, *256*, 169 (1948).
814. W. P. Griffith, J. Lewis, and G. Wilkinson, *J. Chem. Soc.*, 872 (1959).
815. D. H. Svedung and N-G. Vannerberg, *Acta Chem. Scand.*, *22*, 1551 (1968).
816. T. S. Piper and G. Wilkinson, *J. Inorg. Nucl. Chem.*, *3* 104 (1956).
817. F. A. Cotton and P. Legzdins, *J. Am. Chem. Soc.*, *90*, 6232 (1968).
818. N. A. Bailey, W. G. Kita, J. A. McCleverty, A. J. Murray, B. E. Mann, and N. W. Walker, *J. Chem. Soc. Chem. Commun.*, 592 (1974).
819. W. R. Robinson and M. E. Swanson, *J. Organometal. Chem.*, *35*, 315 (1972).
820. F. A. Cotton and B. F. G. Johnson, *Inorg. Chem.*, *3*, 1609 (1964).
821. F. Canziani, U. Sartorella, and F. Cariati, *Ann. Chem.* (*Rome*), *54*, 1354 (1964) through *Chem. Abst.*, *62*; 14159g.
822. B. F. G. Johnson, *J. Chem. Soc.* (*A*), 475 (1974).
823. J. A. Bowden, R. Colton, and C. J. Commons, *Austral. J. Chem.*, *25*, 1393 (1972).
824. L. Bencze, *J. Organomet. Chem.*, *56*, 303 (1973).
825. W. B. Hughes and E. A. Zuech, *Inorg. Chem.*, *12*, 471 (1973).
826. M. O. Visscher and K. G. Caulton, *J. Am. Chem. Soc.*, *94*, 5293 (1972).
827. S. J. Lippard and G. J. Palenik, *Inorg. Chem.*, *10*, 1322 (1971).
828. W. B. Hughes and B. A. Baldwin, *Inorg. Chem.*, *13*, 1531 (1974).
829. M. W. Anker, R. Colton, and I. B. Tomkins, *Austral. J. Chem.*, *21*, 1149 (1968).
830. M. W. Anker, R. Colton, and I. B. Tomkins, *Austral. J. Chem.*, *20*, 9 (1967).
831. L. Bencze, J. Kohan, B. Mohai, and L. Markó, *J. Organometal. Chem.*, *70*, 421 (1974).
832. U. Sartorelli, F. Zingales, and F. Canziani, *Chim. Ind.* (*Milan*), *49*, 751 (1967) through *Chem. Abst.*, *68*, 35443C (1968).
833. R. D. Feltham. W. Silverthorn, and G. McPherson, *Inorg. Chem.*, *8*, 344 (1969).
834. R. Davis, B. F. G. Johnson, and H. H. Al-Obaidi, *J. Chem. Soc. Dalton*, 508 (1972).
835. B. F. G. Johnson, K. H. Al-Obaidi, and J. A. McCleverty, *J. Chem. Soc.* (*A*), 1668 (1969).
836. M. Green and S. H. Taylor, *J. Chem. Soc. Dalton*, 2629 (1972).
837. B. F. G. Johnson, A. Khair, C. G. Savory, and R. H. Walter, *J. Chem. Soc. Chem. Commun.*, 744 (1974).
838. W. G. Kita, J. A. McCleverty, B. Patel, and J. Williams, *J. Organometal. Chem.*, *74*, C9 (1974).
839. E. A. Zeuch, *Chem. Commun.*, 1182 (1968).
840. E. A. Zuech, W. B. Hughes, D. H. Kubicek, and E. T. Kittleman, *J. Am. Chem. Soc.*, *92*, 528 (1970).
841. R. Taube and K. Seyferth, *Z. Chem.*, *14*, 284 (1970).
841a. R. J. Haines and G. J. Leigh, *Chem. Soc. Rev.*, *4*, 155 (1975).
842. W. B. Hughes, *J. Am. Chem. Soc.*, *92*, 532 (1970).
843. W. B. Hughes, *Chem. Commun.*, 431 (1969).
843a. T. J. Katz and J. McGinniss, *J. Am. Chem. Soc.*, *97*, 1594 (1975).
843b. R. H. Grubbs, P. L. Burk, and D. D. Carr, *J. Am. Chem. Soc.*, *97*, 326(1975)5 .
844. A. P. Ginsberg, *Transition Metal Chem.*, *1*, 112 (1965).
845. M. L. H. Green and D. J. Jones, *Adv. Inorg. Chem. Radiochem.*, *7*, 115 (1965).
846. B. L. Shaw, *Inorganic Hydrides*, Pergamon Press, New York, 1967.
847. H. D. Kaesz and R. B. Saillant, *Chem. Rev.*, *72*, 231 (1972).
848. E. L. Muetterties, Ed., *Transition Metal Hydrides*, Dekker, New York, 1971.
849. J. P. McCue, *Coord. Chem. Rev.*, *10*, 265 (1973).

850. E. O. Fischer and Y. Hristidu, *Z. Naturforsch.*, *156*, 135 (1960).
851. M. L. H. Green, L. C. Mitchard, and W. E. Silverthorn, *J. Chem. Soc. Dalton*, 1361 (1974).
852. F. Penella, *Chem. Commun.*, 158 (1971).
853. A. Frigo, G. Puosi, and A. Turco, *Gazz. Chim. Ital.*, *101*, 637 (1971).
854. T. Ito, T. Kokubo, T. Yamamoto, A. Yamamoto, and S. Ikeda, *J. Chem. Soc. Chem. Commun.*, 136 (1974).
855. J. P. Jesson, E. L. Muetterties, and P. Meakin, *J. Am. Chem. Soc.*, *93*, 5261 (1971).
856. P. Meakin, L. J. Guggenberger, W. G. Peet, E. L. Muetterties, and J. P. Jesson, *J. Chem. Soc.*, *95*, 1467 (1973).
857. B. Bell, J. Chatt, G. J. Leigh, and T. Ito, *J. Chem. Soc. Chem. Commun.*, 34 (1972).
858. T. Kruck and A. Prasch, *Z. Naturforsch.*, *19b*, 669 (1964); R. J. Clark and P. I. Hoberman, *Inorg. Chem.*, *4*, 1771 (1965).
859. F. W. S. Benfield and M. L. H. Green, *Chem. Commun.*, 1274 (1971).
860. F. W. S. Benfield and M. L. H. Green, *J. Chem. Soc. Dalton*, 1244 (1974).

The Derivation and Application of Normalized Spherical Harmonic Hamiltonians

by J. C. DONINI, B. R. HOLLEBONE

Dept. of Chemistry (JCD)
St. Francis Xavier Univesity
Antigonish, Nova Scotia,
Canada

Dept. of Chemistry (BRH)
University of Alberta
Edmonton, Alberta
Canada

and A. B. P. LEVER

Department of Chemistry
York University
Downsview (Toronto), Ontario, Canada

Part 1

Part 2 Calculation Techniques

PART 1

I. INTRODUCTION

At the present time there is essentially no standardization in the Hamiltonians used to describe molecules of low symmetry. The use of diverse Hamiltonians and various coordinate frames of reference has made it extremely difficult to compare data obtained from one molecule with that obtained from another, particularly if they are of different symmetry. As a consequence it is not surprising that while we have a good working knowledge of the electronic spectra of octahedral and tetrahedral derivatives, our command of lower symmetry systems is relatively dismal. Indeed, few authors attempt to extract chemically useful information from the spectra of low-symmetry molecules, even when single-crystal polarized spectra are available.

Recently, we (1, 2) have demonstrated the utility of normalized spherical harmonic (NSH) Hamiltonians as a means of analyzing the electronic spectra (and magnetism, etc.) of both cubic and noncubic metal complexes. This procedure provides a number of advantages, including the necessary standardization alluded to above. In this review article we explore, in depth, the applicability of this approach and its relationship with alternate procedures.

We have chosen to present this material in a pedagogical style at a level appropriate to those who might wish to analyze their spectra, but who lack a detailed theoretical background. A knowledge of group theory to the level of Cotton's book (3) is presumed. Moreover, the article is divided into two sections. The first section describes the procedures and provides the tools necessary to carry out an analysis of an electronic spectrum, without a great deal of formal proof. It also discusses in some detail the value of using the NSH approach in relation to other current methods, and hints at valuable extensions of the procedure, such as coupling coefficients and symmetry-ascent selection rules. The second part provides some of the theoretical background in more detail and illustrates how, with the help of the literature, the reader can become proficient in this area. In addition, we work through a complete calculation from start to finish.

II. HISTORICAL PERSPECTIVE

The original work by Bethe (4) and Van Vleck (5), dealing with crystal-field theory, magnetic phenomena, and atomic spectroscopy, was based on a firm understanding of group theoretical principles available at the time. Recent work in the area of transition-metal electronic spectroscopy and in particular the two most often used texts, by Figgis (6) and Ballhausen (7) published in the 1960s, do not take full advantage of the currently available knowledge in group theory and vector algebra. These texts, while presenting an excellent introduction to the classical approach, fail to realize the full power of group theory. The more recent text by Lever (8) must be subject to the same criticism. In particular, apart from notable exceptions of Sugano, Tanabe, and Kamimura (9) and J.S. Griffith (10, 11), comparatively little use has been made of the powerful techniques developed by Wigner (12, 13) and Racah (14–17) in dealing with the spectra of complex ions, and later extended by Judd (18) and Wybourne (19) to the spectra of complexes of the *f* block. Many early crystal-field calculations (e.g., *Ref.* 20), while important at the time, distinguish themselves by the enormous complexity of the algebraic manipulation needed and the lack of meaningful comments on the significance of the radial parameters arising in the lower symmetries (21). Even today such radial parameters have little value to the chemist because of the utilization of poorly constructed Hamiltonians to describe these lower symmetries. Thus chemically useful concepts such as the spectrochemical and nephelauxetic series have developed from investigations of cubic systems, but similar correlations have not been generally forthcoming, or indeed possible, in the lower symmetries until recently (1, 22).

The advent of the computer in conjunction with vector coupling techniques has made possible calculations of considerable complexity (e.g., Ref. 23); however, there has not been a parallel development in the definition of the Hamiltonians. The NSH formalism, firmly based on group theoretical arguments of general validity, fills this vacuum. The principles of group theory are central to formulation of the various theories which aim to rationalize and explain the empirical parameters resulting from use of the new Hamiltonians. Thus, following and expanding on the original investigations of Griffith, Tanabe, Sugano, and others, the definitions of all operators, wave functions, and their coupling coefficients (24, 25) have been explored in terms of exclusively group theoretical arguments.

III. THE HAMILTONIAN OPERATORS OF COMPLEX IONS

In the weak-field approach the energy levels of a complexed ion are calculated as a first-order perturbation of the levels of the spherical free ion (6–8). The energies are the solution of the matrix element

$$\int \psi^* H \psi d\tau = E \qquad (1)$$

in which

$$H = H_o + H_p \qquad (2)$$

The unperturbed operator H_o is the spherical Hamiltonian of the free ion, whereas H_p reduces this symmetry from spherical to that appropriate to the geometry of the environment. This perturbing geometry can always be uniquely described by a catalog of symmetry operations, which when used leave the environment unchanged in appearance. Therefore, any analytical function used to describe the geometry of the environment must be unchanged by the same set of symmetry operations. *In group theoretical language this mathematical function is said to transform as the fully symmetric representation of the finite point group which describes the physical system.*

The shape of the environment of the ion can be regarded in a second way. The solid geometric figure that is formed by joining the ligands, such as octahedron for six equivalent ligands, can be described by a set of vectors which point toward the unique equivalent symmetry features, such as the centers of the faces of the solid body. Such a set of vectors is called a tensor. Its form can be described by a reducible set of analytical polynomial functions of degree q. The $(2q+1)$ vector components must be orthogonal and normalized and can be degenerate, under an appropriate operator, only if they exist in a space containing at least $(2q + 1)$ dimensions (17). The highest order tensor, therefore, that can retain independent and degenerate vector components in three-dimensional nonspherical space is of order 1. If tensors of higher order are forced into three dimensions, some vector components must become different in energy from others. This occurs if a tensor of order 2, or more (e.g., the set of five d orbitals), is constrained into a nonspherical environment.

There are many types of polynomial function which can be used to describe the shape of the environment. However, spherical harmonics (26) are usually chosen because H_p is a perturbation of the spherical ion whose original Hamiltonian H_o and basis states are almost invariably formulated with these functions. Spherical harmonics map directly onto the properties of tensors since, for the set of harmonics of degree l, there are $(2l + 1)$ degenerate values of m_l. Thus each value m_l represents an orthogonal vector component of the complete tensor. In the group theoretical sense, the tensor of degree l is a basis for the lth representation of degeneracy $(2l + 1)$ of the spherical point group R_3. For example, the functions acting as a basis for the d representations of R_3 are the components of a tensor of a degree 2 (i.e., $l = 2$). The spherical harmonics may be thought of as functions which can be used to describe the standing waves (normal vibrations) set up on a flooded planet whose ocean is of uniform depth (26). The harmonics and their components have both magnitude and direction, and by appropriate combination any geometry, being an angular distortion from spherical symmetry, can be simulated as a standing wave in this ocean. The description of this geometry, and therefore of H_p (27), will not necessarily require all components of a particular harmonic, but only a limited number, or none at all. Moreover, these functions will be multiplied by a scalar magnitude which may be different in different orthogonal directions. The first task of any theory of complex ions (6–10, 28–30), therefore, is to define correctly the combination of spherical vector components which, together with their appropriate magnitude, transforms as the totally symmetric representation of the desired environment point group.

In the development of such theories, three distinct procedures have evolved.

1. *Bond addition.* In this model, generally referred to as the orbital angular overlap approach (25, 29, 31–37), each bond is assigned as a separate vector component. These are then added vectorially to form H_p as if the complex were assembled from separate bond entities. This model has attraction because of the relative ease with which some parameters can

apparently be related to the chemistry of the metal and ligands involved. It has the important disadvantage that it must assume that bond properties are linearly additive.

2. *Symmetry adaptation.* In this approach a combination of spherical vector components is derived by subjecting the tensors in turn to the transformation matrices appropriate to the generating operators of the group. This procedure selects multiples of the correct linear combinations of vector components forming bases of H_p, but as conventionally applied, the normalization relationships are ignored (6, 7, 10, 38). This scheme must be uniquely applied to every different complex geometry. Moreover, the neglect of normalization relationships results in different parameters being derived for each different axis chosen as the principal generator (often called axis of quantization). Thus two groups of workers dealing with the same compound, but using different axes of quantization, would produce different parameter values. This is hardly an approach that could lead to a unified treatment of the bonding in noncubic metal complexes. Crystal field Hamiltonians (39) suffer this disadvantage.

3. *Projection and subduction.* This procedure (1, 2) recognizes that most 6-coordinate and many 5- and 4-coordinate complexes can be considered to arise through some kind of distortion, either radial or angular, from an octahedron. Similarly, there are a series of 4- and 3-coordinate systems formally derivable from a tetrahedron. These are regarded as generative or parent groups. The set of symmetry operations common to both the generative group and the less-symmetric subgroup of interest is used to project the basis functions of the generative group into the representations of the subgroup. Indeed, by using the angular momentum l, m_l vectors of the infinite group R_3, this group may be regarded as the generative group for the entire series of complexes. Thus appropriate combinations of the l, m_l vectors comprising the S, P, D, F, and other free-ion atomic terms (R_3) provide bases for the group theoretical representations of the octahedral group 0_h (Table I). Note, however, that for a multidimensional representation of 0_h, such as T_{2g}, the three components can be linearly combined in an infinite number of ways, yet remain a proper basis for T_{2g} (3). Therefore, if we are interested in a specific subgroup of 0_h, such as C_{2v}, we use the projection operator to choose linear combinations of these components which transform correctly as unidimensional representations in C_{2v}. In the example chosen the three components are required to transform as A_2, B_1, and B_2 in C_{2v}. *Indeed the, components of each multidimensional representation in O_h are projected from R_3 such that, in a chain of subgroups (section III. A), they provide normalized bases for all the representations with which they correlate.* The process is similar to symmetry adaptation, but the use of a fully defined projection operator rather than the arbitrarily chosen transformation matrices guarantees several important properties of the projected functions, including the

TABLE I
Mapping of lm_l Quantum Numbers from R_3 into O_h and T_d

l	m_l[a]	L (R_3)	Γ (O_h)	Γ (T_d)
0	0	S	A_{1g}	A_1
1	m_l	P	T_{1u}	T_2
2	m_l	D	$E_g + T_{2g}$	$E + T_2$
3	m_l	F	$A_{2u} + T_{2u} + T_{1u}$	$A_1 + T_2 + T_1$
4	m_l	G	$A_{1g} + E_g + T_{1g} + T_{2g}$	$A_1 + E + T_1 + T_d$
etc.				

[a]Recall that m_l assumes all integer values from $+l$ to $-l$.

preservation of all phase- and coordinate-system relationships. As a consequence, *all basis functions for all finite groups become directly traceable to the conventions in R_3*. Moreover, the use of the complete projection operator imposes normalization conditions on the projected functions which necessarily retain all information concerning the magnitudes as well as the directions of the projected basis vectors. Thus the magnitudes of projected parameters are also directly traceable to the magnitudes of the functions in R_3. The complete solution of the electronic structure requires no external assumptions about the nature of the metal-ligand bonding, and produces the minimal set of empirical parameters.

Of these three systems, the projection technique is the most fully formalized, but at the same time involves in its application the smallest number of arbitrary choices because all relationships are defined by reference to the single generative group R_3. As indicated above, for practical convenience, the octahedral group O_h, the first important subgroup of R_3, can be chosen as the generative group for most geometries ordinarily found in transition metal complexes.

The octahedral group can be quantized on at least four separate types of axis, namely, C_4^z, C_2^z, C_2^{xy} and C_3^{xyz}, producing four apparently different but equivalent sets of basis functions from linear combinations of spherical harmonics. Since in each case the totally symmetric representation is normalized, the perturbation Hamiltonian yields identical magnitudes for the splitting parameter DQ (or any other parameter) along each axis.

To distinguish projected normalized parameters from conventional parameters, upper case nomenclature (e.g., DQ, DS, DT, etc.) will be employed instead of the lowercase (Dq, Ds, Dt, etc.).

A. Chains of Subgroups

Four separate, physically significant chains of subgroups can be defined from the octahedron, one related to each of the previously mentioned axes of quantization. A chain of sugroups is defined (1, 40–44) as a sequence in which each successive point is a proper subgroup (3) of any predecessor, that is, all operators in a given group exist in all previous higher groups in the chain. Series of related complex ions may belong to such chains. For example, consider an octahedral complex ML_6 belonging to the group O_h. If two trans bonds are similarly elongated or compressed, or if there is trans disubstitution, for example, *trans*-ML_4Z_2, a D_{4h} complex is generated. Differences in the two trans ligands, or removal of one of them, for example, to form ML_4Z, provide complexes belonging to the group C_{4v}. If the in-plane angles of such a complex deviate equivalently from 90°, a C_{2v} complex will arise; clearly there are alternate routes. Thus formation of the trans-$MX_2Y_2Z_2$ will yield a D_{2h} complex, which will become C_{2v} if two trans ligands are changed (*trans*-MX_2Y_2ZZ'). If the XY ligands are considered to be bidentate, then the *trans*-$M(XY)_2Z_2$ belongs to C_{2h}. If in this latter trans complex the bidentate groups have a cis arrangement of the XY groups (Fig. 1), then another C_{2v}

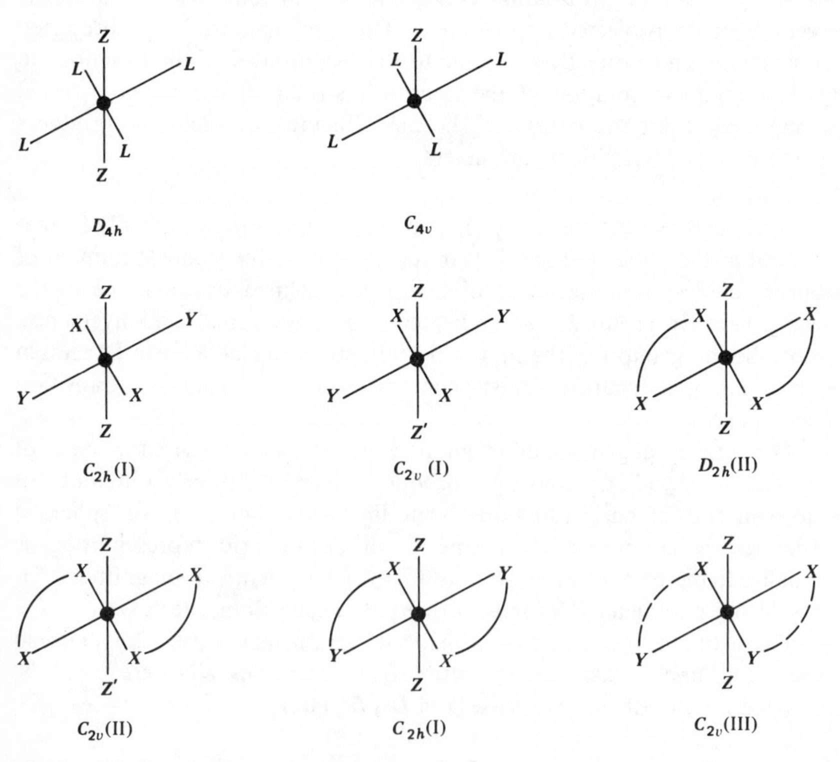

Fig. 1. Some examples of complexes belonging to various D_{4h}, C_{4v}, D_{2h}, C_{2v}, and C_{2h} point groups.

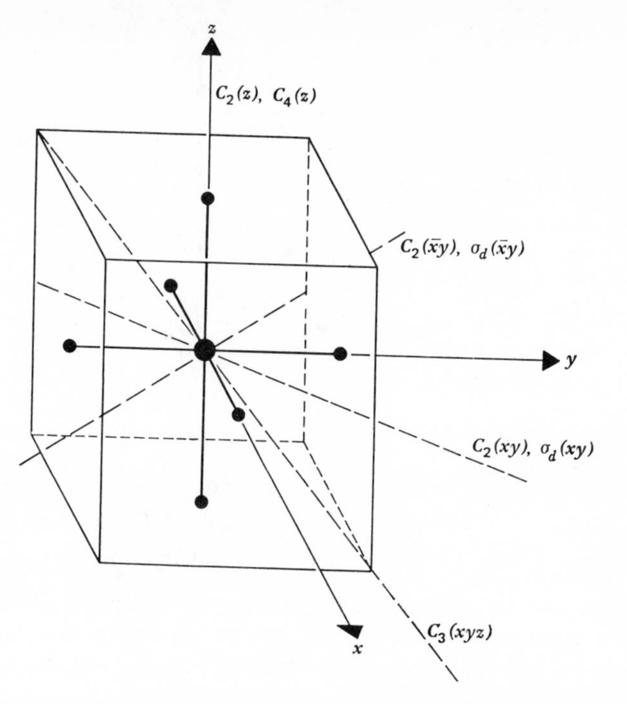

Fig. 2. Coordinate reference frame and some symmetry elements of an octahedron.

group is achieved. Obviously many other groups can be conceived to occur, especially if polyatomic ligands are considered.

It is important to recognize that the several C_{2v} systems mentioned above are not necessarily identical, nor are they necessarily identical with the C_{2v} system which would obtain if we consider substitution of a tetrahedral ML_4 species, to make ML_2X_2. Depending upon the specific operators which survive from the cubic parent, three distinguishable C_{2v} systems can be designated. These are defined (1) by $C_{2v}(\mathrm{I})$, $C_{2v}(\mathrm{II})$, and $C_{2v}(\mathrm{III})$, where the roman numerals designate the surviving operators (see Fig. 2), namely:

$$\begin{aligned} C_{2v}(S) \quad S &= \mathrm{I} = [E, C_2(z), \sigma_h(xz), \sigma_h(yz)] \\ S &= \mathrm{II} = [E, C_2(z), \sigma_d(xy), \sigma_d(\bar{x}y)] \\ S &= \mathrm{III} = [E, C_2(xy), \sigma_h(xy), \sigma_d(xy)] \end{aligned} \tag{3}$$

In a similar fashion, two different D_{2h} groups may arise [specified by $D_{2h}(\mathrm{I})$ and $D_{2h}(\mathrm{II})$], the relevant operators being

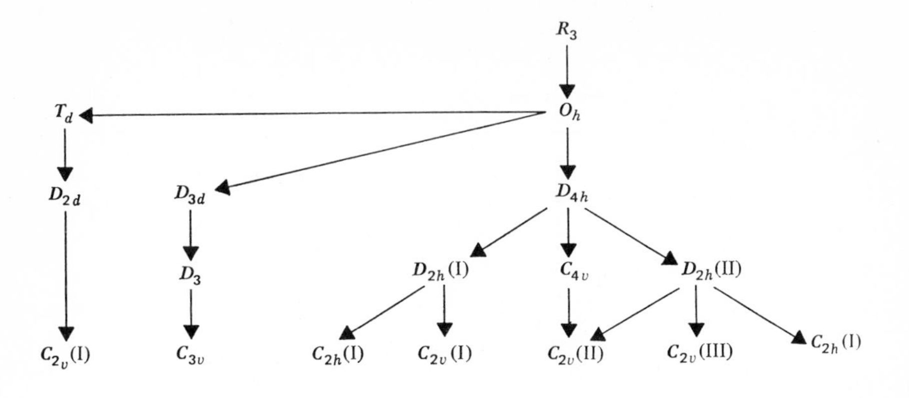

Fig. 3. Various fourfold, threefold, and twofold subduction chains.

$$\begin{aligned} D_{2h}(S) \quad S &= \text{I} = [E, 3C_2, i, 3\sigma_h] \\ S &= \text{II} = [E, C_2(z), C_2(xy), C_2(\bar{x}y), i, \\ &\qquad \sigma_h(xy), \sigma_d(xy), \sigma_d(\bar{x}y)] \end{aligned} \tag{4}$$

The above discussion is summarized in Fig. 3 (see also Tables II-IV). Examples of molecules belonging to these various groups are discussed in Section IV. A–C). These chains have a number of relevant properties (45, 46), namely:

1. The totally symmetric representation of any group is also the totally symmetric representation for all of its subgroups.
2. One or more of the functions transforming as the totally symmetric representation of a group will derive from functions which are not fully symmetric in the groups preceding it in the chain.
3. Multidimensional representations in the parent group might be reducible in its subgroups.

B. The Subduction Criterion

We demonstrated above that it is possible to project out of R_3 a set of basis functions which will transform correctly as bases for the various representations in all the groups of a chain. The special operator which achieves this projection is called a *subduction operator*, and the basis functions of the subgroup are said to be *subduced* from the generative group bases, whereas the basis functions of the generative group are said to be *reduced* with respect to the subgroup.

The vector components of the representations of R_3 can each be labeled with a specific numerical value of m_l. Because the projection operator preserves all the conventions of R_3 on projection into O_h, the labeling scheme of

TABLE II
Basis Functions and Correlation Table for the Gerade Irreducible Representations of O_h Relevant to Chains Terminating in C_{2v}(I) or C_{2h}(I)

γ	O_h	D_{4h}	C_{4h} or	D_{2d}(I)[a] or	D_{2h}(I)[b]	C_{2v}(I)[c] or	C_{2h}(I)[d]	Bases[e]
0	A_{1g}	A_{1g}	A_1	A_1	A_g	A_1	A_g	R
0	$E_g\theta$	A_{1g}	A_1	A_1	A_g	A_1	A_g	$(2z^2 - x^2 - y^2)$
2+	$E_g\varepsilon$	B_{1g}	B_1	B_2	A_g	A_1	A_g	$\sqrt{3}(x^2 - y^2)$
0	A_{2g}	B_{1g}	B_1	B_2	A_g	A_1	A_g	$(x^2 - y^2)(y^2 - z^2)(z^2 - x^2)$
2−	$T_{2g}(xy)$	B_{2g}	B_2	B_1	B_{1g}	A_2	A_g	xy
0	$T_{1g}(z)$	A_{2g}	A_2	A_2	B_{1g}	A_2	A_g	S_z
1+	$T_{2g}(xz)$	$E_g(xz)$	$E(xz)$	$E(xz)$	B_{2g}	B_1	B_g	xz
1+	$T_{1g}(y)$	$E_g(y)$	$E(y)$	$E(y)$	B_{2g}	B_1	B_g	S_y
1−	$T_{2g}(yz)$	$E_g(yz)$	$E(yz)$	$E(yz)$	B_{3g}	B_2	B_g	yz
1−	$T_{1g}(x)$	$E_g(x)$	$E(x)$	$E(x)$	B_{3g}	B_2	B_g	S_x

[a] I = $\{E, 2S_4, C_2(z), 2C_2', 2\sigma_h\}$.
[b] I = $\{E, 3C_2, i, 3\sigma_h\}$.
[c] I = $\{E, C_2(z), \sigma_h(xz), \sigma_h(yz)\}$.
[d] I = $\{E, C_2(z), i, \sigma_h\}$.
[e] The bases transforming as each set of representations in a chain are given in the extreme right-hand column. This information includes the transformation properties of the d orbitals. Readers should take note that the transformation property of a given d orbital (or function) shown in Tables II to IV, for say a C_{2v} group, will not necessarily be the same as that in the right-hand column of a standard character table for C_{2v}. This situation arises because the surviving operators differ in the various C_{2v} groups and are not necessarily the same as those in the standard character table. S_z denotes a function that transforms like z but does not change sign under inversion (1).

TABLE III
Basis Functions and Correlation Table for the Gerade Irreducible Representations of O_h Relevant to Chains Terminating with C_{2v} (II) or with C_{2v} (III)

γ[e]	O_h	D_{4h}	C_{4v}	or D_{2h}(II)[a]	C_{2v}(II)[b]	C_{2v}(III)[c]	Bases[d]
0	A_{1g}	A_{1g}	A_1	A_g	A_1	A_1	R
0	$E_g\theta$	A_{1g}	A_1	A_g	A_1	A_1	$(2z^2 - x^2 - y^2)$
2−	$T_{2g}(xy)$	B_{2g}	B_2	A_g	A_1	A_1	xy
2−	A_{2g}	B_{1g}	B_1	B_{1g}	A_2	B_1	$(x^2 - y^2)(y^2 - z^2)(z^2 - x^2)$
2+	$E_g\varepsilon$	B_{1g}	B_1	B_{1g}	A_2	B_1	$\sqrt{3}(x^2 - y^2)$
0	$T_{1g}(z)$	A_{2g}	A_2	B_{1g}	A_2	B_1	S_z
1+	$T_{1g}\tau_1$	$E_g\tau_1$	$E\tau_1$	B_{2g}	B_1	B_2	$\left(\frac{1}{\sqrt{2}}\right)(S_x - S_y)$
1+	$T_{2g}\tau_2$	$E_g\tau_2$	$E\tau_2$	B_{2g}	B_1	B_2	$\left(\frac{1}{\sqrt{2}}\right)(yz + xz)$
1−	$T_{1g}\tau_1'$	$E_g\tau_1'$	$E\tau_1'$	B_{3g}	B_2	A_2	$\left(\frac{1}{\sqrt{2}}\right)(S_x + S_y)$
1−	$T_{2g}\tau_2'$	$E_g\tau_2'$	$E\tau_2'$	B_{3g}	B_2	A_2	$\left(\frac{1}{\sqrt{2}}\right)(yz - xz)$

[a]II = $\{E, C_2(z), C_2(xy), C_2(\bar{x}y), i, \sigma_h(xy), \sigma_d(xy), \sigma_d(\bar{x}y)\}$.
[b]II = $\{E, C_2(z), \sigma_d(xy), \sigma_d(\bar{x}y)\}$
[c]III = $\{E, C_2(xy), \sigma_h(xy), \sigma_d(xy)\}$. *Note.* C_{2v}(III) is not subduced by C_{4v}.
[d]Ref. 1, see footnote to Table II.
[e]Appropriate to C_{2v}(II).

TABLE IV
Basis Functions and Correlation Table for the Gerade Irreducible Representations of T_d and Its Subgroups

γ	O_h	T_d	D_{2d}(II)[a]	C_{2v}(II)[b]	Bases[c]
0	A_{1g}	A_1	A_1	A_1	R
0	$E_g\theta$	E	A_1	A_1	$(2z^2 - x^2 - y^2)$
2−	$T_{2g}(xy)$	$T_2(xy)$	B_2	A_1	xy
2−	A_{2g}	A_2	B_1	A_2	$(x^2 - y^2)(y^2 - z^2)(z^2 - x^2)$
2+	$E_g\varepsilon$	E	B_1	A_2	$\sqrt{3}(x^2 - y^2)$
0	$T_{1g}(z)$	$T_1(z)$	A_2	A_2	S_z
1+	$T_{1g}\tau_1$	$T_1\tau_1$	$E\tau_1$	B_1	$\left(\frac{1}{\sqrt{2}}\right)(S_x - S_y)$
1+	$T_{2g}\tau_2$	$T_2\tau_2$	$E\tau_2$	B_1	$\left(\frac{1}{\sqrt{2}}\right)(yz + xz)$
1−	$T_{1g}\tau_{1'}$	$T_1\tau_{1'}$	$E\tau_{1'}$	B_2	$\left(\frac{1}{\sqrt{2}}\right)(S_x + S_y)$
1−	$T_{2g}\tau_{2'}$	$T_2\tau_{2'}$	$E\tau_{2'}$	B_2	$\left(\frac{1}{\sqrt{2}}\right)(yz - xz)$

[a]II = $\{E, 2S_4, 3C_2, 2\sigma_d\}$
[b]II = $\{E, C_2(z), \sigma_d(xy), \sigma_d(\bar{x}y)\}$.
[c]Ref. 1, see footnote to Table II.
Note. Three fold chains of O_h and T_d. The A_{1g}, A_{2g}, and E_g representations of O_h correlate simply with A_1, A_2, and E down chain (Eq. 14). Similarly, T_{1g} correlates with $A_2 + E$, and T_{2g} with $A_1 + E$.

vector components in R_3 can be projected into a finite set of labels appropriate to the components of the O_h representation. The mapping creates a finite set of labels whose size is limited by the order of the quantization axis. While many useful consequences of this quantitative labeling are described in Part 2, its most important property is the assignment of a label $|\Gamma O|$ only to those components of any representation L with no net azimuthal angular momentum.*

The set of components of any finite group labeled $|\Gamma O|$ must include the totally symmetric representation since the Hamiltonian remains unaffected by any symmetry operation. This must remain true for all proper subgroups; but during subduction, parts of representations of the generative group other than the totally symmetric may correlate to the totally symmetric representation of the subgroup. In physical terms the descent in symmetry may require the addition of more distortion vectors to the perturbation Hamiltonian. For example, the two components of the E_g representation in O_h transform as A_{1g} and B_{1g} in the subgroup D_{4h}. Thus to produce the D_{4h} Hamiltonian, we must add to the octahedral Hamiltonian the vector component of E_g which transforms as A_{1g} in D_{4h}. In general, therefore, the Hamiltonian for any subgroup is obtained by adding together those octahedral representation components which transform as the totally symmetric representation in the subgroup concerned. Any ambiguity in this choice is easily removed by choosing only those components bearing the label $|\Gamma O|$ from the correlating representations of the generative group. This rule is called the *subduction criterion*, and it provides a direct unambiguous method of formulating the perturbation Hamiltonian of any physically significant subgroup if the bases of the generative group have been projected from R_3 onto the appropriate axis.

IV. EXAMPLES OF THE USE OF THE SUBDUCTION CRITERION

According to the formation of physically significant chains, the Hamiltonian operator for any subgroup is constructed by addition of any necessary additional vector components of O_h to the Hamiltonian of its immediate predecessor in the chain. Then

$$H_p = H_o + \sum_G H_G \tag{5}$$

*Namely eigenstates for which $\hat{L}_z|\Gamma\gamma> = 0|\Gamma\gamma>$. Note, as discussed in part 2, that certain components which carry the 2+label have no azimuthal angular momentum and can contribute to the Hamiltonian operator. This also applies to the $|T_{2g}2-(xy)|$ operator quantized along C_4^2. It will always contain functions with the imaginary i, but can be avoided by re-quantization along another axis.

where H_G are the necessary additional components carrying no net angular momentum which transform as the totally symmetric representation in the subgroup G. Each term in this sum is a projected octahedral vector component and therefore is normalized for all finite three-dimensional geometries obtained from O_h. Consequently H_p is called an NSH Hamiltonian (1). This normalization guarantees that the magnitudes of each term in the perturbation Hamiltonian are entirely independent of the magnitudes of any other term. Thus the various perturbations may be very different in size, with these differences reflected in the magnitudes of the distortion parameters DQ, DS, DT, DU, DV, DM, DN, and so forth. Correlation tables of symmetry descent (e.g., Ref. 47, Table X14) from O_h to its subgroups can be used to find the reduction of octahedral representations to the sum of irreducible representations of the subgroup. Since these tables do not distinguish the correlation of the various multidimensional representations of the octahedral group into the lower groups, expanded correlations tables are presented here (Tables II-IV). For the readers' convenience the vector components of the subgroup Hamiltonian are labeled both with the numerical label $|\Gamma O|$ and, where possible, the more familiar cartesian coordinate label appropriate to the real form of the required linear combination of spherical harmonics. In Part 2 this latter terminology is omitted. Throughout this development the spherical harmonics are expressed in their rational form which is related to the more conventional functions by the following relationship (17):

$$C^l_{m_l} = \frac{4\pi}{(2l + 1)^{1/2}} Y^l_{m_l} \tag{6}$$

The required bases projected from R_3 onto O_h along the C_2^z, C_2^{xy}, and C_3^{xyz} axes are presented in Table V in terms of *subduction coefficients*, which are the coefficients of the spherical harmonics required for each vector component. The means by which such combinations are derived has been discussed at length in the literature (7, 10, 48, 49) and will not therefore be repeated here. Not all representations are included. The complete set, up to $l = 7$, is shown in Table VII. Table V, for simplicity, only includes those representations up to $l = 4$ which are suitable for the construction of Hamiltonian operators for d^n configurations in the various octahedral subgroups containing at least one reflection plane and at least one rotation axis. It is convenient to make this last restriction because the number of components required to construct an operator in, for example, C_2 would be so large as to make the procedure cumbersome. Thus for the Hamiltonian of the octahedron itself, we simply require the totally symmetric representation, namely, for a d^n basis:

$$H_p = DQ \,|\, A_{1g}\, O(i) \,|\, O_h^4$$
$$= DQ\left[\left(\frac{\sqrt{7}}{\sqrt{12}}\right)C_0^4 + \left(\frac{\sqrt{5}}{\sqrt{24}}\right)(C_4^4 + C_{-4}^4)\right] \tag{7}$$

For an f^n basis, the A_{1g} component projected from $l = 6$ must also be included (see Table VII). The symmetry descent correlation tables (Tables II-IV) together with the subduction criterion can now be applied to identify the desired Hamiltonian operators for subgroups occurring in the chains arising through distortion along the C_4^z, C_2^z or C_2^{xy} axes. The operators for these symmetries are obtained by *direct inspection* of the relevant tables (tables II-IV), identifying those octahedral (or tetrahedral) representations which subduce A_1 in the subgroup concerned. Thus for the point group D_{4h}, we find that the totally symmetric operator is subduced by $A_{1g}O(i) + E_gO(\theta)$ in O_h. Referring to the table of operator eigenstates (Table V, subduction coefficients) and noting (Table I) that tensorial sets $l = 2$ and $l = 4$ contribute to $E_gO(\theta)$, the Hamiltonian may be written

$$H_{D_{4h}} = DQ \,|\, A_{1g}O(i) \,|\, _{O_h}^4 + DS \,|\, E_gO(\theta) \,|\, _{O_h}^2 + DT \,|\, E_gO(\theta) \,|\, _{O_h}^4$$
$$= DQ\left[\left(\frac{7}{12}\right)^{1/2} C_0^4 + \left(\frac{5}{24}\right)^{1/2}\left(C_4^4 + C_{-4}^4\right)\right] \tag{8}$$
$$+ DS\, C_0^2 + DT\left[\left(\frac{5}{12}\right)^{1/2} C_0^4 - \left(\frac{7}{24}\right)^{1/2}\left(C_4^4 + C_{-4}^4\right)\right]$$

In summary, inspection of Tables I through IV reveals that those compounds characterized by the point group symmetry of a member of the chains terminating with C_{2v} require Hamiltonians of the following form:

$$\begin{aligned} H_G &= V && \text{for } G = O_h \text{ and } T_d \\ H_G &= V + V' && \text{for } G = D_{4h}, D_{4v}, D_{2d}(\mathrm{I}), D_{2d}(\mathrm{II})\ldots \\ H_G &= V + V' + V'' && \text{for } G = D_{2h}(\mathrm{I}), C_{2v}(\mathrm{I})\ldots \\ H_G &= V + V' + V''' && \text{for } G = D_{2h}(\mathrm{II}), C_{2v}(\mathrm{II}), C_{2v}(\mathrm{III})\ldots \end{aligned} \tag{9}$$

where

$$\begin{aligned} V &= DQ \,|\, A_{1g}O(i) \,|\, _{O_h}^4 \\ V' &= DS \,|\, E_gO(\theta) \,|\, _{O_h}^2 + DT \,|\, E_gO(\theta) \,|\, _{O_h}^4 \\ V'' &= DU \,|\, E_g2{+}(\varepsilon) \,|\, _{O_h}^2 + DV \,|\, E_g2{+}(\varepsilon) \,|\, _{O_h}^4 \\ V''' &= DM \,|\, T_{2g}O(xy) \,|\, _{O_h}^2 + DN \,|\, T_{2g}O(xy) \,|\, _{O_h}^4 \end{aligned} \tag{10}$$

which may be expanded by reference to Table V*. Readers should have no

*The $|\,T_{2g}0(xy)\,|$ label is appropriate for C_{2v}(III). For C_{2v}(II) the component label differs, viz: $|\,T_{2g}\ 2-(xy)\,|$.

TABLE V

Operator Eigenstates Quantized in the Point Group 0_h[a]

Quantized along $C_2(z)$[b]

$$|A_{1g}0(i)|^{4}_{0_h} = C^0_0$$

$$|E_g0(\theta)|^{2}_{0_h} = C^2_0$$

$$|E_g2+(\varepsilon)|^{2}_{0_h} = \left(\frac{1}{\sqrt{2}}\right)(C^2_2 + C^2_{-2})$$

$$|T_22-(xy)|^{2}_{0_h} = \left(\frac{-i}{\sqrt{2}}\right)(C^2_2 - C^2_{-2})$$

$$|A_{1g}0(i)|^{4}_{0_h} = \left(\frac{\sqrt{7}}{\sqrt{12}}\right)C^4_0 + \left(\frac{\sqrt{5}}{\sqrt{24}}\right)(C^4_4 + C^4_{-4})$$

$$|E_g0(\theta)|^{4}_{0_h} = \left(\frac{\sqrt{5}}{\sqrt{12}}\right)C^4_0 - \left(\frac{\sqrt{7}}{\sqrt{24}}\right)(C^4_4 + C^4_{-4})$$

$$|E_g2+(\varepsilon)|^{4}_{0_h} = \left(\frac{1}{\sqrt{2}}\right)(C^4_2 + C^4_{-2})$$

$$|T_{2g}2-(xy)^{4}_{0_h} = \left(\frac{-i}{\sqrt{2}}\right)(C^4_2 - C^4_{-2})$$

$$|T_{1g}0(z)|^{4}_{0_h} = \left(\frac{-i}{\sqrt{2}}\right)(C^4_{-4} - C^4_4)$$

Quantization along $C_2(xy)$

$$|A_{1g}0(i)|^{L=0}_{0_h} = C^0_0$$

$$|E_g0(\theta)|^{2}_{0_h} = \frac{1}{2}C^2_0 + \left(\frac{\sqrt{3}}{\sqrt{8}}\right)(C^2_2 + C^2_{-2})$$

$$|E_g0(\theta)|^{4}_{0_h} = \left(\frac{\sqrt{125}}{\sqrt{192}}\right)C^4_0 - \left(\frac{1}{\sqrt{96}}\right)(C^4_2 + C^4_{-2}) + \left(\frac{\sqrt{63}}{\sqrt{384}}\right)(C^4_4 + C^4_{-4})$$

$$|T_{2g}0(xy)|^{2}_{0_h} = \left(\frac{\sqrt{3}}{\sqrt{4}}\right)C^2_0 - \left(\frac{1}{\sqrt{8}}\right)(C^2_2 + C^2_{-2})$$

$$|T_{2g}0(xy)|^{4}_{0_h} = \left(\frac{\sqrt{5}}{\sqrt{16}}\right)C^4_0 + \left(\frac{1}{\sqrt{8}}\right)(C^4_2 + C^4_{-2}) - \left(\frac{\sqrt{7}}{\sqrt{32}}\right)C^4_4 + C^4_{-4})$$

$$|A_{1g}0(i)|^{4}_{0_h} = \left(\frac{\sqrt{7}}{\sqrt{192}}\right)C^4_0 - \left(\frac{\sqrt{35}}{\sqrt{96}}\right)(C^4_2 + C^4_{-2}) - \left(\frac{\sqrt{45}}{\sqrt{384}}\right)(C^4_4 + C^4_{-4})$$

Quantization along $C_3\ (xyz)$[c]

$$^3|A_{1g}0(i)|^{L=0}_{0_h} = C^0_0$$

$$^3|A_{1g}0(i)|^{4}_{0_h} = \left(\frac{\sqrt{7}}{\sqrt{27}}\right)C^4_0 + \left(\frac{\sqrt{10}}{\sqrt{27}}\right)(C^4_3 - C^4_{-3})$$

$$^3|T_{2g}0|^{2}_{0_h} = C^2_0$$

$$^3|T_{2g}0|^{4}_{0_h} = \left(\frac{\sqrt{20}}{\sqrt{27}}\right)C^4_0 - \left(\frac{\sqrt{7}}{\sqrt{54}}\right)(C^4_3 - C^4_{-3})$$

---

[a]The coefficients in this table are subduction coefficients extracted from Table VII. The parameters are independent of the quantization axis provided that both operator and basis f unctions are quantized along the same axis. However, for simplicity in computation, and particularly to avoid the occurrence of the imaginary i appearing in an operator component, it is convenient to use functions quantized along $C_3(xyz)$ for the threefold groups and those down $C_2(xy)$ for groups in chains terminating in C_{2v}(III). Note further that the operator quantized along C^z_2 can also be used for groups which contain a C^z_4 axis (either

alone or with a collinear C_2^z). The 2+ labels on the $E_g\varepsilon$ and $T_{2g}(xy)$ tensor components reflect the dual nature of the z axis carrying both C_4^z and C_2^z elements; functions with these representations do not carry any net angular momentum.

[b]In our previous publications (1, 2, 21) the phases of the subduction coefficients followed those prepared by Griffith (10). The phases in Table VII follow the Condon and Shortley convention, which is more generally accepted. The primary consequence is a change in the sign of the $|E_g(0)\theta|^4_{0_h}$ tensor component which results in a change in sign of DT. Thus our previously published equations involving DT must also be multiplied by -1.

[c]The three fold quantized operator is used in conjunction with a *negative* value of DQ (52).

TABLE VI

Mixed Ligand Systems Which Can Exhibit C_{2v} Symmetry

	Highest available point-group symmetry[a]		Possible C_{2v} surviving operators (S)[b]
6-coordinate			
ML_6^c		O_h	I, II, III
ML_4ZZ'	trans	C_{4v}	I, II
ML_4Z_2	trans	D_{4h}	I, II, III
	cis	C_{2v}	III
MX_2Y_2ZZ'	trans	C_{2v}	I
$MX_2Y_2Z_2$	trans	D_{2h}	I
	cis	C_{2v}	III
$ML_3L'_3$	equatorial	C_{2v}	I
$M(L\text{-}L)_2ZZ'$	trans	C_{2v}	II
$M(L\text{-}L)_2Z_2$	trans	D_{2h}	II, III
$M(L\text{-}L')_2Z_2$	trans[d]	C_{2v}	III
5-coordinate			
$M(L\text{-}L)_2Z$	trans	C_{2v}	II
MX_2Y_2Z	trans	C_{2v}	I
4-coordinate planar			
MX_2Y_2	cis[e]	C_{2v}	III
$M(L\text{-}L)L'_2$		C_{2v}	III
$M(L\text{-}L'\text{-}L)L''$		C_{2v}	I
ML_3L'		C_{2v}	I
$ML_2L'L''$	trans	C_{2v}	I
4-coordinate tetrahedral			
$ML_2L'_2$		C_{2v}	II
$M(L\text{-}L)L'_2$		C_{2v}	II

[a]Corresponds to the symmetry of an octahedral (or tetrahedral in the last case) angular configuration.

[b]C_{2v} point groups that may arise through an angular distortion.

[c]L represents a general ligand along any axis; X, Y, and Z refer to ligands along the x, y, and z axes, respectively, assuming a standard left-handed coordinate scheme. This arrangement is not essential, but is an aid to visualization of the structure.

[d]This structure is trans with respect to the Z ligands, but cis with respect to the L-L′ in-plane ligands. An all-trans structure would be long to C_{2h}(I).

[e]The trans isomer belongs to D_{2h} which can be C_{2h} (I) with angular distortion.

TABLE VII

Subduction Coefficients[a,b]

Fourfold Adapted Functions (C_4^z)									
α	Γ	γ	J	MJ	$S(MJ)^2$	$S(MJ+4)^2$	$S(MJ+8)^2$	$S(MJ+12)^2$	$(\text{Den})^2$
	A_1	0	0	1					1
			4	−4	5	14	5		24
			6	−4	*7	2	*7		4
	A_2	2−	3	−2	*1	1			2
			6	−6	*5	11	11	*5	32
			7	−6	*11	*13	13	11	48
	E	0	2	0	1				1
			4	−4	*7	10	*7		24
			5	−4	1	0	*1		2
			6	−4	1	14	1		16
			7	−4	1	0	*1		2
	E	2+	2	−2	1	1			2
			4	−2	1	1			2
			5	−2	*1	1			2
			6	−6	11	5	5	11	32
			7	−6	13	*11	11	*13	48
	T_1	1	1	1	1				1
			3	−3	5	3			8
			4	−3	1	7			8
a			5	−3	35	30	63		128
b			5	−3	*81	42	5		128
			6	−3	*15	6	*11		32
a			7	−7	429	189	175	231	1,024
b			7	−7	91	*11	297	*625	1,024
	T_1	0	1	0	1				1
			3	0	1				1
			4	−4	1	0	*1		2
a			5	0	1				1
b			5	−4	1	0	1		2
			6	−4	1	0	*1		2
a			7	0	1				1
b			7	−4	1	0	1		2
	T_1	−1	1	−1	1				1
			3	−1	3	5			8
			4	−1	7	1			8
a			5	−5	63	30	35		128
b			5	−5	5	42	*81		128
			6	−5	*11	6	*15		32
a			7	−5	231	175	189	429	1,024
b			7	−5	*625	297	*11	91	1,024
	T_2	1	2	1	1				1
			3	−3	*3	5			8

Subduction Coefficients[a,b] (Continued)

α	Γ	γ	J	MJ	$S(MJ)^2$	$S(MJ+4)^2$	$S(MJ+8)^2$	$S(MJ+12)^2$	$(\text{Den})^2$
			4	−3	*7	1			8
			5	−3	3	14	*15		32
a			6	−3	*81	10	165		256
b			6	−3	55	198	3		256
a			7	−7	*1,001	361	675	11	2,048
b			7	−7	*7	*1,287	429	325	2,048
	T_2	2−	2	−2	*1	1			2
			3	−2	1	1			2
			4	−2	*1	1			2
			5	−2	1	1			2
a			6	−2	*1	1			2
b			6	−6	1			*1	2
a			7	−2	1	1			2
b			7	−6	1			1	2
	T_2	−1	2	−1	1				1
			3	−1	5	*3			8
			4	−1	1	*7			8
			5	−5	*15	14	3		32
a			6	−5	165	10	*81		256
b			6	−5	3	198	55		256
a			7	−5	11	675	361	*1,001	2,048
b			7	−5	325	429	*1,287	*7	2,048

Threefold Adapted Functions (C_3^{xyz})

ν	Γ	γ	J	MJ	$S(MJ)^2$	$S(MJ+3)^2$	$S(MJ+6)^2$	$S(MJ+9)^2$	$S(MJ+12)^2$	$(\text{Den})^2$
	A_1	0	0	0	1					1
			4	−3	*10	7	10			27
			6	−6	77	70	192	*70	77	486
	A_2	0	3	−3	2	5	*2			9
			6	−6	*5	22	0	22	5	54
			7	−6	*176	*52	273	52	*176	729
	E	1	2	−2	*1	2				3
			4	−2	16	4	7			27
			5	−5	*20	*3	42	*16		81
			6	−5	22	*15	12	32		81
			7	−5	128	352	66	1	*182	729
	E	−1	2	−1	2	1				3
			4	−4	*7	4	*16			27
			5	−4	*16	42	3	*20		81
			6	−4	*32	12	15	22		81
			7	−7	*182	*1	66	*352	128	729
	T_1	1	1	1	1					1
			3	−2	*5	1				6
			4	−2	*7	7	4			18

Three fold Adapted functions C_3^{xyz} (Continued)

α	Γ	γ	J	MJ	$S(MJ)^2$	$S(MJ+3)^2$	$S(MJ+6)^2$	$S(MJ+9)^2$	$S(MJ+12)^2$	(Den)^2
a			5	−5	448	420	30	*560		1 458
b			5	−5	245	12	672	529		1 458
			6	−5	11	30	96	*25		162
a			7	−5	*1,232	3,703	6,069	*7,546	8,152	27,702
b			7	−5	525	*176	132	2	91	1026
	T_1	0	1	1	1					1
			3	−3	*5	8	5			18
			4	−3	1	0	1			2
a			5	0	1	1				2
b			5	−3	1	0	*1			2
			6	−6	22	5	0	*5	*22	54
a			7	−6	4,004	1,183	17,328	*1,183	4,004	27,702
b			7	−6	*13	44	0	*44	*13	114
	T_1	−1	1	−1	1					1
			3	−1	*1	*5				6
			4	−4	*4	7	7			18
a			5	−4	560	30	*420	448		1,458
b			5	−4	*529	642	*12	245		1,458
			6	−4	25	96	*30	11		162
a			7	−7	8,152	7,546	6,069	*3,703	*1,232	27,702
b			7	−7	91	*2	132	176	525	1,026
	T_2	1	2	−2	2	1				3
			3	−2	1	5				6
			4	−2	1	25	*28			54
			5	−5	*5	12	0	1		18
a			6	−5	*352	*60	147	8		567
b			6	−5	*5	66	0	55		126
a			7	−5	*44	*2,209	507	*6,358	*4,004	13,122
b			7	−5	*2,197	*572	6,864	3,146	*343	13,122
	T_2	0	2	0	1					1
			3	−3	1	0	1			2
			4	−3	7	40	*7			54
			5	−3	1	0	1			2
a			6	−6	*176	*160	1,029	160	*176	1,701
b			6	−6	*10	11	0	*11	*10	42
a			7	−3	1	0	1			2
b			7	−6	*1	0	0	*1		2
	T_2	−1	2	−1	1	*2				3
			3	−1	5	*1				6
			4	−4	8	25	*1			54
			5	−4	1	0	5	5		18
a			6	−4	*8	147	60	*352		567
b			6	−4	55	0	66	5		126
a			7	−7	*4,004	6,358	507	2,209	*44	13,122
b			7	−7	*343	*3,146	6,364	572	*2,197	13,122

Twofold Adapted Functions(C_2^z)

α	Γ	γ	J	MJ	$S(MJ)^2$	$S(MJ+2)^2$	$S(MJ+4)^2$	$S(MJ+6)^2$	$S(MJ+8)^2$	$S(MJ+10)^2$	$S(MJ+12)^2$	$S(MJ+14)^2$	$(\text{Den})^2$
	A_1	0	0	0	1								1
			4	−4	5		14		5				24
			6	−4	*7		2		*7				16
	A_2	2−	3	−2	*1		1						2
			6	−6	*5		11		11		*5		32
			7	−6	*11		*13		13		11		48
	E	0	2	0	1								1
			4	−4	*7		10		*7				24
			5	−4	1				*1				2
			6	−4	1		14		1				16
			7	−4	1				*1				2
	E	2+	2	−2	1		1						2
			4	−2	1		1						2
			5	−2	*1		1						2
			6	−6	11		10		10		11		32
			7	−6	13		*11		11		*13		48
	T_1	1+	1	−1	1	1							2
			3	−3	5	3	3	5					16
			4	−3	1	*7	7	*1					16
a			5	−5	63	35	30	30	35	63			256
b			5	−5	5	*81	42	42	*81	5			256
			6	−5	11	*15	*6	6	15	*11			64
a			7	−7	429	231	189	175	175	189	231	429	2,048
b			7	−7	91	*625	*11	297	297	*11	*625	91	2,048
	T_1	0	1	0	1								1
			3	0	1								1
			4	−4	1				*1				2
a			5	0	1								1
b			5	−4	1				1				2

Twofold Adapted Functions (C_2^*) (Continued)

			6	−4	1				*1				2
a			7	0	1								1
b			7	−4	1				1				2
	T_1	1−	1	−1	1	*1							2
			3	−3	*5	3	*3	5					16
			4	−3	1	7	7	1					16
a			5	−5	63	*35	30	*30	35	*63			256
b			5	−5	5	81	42	*42	*81	*5			256
			6	−5	*11	*15	6	6	*15	*11			64
a			7	−7	*429	231	*189	175	*175	189	*231	429	2,048
b			7	−7	*91	*625	11	297	*297	*11	625	91	2,048
	T_2	1+	2	1	1	1							2
			3	−3	*3	*5	5	3					16
			4	−3	*7	1	1	*7					16
			5	−5	60	12	*56	56	*12	*60			256
a			6	−5	165	*81	10	10	*81	165			512
b			6	−5	3	55	198	198	55	3			512
a			7	−7	*1,001	*11	361	*675	675	*361	11	1,001	4,096
b			7	−7	*7	*325	*1,287	*429	429	1,287	325	7	4,096
	T_2	2−	2	−2	*1		1						2
			3	−2	1		1						2
			4	−2	*1		1						2
			5	−2	1		1						2
a			6	−2	*1		1						2
b			6	−6	1						*1		2
a			7	−2	1		1						2
b			7	−6	1						1		2
	T_2	1−	2	−1	1	*1							2
			3	−3	*3	5	5	*3					16
			4	−3	7	1	*1	*7					16
			5	−5	*60	12	56	56	12	*60			256
a			6	−5	165	81	10	*10	*81	*165			512
b			6	−5	3	*55	198	*198	55	*3			512
a			7	−7	*1,001	11	361	675	675	361	11	*1,001	4,096
b			7	−7	*7	325	*1,287	429	429	*1,287	325	*7	4,096

α	Γ	γ	J	MJ	$S(MJ)^2$	$S(MJ+2)^2$	$S(MJ+4)^2$	$S(MJ+6)^2$	$S(MJ+8)^2$	$S(MJ+10)^2$	$S(MJ+12)^2$	$(\text{Den})^2$
					Twofold Adapted Functions (C_2^{xy})							
	A_1	0	0	0	1							1
			4	−4	*45	*140	14	*140	*45			384
			6	−6	231	*350	105	676	105	*350	231	2,048
	A_2	1+	3	−3	*3	5	5	*3				16
			6	−5	*135	*11	*110	110	11	135		512
	E	0	2	−2	3	2	3					8
			4	−4	63	*4	250	*4	63			384
			5	−4	*1	3		*3	1			8
			6	−6	297	578	135	28	135	578	297	2,048
	E	1−	2	−1	*1	1						2
			4	−3	*7	*1	1	7				16
			5	−5	*60	12	56	56	12	*60		256
			6	−5	132	500	*392	392	*500	*132		2,048
	T_1	1+	1	1	1	1						2
			3	−3	5	3	3	5				16
			4	−3	1	*7	7	*1				16
a			5	−5	7,623	*875	10,830	10,830	*875	7,623		38,656
b			5	−5	*2,645	*1,664	42	42	*1,664	*2,645		38,656
			6	−5	55	*75	*30	30	75	*55		320
	T_1	0	0	1	1							1
			3	−2	*45	2	*15					62
			4	−4	9	7		*7	*9			32
a			5	−4	*15,435	420	45,602	420	*15,435			77,312
b			5	−4	4	147		147	4			302
			6	−6	297	*200	15		*15	200	*297	1,024
	T_1	1−	1	1	1	*1						2
			3	−3	5	27	*27	*5				64
			4	−3	24	7	7	24				64

Twofold Adapted Functions(C_2^{xy}) (Continued)

a			5	−5	*18.207	58,835	270	*270	*58,835	18,207		154,624
b			5	−5	*605	*225	378	*378	225	605		2,416
			6	−5	*11	*15	486	486	*15	*11		1,024
	T_2	0−	2	−2	*1		1					2
			3	−2	1	30	1					32
			4	−4	7	*9		9	*7			32
			5	−4	*147	4	*210	4	*147			512
a			6	−6	*2,475	*4,056	*125		125	4,056	2,475	13,312
b			6	−6	5		*99		99		*5	208
	T_2	1+	2	−1	1	1						2
			3	−3	27	*5	5	*27				64
			4	−3	*7	25	25	*7				64
			5	−5	*135	*27	*350	350	27	135		1,024
a			6	−5	2,673	3,645	338	338	3,645	2,673		13,312
b			6	−5	15	11			11	15		52
	T_2	0	2	−2	*1	6	*1					8
			3	−2	1	0	*1					2
			4	−4	*7	4	10	4	*7			32
			5	−4	3	1	0	*1	*3			8
a			6	−6	99	294	*4,205	4,116	*4,205	294	99	13,312
b			6	−6	*3,125	330	891	4,620	891	330	*3,125	13,312

[a]Condon and Shortley (54) phase convention.

[b]Squares of the subduction coefficients are expressed as fractions, with the square of the numerator in the body of the table and the square of the denominator in the extreme right-hand column. Starred numbers carry negative signs. For example, in the fourfold adapted set, the A_2 function with $J = 6$ is given by reading accross the table as

$$| A_2 2- > = -(5/32)^{\frac{1}{2}} C^6_{-6} + (11/32)^{\frac{1}{2}} C^6_{-2} + (11/32)^{\frac{1}{2}} C^6_2 - (5/32)^{\frac{1}{2}} C^6_6$$
$$= (11/32)^{\frac{1}{2}} (C^6_2 + C^6_{-2}) - (5/32)^{\frac{1}{2}} (C^6_6 + C^6_{-6})$$

difficulty in assigning appropriate Hamiltonians to two fold or four fold groups not considered above. Note that the octahedral representation $A_{2g}O(i)$ is not included in Table V, even though A_{2g} does subduce A_1 in some subgroups. The A_{2g} representation is projected from $l = 3$ in R_3, but not from $l = 2$ or 4. A harmonic of odd degree is odd with respect to inversion; an odd operator cannot couple d orbitals to themselves and is therefore of no value as a ligand field operator. Note that A_{2g} is subduced by the even $l = 6$ and would be necessary for an f-orbital basis. Odd representations of O_h, even where they subduce A_1 in noncentrosymmetric subgroups, cannot contribute to the ligand field Hamiltonian. However, *these components would be necessary in noncentrosymmetric groups if configurational interaction between states of different parity were to be considered.* This is a useful example of the conventions of R_3 and O_h carrying down into the lower groups. This general model can be applied to a wide range of specific complex geometries. Sections IV.A through C note a few examples.

A. Six-Coordinate Derivatives Belonging to Twofold or Fourfold Groups

Regular octahedral complexes such as the $Cr(NH_3)_6^{3+}$ ion are described by Hamiltonian (Eq. 7), whereas derivatives such as $Cr(NH_3)_5Cl^{2+}$ belonging to C_{4v} and *trans*-$Cr(NH_3)_4Cl_2^+$ belonging to D_{4h} require Hamiltonian (Eq. 8).

The *cis*-$Cr(NH_3)_4Cl_2^+$ ion belongs to a C_{2v} group in which the C_2^{xy} operator is retained, that is, C_{2v}(III). Hence from Eq. 9

$$
\begin{aligned}
H_G &= V + V' + V''' \\
&= DQ \mid A_{1g}O(i) \mid^4_{O_h} + DS \mid E_gO(\theta) \mid^2_{O_h} + DT \mid E_gO(\theta) \mid^4_{O_h} \\
&\quad + DM \mid T_{2g}O(xy) \mid^2_{O_h} + DN \mid T_{2g}O(xy) \mid^4_{O_h}
\end{aligned} \tag{11}
$$

which can be expanded using Table V. Note that in expanding the Hamiltonian using Eq. 11 and the spherical harmonic functions listed for each representation in Table V, it does not matter whether you use such functions quantized along $C_2(z)$ or $C_2(xy)$; provided that the basis wave functions (see Section V) are similarly quantized the magnitudes of the empirical parameters obtained will be identical. However, as will be discussed in some detail in Part 2, there are definite computational advantages to using functions quantized along the principle axis of the molecule concerned, in this case using those down $C_2(xy)$. In particular, the Hamiltonian expressed this way will not contain the imaginary i.

If the angles in the CrN_2Cl_2 plane are 90°, the $T_{2g}O(xy)$ operator eigenstates both vanish, that is, $DM = DN = 0$. This is most easily appreciated by visualizing the $T_{2g}xy$ operator eigenstate as the d_{xy} orbital. For the specific 90° case, the metal ligand bonds lie in the nodal planes of this orbital, and under such a circumstance these components vanish. Thus such a *cis*-C_{2v}(III) complex has the same Hamiltonian as the trans D_{4h} isomer, that is, it has pseudotetragonal symmetry, as has been explained in the crystal field model by Ballhausen (7). If the angles are not 90°, then DM and DN do not vanish and, given sufficient experimental information, can be calculated. However, it is clear that the system is overdefined since the relative energies of the five d orbitals under C_{2v} symmetry can be described by, at most, four-orbital energy separations. Hence the five NSH empirical parameters cannot be linearly independent; indeed, DM and DN are related through the in-plane angle, for which an effective value can be calculated. The same Hamiltonian and arguments apply for a D_{2h}(II) species such as the bidentate complex *trans*-Ni(ethylenediamine)$_2$I$_2$.

This is an example of "intermediate symmetry" where the effective Hamiltonian possesses a higher symmetry than suggested by the physical environment (1, 28). As a consequence degeneracies occur, in the electronic states, which at first sight might not have been expected. For example, the E_g components of the T_{1g} and T_{2g} states in O_h will not split if a 90° angle is retained. In practice this is seen to be the case, although careful work (50) reveals very small deviations in the energies of the two E components as a consequence of distortion. In general, however, unless the distortion is great, small or zero values of DM and DN can be anticipated.

Consider now an all-trans species MX_2Y_2WZ; this will belong to C_{2v}(I). The appropriate Hamiltonian is seen (Eq. 9) to be the following:

$$\begin{aligned} H_G &= V + V' + V'' \\ &= DQ \mid A_{1g}O(i) \mid^4_{O_h} + DS \mid E_gO(\theta) \mid^2_{O_h} + DT \mid E_gO(\theta) \mid^4_{O_h} \\ &\qquad + DU \mid E_g\, 2+(\varepsilon) \mid^2_{O_h} + DV \mid E_g 2+(\varepsilon) \mid^4_{O_h} \end{aligned} \tag{12}$$

which may be expanded with Table V.

The form of the $E_g 2+(\varepsilon)$ operator eigenstate (Table V) is that of the $d_{x^2-y^2}$ orbital. Since the metal ligand bonds lie along the lobes of this orbital, DU and DV do not vanish in the 90° case. They will, however, vanish for an admittedly unlikely angular arrangement in the plane (1). The meridional isomer of a species MX_3Y_3, such as *mer*-Co(NH$_3$)$_3$Cl$_3$, will also belong to C_{2v}(I). It also exhibits intermediate symmetry with DS and DT vanishing if the octahedral angular configuration is retained. These special cases are best detected by a comparison of the classical crystal field Hamiltonian, inclusive of its angular dependence against the NSH Hamiltonian. Many such special cases exist and we mention here only the more common.

A complex such as *trans*-Co(ethylenediamine)$_2$ClBr$^+$ belongs to C_{2v}(II). Reference to Table III or Eq. 9 reveals that the Hamiltonian will have the same form as that for a C_{2v}(III) complex, namely, Hamiltonian Eq. 11. $C_2(z)$ quantized functions should preferably be used. DM and DN will be small and will vanish if the octahedral angular configuration is retained.

Note that a clear distinction has been made between the three C_{2v} examples, placing the complexes in C_{2v}(III), C_{2v}(I), and C_{2v}(II), respectively. Many complexes belong to C_{2v} point groups (Table VI). It is emphasized that they can be readily distinguished by consideration of the surviving operators from the parent octahedron. Thus if the surviving twofold operator lies between the bonding axes, that is, $C_2(xy)$, C_{2v}(III) is indicated. C_{2v}(I) and C_{2v}(II) are distinguished by the surviving planes of symmetry, being horizontal in the former and dihedral in the latter (Eq. 3). This decision may be less clear if considerable angular distortion occurs at the octahedron. An unambiguous decision may not be possible. It is part of the beauty of this approach that this ambiguity does not matter. Thus if the same molecule is treated first as C_{2v}(I), then as C_{2v}(II) identical values of DQ, DS, and DT would be obtained. In the former case, nonzero values of DU and DV occur, whereas in the latter, different nonzero values of DM and DN result. If it is intended, for example, to compare these data against those obtained from a molecule belonging to D_{2h}(I), then C_{2v}(I) should be chosen. Thus the reason for the collection of data may dictate the preferred choice in cases of ambiguity.

B. Four- and Five-Coordinate Complexes of the Fourfold and Twofold Groups

Square pyramidal 5-coordinate complexes of C_{4v} symmetry also share Hamiltonian Eq. 8 and present no problems. Complexes such as the square pyramidal $VO(L\text{-}L)_2$ are readily discerned to belong to C_{2v}(II) and therefore to require Hamiltonian Eq. 11. The special case of the 5-coordinate trigonal bipyramid is discussed below. Four-coordinate derivatives can generally be classified into square planar and tetrahedral complexes. It is clearly convenient to consider the former as octahedra in which two trans ligands have been removed. Thus square planar $Ni(CN)_4^{2-}$ of D_{4h} symmetry would require Hamiltonian Eq. 8. The square planar *trans*-$Pt(NH_3)_2Cl_2$ belongs to D_{2h}(I), and since it retains the octahedral horizontal reflection planes it utilizes Hamiltonian Eq. 13. In both cases, a relationship between DQ and DT can be demonstrated($DT = -\frac{1}{2}(\sqrt{5}/\sqrt{7})DQ$)(2), and is theoretically valid in the absence of any perturbation along the z axis, reducing the number of required parameters by one. However, "absence of any perturbation along the z axis" seems to be an extremely stringent requirement and no definite examples of such a relationship have been demonstrated yet. *Cis*-$Pt(NH_3)_2Cl_2$ has C_{2v}(III) symmetry (Hamiltonian Eq. 11) and will exhibit intermediate symmetry with only DQ, DS, and DT as variables if 90° angles are retained. The same comments and Hamiltonian apply to a complex such as Pt(ethylenediamine)$_2^{+2}$ belonging to D_{2h}(II). ML_3X systems such as $Pt(NH_3)_3Cl^+$ fall into C_{2v}(I) (see Table VI). If consideration is given to the positions of the hydrogen atoms in some of these complexes, other point group symmetries may be realized, especially C_{2h}. Like C_{2v} this arises with varying surviving operators. Generally speaking, it requires a considerable number of parameters for its description, some of which are likely to be small, or zero. High-quality experimental data would be necessary before such analysis could be justified.

Regular tetrahedral complexes such as the $NiCl_4^{2-}$ ion utilize exactly the same Hamiltonian as a regular octahedron. The groups O and T_d are isomorphous, and functions quantized down S_4 of T_d are identical to the C_2^z functions in Table V. Since the tetrahedron lacks a center of symmetry, an additional A_1 term can be subduced, namely, A_1 in T_d from A_{2u} in O_h. As indicated above, this representation, being odd, does not contribute to the Hamiltonian.

The many complexes of general formula ML_2X_2 based on a tetrahedron belong to C_{2v}(II), Hamiltonian Eq. 11.* If the tetrahedral angular configuration is maintained, DS and DT should vanish (1), although this has yet to be conclusively demonstrated experimentally. It is worthwhile noting that the various cases of intermediate symmetry cited above are strictly valid only when ligand additivity is assumed, that is, the individual effects of each metal ligand bond can be added to generate a picture of the whole (1, 28).

C. Threefold Groups

We require only one kind of quantization, namely, along C_3^{xyz}. Table V also lists the operator eigenstates quantized along this axis. The Hamiltonian for the octahedron can be written as

$$H_P = DQ^3|A_{1g}O(i)|^4_{o_h} = DQ\left[\left(\frac{\sqrt{7}}{\sqrt{27}}\right)C_0^4 + \left(\frac{\sqrt{10}}{\sqrt{27}}\right)(C_3^4 - C_{-3}^4)\right] \quad (13)$$

Using basis states also quantized along C_3^{xyz}, it will of course yield an identical value of DQ for a given data base as Hamiltonian Eq. 7.

This quantization leads to a single chain of trigonal subgroups:

$$O_h \rightarrow D_{3d} \rightarrow D_3 \rightarrow C_{3v} \rightarrow C_3 \quad (14)$$

The A_1 representation in D_{3d}, D_3, and C_{3v} is subduced by the even representations $A_{1g}O$ and $T_{2g}O$ in O_h (Table IV); hence within a given d^n configuration, all these groups require the same Hamiltonian:

* See footnote, p. 239.

$$H_G = W + W' \qquad \text{for } G = D_{3d}, D_3, C_{3v}$$

where

$$W = DQ^3 \left| A_{1g}(i) \right|^4_{O_h}$$
$$W' = DM^3 \left| T_{2g}O \right|^2_{O_h} + DN^3 \left| T_{2g}O \right|^4_{O_h} \tag{15}$$

and the superfix 3 refers to quantization along the threefold axis (Table V).

It may seem curious that the D_{3d}, D_3, and C_{3v} subgroups of O_h (or T_d) possess the same Hamiltonian. It is emphasized that this is only true for a d^n basis. If configurational interaction with odd atomic orbital states is included, and the full Hamiltonians inclusive of odd contributions are written, the three groups above can be distinguished, the D_3 group requiring $A_{1u}O(i)$ + $T_{2u}O$ tensor contributions and the C_{3v} group requiring contributions from $A_{2u}O(i)$ and $T_{1u}O$ (47).

The only group of chemical significance which is not covered within the above treatment is D_{3h}; this presents certain difficulties because it is not a subgroup of O_h. However, it can be handled within the same framework by noting that it is isomorphous with D_{3d}, which is a subgroup of O_h. Two isomorphous groups (3) can always be represented by a common character table and will contain common representations. The difference between D_{3d} and D_{3h} is that the inversion operator i of the former becomes the horizontal reflection plane of the latter. Thus Γ_g and Γ_u of D_{3d} become Γ' and Γ''' of the latter. However, there is no one-to-one mapping in this respect since Γ' and Γ''' can be subduced by the same l value in R_3, whereas Γ_g and Γ_u neatly divide themselves among the even and odd l values of R_3, respectively. All generators of the D_{3h} group appear in O_h, except for the σ_h plane perpendicular to C_3. Any representation of O_h which is not scrambled into another function by this reflection operator form an appropriate basis for D_{3h}. Thus a connection between O_h and D_{3h} can be made, even though the latter is not a proper subgroup of the former. Parallel to D_{3d}, therefore, we note that A_1 of D_{3h} is subduced by $A_{1g}O(i) + T_{2g}O$. However, the Hamiltonian, in this special case, has an unusual form, the reasons for which have been fully described (51). It can be written as the following:

$$H_G = DQ\left[\left(\frac{\sqrt{7}}{3\sqrt{3}}\right)^3 \left| A_{1g}O(i) \right|^4_{O_h} + \left(\frac{\sqrt{10}}{3\sqrt{3}}\right)^3 \left| T_{2g}O \right|^4_{O_h}\right] + DM^3 \left| T_{2g}O \right|^2_{O_h} \tag{16}$$

The commonly occurring threefold complexes are those such as Co(ethylenediamine)$_2^{+3}$, trigonally distorted octahedral complexes (often obtained by doping metal ions into nontransition metal trigonal lattices), and "tetrahedral" derivatives of the MLX_3-type belonging to the groups D_3, D_{3d} and C_{3v}, respectively.

V. BASIS FUNCTIONS AND SELECTION RULES

Having established the Hamiltonian to use, we should now consider the nature of the basis wave functions to use. It is also convenient here to consider ways in which the existence of a given matrix element may be detected (selection rules). The energies of the states of complexed ions are calculated as the solutions of the matrices obtained (Eq. 1) by operating with the Hamiltonian H_P on all possible basis functions ψ of the ion. In the lower groups this leads to a large number of calculations (i.e., a large number of elements in the matrix generated) because of the size of the Hamiltonian. The problem can be greatly facilitated if there is a ready procedure to detect those matrix elements which vanish, that is, those which are zero. This is accomplished by means of selection rules which take full advantage of the group theoretical nature of both the Hamiltonian operator and the wave functions, and the fact that they are subduced from octahedral representations. In order to pursue these simplified models, the relation of the properties of the Hamiltonian operator and the basis functions to vector and group theoretical terms needs to be defined, and some indication given here to the means of evaluating matrix elements.

Any of the three procedures which have been developed to derive the perturbation Hamiltonian can be used to define a set of consistent basis wave functions for a particular point group. Of these three techniques, however, only the projection and subduction method guarantees the availability of a basis set of wave functions which are self-consistent within an entire chain of groups (52). This is an important requirement because inconsistencies in either phase definitions or coordinate-system convention can quickly lead to confusion in the assignment of states. Using the projection operator, quantization on the appropriate symmetry axis of O_h yields sets of linearly combined components of the basis functions of R_3. The coefficients of these linear combinations of spherical harmonic components are indeed the *subduction coefficients*. For example, the transformation properties of the d orbitals are included in Tables II through IV. Thus the $d_{x^2-y^2}$ orbital transforms as a_2 in C_{2v}(II) (see footnote to Table II) and is subduced by $E_g2+(\varepsilon)$ (Table III). This latter representation, quantized along $C_2(z)$, as shown in Table V, can therefore be used as a basis wave function for the $d_{x^2-y^2}$ orbital in this group. Bases for other d orbitals can be ascertained from Tables II to IV, whereas bases for the various terms arising from the d^n configurations are presented in Table VII. It is emphasized that the functions used as operator eigenstates are identical in derivation to the wave-functions and can therefore be used as such.

Because the components of the spherical harmonics can be recognized as vector components of a tensor of degree l (17), it is clear that projected or subduced linear combinations of these components must also be vectors (11). Any such quantized combination is called an eigenvector, and in the notation used by Dirac (53), the wavefunction ψ_γ^Γ is mapped onto the eigenevector $|\Gamma\gamma>$ in which the symbol $| >$ is called a *ket*. The symbol Γ stands for a representation label in either infinite (R_3) or finite groups (O_h, etc.), and γ denotes the vector component. The complex conjugate of the eigenvector is denoted as a *bra* $<\Gamma\gamma|$.

In parallel with these definitions the vectorial operator H_G can be described by a symbol characterizing its representation Γ_H and component γ_H (usually O). Thus H_G is mapped into $|P\Gamma_H O|$, the operator eigenstate (10), in which P is the scalar magnitude of the vector component and represents the observable, such as the energy of a state.

Using this nomenclature, for example, the first-order energy of the $d_{x^2-y^2}$ orbital energy in the group C_{2v}(II) can be written as

$$E(d_{x^2-y^2}) = < A_2 | PA_1 | A_2 > \tag{17}$$

Using C_{2v} group representations, in expanded form using the octahedral parent representations, and recalling Eq. 11, the equation can be expressed as (recall footnote on p. 239).

$$\begin{aligned} E(d_{x^2-y^2}) = \; & < E_g 2+(\varepsilon) | PA_{1g}O(i) + P'E_gO(\theta) \\ & + P''T_{2g}2-(xy) | E_g 2+(\varepsilon) > \end{aligned} \tag{18}$$

which may be further expanded into

$$\begin{aligned} E(d_{x^2-y^2}) = \; & DQ < E_g 2+(\varepsilon) | A_{1g}O(i) | E_g 2+(\varepsilon) > \\ & + DS < E_g 2+(\varepsilon) | E_gO(\theta) | E_g 2+(\varepsilon) > \\ & + DT < E_g 2+(\varepsilon) | E_gO(\theta) | E_g 2+(\varepsilon) > \\ & + DM < E_g 2+(\varepsilon) | T_{2g}2-(xy) | E_g 2+(\varepsilon) > \\ & + DN < E_g 2+(\varepsilon) | T_{2g}2-(xy) | E_g 2+(\varepsilon) > \end{aligned} \tag{19}$$

which admirably demonstrates the apparent complexity of low-symmetry calculations. However, as the selection rules will demonstrate, a number of these matrix elements are zero.

The wave functions ψ_γ^Γ now representing an eigenvector, or the operator eigenstate, must, as a minimum condition, meet the symmetry requirements of the group G in which Γ is defined. In the present model, however, these functions have all been projected out of one generative group and are written as linear combinations, specified by the subduction coefficient, of spherical harmonics. These harmonics, in polar coordinates, are product functions of radial and angular functions (26, 54, 55):

$$\Psi = \sum_l \sum_m \psi^Z_{n,l,m} = \sum_l \sum_m R^Z_{n,l}(r)\, Y_{l,m}(\theta, \phi) \tag{20}$$

where Z is the effective nuclear charge and n is the principal quantum number.

The expectation value can be written in a fully expanded form *and then factored to reflect the preservation of l and m_l as good quantum numbers for R_3* (26).

$$\begin{aligned} &\int \psi^{Z\,*}_{n,l,m}(r, \theta, \phi)\, H_{qk}(r, \theta, \phi)\, \psi^{Z'}_{n',l',m'}(r, \theta, \phi)\, r^2\, dr \sin\theta d\theta\, d\phi \\ &= R(Znl)(r)^*\, H^R_q(r)\, R(Z'n'l')(r)\, r^2 dr \\ &\times \int Y^*_{l,m}(\theta, \phi)\, C^q_k\, Y_{l',m'}(\theta, \phi) \sin\theta d\theta\, d\phi \end{aligned} \tag{21}$$

The second factorization has been written explicitly in terms of the quantum numbers which are derived from the geometric coordinate functions. The first term is usually called the *radial integral*. The second term is called the *angular integral* and describes the behavior of the wave functions and operators in two-dimensional space in terms of spherical harmonics. Projection of these spherical wave functions into finite groups cannot affect this basic factorization. Note that q and k are used in this and following equations to designate the operator values of l and m_l, respectively.

The unfactorized energy can be rewritten in the Dirac vector formalism. The factorization produces a parallel separation of radial and angular terms, and as before for R_3, the angular symbol depends only on the quantum numbers l and m_l. Thus

$$< Z'n'l'm_l' \mid Hqk \mid Znlm_l > \; = \; < Z'n'l' \mid Hq \mid Znl > \; < l'm_l' \mid qk \mid lm_l > \tag{22}$$

The first term, on the right-hand side, remains the radial integral which can either be calculated in principle, or can be fitted as the empirical parameter P, that is, DQ, DS, and so on, in an experimental measurement. Each matrix element in Eq. 19, for example, comprises a sum of angular terms, summed over the values of l and m_l, l' and m_l', contained within the wave function, and the values of q and k appropriate to the operator. The angular term can be further factored to separate the metric or scaling properties of the eigenvector from their directional properties. The full factorization of the matrix element becomes (18)

$$< Z'n'l'm_l' \mid Hqk \mid Znlm_l > \; = P < l' \| q \| l > < qlkm_l \mid qll'm_l' > \tag{23}$$

Here the radial integral is abbreviated P. The angular term is expanded into a *reduced matrix element* representing the scalar factors, and a *vector coupling coefficient* related only to the directional properties of the eigenvector. This separation of the energy of a state into three terms is achieved using the Wigner—Eckart theorem (12, 17, 55), and it reduces quantum mechanical problems to finding a product of a geometry factor which is completely defined by the environment and a scalar that can, in principle, be calculated.

In the matrix element the angular integral is defined in a triple product of normalized spherical harmonics. These harmonics are functions of sin θ and cos θ (6), the and integration in real space either vanishes or yields a finite value equal to, or less than, the normalization condition. Thus with NSH Hamiltonians the angular integral value ranges from -1 to 1. The integral may be evaluated using *closed analytical formulae* (56). The relationship between the Schrödinger form and the Dirac notation described earlier emphasizes that these integrated values can be reproduced by a *simple vector combination model* (53) without *apparent* integration (14–16), in a properly defined vector space.

In group theoretical language any such vector combination is interpreted as the formation of a direct product of representations and their components (13), a process similar to the more familiar multiplication of representations Γ alone (3). In that case, the result is generally a reducible representation containing integral numbers of other irreducible representations of the group. These integers are called *reduction coefficients* (12). The result of direct products involving *components* is a reducible representation which may contain nonintegral numbers of the components of the group because of normalization conditions. These nonintegral numbers are the coupling coefficients (Eq. 18), and they are expressed in a generalized notation that illustrates the total component $|\Gamma_T \gamma_T >$ formed by the multiplication of two initial components $|\Gamma\gamma >$ and $|\Gamma'\gamma' >$:

$$|\Gamma\gamma > \times \; |\Gamma'\gamma' > = \sum_{\Gamma_T} < \Gamma\Gamma'\Gamma_T\gamma_T \,|\, \Gamma\Gamma'\gamma\gamma' > \; |\Gamma\Gamma'\Gamma_T\gamma_T > \qquad (24)$$

These coefficients are often more conveniently written as 3Γ symbols, defined in the following relationship:

$$< \Gamma\Gamma'\gamma\gamma' \,|\, \Gamma\Gamma'\Gamma_T\gamma_T > \; = (-1)^{\Gamma+\Gamma'+\gamma_T} [2\Gamma_T + 1]^{1/2} \begin{pmatrix} \Gamma & \Gamma' & \Gamma_T \\ \gamma & \gamma' & \gamma_T \end{pmatrix} \qquad (25)$$

Knowledge of the magnitudes of these coupling coefficients or 3Γ symbols provides the means for evaluating matrix elements such as Eq. 23 and therefore Eq. 19. In the special case of the group R_3, the representation labels Γ and components γ are simply the values of j and m_j, where j may be j, l, or s, and the symbol is the well known $3j$ symbol (10, 12, 13, 18, 55, 56).

When the angular integral is expressed in this 3Γ (or $3j$) form, the conditions under which it vanishes are particularly easy to define. These conditions are called *selection rules.* Since they arise solely from the angular integrals, the selection rules governing the existence of any quantum mechanical property are *entirely dependent on the geometric functions.* Additional selection rules may arise from the reduced matrix elements. Any reduced matrix element must be the scalar of its point group, or otherwise it would vary with the conditions of observation and be undefinable. The vector product must therefore be a scalar, and this is reflected in the allowed combinations of representation j and component m_j in the $3j$ symbol, the existence of which is easily determined. Briefly,

$$\text{in the symbol} \quad \begin{pmatrix} j_1 & j_2 & j_3 \\ m_1 & m_2 & m_3 \end{pmatrix}$$

$$\begin{aligned} j_1 + j_2 - j_3 &\geq 0 \\ j_1 - j_2 + j_3 &\geq 0 \quad \text{for representations} \\ j_1 + j_2 + j_3 &\geq 0 \end{aligned} \tag{26}$$

known as the "triangle rule" since it infers that the magnitudes of the three j values should be capable of forming a triangle, in their number space, and

$$m_1 + m_2 + m_3 = 0 \qquad \text{for components} \tag{27}$$

If any other combinations of j or m_j occur in the $3j$ symbol, the symbol and matrix element must vanish.

The projection operator that produces the wave functions of finite groups from R_3 preserves much of the multiplication behavior of the vector components. Hence the coupling coefficients of finite groups are expected to resemble closely those of R_3, and the 3Γ symbols *retain the selection rules defined by the 3j symbols.*

The representation and component nomenclature in finite groups can indeed be defined to resemble closely those of R_3 (52). This has been intimated in the assignment of $|\Gamma_H O|$ for Hamiltonian eigenstates, but it extends to all eigenvectors (Table VII). Thus specific numerical values can be assigned to each component of Γ in the finite groups. Since the label designation depends on the axis of quantization, it will not be considered in detail here, but will be dealt with in Part 2. We summarize below the existence rules for the finite group 3Γ symbols to emphasize their relationship to those of the $3j$ symbols. Thus the finite group representation condition is (3)

$$\Gamma_1 \times \Gamma_2 \times \Gamma_3 \,\varepsilon\, A_1 \tag{28}$$

That is, the triple direct product must contain the totally symmetric representation of the group. There is, however, a second requirement. We can project from R_3 a value of $j(L)$ for each representation of O_h. For example, a T_{2g} wave function arising from the ^3F term ($L = 3$) of a d^2 configuration can naturally be assigned $j = 3$. Similarly, the second- and fourth-order $E_gO(\theta)$ operator components can be assigned $j = 2$ and 4, respectively. Substituting such numerical values for each Γ, *the triangular rules for the finite groups become identical to the 3j-symbol rules* (*Eq. 26*) (*52*), and they can be readily programmed for a computer. For any particular quantization on an axis of order n, a finite set of at least n-component labels can be defined.The component selection rules must be modified for these finite sets to

$$\gamma_1 + \gamma_2 + \gamma_3^* = 0 \qquad \text{or } |n| \tag{29}$$

That is, the sum of the component values of two interacting vectors along with the complex conjugate value of the total component must vanish or equal the modulus of the order of the quantizing axis. A matrix element failing to meet this requirement vanishes.

The use of these numerical values of finite component labels is clearly of considerable advantage. Since they are obtained by the projection operator, which preserves self-consistency in any quantization, the finite group labels of O_h used in the lower subgroups retain the powerful selection-rule meaning they acquired in R_3. For this reason and because numerical values lead directly to the subduction criterion, the tabulated data and the more rigorous development noted in Part 2 use this nomenclature exclusively. The real sets of components labeled with cartesian symbols can always be recombined to their complex equivalent forms in which the numerical labeling scheme is defined (11).

It is now clear that if the representations in the subgroups are subduced along a physical chain from R_3 via O_h, the infinite group selection rules are also maintained in the finite subgroups, since subduction leaves the mathematical form of the wave functions unchanged. This may not be reflected in the direct product rules of the subgroup itself (*vide infra*). Thus to define the selection rules completely those effective in O_h (or R_3) must act as a Kronecker delta function on the selection rules of the subgroup. These augmented coupling coefficients are called *symmetry-ascent coupling coefficients* (57). They are particularly important in groups of low symmetry as a means of simplifying calculations by reducing the number of matrix elements which must be calculated. Consider, for example, the expansion of the matrix element (Eq.17). Application of the direct product rules simply infers that the element (Eq.17) is nonzero because the triple direct product contains A_1. Expansion into

TABLE VIII

Nonzero Matrix Elements for Normalized Spherical Harmonic Operators on the d^1 Basis Symmetry Adapted to C_{2v}(I), C_{2v}(II), and C_{2v}(III)

$a(C_{2v}(S))$						$<{}^2a(O_h);{}^2a(C_{2v}(S))\mid \text{operator} \mid {}^2a'(O_h);{}^2a(C_{2v}(S))>$						
I	II	III	$a(O_h)$	Operator[a]	$a'(O_h)$	DQ	DS	DT	DU	DV	DM	DN
A_1	A_1	A_1	$E_g\theta$	$V + V'$	$E_g\theta$	$\left(\frac{1}{\sqrt{21}}\right)$	$\frac{2}{7}$	$\frac{5}{7}\left(\frac{1}{\sqrt{15}}\right)$				
A_1			$E_g\theta$	V''	$E_g\varepsilon$				$\frac{1}{7}(\sqrt{2})$	$-\frac{5}{14}\left(\frac{\sqrt{2}}{\sqrt{15}}\right)$		
	A_1	A_1	$E_g\theta$	V'''	$T_{2g}(xy)$						$\frac{1}{7}(\sqrt{2})$	$-\frac{5}{14}\left(\frac{\sqrt{2}}{\sqrt{15}}\right)$
A_1	A_2	B_1	$E_g\varepsilon$	$V + V'$	$E_g\varepsilon$	$\left(\frac{1}{\sqrt{21}}\right)$	$-\frac{2}{7}$	$-\frac{5}{7}\left(\frac{1}{\sqrt{15}}\right)$				
A_2	A_1	A_1	$T_{2g}(xy)$	$V + V'$	$T_{2g}(xy)$	$-\frac{2}{3}\left(\frac{1}{\sqrt{21}}\right)$	$-\frac{2}{7}$	$\frac{20}{21}\left(\frac{1}{\sqrt{15}}\right)$				
B_1			$T_{2g}(xz)$	$V + V' + V''$	$T_{2g}(xz)$	$-\frac{2}{3}\left(\frac{1}{\sqrt{21}}\right)$	$\frac{1}{7}$	$-\frac{10}{21}\left(\frac{1}{\sqrt{15}}\right)$	$\frac{1}{7}(\sqrt{3})$	$\frac{10}{21}\left(\frac{1}{\sqrt{5}}\right)$		
B_2			$T_{2g}(yz)$	$V + V'$ V''	$T_{2g}(yz)$	$-\frac{2}{3}\left(\frac{1}{\sqrt{21}}\right)$	$\frac{1}{7}$	$-\frac{10}{21}\left(\frac{1}{\sqrt{15}}\right)$	$-\frac{1}{7}(\sqrt{3})$	$-\frac{10}{21}\left(\frac{1}{\sqrt{5}}\right)$		
	B_1	B_2	$T_{2g}\tau_2$	$V + V' + V'''$	$T_{2g}\tau_2$	$-\frac{2}{3}\left(\frac{1}{\sqrt{21}}\right)$	$\frac{1}{7}$	$-\frac{10}{21}\left(\frac{1}{\sqrt{15}}\right)$			$\frac{1}{7}(\sqrt{3})$	$\frac{10}{21}\left(\frac{1}{\sqrt{5}}\right)$
	B_2	A_2	$T_{2g}\tau_2'$	$V + V' + V'''$	$T_{2g}\tau_{2'}$	$-\frac{2}{3}\left(\frac{1}{\sqrt{21}}\right)$	$\frac{1}{7}$	$-\frac{10}{21}\left(\frac{1}{\sqrt{15}}\right)$			$-\frac{1}{7}(\sqrt{3})$	$-\frac{10}{21}\left(\frac{1}{\sqrt{15}}\right)$

[a]Defined in equations (10).

octahedral representations is shown in Eq. (19). The last two elements are immediately seen to vanish because $E_g \times T_{2g} \times E_g$ in O_h does not contain A_{1g}. The first three elements obey the triple direct product rule (Eq. 28). The triangle rule may be applied by noting that the value of l which projects $E_g2+(\varepsilon)$ is 2, and that the values of l projecting the operator components $A_{1g}O(i)$ and $E_gO(\theta)$ are 4 and 2, respectively; thus the triangle rules (Eq. 26) are obeyed. Consideration of the component rules (Eq. 29) (see Part 2 for further detail) reveals that the first three elements in the expansion (Eq. 18) are indeed nonzero (see Table VIII).

The use of numerical selection rules is not limited to the definition of energy levels alone. Peturbation operators due to externally applied fields of all kinds can be formulated by defining appropriate point group representations and components of the field. For example, a Zeeman perturbation by an external, parallel, magnetic field affects orbital energies of atoms or ions by coupling with orbital L or spin S angular momentum. Both angular momenta have representations T_{1u} in O_h, and the selection rules for the matrix elements of each component operator eigenstate $|T_1u\gamma|$, where γ is a numerical component, can be derived by inspection.

The assignment of numerical component values and their unequivocal algebra arise only through use of the NSH Hamiltonians because both wave functions and operators are orthonormal representations of both the subgroup and the parent group. Such a procedure can be programmed and provides facility of calculation which does not occur for any other previously used ligand-field Hamiltonian.

VI. MATRIX ELEMENTS OF THE HIGH SPIN STATES OF THE d^n CONFIGURATIONS

Using the NSH Hamiltonians and the projected wave functions (Table VII), matrix elements for any d^n (or f^n) configuration in any geometry can be simply constructed. The detailed expansion of these matrix elements is left until Part 2. However, the results of these calculations for the high spin states of the d^n configurations are presented here. Using Tables VIII through XII, numerical values of the empirical parameters can be derived from a given data base.

Consider first the d^1 case. We require the energies of the individual d orbitals in a complex of given symmetry. In general, this requires calculation of matrix elements $< \psi_i|H|\psi_j >$, where ψ_i and ψ_j are d-orbital functions and H is an NSH operator. If the d-orbital functions are symmetry adapted to the group concerned, the matrix so generated will be block diagonal and present no problem in solution. The results of such calculations for a d^1

TABLE IX
Nonzero Matrix Elements for the Normalized Spherical Harmonic Operator on the d^1 Basis Symmetry Adapted to the Threefold Axis C_3^{xyz}

aC_{3v}	$\Gamma\gamma$	$\Gamma'\gamma'$	$<\Gamma^3 aC_{3v}\|\text{operator}\|\Gamma'^3 aC_{3v}>$ DQ	DM	DN
A_1	$T_2 0$	$T_2 0$	$\frac{2}{3}\left(\frac{1}{\sqrt{21}}\right)$	$\frac{2}{7}$	$\frac{20}{21}\left(\frac{1}{\sqrt{15}}\right)$
$E \pm 1$	$T_2 \pm 1$	$T_2 \pm 1$	$\frac{2}{3}\left(\frac{1}{\sqrt{21}}\right)$	$-\frac{1}{7}$	$-\frac{2}{21}\left(\frac{\sqrt{5}}{\sqrt{3}}\right)$
	$T_2 \pm 1$	$E \pm 1$		$\frac{\sqrt{2}}{7}$	$-\frac{5}{14}\left(\frac{\sqrt{2}}{\sqrt{15}}\right)$
	$E \pm 1$	$E \pm 1$	$-\frac{1}{\sqrt{21}}$		

configuration along both the C_2 and C_3 axes is given in Tables VIII and IX. Obviously these may also be utilized for the high spin states of d^4, d^6, and d^9 complexes.

Thus for a D_{4h} molecule the d-orbital energies are the following:

$$
\begin{aligned}
&E_g O(\theta) \rightarrow a_{1g} && \left(\frac{1}{\sqrt{21}}\right) DQ + \left(\frac{2}{7}\right) DS + \left(\frac{5}{7\sqrt{15}}\right) DT \\
&E_g 2+(\varepsilon) \rightarrow b_{1g} && \left(\frac{1}{\sqrt{21}}\right) DQ - \left(\frac{2}{7}\right) DS - \left(\frac{5}{7\sqrt{15}}\right) DT \\
&T_{2g} 2-(xy) \rightarrow b_{2g} && -\left(\frac{2}{3\sqrt{21}}\right) DQ - \left(\frac{2}{7}\right) DS + \left(\frac{20}{21\sqrt{15}}\right) DT \\
&\left.\begin{matrix} T_{2g} 1+(xz) \\ T_{2g} 1-(yz) \end{matrix}\right. \rightarrow e_g && -\left(\frac{2}{3\sqrt{21}}\right) DQ + \left(\frac{1}{7}\right) DS - \left(\frac{10}{21\sqrt{15}}\right) DT
\end{aligned} \tag{30}
$$

Notice that unlike the classical D_{4h} Hamiltonian these levels obey the center of gravity rule (7, 29, 44). Similarly, the energy levels under a threefold field can be obtained from Table IX, but they include an off-diagonal contribution since there are two e levels. Off-diagonal contributions also occur under the C_{2v} groups since the d-orbital transformation includes at least two a_1 states. Tables X through XII contain the matrix elements evaluated for all the operators connecting the various maximum spin F and P terms of a d^2 basis. The tables can also be used for the maximum spin states of d^3, d^7, and d^8. The components of these terms are distinguished according to their subduction into the various C_{2v} groups. Tables VIII to XII can be used to determine DQ, DS, DT, and so on, from an experimental spectrum. The calculated energies are fitted to the experimental energies once an assignment has been made, and the empirical parameters are thereby deduced.

TABLE X

Nonzero Matrix Elements of the C_{2v}(I) Hamiltonian for the Triplet States of a d^2 Basis[a]

$a(C_{2v}(\mathrm{I}))$	$\Gamma\gamma$	$\Gamma'\gamma'$	B	DQ	DS	DT	DU	DV
					$\langle \Gamma^3 a(C_{2v}(\mathrm{I}))\|H^0 + H_{c_{2v}}(\mathrm{I})\|\Gamma'^3 a(C_{2v}(\mathrm{I}))\rangle$			
A_1	$A_2 2-$	$A_2 2-$		0.43644				
A_2	$T_1 O(P)$	$T_1 O(P)$	15		−0.40000			
	$T_1 O(P)$	$T_1 O$		0.14548	−0.34286	0.12295		
	$T_1 O(P)$	$T_2 2-$					−0.25555	0.16496
	$T_1 O$	$T_1 O$		−0.21822	0.11429	−0.18443		
	$T_1 O$	$T_2 2-$					−0.12778	0.08248
	$T_2 2-$	$T_2 2-$		0.07274		0.43033		
B_1[b]	$T_1 1-\ (P)$	$T_1 1-\ (P)$	15		0.20000		0.34641*	
(B_2)[b]	$T_1 1-\ (P)$	$T_1 1-$		0.14548	0.17143	−0.06148	0.29692*	0.10648*
	$T_1 1-$	$T_2 1+$			−0.22131*	−0.14286*	0.12778	−0.08248
	$T_1 1-$	$T_1 1-$		−0.21822	−0.05714	0.09221	−0.09897*	−0.15972*
	$T_1 1-$	$T_2 1+$			−0.11066*	−0.07143*	0.06389	−0.04124
	$T_2 1+$	$T_2 1+$		0.07274		−0.21517		0.37268*

[a]Symmetry adapted to C_{2v}(I).

[b]The nonzero matrix elements for the $^3B_2$ states are obtained from those of the $^3B_1$ matrix elements using the negative of the starred (*) multipliers and the other multipliers as tabulated. The component labels (in O_h) for the B_2 states are the reverse of those for the B_1 states.

TABLE XI

Nonzero Matrix Elements of the $C_{2v}(S)$ Hamiltonian for the Triplet States of a d^2 Basis (for S = II and III)[a]

$aC_{2v}(S)$				$\langle \Gamma^3 a(C_{2v}(S)) \mid H^0 + H_{C_{2v}}(S) \mid \Gamma'^3 a(C_{2v}(S)) \rangle$					
II	III	$\Gamma\gamma$	$\Gamma'\gamma'$	B	DQ	DS	DT	DM	DN
A_1	A_1	$T_2 2-$	$T_2 2-$		0.07274		0.43033		
$B_1{}^b$	$A_2{}^c$	$T_1 1+ (P)$	$T_1 1+ (P)$	15		0.20000		−0.34641*	
		$T_1 1+$	$T_1 1+$		0.14548	0.17143	−0.06148	0.19795	−0.05324*
		$T_1 1+$	$T_2 1-$			−0.22131*	−0.14286*	0.25555	0.12372
		$T_1 1+$	$T_1 1+$		−0.21822	−0.05714	0.09221	−0.02474*	−0.31944*
		$T_1 1+$	$T_2 1-$			−0.11066*	−0.07143*	−0.03194	0.16496
		$T_2 1-$	$T_2 1-$		0.07274		−0.21517	0.12372*	0.10648*
A_2	B_1	$T_1 O(P)$	$T_1 O(P)$	15		−0.40000			
		$T_1 O(P)$	$T_1 O$		0.14548	−0.34286	0.12295		
		$T_1 O$	$A_2 2-$					−0.25555	0.16496
		$T_1 O$	$T_1 O$		−0.21822	0.11429	−0.18443		
		$T_1 O$	$T_2 2-$					−0.12778	0.08248
		$T_2 2-$	$A_2 2-$		0.43644				
$B_2{}^b$	$B_2{}^c$								

[a]Basis symmetry adapted to C_{2v}(II). Hence component labels appropriate to C_{2v}(II)

[b]The nonzero matrix elements for the $^3B_2$ states of C_{2v}(II) are obtained from the $^3B_1$ matrix elements using the negative of the starred (*) multipliers and the other multipliers as tabulated.

[c]The same procedure as footnote b, but using the matrix elements of the $^3A_2$ states of C_{2v}(III). The $B_2(B_2)$ component labels are the reverse of those of $B_1(A_2)$.

TABLE XII
Nonzero Matrix Elements of the C_{3v} Hamiltonian for the Triplet States of a d^2 Basis Quantized along C_3^{xyz} in O_h[a]

$\alpha(C_{3v})$	$\Gamma\gamma$	$\Gamma'\gamma'$	$<\Gamma\gamma^3 a(C_{3v})\|$operator$\|\Gamma'\gamma'^3 a(C_{3v})>$			
			B	DQ	DM	DN
A_1	$T_2 0$	$T_2 0$	15	$-\left(\frac{1}{3\sqrt{21}}\right)$	$-\frac{1}{7}$	$-\left(\frac{2\sqrt{5}}{21\sqrt{3}}\right)$
A_2	$A_2 0$	$A_2 0$		$-\left(\frac{2}{\sqrt{21}}\right)$		
	$T_1 0$	$A_2 0$			$\left(\frac{2}{7\sqrt{5}}\right)$	$-\left(\frac{1}{7\sqrt{3}}\right)$
	$A_2 0$	$T_1 0(P)$			$-\left(\frac{4}{7\sqrt{5}}\right)$	$\left(\frac{2}{7\sqrt{3}}\right)$
	$T_1 0$	$T_1 0$		$\left(\frac{1}{\sqrt{21}}\right)$	$-\frac{1}{35}$	$-\left(\frac{2\sqrt{5}}{7\sqrt{3}}\right)$
	$T_1 0$	$T_1 0(P)$		$\left(\frac{2}{3\sqrt{21}}\right)$	$-\frac{8}{35}$	$\left(\frac{5}{21\sqrt{3}}\right)$
	$T_1 0(P)$	$T_1 0(P)$	15		$-\frac{2}{5}$	
E	$T_2 \pm 1$	$T_2 \pm 1$		$-\left(\frac{1}{3\sqrt{21}}\right)$	$\frac{1}{14}$	$\left(\frac{\sqrt{5}}{21\sqrt{3}}\right)$
	$T_1 \pm 1$	$T_2 \pm 1$			$-\left(\frac{1}{14\sqrt{5}}\right)$	$\left(\frac{2}{7\sqrt{3}}\right)$
	$T_2 \pm 1$	$T_1 \pm 1(P)$			$-\left(\frac{4}{7\sqrt{5}}\right)$	$-\left(\frac{\sqrt{3}}{14}\right)$
	$T_1 \pm 1$	$T \pm 1$		$\left(\frac{1}{\sqrt{21}}\right)$	$\frac{1}{70}$	$\left(\frac{\sqrt{5}}{7\sqrt{3}}\right)$
	$T_1 \pm 1$	$T_1 \pm 1(P)$		$\left(\frac{2}{3\sqrt{21}}\right)$	$\frac{4}{35}$	$-\left(\frac{\sqrt{5}}{42\sqrt{3}}\right)$
	$T_1 \pm 1(P)$	$T_1 \pm 1(P)$	15		$\frac{1}{5}$	

[a]Operator (Eq. 15) appropriate for D_{3d}, D_3, C_{3v}.

Consider, for example, a C_{2v}(I) complex whose energy-level correlation diagram is shown in Fig. 4. There are three A_2 states, and thus the matrix describing their energies is 3 × 3. Since the matrix must be hermitian, that is, $< T_1O|H|T_22- >^* = < T_22-|H|T_1O >$, only the six upper diagonal elements are stated in Table X, the lower diagonal being obtained upon reflection. A detailed study, preferably a single-crystal, polarized light study, of such a complex should provide three experimental A_2-state energies. These must be fitted to the roots of the 3 × 3 matrix (*vide infra*). Similar fitting of the A_1, B_1, and B_2 energies provides the magnitudes of the various empirical parameters.

Using Tables VIII to XII readers can analyze the spin-allowed spectra

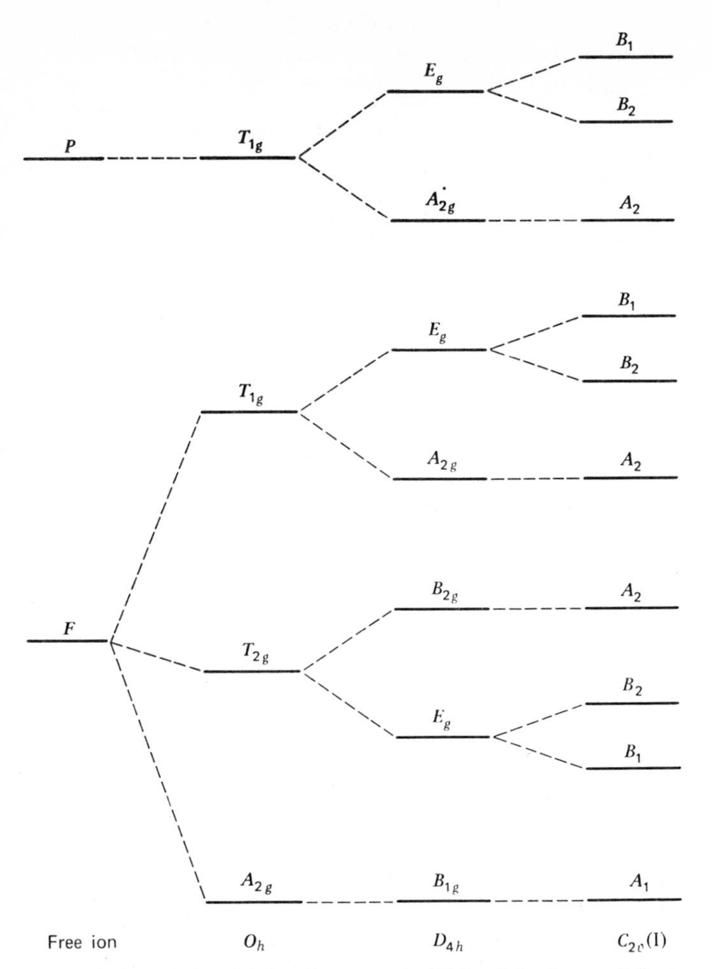

Fig. 4. Splitting of F and P free ion terms in fields of O_h, D_{4h}, and C_{2v} symmetry.

arising from the maximum spin states of all the d^n configurations in the point groups discussed. Once the appropriate point group and Hamiltonian have been deduced, no further calculations are necessary, other than the fitting of the experimental spin-allowed data to the matrices presented here.

A. Trace Method for Fitting Spectra

Various iterative programs exist to fit experimental data to calculated energies. Clearly the effort is minimized by reducing the number of variables. This can be achieved by noting that the roots of any matrix added together yield a trace of the matrix, that is, the sum of the roots equals the sum of the diagonal elements. Thus in the C_{2v}(I) example cited above, reference to Table X shows that

$$-0.14548\ DQ - 0.28571\ DS + 0.2459\ DT + 15B = Tr(A_2) - 3GS \quad (31)$$

where $Tr(A_2)$ signifies the sum of the three observed transition energies to the A_2 states, which are relative (of course) to the ground state (GS) as zero. Since the matrix roots are relative to the ^3F term as zero, this ground-state energy, which is also relative to the ^3F as zero, must be subtracted out. Similarly, we derive from Table X the following:

for the three B_1 states,

$$\begin{aligned}&-0.14548\ DQ + 0.14286\ DS - 0.12296\ DT + 0.24744\ DU \\ &+ 0.21296\ DV + 15B = Tr(B_1) - 3\ \mathrm{GS}\end{aligned} \quad (32)$$

for the three B_2 states,

$$\begin{aligned}&-0.14548\ DQ + 0.14286\ DS - 0.12296\ DT - 0.24744\ DU \\ &-0.21296\ DV + 15B = Tr(B_2) - 3\mathrm{GS}\end{aligned} \quad (33)$$

for the A_1 state,

$$0.43644\ DQ = Tr(A_1) - \mathrm{GS} \quad (34)$$

Rearrangement eliminates the unknown ground-state energy to give expressions of the following type:

$$\begin{aligned}0.49488\ DU + 0.42592\ DV &= Tr(B_1) - Tr(B_2) \\ 0.42857\ DS - 0.36886\ DT &= \tfrac{1}{2}(Tr(B_1) + Tr(B_2)) - Tr(A_2) \\ 15B - 1.4548\ DQ &= (\tfrac{1}{3})(Tr(B_1) + Tr(B_2) + Tr(A_2)) - 3Tr(A_1)\end{aligned} \quad (35)$$

Thus if all possible transitions can be resolved, three independent variables must be fitted (1). Similar expressions can be derived for other point group symmetries from Table VII to XI. Note, for example, that if one of the A_2 transition energies is not observed, it may be worthwhile to use $Tr(A_2)$ itself as a variable in the iteration, since it is likely that one will have a reasonably good idea of its magnitude. One may have less idea of the magnitudes of other parameters.

VII. THE ANALYSIS OF SCALAR PARAMETERS

The chemical significance of parameters obtained from classical Hamiltonians, be they crystal field or semiempirical molecular orbital, has been a contentious issue in the past. Gerloch and Slade discussed this problem at length (22), and the reader is referred to the literature for general background information. In this section we pursue the scalar parameters from the viewpoint of the NSH approach.

The Wigner—Eckart theorem removes all of the formal geometry which can be described by the point group from the value of the matrix element. The remaining part of the value is the yet unspecified scalar parameter P which may represent DQ or any of the distortion parameters in the subgroup.

Within any group of complexes of the same symmetry, a meaningful comparison of the scalar-parameter magnitudes can be made. Thus we may compare the DT values generated in a series of D_{4h} complexes. The comparison is direct, although the chemical significance we place on the comparison depends on the model which we choose to extract such chemical information. Indeed, we may choose to compare DT values for a series of complexes of different symmetries belonging to various subgroups which have the $E_gO(\theta)$ tensor component in common. The comparison is precise in that it measures the magnitude of the $E_gO(\theta)$ tensor component to the field experienced by the complex concerned; its chemical significance, however, requires further analysis of the form of the parameters in order to extract their common features. Scalar parameters of all finite groups are functions of the radial integral G^l, defined by a single-center expansion of contributions both within and beyond the distance to the off-center atoms (22, 53):

$$G^l = \int_0^a R^*(Znl)\,\frac{r^l}{a^{l+1}}R(Znl)r^2dr + \int_a^\infty R^*(Znl)\frac{a^l}{r^{l+1}}\,R(Znl)r^2dr \tag{36}$$

The two contributions to the magnitude of the observable arise from different parts of the perturbation potential, namely, the perturbation in the interior region, within the metal-ligand distance, and in the exterior region beyond the ligands. In similar fashion to the classical crystal-field Dq parameter (22), the NSH DQ parameter in an octahedron can be expressed as

$$DQ = \sqrt{21}\;ze^2G^4 \tag{37}$$

where G^4 is the fourth-order radial parameter. Using the approximation in which the perturbed orbitals are compact, this parameter can be expressed as

$$DQ = \sqrt{21}\;ze^2\left(\frac{r^4}{a^5}\right) \tag{38}$$

where ze is the charge on a ligand, and a is the length of the electric-gradient vector created by the presence of six identical ligands.

In any subgroup down a physically significant chain, the deviations from octahedral geometry change the electric-field gradients in different orthogonal directions. This is manifested in the presence of nonzero distortion parameters $D(l)$, which can be expressed as a difference between two octahedral parameters (22), namely,

$$DS = -2ze^2\left(\left(\frac{\overline{r^2}}{a^3}\right) - \left(\frac{\overline{r^2}}{b^3}\right)\right) \tag{39}$$

$$DT = -\left(\frac{7\sqrt{15}}{21}\right)ze^2\left(\left(\frac{\overline{r^4}}{a^5}\right) - \left(\frac{\overline{r^4}}{b^5}\right)\right) \tag{40}$$

where a and b are two different orthogonal electric-field-gradient vectors. The DQ parameter must be similarly factored when symmetry lower than octahedral is considered, since this parameter is a "global" parameter reflecting the average field experienced by the metal ion. In D_{4h} symmetry, DQ can be expressed as

$$DQ = ze^2\frac{\sqrt{21}}{3}\left\{2\frac{\overline{r^4}}{a^5} + \frac{\overline{r^4}}{b^5}\right\} \tag{41}$$

General expressions for these parameters, taking into account both angular as well as radial distortion, have been developed (59).

If we may assume in Eq. 36 that the numerator and denominator of the radial integral can be separated, further factorization of Eqs. 39 to 41 is possible:

$$DS = -2e\bar{r}^2\left[\frac{(x)e}{\mathrm{a}^3} - \frac{(z)e}{\mathrm{b}^3}\right] \tag{42}$$

$$DT = -\left(\frac{7\sqrt{15}}{21}\right)e\bar{r}^4\left[\frac{(x)e}{\mathrm{a}^5} - \frac{(z)e}{\mathrm{b}^5}\right] \tag{43}$$

$$DQ = e\bar{r}^4\frac{\sqrt{21}}{3}\left\{2\frac{(x)e}{\mathrm{a}^5} + \frac{(z)e}{\mathrm{b}^5}\right\} \tag{44}$$

where (z) and (x) are chemical-site parameters in two orthogonal directions, and a and b are now regarded as geometric quantities. The chemical-site parameters are numbers modifying the radial integrals in different orthogonal directions. They do not refer to specific ligands along any particular axis, but rather to the total potential, from whatever source, experienced by the metal ion in that orthogonal direction. At this stage, no model is imposed with which to obtain chemical information about the ligands themselves.

The advantages to the factorization are clear; it allows a separation between the chemical and geometric contributions to the distortion. Thus we can consider a limiting case in which there is no geometric distortion (a = b), but different ligands in two orthogonal directions generate a molecule exhibiting low symmetry. In cubic molecules there is much less need to consider such a distinction, so that the question of the physical significance of the terms making up DQ is less relevant. If we are to obtain useful chemical information from the electronic properties of low-symmetry complexes, some acceptable procedure for separating chemical from geometric contributions would be beneficial. *If* a and b are associated with the geometry of the complex as deduced by x-ray structural analysis, and *if* the compact orbital approximation of the radial integral is permitted, then chemical information through evaluation of (z) and (x) can be extracted. *However, there is no theoretical justification for these conditions*; indeed, there are reasons, as argued by Gerloch (60), for not accepting either of them. The realization that covalency must contribute to the magnitude of crystal-field parameters, even in systems regarded as primarily ionic (8), shows that a too specific interpretation of the terms in the crystal-field expansions of DQ, DS, DT, and so on, is naive. Nevertheless, the collection of the appropriate spectroscopic and crystallographic data would prove highly illuminating, and it is the purpose, in part, of this admittedly speculative treatment to spur the collection of such data, which hardly exist at the moment. The approach can then be seriously tested. If the factorization fails, then at least we can hope that the data will point to a more profitable theoretical direction.

VIII. RELATIONSHIPS OF THE NSH HAMILTONIAN PARAMETERS WITH THOSE OF OTHER MODELS

Mathematical relationships between the NSH parameters and those of other models can, of course, be established. However, in establishing these relationships we are, in effect, imposing the chemical assumptions inherent on the model being compared, and this fact should be borne in mind. Thus there is a simple numerical factor relating the classical octahedral, fourfold axis, quantized Hamiltonian to that used here such that

$$DQ = (6\sqrt{21})\ Dq \tag{45}$$

Comparison of the NSH Hamiltonian for D_{4h} molecules with that classically used (7) provides the relationships

$$\begin{aligned} DS &= -7Ds \\ DT &= -\left(\frac{7\sqrt{15}}{2}\right)Dt^* \\ DQ &= \left(6\sqrt{21}\right)Dq_E - \left(\frac{7\sqrt{21}}{2}\right)Dt \end{aligned} \tag{46}$$

*Recall (see footnote to Table V) that the $|E_gO(\theta)|_{O_h}^{4}$ tensor component is defined with opposite sign in this review, relative to our earlier publications.

The last relationship clearly reveals how the classical D_{4h} Hamiltonian scrambles the DQ and DT parameters. In fact,the classical Dq in a tetragonal complex is a measure of the in-plane crystal-field strength (Dq_E), whereas DQ is a measure of the average ligand field experienced by the metal ion. The factorization of the crystal field in tetragonal complexes into equatorial, Dq_E, and axial, Dq_A, components can be mimicked in the NSH approach through

$$\begin{aligned} DQ_A &= DQ + 2\left(\frac{\sqrt{7}}{\sqrt{5}}\right)DT \\ DQ_E &= DQ - \left(\frac{\sqrt{7}}{\sqrt{5}}\right)DT \end{aligned} \tag{47}$$

While DQ_E and DQ_A have rather artificial significance, they can be conveniently used to establish limiting values of DT. Thus in a tetragonal complex in which the axial field becomes vanishingly small, DQ_A approaches zero and the ratio DT/DQ approaches $-\frac{1}{2}(\sqrt{5}/\sqrt{7})$ (2). This should provide a limit for DT in a truly square planar complex. Similarly, if the equatorial field vanishes, a limiting value of the ratio $DT/DQ = (\sqrt{5}/\sqrt{7})$ is obtained as the molecule approaches linearity (2).

Orbital energy separations in D_{4h} symmetry (Fig. 5) are related to the NSH parameters through

$$\Delta_1 = -\left(\frac{3}{7}\right)DS + \left(\frac{10}{7\sqrt{15}}\right)DT$$

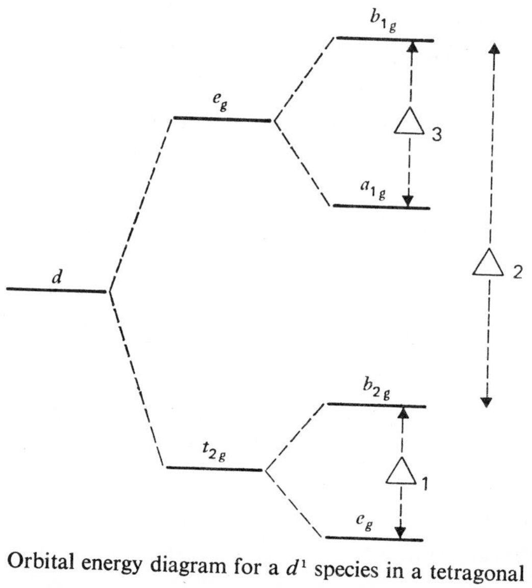

Fig. 5. Orbital energy diagram for a d^1 species in a tetragonal (D_{4h}) field.

$$\Delta_2 = \left(\frac{5}{3\sqrt{21}}\right)DQ - \left(\frac{\sqrt{5}}{\sqrt{27}}\right)DT \tag{48}$$

$$\Delta_3 = -\left(\frac{4}{7}\right)DS - \left(\frac{10}{7\sqrt{15}}\right)DT$$

Gerloch (22) utilized a crystal-field parameterization using Dq and the empirical second-order parameter Cp, defined by

$$\begin{aligned} Cp &= \left(\frac{2}{7}\right)Ze^2G^2 \\ &= \left(\frac{2}{7}\right)Ze^2\left(\frac{\overline{r^2}}{a^3}\right) \end{aligned} \tag{49}$$

where G^2 is the second-order radial integral, as defined in Eq.36. The Cp parameter refers to a given orthogonal direction and is related to DS (Eq.39) through

$$DS = -7Ds = -7(Cp_a - Cp_b) \tag{50}$$

All these models are single-center theories of varying sophistication. A typical multicenter L. C. A. O. theory is the semiempirical molecular orbital approach. The NSH Hamiltonian in D_{4h} symmetry is related through the McClure parameters (31) via

$$\begin{aligned} d\sigma &= \left(\frac{1}{28}\right)(6DS + (\sqrt{15})DT) \\ d\pi &= \left(\frac{1}{14}\right)\left(3DS - 2\left(\frac{\sqrt{5}}{\sqrt{3}}\right)DT\right) \end{aligned} \tag{51}$$

and to the semiempirical, orbital-angular-overlap parameters (35) via expressions of the type

$$DQ_L = \left(\frac{3\sqrt{21}}{5}\right)\left(3e'_{\sigma L} - 4e'_{\pi L}\right) \tag{52}$$

$$DS = -2\,(e'_{\sigma L} + e'_{\pi L} - e'_{\sigma Z} - e'_{\pi Z}) \tag{53}$$

$$DT = -\left(\frac{\sqrt{15}}{5}\right)\left(3e'_{\sigma L} - 4e'_{\pi L} - 3e'_{\sigma Z} + 4e'_{\pi Z}\right) \tag{54}$$

These refer specifically to a tetragonal *trans*-ML_4Z_2 complex; similar expressions can be readily derived for other complexes.

It may be convenient to use such expressions to abstract chemical information from NSH parameters by converting them into parameters which have apparent greater intuitive significance. However, this factorization is similar to that proposed in Eqs. 42 to 44, though perhaps more justifiable, theoretically, in that it imposes a chemical model upon the NSH parameterization. The NSH parameters as they stand have precise group theoretical significance. The orbital-angular-overlap parameters, designed to be transferable from one complex to another, have a ligand additivity model imposed upon them. In other words, we can describe a molecule A-M-B by addition of two components described by A-M and B-M; no account is taken directly of the possible perturbation of the A-M bond by the B-M bond in the total description of A-M-B. If such interactions are important, the significance of the orbital-angular-overlap parameters becomes less sharply defined as will, indeed, the factorization described in Eqs. 42 to 44.

Schäffer developed (29, 61) a nonadditive ligand field approach involving a spherical harmonic parameterization which is conceptually the same as the NSH procedure. His parameters are related to ours through

$$I^g_{a_1} = DQ; \qquad I^d_{e\theta} = DS; \qquad I^g_{e\theta} = DT \tag{55}$$

Because of the lack of standardization in the past, it is not practical to include further relationships of the type described in this section for other stereochemistries. Similar expressions be can derived by direct mapping of the classically used Hamiltonian with the NSH Hamiltonian. Where angular distortion is present, information about such distortion can be obtained through a comparison of the crystal-field and NSH parameterization. Thus ratios of NSH parameters of the same order lead to expressions in which the radial part has canceled out, leaving a number related directly to the magnitude of the angular distortion. While angles determined this way may not, and indeed are unlikely to, agree perfectly with crystallographic data, the deviation between crystallographic and calculated values should be small.

IX. GROUND-STATE ENERGY SPACE DIAGRAMS

The NSH approach and its associated algebra have been developed to deal with complexes of comparatively low symmetry. Uncertainty concerning the identity of the ground state is a problem that does not generally arise when dealing with cubic complexes; it is frequently a problem in low-symmetry systems. To alleviate this we have introduced *ground-state energy space diagrams* (2, 21, 62, 63). The energy levels in, for example, a D_{4h} complex are defined by values of DQ, DS, DT, B, and C (neglecting spin-orbit coupling for the moment). To display the energies of these states as continuous functions of the five variables we would need a six-dimensional surface. It is, of course, this complexity that leads to ambiguity in the identification of

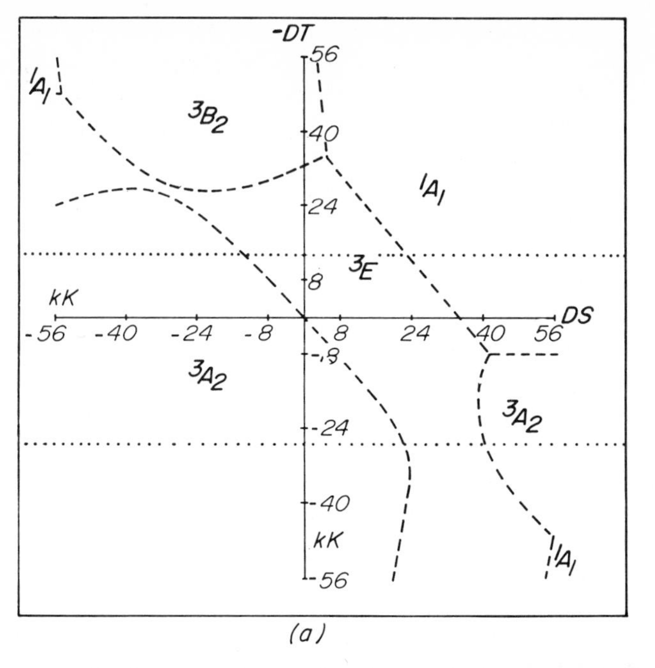

(a)

the ground state if other states lie nearby. Any excited state which, for some specified values of the distortion parameters, can become the ground state, will intersect the energy surface at E = O. A slice of the multidimensional space at E = O reveals contours showing the values of each parameter at which particular states become the ground state. There are still too many surfaces to depict conveniently. However, if we consider a series of D_{4h} complexes of similar DQ and B, and use a conventional C/B ratio, a two-dimensional diagram can be constructed that shows the values of DT and DS required to generate different ground states (e.g., see Fig. 6*a*). Similar diagrams can be constructed using $d\sigma$ and $d\pi$ as variables (2, 21) (Fig. 6*b*), and they provide a challenge to the synthetic chemist to design molecules with hitherto unobserved ground states.

In a given complex, the sign of each distortion parameter can usually be deduced using relationships such as Eq. 40 or 47 and frequently leads to the correct identification of the ground state. That this procedure indeed works provides credence for the chemical significance generally placed on the various classical parameters. As indicated briefly above (Section VIII), limiting values of certain parameters can sometimes be deduced, implying that real complexes cannot exist in some regions of the energy space diagrams. This problem was also explored earlier in terms of the parameter $A(l)$ (21). Diagrams have also been published for linear (62) and trigonal groups (63)

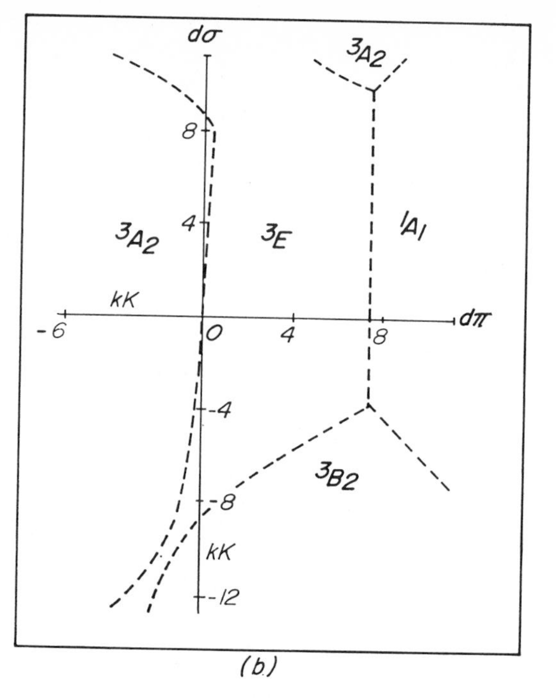

Fig. 6. Ground-state energy space diagrams for a d^2 species in a tetragonal (D_{4h}) field. (*a*) DS versus DT, and (*b*) $d\sigma$ versus $d\pi$. In both diagrams DQ = 31,800 cm^{-1}, B = 800 cm^{-1}, and C = 3,000 cm^{-1}. The ordinate in Fig. 6*a* is defined as $-DT$ for ease of comparison with previous diagrams (2).

and for systems inclusive of spin-orbit coupling where the boundary region between ground states becomes diffuse (64).

X. DATA ACQUISITION

So far it has been assumed that there exists a body of experimental information which can be used, via a fitting process, to generate values for the various parameters. Two factors have not been mentioned yet—vibronic coupling and spin-orbit coupling. The room-temperature, broad, spectra of low-symmetry complexes contain contributions from these processes which have to be factored out if a meaningful analysis is to be completed. In published crystal-field analyses, both these processes are generally ignored. In effect, values of the transition energies (band maxima) are being used which deviate from the true electronic excitation energies by, possibly, several hundreds of wave numbers. This can lead to *considerable percentage errors* in the magnitudes of the parameters so produced. In addition, room-temperature spectra are generally so broad that even the identification of the band maximum has considerable uncertainty associated therewith. The apparent success of previous treatments of D_{4h} complexes (in particular, Refs. 2 and 65–69) may arise through a fortuitous cancellation of some of these errors. Even here, however, there have been disagreements concerning assignments (e.g., Refs. 65 and 67), highlighting the difficulty, in low-symmetry complexes, of being sure of the identity of the excited states. In this respect, powder or solution spectra have decided disadvantages.

For the future study of low-symmetry complexes such approximations and ambiguities must be eliminated as far as possible. High-quality data must be obtained so that both the spin-orbit and vibronic coupling Hamiltonians can be included in the analysis. Such high-quality data can only be obtained through single-crystal studies carried out at, or close to, liquid-helium temperatures, using polarized light. In this way, band energies can be measured with highly accuracy, and their assignments can be confirmed by the dichroism data. The commercial availability of closed-circuit Joule—Thompson refrigerators, at relatively low cost, should facilitate studies of this type. It is, of course, important that the x-ray structures of these crystals be available. The phenomenal improvement in x-ray structural analysis, which has taken place over the last ten years, should ensure that these structures will be available in the future.

Magnetic circular dichroism (MCD) can also play an important role in the study of low-symmetry systems in that bands poorly resolved in the absorption spectrum can be well resolved in the MCD spectrum. Finally, there has been growing interest in the photochemistry of low-symmetry complex, confined for the moment primarily to complexes of D_{4h} symmetry (70). Expansion and integration of this work with the absorption, emission, and MCD spectra of low-symmetry complexes will ultimately lead to a much more detailed understanding of metal ligand interactions.

PART 2 CALCULATION TECHNIQUES

XI. INTERMEDIATE-FIELD MATRIX ELEMENTS

This section deals with procedures for the evaluation of the matrix elements presented in Part 1, and it discusses the properties of the 3Γ symbols in more detail with special relevance to their algebra and selection rules. The section is presented in a pedagogical fashion, but at a higher level than Part 1. The reader is expected to have familiarity with classical techniques, as discussed, for example, by Ballhausen (7), and to have had some exposure to calculations involving vector algebra (18). It is not intended to present an exhaustive survey of the methods available to calculate energy matrices. Instead, we present sufficient background material to show the best procedures for the computation of data using the NSH Hamiltonian, and to illustrate how this approach simplifies such calculations.

A purpose of any calculation of the electronic structure of a complexed transition-metal ion is the precise prediction of the observed transition energies. In some cases, the weak- or strong-field approximations provide an adequate model, but more often second-order interactions between states become important. This more general situation is called the intermediate field. Its solution requires the calculation of many off-diagonal matrix elements, which if pursued by traditional techniques is a long and tedious procedure. The calculation can be simplified, however, by the definition and full understanding of the behavior of *intermediate-field matrix elements.*

These matrix elements, presented by König et al. (71), have the following form:

$$\begin{aligned}&<\alpha_1 j_1 \gamma_1 a_1 \mid T\gamma^{k\Gamma a} \mid \alpha_2 j_2 \Gamma_2 \gamma_2 a_2>\\ &=(-1)^{2k}[2j_1+1]^{-1/2}<j_1\Gamma_1\gamma_1 a_1 \mid k\Gamma\gamma a j_2 \Gamma_2\gamma_2 a_2>\\ &<\alpha_1 j_1 \parallel T^{(k)} \parallel \alpha_2 j_2>\end{aligned}$$

in which

$$\mid \alpha j\Gamma\gamma a> = \sum_m \mid jm> <jm \mid j\Gamma\gamma a>$$

and

$$\mid T\gamma^{k\Gamma a} \mid = \sum_q \mid T_q^k \mid <kq \mid k\Gamma\gamma a> \tag{56}$$

Thus the intermediate-field eigenvectors are defined as linear combinations of components of $R_3 \mid jm_j>$ and the combination coefficients are the subduction coefficients $<jm|j\Gamma\gamma a>$ of Table VII. Similarly, the tensor operators are linear combinations of tensor components of R_3, and the combination coefficients are again the appropriate subduction coefficients written in tensor nomenclature $<kq|k\Gamma\gamma a>$ from the same tabulations. They are indeed the NSH operators.

The factorization of the intermediate matrix element is accomplished using the Wigner—Eckart theorem. This yields a product of an intermediate coupling coefficient, $<j_1\Gamma_1\gamma_1 a_1 \mid k\Gamma\gamma a,\ j_2\Gamma_2\gamma\gamma_2 a_2>$, and a reduced matrix element specified only in terms of representation labels of the infinite group R_3, $<\alpha_1 j_1 \parallel T^{(k)} \parallel \alpha_2 j_2>$.

If the development of the intermediate-field formalism is stopped at this point, every matrix coupling coefficient must be expanded in terms of a summation of coupling coefficients of R_3 multiplied by the product of the three relevant subduction coefficients:

$$\begin{aligned} &\langle j_1\Gamma_1\gamma_1 a_1 \,|\, k\Gamma\gamma a, j_2\Gamma_2\gamma_2 a_2\rangle \\ &= \sum_{m_1 q_2 m_2} \langle j_1 m_1 \,|\, j_1\Gamma_1\gamma_1 a_1\rangle^* \langle kq \,|\, k\Gamma\gamma a\rangle \ \langle j_2 m_2 \,|\, j_2\Gamma_2\gamma_2 a_2\rangle \\ &\qquad \langle j_1 m_1 \,|\, kq j_2 m_2\rangle \\ &= \sum_{m_1 q_2 m_2} \langle j_1 m_1 \,|\, j_1\Gamma_1\gamma_1 a_1\rangle^* \langle kq \,|\, k\Gamma\gamma a\rangle \ \langle j_2 m_2 \,|\, j_2\Gamma_2\gamma_2 a_2\rangle \\ &\qquad (-1)^{j_1-j_2+m_1}[2j_1+1]^{+1/2}\begin{pmatrix} j_1 & k & j_2 \\ -m_1 & q & m_2 \end{pmatrix} \end{aligned} \tag{57}$$

Conventionally, the interpretation of the intermediate matrix element is terminated with this formula (Eq. 57). Such an expansion has both advantages and disadvantages. The Hamiltonian operator for d electrons, $T(^{(k\Gamma a)}_{\gamma})$, *of any point group* is a linear combination of physical harmonics (72) which may include any of the following: (C^4_0), $(C^4_4 + C^4_{-4})$, $(C^4_2 + C^4_{-2})$, $(C^4_3 - C^4_{-3})$, (C^2_0), and $(C^2_2 + C^2_{-2})$, but no others. The matrices of these operators, involving the elements

$$\langle j_1 m_1 \,|\, C^k_0 \,|\, j_2 m_2\rangle \qquad \text{and} \qquad \langle j_1 m_1 \,|\, C^k_q \pm C^k_{-q} \,|\, j_2 m_2\rangle$$

can be very rapidly calculated in the R_3 group using closed formulae (18). Once calculated they can be stored and used in any desired linear combination without recalculation for each specific case. Because of this, expansion (Eq. 56) is particularly useful in programming central-field calculations into computer languages.

However, many of the summations of expansion (Eq. 57) are identically zero. Some of these vanishing intermediate matrix elements can be identified by the direct product rules of the finite point group, but many more are unpredictable using conventional rules without performing the full summation. An additional difficulty of expansion (Eq. 56) is that a completely separate set of subduction coefficients is required for the basis set of each different point group. Moreover, if several $|j_2 m_2\rangle$ map onto one $|\Gamma\gamma\rangle$ during subduction, the $3j$ symbol of expansion (Eq. 57) must be repeatedly calculated as often as it appears.

These tedious procedures can be avoided with the use of the previously discussed basis sets projected into O_h and subduced into the physically significant subgroups, providing a single basis set for each chain. The expansion of the intermediate matrix elements in terms of these projected eigenvectors is (52)

$$\begin{aligned} &\langle \alpha_1 j_1\Gamma_1\gamma_1 a_1 \,|\, T^{k\Gamma a}_{\gamma} \,|\, \alpha_2 j_2\Gamma_2\gamma_2 a_2\rangle \\ &= (-1)^{\Gamma_1+\gamma_1^*}\begin{pmatrix} \Gamma_1 & \Gamma & \Gamma_2 \\ \gamma_1^* & \gamma & \gamma_2 \end{pmatrix} \langle \alpha\Gamma_1 \,\|\, T^{k\Gamma} \,\|\, \alpha_2\Gamma_2\rangle \end{aligned} \tag{58}$$

in which

$$\begin{pmatrix} \Gamma_1 & \Gamma & \Gamma_2 \\ \gamma_1^* & \gamma & \gamma_2 \end{pmatrix}$$

is the 3Γ vector coupling symbol for the finite group (usually O_h) and $<\alpha\Gamma_1 \| T^{k\Gamma} \| \alpha_2\Gamma_2>$ is the accompanying reduced matrix element. γ_1^* is the congugate of γ_1.

This factorization has the distinct advantage that the finite group component algebra, which was defined by the projection operators discussed in Part 1, can immediately be used to discover the vanishing 3Γ symbols. In most cases, the summations necessary with the conventional expansion can thus be avoided.

The main difficulty with this finite-group expansion is the evaluation of the reduced matrix element (71). If the projection conditions are ignored, the reduced matrix elements must be evaluated within an arbitrary set of orthonormalization conditions imposed for the specific finite group. Several tabulations of this kind have been published (11). However, paying regard to the projection conditions, the infinite-and finite-group expansions can be equated. Thus, from Eg. 56.

$$\sum_{m_1 q m_2} <j_1 m_1 | j_1 \Gamma_1 \gamma_1 a_1>^* \quad <k_q | k\Gamma\gamma a> \quad <j_2 m_2 | j_2 \Gamma_2 \gamma_2 a_2>$$
$$\times (-1)^{j_1+2k-j_2+m_1} \begin{pmatrix} j_1 & k & j_2 \\ -m_1 & q & m_2 \end{pmatrix} <\alpha_1 j_1 \| T^k \| \alpha_2 j_2>$$
$$= (-1)^{\Gamma_1+\gamma_1^*} \begin{pmatrix} \Gamma_1 & \Gamma & \Gamma_2 \\ \gamma_1^* & \gamma & \gamma_2 \end{pmatrix} <\alpha j_1 \Gamma_1 \| T^{k\Gamma} \| \alpha_2 j_2 \Gamma_2> \tag{59}$$

$$\therefore <\alpha_1 j_1 \Gamma_1 \| T^{k\Gamma} \| \alpha_2 j_2 \Gamma_2>$$
$$= \frac{\sum_{m_1 q m_2} <j_1 m_1 | j_1 \Gamma_1 \gamma_1 a_1> <kq | k\Gamma\gamma a> <j_2 m_2 | j_2 \Gamma_2 \gamma_2 a_2> \begin{pmatrix} j_1 & k & j_2 \\ -m_1 & q & m_2 \end{pmatrix}}{(-1)^{\Gamma_1+\gamma^*}}$$
$$\times \frac{(-1)^{j_1+2k-j_2+m_1} <\alpha_1 j_1 \| T^k \| \alpha_2 j_2>}{\begin{pmatrix} \Gamma_1 & \Gamma & \Gamma_2 \\ \gamma_1^* & \gamma & \gamma_2 \end{pmatrix}}$$
$$= \frac{F^{\Gamma_1 \Gamma \Gamma_2}_{j_1 k j_2} <\alpha_1 j_1 \| T^k \| \alpha_2 j_2>}{(-1)^{\Gamma_1+\gamma_1^*} \begin{pmatrix} \Gamma_1 & \Gamma & \Gamma_2 \\ \gamma_1^* & \gamma & \gamma_2 \end{pmatrix}} \tag{60}$$

where the F symbol collects all the R_3 coupling and subduction coefficients together. It is clear from either Eqs. 59 or 60 that the existence of the intermediate-field matrix element is dependent directly on the existence of the 3Γ symbol; the former can exist only if the latter does not vanish. Thus the selection rules for the existence of the 3Γ symbol, discussed fully below, can be used to determine which intermediate-field matrix ele ments do not vanish. The problem of summation over m_{j_1}, q, and m_{j_2}, contained within the F symbol, still persists. However, note that rearrangement of Eq. 59 yields

$$\frac{F^{\Gamma_1\Gamma\Gamma_2}_{j_1kj_2}}{(-1)^{\Gamma_1+\gamma_1^*}\begin{pmatrix}\Gamma_1 & \Gamma & \Gamma_2\\ \gamma_1^* & \gamma & \gamma_2\end{pmatrix}} = \frac{<\alpha_1 j_1\Gamma_1 \parallel T^{k\Gamma} \parallel \alpha_2 j_2\Gamma_2>}{<\alpha_1 j_1 \parallel T^k \parallel \alpha_2 j_2>} \tag{61}$$

Since a reduced matrix element is an observable of the system, it must be independent of the axis of quantization and components. Therefore, the ratio of coupling functions of the two factorization schemes must also be independent of these variables, permitting the definition of a *partition symbol* which is written as (25, 28)

$$\begin{pmatrix}j_1 & k & j_2\\ \Gamma_1 & \Gamma & \Gamma_2\end{pmatrix} = \frac{F^{\Gamma_1\Gamma\Gamma_2}_{j_1kj_2}}{(-1)^{\Gamma_1+\gamma_1^*}\begin{pmatrix}\Gamma_1 & \Gamma & \Gamma_2\\ \gamma_1^* & \gamma & \gamma_2\end{pmatrix}} \tag{62}$$

These symbols are identical to the "V- coefficients" of Tang Au-Chin et al. (73). In spite of derivations of these latter coefficients, solely in terms of a tetragonal quantization axis, they have general applicability for all projected basis sets of an octahedron. The partition symbols act as normalization constants and are denoted N in a recent treatment of trigonally quantized bases (51).

The final factorization of the intermediate element can be derived by substitution for the reduced matrix element of the finite group (Eq. 58) in terms of the partition symbol (Eq. 62) and the reduced matrix element of the in finite group. Thus

$$\begin{aligned}&<\alpha_1 j_1\Gamma_1\gamma_1 a_1 \mid T^{k\Gamma a}_{\gamma} \mid \alpha_2 j_2\Gamma_2\gamma_2 a_2>\\ &= (-1)^{\Gamma_1+\gamma_1^*}\begin{pmatrix}\Gamma_1 & \Gamma & \Gamma_2\\ \gamma_1^* & \gamma & \gamma_2\end{pmatrix}\begin{pmatrix}j_1 & k & j_2\\ \Gamma_1 & \Gamma & \Gamma_2\end{pmatrix} <\alpha_1 j_1 \parallel T^k \parallel \alpha_2 j_2>\end{aligned} \tag{63}$$

In this formulation, the intermediate matrix elements can be more readily handled by manual calculations. The evaluation of the nonvanishing elements requires knowledge of the 3Γ and partition-symbol values. The 3Γ symbols can be calculated on a one-time basis for each quantization axis of O_h and tabulated for this purpose, and the partition symbols can be tabulated and used in all projections. In subgroups on each chain, symmetry-ascent 3Γ symbols can easily be derived since the basis vectors remain unchanged by the subduction technique (57).

The two expansions noted in Eqs. 56 and 63 thus have complimentary properties. In a computer environment, the former is very useful because 3j symbols can be calculated rapidly using closed formulae, and the small number of universally applicable matrices needed for the NSH Hamiltonians are easily stored. On the other hand, the expansion noted in Eq. 63 is very suitable for manual calculations because it minimizes the number of necessary mathematical steps. In the next two sections, the detailed properties of each technique are derived and exemplified.

XII. THE INFINITE-GROUP EXPANSION OF INTERMEDIATE-FIELD MATRIX ELEMENTS

In the factorization shown in Eq. 56, the use of the Wigner—Eckart theorem directly on the representations of R_3 produces a sum of products of $3j$ symbols and reduced matrix elements of the full rotation group. All basis vectors of the expansion are therefore labeled $|l^nLSM_LM_S>$ in any many-electron configuration l^n. In this form, the system is ideal for programming and easily extends to include the matrix elements of other operators such as spin-orbit coupling $|\lambda L.S|$ and the Zeeman effect $|\beta H(L+2S)|$. If necessary, fully general calculations can be carried out on the totally combined bases $|l^nLSJM_J>$. These latter bases are obtained as linear combinations of the $|l^nLSM_LM_S>$ bases, using coefficients defined by the appropriate $3j$ symbols. Examples of such calculations are available (74, 75).

The programming of the expansion (Eq. 56) of intermediate matrix elements is greatly simplified if full use is made of the symmetric properties of the $3j$ symbols. These were discussed briefly in Part 1, but more particular attention to their behavior under conjugation and permutation (or commutation) is necessary.

1. *Conjugation.* The complex conjugate of an eigenvector described as a spherical harmonic component is

$$<m| = (m^*) = (-1)^m(-m) = (-1)^m|-m> \quad (64)$$

Therefore, formation of the complex conjugate of a $3j$ symbol is equivalent to negation of all m_i:

$$\begin{pmatrix} j_1 & j_2 & j_3 \\ m_1 & m_2 & m_3 \end{pmatrix}^* = (-1)^{j_1+j_2+j_3}\begin{pmatrix} j_1 & j_2 & j_3 \\ -m_1 & -m_2 & -m_3 \end{pmatrix} \quad (65)$$

2. *Permutation.* The $3j$ symbol is defined to be invariant under permutation (commutation) within a phase factor:

$$\begin{pmatrix} j_1 & j_2 & j_3 \\ m_1 & m_2 & m_3 \end{pmatrix} = (-1)^{j_1+j_2+j_3} \begin{pmatrix} j_2 & j_1 & j_3 \\ m_2 & m_1 & m_3 \end{pmatrix}$$
$$= (-1)^{j_1+j_2+j_3} \begin{pmatrix} j_1 & j_3 & j_2 \\ m_1 & m_3 & m_2 \end{pmatrix} \text{ etc.} \tag{66}$$

These two symmetry properties dictate two selection rules in addition to the sum rules.

1. If $m_1 = m_2 = m_3 = 0$ and $j_1 + j_2 + j_3$ is odd,

$$\begin{pmatrix} j_1 & j_2 & j_3 \\ m_1 & m_2 & m_3 \end{pmatrix} = (-1) \begin{pmatrix} j_1 & j_2 & j_3 \\ -m_1 & -m_2 & -m_3 \end{pmatrix} = 0 \tag{67}$$

2. If $m_1 = m_2$, $j_1 + j_2 + j_3$ is odd, and $j_1 = j_2$

$$\begin{pmatrix} j_1 & j_2 & j_3 \\ m_1 & m_2 & m_3 \end{pmatrix} = (-1) \begin{pmatrix} j_2 & j_1 & j_3 \\ m_2 & m_1 & m_3 \end{pmatrix} = 0 \tag{68}$$

In both cases, the only number that is equal to its negative must be zero.

The $3j$ symbols were designed to express not only the coupling of different vectors, for example, l and s for a single electron, but also to express fully the antisymmetrized conditions for two-electron systems. Because of this their use eliminates the need for the Slater-determinant form of electron basis sets.

NSH and crystal-field Hamiltonians are one-electron operators, but may act upon the individual parts of a many-electron basis set. The $|l^nLSM_L M_S>$ bases must therefore be uncoupled into the sum of products of single-electron l sets. This is a simple matter for l^2 configurations, but for l^3 and higher configurations, the uncoupling requires the use of *fractional-parentage coefficients* (15, 16). These are difficult to calculate and extensive tabulations are necessary. In both cases, their adaptation to programmed formats is very inconvenient. The alternative is the use of tables of reduced matrix elements of *unitary operators*. These are much more economical of computer space and are readily available.

A. Unitary Operators

The techniques of using these matrix elements can be described briefly without any development of the supporting theory of coefficients of fractional parentage. If $|j_i m_i>$ are single-electron l sets, a unit tensor operator u can be defined using the Wigner—Eckart theorem:

$$<j_1 m_1 | u_q^k | j_2 m_2> = (-1)^{j_1 - m_1} \begin{pmatrix} j_1 & k & j_2 \\ -m_1 & q & m_2 \end{pmatrix} <j_1 \| u^k \| j_2> \tag{69}$$

which is identical to the general expansion for an arbitrary operator X_q^k (such as C_q^k).

The reduced unitary matrix element

$$<j_1\|u^k\|j_2> \;=1 \qquad \text{if } j_1+j_2 \geq k \geq |j_1-j_2|$$

and vanishes otherwise. Furthermore,

$$U_q^k = u_{1q}^k + u_{2q}^k + \cdots u_{nq}^k = \sum_{i=1}^{n} u_{iq}^k \tag{70}$$

in which n is the number of electrons of the configuration l^n, and u_i is the operator acting on the ith electron alone.

The application of the Wigner—Eckart theorem to a unitary matrix element for states of a many-electron configuration yields

$$\begin{aligned}&\langle l^n LSM_LM_S|U_q^k|l^nL'S'M_L'M_S'\rangle \\ &= \delta(SS')\cdot\delta(M_SM_S')(-1)^{L-M_L}\begin{pmatrix} L & k & L' \\ -M_L & q & M_L'\end{pmatrix}\langle l^nSL\|U^k\|l^nS'L'\rangle\end{aligned} \tag{71}$$

The spin function, unaffected by the operator, is redundant and dropped from further development. The reduced matrix element can be expanded:

$$\begin{aligned}&(-1)^L\begin{pmatrix} L & k & L' \\ 0 & 0 & 0\end{pmatrix}<l^nL\|U^k\|l^nL'> \;=\; <l^nL0|U_0^k|l^nL'0> \\ &= <\sum_{z=1}^{g} a(z) <l_1m_1|<l_2m_2|\cdots<l_nm_n\|\sum_{i=1}^{n} U_i^k0| \\ &\sum_{z'=1}^{g'} a'(z')|l_1'm_1'>|l_2'm_2'>\cdots|l_n'm_n'\gg\end{aligned} \tag{72}$$

where the values of m, m', $a(z)$, $a(z')$, g, and g' are dictated by the coefficients of fractional parentage. Rearrangement and application of the Wigner—Eckart theorem again yields

$$\begin{aligned}&(-1)^L\begin{pmatrix} L & k & L' \\ 0 & 0 & 0\end{pmatrix} <l^nL\|U^k\|l^nL'> \\ &= \sum_{i=1}^{n}\sum_{z=1}^{g}\sum_{z'=1}^{g'} a(z)\cdot a'(z')\cdot(-1)^{l_z-m_z}\,\delta(m_1,m_1')\delta(m_2m_2') \\ &\quad\cdots\delta(m_{z\pm'1},m_{z'\pm1})\cdots\delta(m_zm_z')\begin{pmatrix} l_z & k_i & l_z' \\ -m_z & 0 & m_{z'}\end{pmatrix}<l_z\|u^k\|l_{z'}^1> \\ &= B\end{aligned} \tag{73}$$

in which B is a constant. The value of the reduced matrix elements $<l\|u^k\|l'>$ equals 1 or 0. Thus for nonvanishing cases,

$$<l^nL\|U^k\|l^nL'> \;=\; \frac{B}{(-1)^L\begin{pmatrix} L & k & L' \\ 0 & 0 & 0\end{pmatrix}} \tag{74}$$

Since they rely directly on B, the many-electron, unitary-reduced matrix elements contain both the coefficients of fractional parentage and the results of application of the Wigner—Eckart theorem. These unitary elements are readily related to the elements of a spherical harmonic operator. Thus

$$<l^nL|C_q^k|l^nL> = <l\|C^k\|l> \cdot <l^nL|U_q^k|l^nL> \tag{75}$$

The reduced matrix elements of these harmonic operators are (18)

$$<l\|C^k\|l'> = (-1)^l[(2l+1)(2l'+1)]^{1/2}\begin{pmatrix} l & k & l' \\ 0 & 0 & 0 \end{pmatrix} \tag{76}$$

which upon substitution into Eq. 71 yields (assuming $l = l'$)

$$\begin{aligned} &<l^nLM_L|C_q^k|l^nL'M_L'> \\ &= (-1)^{L-M_L+l}(2l+1)\begin{pmatrix} l & k & l \\ 0 & 0 & 0 \end{pmatrix}\begin{pmatrix} L & k & L' \\ -M_L & q & M_L' \end{pmatrix}<l^nL\|U^k\|l^nL'> \end{aligned} \tag{77}$$

Thus matrix elements whose eigenvectors and operator eigenstates are specified as spherical harmonics can be calculated as products of $3j$ symbols and reduced unitary matrix elements. The former can be calculated directly and the latter is easily stored in a computer program of convenient size.

This completes the formalism which is needed to perform calculations of eigenvalues. As an example of its use, the energy levels originating from the ^3F and ^3P states of d^2 configuration in a D_{3d} central field will be derived.

B. Sample Calculations Assuming Either a Weak or Intermediate Field

The complete Hamiltonian has the following general form:

$$H = H_o + H_{ir} + H_{lf}$$

where H_o is the single-electron nucleus attraction;
H_{ir} is the interelectron repulsion potential;
H_{lf} is the ligand field perturbation.

The energies of the two-electron states of the free ion are found by expansion of the interelectron repulsion operator e^2/r_{ij}. They can be calculated by vector coupling techniques and are readily available in many tabulated sources. They are omitted from the following discussion.

1. The Computer Optimized Approach

The basis sets used for the two states are in terms of a notation $|LM_L>$:

^3P $|11>, |10>, |1-1>$

^3F $|33>, |32>, |31>, |30>, |3-1>, |3-2>, |3-3>$.

The electrostatic matrix elements of these two states are (7)

$$E(^3P) = F_0 + 7F_2 - 84F_4 = A + 7B$$
$$E(^3F) = F_0 - 8F_2 - 9F_4 = A - 8B$$

If only relative energies are needed, then setting the energy of the ground state ^3F as an arbitrary zero, $E(^3P) - E(^3F) = 15B$.

Superimposed on these states, the ligand-field Hamiltonian cause further splitting of both the ^3P and ^3F states. In expanded form the Hamiltonian for trigonal symmetry is (see Eq. 15)

$$H_{D_{3d}} = \frac{DQ}{\sqrt{27}}(\sqrt{7}\, C_0^4 + \sqrt{10}\,(C_3^4 - C_{-3}^4)) + DM\, C_0^2 + \frac{DN}{\sqrt{54}}(\sqrt{40}\, C_0^4 - \sqrt{7}(C_3^4 - C_{-3}^4)) \qquad (78)$$

Inspection of this operator will show that individual parts, such as C_0^4, may appear repeatedly in different linear combinations. Thus it is advantageous to operate with these individual operator elements on the basis set. Later, taking advantage of the linear properties of the spherical harmonic operators, the appropriate matrices can be found by addition of the individual matrices arising from the individual spherical harmonic operators.

Equation 77 clearly shows that the quantities $\langle l^nL \| U^k \| l^nL' \rangle$, $\begin{pmatrix} l & k & l \\ 0 & 0 & 0 \end{pmatrix}$, and $(2l + 1)$ will often be used. It is therefore useful to define them immediately. Using the tabulated unitary-reduced matrix elements for the d^2 configuration (40),

$$\begin{aligned}
\langle d^2\, {}^3P \| U^2 \| d^2\, {}^3P \rangle &= -\frac{\sqrt{21}}{\sqrt{25}} \\
\langle d^2\, {}^3F \| U^2 \| d^2\, {}^3F \rangle &= \frac{\sqrt{6}}{\sqrt{25}} \\
\langle d^2\, {}^3P \| U^2 \| d^2\, {}^3F \rangle &= \frac{\sqrt{24}}{\sqrt{25}} \\
\langle d^2\, {}^3P \| U^4 \| d^2\, {}^3F \rangle &= \frac{\sqrt{6}}{\sqrt{5}} \\
\langle d^2\, {}^3F \| U^4 \| d^2\, {}^3F \rangle &= -\frac{\sqrt{11}}{\sqrt{5}}
\end{aligned} \qquad (79)$$

Similarly, tabulations of $3j$ symbols yield the desired values (18, 55, 56)

$$\begin{pmatrix} 2 & 2 & 2 \\ 0 & 0 & 0 \end{pmatrix} = -\frac{\sqrt{2}}{\sqrt{35}}$$

$$\begin{pmatrix} 2 & 4 & 2 \\ 0 & 0 & 0 \end{pmatrix} = \frac{\sqrt{2}}{\sqrt{35}} \tag{80}$$

Following Eq. 69, matrix elements of the C_q^k operator are given by

$$<LM_L|C_q^k|L'M_L'> = (-1)^{L-M_L}\begin{pmatrix} L & k & L' \\ -M_L & q & M_L' \end{pmatrix} <L\|C^k\|L'> \tag{81}$$

Substitution of the above into Eq. 77 provides the reduced matrix elements

$$<l^nL\|C^k\|l^nL'> = (-1)^l(2l+1)\begin{pmatrix} l & k & l \\ 0 & 0 & 0 \end{pmatrix} <L\|U^k\|L'>$$

$$<d^2\,{}^3P\|C^2\|d^2\,{}^3P> = (-1)^2((2\times 2)+1)\begin{pmatrix} 2 & 2 & 2 \\ 0 & 0 & 0 \end{pmatrix}\frac{-\sqrt{21}}{\sqrt{25}}$$

$$= 5 \times \frac{-\sqrt{2}}{\sqrt{35}} \times \frac{-\sqrt{21}}{\sqrt{25}} = \frac{\sqrt{6}}{\sqrt{5}} \tag{82}$$

$$<d^2\,{}^3P\|C^2\|d^2\,{}^3F> = <1\|C^2\|3> = \frac{-\sqrt{48}}{\sqrt{35}}$$

$$<d^2\,{}^3F\|C^2\|d^2\,{}^3F> = <3\|C^2|3> = \frac{-\sqrt{12}}{\sqrt{35}}$$

$$<d^2\,{}^3P\|C^4\|d^2\,{}^3F> = <1\|C^4\|3> = \frac{\sqrt{12}}{\sqrt{7}}$$

$$<d^2\,{}^3F\|C^4\|d^2\,{}^3F> = <3\|C^4\|3> = \frac{-\sqrt{22}}{\sqrt{7}}$$

Using these data we can solve Eq. 81. Thus

$$<11|C_0^2|11> = (-1)^{1-1}\begin{pmatrix} 1 & 2 & 1 \\ -1 & 0 & 1 \end{pmatrix}<1\|C^2\|1> = \frac{1}{\sqrt{30}} \times \frac{\sqrt{6}}{\sqrt{5}} = \frac{1}{5} \tag{83}$$

Applying the sum rule (Eq. 27) for existence of $3j$ symbols, and noting Eq. 81 a C_0^k-type operator will only connect elements with the same M_L value. Since $(1 + 2 + 1) = 4$ is an even number, the $<1-1\|C_0^2\|1-1>$ element will be identical to Eq. 83. Taking another example,

$$<31|C_0^2|11> = (-1)^{3-1}\begin{pmatrix} 3 & 2 & 1 \\ -1 & 0 & 1 \end{pmatrix}<3\|C^2\|1> = \frac{\sqrt{2}}{\sqrt{35}} \times \frac{-\sqrt{48}}{\sqrt{35}}$$

$$= \frac{-4\sqrt{6}}{35} \tag{84}$$

	$\|11>$	$\|10>$	$\|1-1>$	$\|33>$	$\|32>$	$\|31>$	$\|30>$	$\|3-1>$	$\|3-2>$	$\|3-3>$
$<11\|$	$\frac{1}{5}$					$\frac{-4\sqrt{6}}{35}$				
$<10\|$		$\frac{-2}{5}$					$\frac{-12}{35}$			
$<1-1\|$			$\frac{1}{5}$					$\frac{-4\sqrt{6}}{35}$		
$<33\|$				$\frac{-1}{7}$						
$<32\|$					0					
$<31\|$	$\frac{-4\sqrt{6}}{35}$					$\frac{3}{35}$				
$<30\|$		$\frac{-12}{35}$					$\frac{4}{35}$			
$<3-1\|$			$\frac{-4\sqrt{6}}{35}$					$\frac{3}{35}$		
$<3-2\|$									0	
$<3-3\|$										$\frac{-1}{7}$

$$= [C_0^2] \quad (85)$$

A similar matrix derived for C_0^4 is the following:

	$1\rangle$	$\|10\rangle$	$\|1-1\rangle$	$\|33\rangle$	$\|32\rangle$	$\|31\rangle$	$\|30\rangle$	$\|3-1\rangle$	$\|3-2\rangle$	$\|3-3\rangle$
$\langle 11\|$	0					$\frac{-\sqrt{2}}{7\sqrt{3}}$				
$\langle 10\|$		0					$\frac{4}{21}$			
$\langle 1-1\|$			0					$\frac{-\sqrt{2}}{7\sqrt{3}}$		
$\langle 33\|$				$\frac{-1}{7}$						
$\langle 32\|$					$\frac{1}{3}$					
$\langle 31\|$	$\frac{-\sqrt{2}}{7\sqrt{3}}$					$\frac{-1}{21}$				
$\langle 30\|$		$\frac{4}{21}$					$\frac{-2}{7}$			
$\langle 3-1\|$			$\frac{-\sqrt{2}}{7\sqrt{3}}$					$\frac{-1}{21}$		
$\langle 3-2\|$									$\frac{1}{3}$	
$\langle 3-3\|$										$\frac{-1}{7}$

$$= [C_0^4] \quad (86)$$

Again, the $<3\,-1|C_0^2|1\,-1>$ element will be identical to Eq. 84 as will the $<1\,-1|C_0^2|3\,-1>$ and $<11|C_0^2|31>$ elements. The matrix resulting from the C_0^2 operator acting on the ^3P, ^3F basis set is written $[C_0^2]$ (see Eq. 85). A similar matrix for the C_0^4 operator is shown in Eq. 86. These two matrices exhibit several features which are common to all matrices needed. They are very sparse, that is, very few elements do not vanish and these are easily found using the summation rules for $3j$ symbols.

The traces, that is, the sum of the diagonal elements, are zero (17). This is the case for all operators composed of spherical harmonics other than C_0^0.

Finally, these matrices are always real and Hermitian, that is, they are matrices composed of real numbers and are symmetric across the diagonal. Therefore,

$$M[i,j] = M[j,i] \tag{87}$$

where $M[i,j]$ equals the jth element in the ith row of matrix M.

The matrices for both C_3^4 and C_{-3}^4 which connect off-diagonal l sets can be constructed in the same way. Because M_L is odd there is a change in sign between elements of C_3^4 and C_{-3}^4:

$$\begin{aligned}&<L(M_L+3)|C_3^4|L'M_L> \\ &\quad = (-1)^{L-(M_L+3)}\begin{pmatrix} L & 4 & L' \\ -(M_L+3) & 3 & M_L\end{pmatrix} <L\|C^4\|<L'>\end{aligned}$$

but

$$\begin{aligned}&<LM_L|C_{-3}^4|L'(M_L+3)> \\ &\quad = (-1)^{L-M_L}\begin{pmatrix} L & 4 & L' \\ -M_L & -3 & (M_L+3)\end{pmatrix} <L\|C^4\|L'>\end{aligned}$$

Therefore,

$$<L(M_L+3)|C_3^4|L'M_L> \;=\; -\;<LM_L|C_{-3}^4|L'(M_L+3)> \tag{88}$$

For this reason the off-diagonal Hermitian operator is the linear combination $(C_3^4 - C_{-3}^4)$. A typical matrix element using this combined operator is

$$\begin{aligned}<33|C_3^4 - C_{-3}^4|30> &= (-1)^{3-3}\begin{pmatrix} 3 & 4 & 3 \\ -3 & 3 & 0\end{pmatrix} <3\|C^4\|3> \\ &= (1)\cdot\frac{1}{\sqrt{22}}\cdot\frac{-\sqrt{22}}{\sqrt{7}} \\ &= \frac{-1}{\sqrt{7}}\end{aligned} \tag{89}$$

In Eq. 89 the C^4_{-3} matrix elements vanish. By similar expansions,

$$\begin{aligned} \langle 33|C^4_3 - C^4_{-3}|30\rangle &= \langle 30|C^4_3 - C^4_{-3}|33\rangle \\ &= \langle 30|C^4_3 - C^4_{-3}|3-3\rangle \\ &= \langle 3-3|C^4_3 - C^4_{-3}|30\rangle \\ &= -\frac{1}{\sqrt{7}} \end{aligned} \tag{90}$$

The resulting matrix of off-diagonal elements will be denoted $[C^4_3 - C^4_{-3}]$. Any other matrix is obtained by analogous manipulations.

The matrix of the complete A_{1g} Hamiltonian is constructed by addition of the individual general matrices, $[C^2_0]$, $[C^4_0]$, and $[C^4_3 - C^4_{-3}]$, in the proportions specified by the subduction coefficients of the A_{1g} linear combinations for quantization under C_3^{xyz}, as shown in Eq. 15, namely,

$$\begin{aligned} [A_1]_{D_{3d}} &= [A_{1g}]_{O_h} + [T_{2g}O]_{O_h} \\ &= DQ\left[\frac{\sqrt{7}}{\sqrt{27}}[C^4_0] + \frac{\sqrt{10}}{\sqrt{27}}[C^4_3 - C^4_{-3}]\right] + DM[C^2_0] \\ &\qquad + DN\left[\frac{\sqrt{20}}{\sqrt{27}}[C^4_0] - \frac{\sqrt{7}}{\sqrt{54}}[C^4_3 - C^4_{-3}]\right] \end{aligned} \tag{91}$$

Matrices for operators appropriate to the C_4^z, C_2^z, and C_2^{xy} quantized subgroups can be built up the same way. In expanded form the full $[A_{1g}]_{O_h}$ matrix has the form shown in Eq. 92:

	$\lvert 11\rangle$	$\lvert 10\rangle$	$\lvert 1-1\rangle$	$\lvert 33\rangle$	$\lvert 32\rangle$	$\lvert 31\rangle$	$\lvert 30\rangle$	$\lvert 3-1\rangle$	$\lvert 3-2\rangle$	$\lvert 3-3\rangle$
$\langle 11\rvert$	0					$-\frac{\sqrt{2}}{9\sqrt{7}}$			$\frac{\sqrt{10}}{9\sqrt{7}}$	
$\langle 10\rvert$		0		$-\frac{\sqrt{10}}{9\sqrt{21}}$			$\frac{4}{9\sqrt{21}}$			$-\frac{\sqrt{10}}{9\sqrt{21}}$
$\langle 1-1\rvert$			0		$\frac{\sqrt{10}}{9\sqrt{7}}$			$-\frac{\sqrt{2}}{9\sqrt{7}}$		
$\langle 33\rvert$		$-\frac{\sqrt{10}}{9\sqrt{21}}$		$-\frac{1}{3\sqrt{21}}$			$-\frac{\sqrt{10}}{3\sqrt{21}}$			
$\langle 32\rvert$			$\frac{\sqrt{10}}{9\sqrt{7}}$		$\frac{\sqrt{7}}{9\sqrt{3}}$			$-\frac{2\sqrt{5}}{9\sqrt{21}}$		
$\langle 31\rvert$	$-\frac{\sqrt{2}}{9\sqrt{7}}$					$-\frac{1}{9\sqrt{21}}$			$-\frac{2\sqrt{5}}{9\sqrt{21}}$	
$\langle 30\rvert$		$\frac{4}{9\sqrt{21}}$		$-\frac{\sqrt{10}}{3\sqrt{21}}$			$-\frac{2}{3\sqrt{21}}$			$-\frac{\sqrt{10}}{3\sqrt{21}}$
$\langle 3-1\rvert$			$-\frac{\sqrt{2}}{9\sqrt{7}}$		$-\frac{2\sqrt{5}}{9\sqrt{21}}$			$-\frac{1}{9\sqrt{21}}$		
$\langle 3-1\rvert$	$\frac{\sqrt{10}}{9\sqrt{7}}$					$-\frac{2\sqrt{5}}{9\sqrt{21}}$			$\frac{\sqrt{7}}{9\sqrt{3}}$	
$\langle 3-3\rvert$		$-\frac{\sqrt{10}}{9\sqrt{21}}$					$-\frac{\sqrt{10}}{3\sqrt{21}}$			$-\frac{1}{3\sqrt{21}}$

$= [A_1]^4$(in units of DQ) (92)

If this matrix were multiplied by a value of DQ, added to the appropriate interelectronic repulsion matrix, and diagonalized, the correct set of DQ energies would be obtained. However, the assignment of representation labels to these energies would be very difficult in more complex cases, for example, low symmetries. Therefore, it is convenient to introduce an intermediate step which is to project the basis set on to the trigonally quantized octahedron.

This projection is accomplished by way of a basis transformation. Such a transformation can be expressed as (13)

$$[A'] = [u^{-1}] \times [A] \times [u] \tag{93}$$

where × means matrix multiplication;

[A] is the original matrix,

[u] is the transformation matrix.

The transformation matrix [u] can be easily constructed. The columns of a matrix are labeled with the appropriate O_h component labels, while the rows are labeled with the original unchanged basis-set labels. Then, at the conjunction of the rows and columns, the appropriate subduction coefficients from Table VII are inserted.

Thus, for example, the [u] matrix appropriate for threefold quantization is shown in Eq. 94:

$[u] =$

$(LM)(\Gamma\gamma)$	$(T_1 1)$	$(T_1 0)$	$(T_1 -1)$	$(A_2 0)$	$(T_1 1)$	$(T_1 0)$	$(T_1 -1)$	$(T_2 1)$	$(T_2 0)$	$(T_2 -1)$
(11)	1									
(10)		1								
(1–1)			1							
(33)				$-\frac{\sqrt{2}}{3}$		$\frac{\sqrt{5}}{3\sqrt{2}}$			$\frac{1}{\sqrt{2}}$	
(32)							$-\frac{\sqrt{5}}{\sqrt{6}}$			$-\frac{1}{\sqrt{6}}$
(31)					$\frac{1}{\sqrt{6}}$			$\frac{\sqrt{5}}{\sqrt{6}}$		
(30)				$\frac{\sqrt{5}}{3}$		$\frac{2}{3}$				
(3–1)							$-\frac{1}{\sqrt{6}}$			$\frac{\sqrt{5}}{\sqrt{6}}$
(3–2)					$-\frac{\sqrt{5}}{\sqrt{6}}$			$\frac{1}{\sqrt{6}}$		
(3–3)				$\frac{\sqrt{2}}{3}$		$-\frac{\sqrt{5}}{3\sqrt{2}}$			$\frac{1}{\sqrt{2}}$	

(94)

Note that the matrix shown in Eq. 94 presents the wave function for a given octahedral representation in terms of the LM_L functions; for example, by reading vertically downward, one derives

$$|A_2O> = \left(\frac{\sqrt{5}}{3}\right)|30> + \left(\frac{\sqrt{2}}{3}\right)(|3-3> - |33>) \qquad (95)$$

The $[u]$ is a real unitary matrix, and its transpose is its inverse. Thus $[u^{-1}]$ will have the row labeled with octahedral representations and the columns labeled by the unchanged basis set.

Operating on the $[A_1]^4$ matrix with the $[u]$ and $[u^{-1}]$ matrices results in a transformed matrix $[A_1O\psi_{O_h}]$ whose rows and columns are labeled by O_h components. That is, $[A_1O\psi_{O_h}]$ is the matrix that is obtained by operating with $[A_1]O_h$ directly on the ψ_{O_h} basis set of octahedrally projected basis functions. The form of the matrix $[A_1O\psi_{O_h}]$ is shown in Eq. 96:

$$[A_1 0 \psi_{o_s}] =$$

	$\lvert T_1 1 \rangle$†	$\lvert T_1 0 \rangle$†	$\lvert T_1 -1 \rangle$†	$\lvert A_2 0 \rangle$	$\lvert T_1 1 \rangle$	$\lvert T_1 0 \rangle$	$\lvert T_1 -1 \rangle$	$\lvert T_2 1 \rangle$	$\lvert T_2 0 \rangle$	$\lvert T_2 -1 \rangle$
$\langle T_1 1 \rvert$†	0				$\frac{2}{3\sqrt{21}}$					
$\langle T_1 0 \rvert$†		0				$\frac{2}{3\sqrt{21}}$				
$\langle T_1 -1 \rvert$†			0				$\frac{2}{3\sqrt{21}}$			
$\langle A_2 0 \rvert$				$-\frac{2}{\sqrt{21}}$						
$\langle T_1 1 \rvert$	$\frac{2}{3\sqrt{21}}$				$\frac{1}{\sqrt{21}}$					
$\langle T_1 0 \rvert$		$\frac{2}{3\sqrt{21}}$				$\frac{1}{\sqrt{21}}$				
$\langle T_1 -1 \rvert$			$\frac{2}{3\sqrt{21}}$				$\frac{1}{\sqrt{21}}$			
$\langle T_2 1 \rvert$								$-\frac{1}{3\sqrt{21}}$		
$\langle T_2 0 \rvert$									$-\frac{1}{3\sqrt{21}}$	
$\langle T_2 -1 \rvert$										$-\frac{1}{3\sqrt{21}}$

(96)

†are the *P* term representations.

If diagonalized and multiplied by DQ, the matrix illustrated in Eq. 96 provides the energies of the components of the ^3F and ^3P terms under the octahedral crystal-field operator, labeled by their group theoretical representations. As can be seen, the first-order energy of $T_{1\gamma}$ arising from ^3P is zero, since this term is unsplit by a crystal field. Similarly, the energies of the A_2O and $T_{2\gamma}$ levels can be directly read from the matrix since these terms are diagonal. The off-diagonal elements connecting the $T_{1\gamma}$ states of ^3P and ^3F permit configurational interaction between these $T_{1\gamma}$ representations to second order.

We may now use this matrix to begin calculation of the energies of the ^3P and ^3F terms under the D_{3d} operator. If we assign D_{3d} labels to matrix Eq. 96 (see Table IV for subduction from O_h into D_{3d}), we can block this 10×10 matrix into a 1×1 matrix, A_1 in D_{3d}, a 3×3 matrix, A_2 in D_{3d}, and two independent, but identical, 3×3 matrices for the two components of each of the three E levels arising from these terms. These matrices are

$[A_1O\phi_{o_h}]\, A_1$

	$\lvert T_2O$
$<T_2O\rvert$	$\frac{-1}{3\sqrt{21}}$

$[A_1O\phi_{o_h}]\, A_2$

	$\lvert T_1O>\dagger$	$\lvert T_1O>$	$\lvert A_2O>$
$<T_1O\rvert\dagger$	0	$\frac{2}{3\sqrt{21}}$	0
$<T_1O\rvert$	$\frac{2}{3\sqrt{21}}$	$\frac{1}{\sqrt{21}}$	0
$<A_2O\rvert$	0	0	$\frac{-2}{\sqrt{21}}$

(97)

$[A_1O\phi_{o_h}]\, E\pm 1$

	$\lvert T_1\pm 1>\dagger$	$\lvert T_1\pm 1>$	$\lvert T_2\pm 1>$
$<T_1\pm 1\rvert\dagger$	0	$\frac{2}{3\sqrt{21}}$	0
$<T_1\pm 1\rvert$	$\frac{2}{3\sqrt{21}}$	$\frac{1}{\sqrt{21}}$	0
$<T_2\pm 1\rvert$	0	0	$\frac{-1}{3\sqrt{21}}$

For D_{3d} symmetry we also require the T_2O tensor component to construct the operator. Thus a matrix, analogous to matrix Eq. 94, is generated by combining the various matrices (such as Eqs. 86 and 87) for the C_k^q operators according to the subduction coefficients in Table V or Eq. 91. Since there are second- and fourth-order contributions to T_2O, this procedure generates two more 10 × 10 matrices, labeled $[T_2O\psi_{o_h}]^2$ and $[T_2O\psi_{o_h}]^4$. Each of these can be blocked into a 1 × 1, a 3 × 3, and two further identical 3 × 3 matrices, labelled by the A_1, A_2, and $E \pm 1$ D_{3d} representations, respectively, in a fashion identical to that previously illustrated for the $[A_1O\psi_{o_h}]^4$ matrix. For example, the $[T_2O\psi_{o_h}]^4$ blocks are

$[T_2O\psi_{o_h}]^4A_1$

	$\vert T_2O>$
$<T_2O\vert$	$\frac{-2\sqrt{5}}{21\sqrt{3}}$

$[T_2O\psi_{o_h}]^4A_2$

	$\vert A_2O>$	$\vert T_1O>$	$\vert T_1O>\dagger$
$<A_2O\vert$	0	$\frac{-1}{7\sqrt{3}}$	$\frac{2}{7\sqrt{3}}$
$<T_1O\vert$	$\frac{-1}{7\sqrt{3}}$	$\frac{-2\sqrt{5}}{7\sqrt{3}}$	$\frac{5}{21\sqrt{3}}$
$<T_1O\vert\dagger$	$\frac{2}{7\sqrt{3}}$	$\frac{5}{21\sqrt{3}}$	0

(98)

$[T_2O\psi_{o_h}]^4E \pm 1$

	$\vert T_2 \pm 1>$	$\vert T_1 \pm 1>$	$\vert T_1 \pm 1>\dagger$
$<T_2 \pm 1\vert$	$\frac{\sqrt{5}}{21\sqrt{3}}$	$\frac{2}{7\sqrt{3}}$	$\frac{-\sqrt{3}}{14}$
$<T_1 \pm 1\vert$	$\frac{2}{7\sqrt{3}}$	$\frac{\sqrt{5}}{7\sqrt{3}}$	$\frac{-\sqrt{5}}{42\sqrt{3}}$
$<T_1 \pm 1\vert\dagger$	$\frac{-\sqrt{3}}{14}$	$\frac{-\sqrt{5}}{42\sqrt{3}}$	0

The next step is to multiply the three matrices $[A_1O\psi_{o_h}]^4$, $[T_2O\psi_{o_h}]^2$, and $[T_2O\psi_{o_h}]^4$ by trial values of DQ, DM, and DN, respectively. Blocked submatrices with the same D_{3d} label are then added. The result of this addition appears together with the interelectronic repulsion energies in Table XII. Values of DQ, DM, and DN are finally obtained by diagonalizing the matrices and fitting the roots to an experimental spectrum.

In summary, an approach to the calculation of ligand-field eigenvalues has been developed which is very easily adapted to program manipulation. Relatively small libraries of the fully general $[C_q^k]d^n$ matrices can be built up, stored, and used repeatedly for all symmetries. The necessary $3j$ symbols can be rapidly calculated using closed formulae and require no machine storage. Thus the only tabulated input necessary for the calculation of the $[C_q^k]d^n$ matrices is the set of reduced matrix elements $< d^n\ SL \parallel C^k \parallel d^n\ SL' >$. These general many-electron matrices are readily adapted to any desired symmetry using the appropriate transformation matrix of subduction coefficients $[u]$. Only the four different sets of subduction coefficients defined in Part 1 for quantization on C_4^z, C_2^z, C_2^{xy}, and C_3^{xyz} axes are necessary for most spectroscopic problems.

XIII. THE FINITE-GROUP EXPANSION OF INTERMEDIATE-FIELD MATRIX ELEMENTS

A. Further Aspects of 3Γ Selection Rules

For pedagogical purposes, and to simplify the manual calculations of intermediate-field eigenvalues, it is useful to construct the $[\Gamma_{o_h}\psi_{o_h}]\Gamma_G$ matrices by a primary application of the Wigner—Eckart theorem to the finite-group quantum numbers. This approach follows the general procedure noted in Eqs. 58 to 63. The purpose is the construction of the matrices of the Hamiltonian (e.g., Eq. 15).

As an example, the matrix obtained by operating with $DN\,|T_2O|_{o_h}^4$ on the $d^2\psi_{o_h}$ basis has 100 elements of the form $< d^2\Gamma''\,|DN\ T_2O|\, d^2\Gamma >$, but reference to the matrices (Eq. 98) reveals that only 24 of these are nonzero, the remainder vanishing. If these elements are expanded using Eq. 58, the 76 zero elements vanish because the appropriate 3Γ symbol vanishes. Many of these 3Γ symbols vanish because they do not meet the group theoretical condition (Eq. 28) in the group O_h; others, such as the element $< T_1O\,|\,T_2O\,|$-$T_1O >$, meet this condition in O_h, but fail the triangular condition in R_3 (Eq. 26) (assuming fourth-order operator T_2O and ^{3}P basis functions; symmetry-ascent selection rule). Still others, such as $< T_11\,|\,T_2O\,|\,A_2O >$, fail because of the component selection rules (Eq. 29). Clearly a recognition that so few elements exist provides a means of drastically reducing the computation necessary to derive these energy matrices.

A more complete understanding of the detailed selection rules that apply to the 3Γ symbols along the various axes provides rapid means of determining which wave function components can be coupled by the various operators. We require the definition of numerical values for both representation Γ and component γ labels of operators and wave functions and a simple concise combination algebra which reproduces all the properties of the 3Γ symbols. By providing such numerical labels to both operator and wave functions, manual or computer calculations can be carried out with facility similar to that enjoyed in calculations in the infinite R_3 group using $3j$ symbols.

Wigner (12) proved that all coupling symbols belonging to simply reducible groups (SR groups) must obey relationships such as those already discussed for the $3j$ symbols. However, his proof of existence did not indicate the way these coupling symbols could be derived nor did it establish the nomenclature that should be employed to make these relationships obvious. Recent developments (51, 52) have overcome both of these problems.

The important features of finite-group basis functions which lead to the desired definitions are the following:

1). There is a particularly simple behavior of the spherical harmonics under a rotation about the z reference axis. If the quantization axis lying along z has order n, then only those spherical harmonics for which $(m_j - m_j') = fn$ (f is an integer) can be linearly combined in the basis vectors of the finite group.

2. The linear combinations of spherical harmonics thus obtained behave under conjugation *in exactly the same way* as the spherical harmonic with the lowest m_j value in the combination.

3. Linear combinations which are either symmetric or skew symmetric about $m_j = 0$ form a class by themselves. In these linear combinations there is no unique minimum m_j, but instead two harmonics appear for which $|m_j| = fn/2$. In these cases, the harmonics for (m_j) and $(-m_j)$ appear in the same linear combination and have the same absolute magnitude of the subduction coefficients.

Under complex conjugation these particular linear combinations reproduce themselves, rather than producing some other component, up to a possible sign inversion. Their behavior is analogous to that of C_0^k spherical harmonics, which act as their own complex conjugate functions.

4. Since wave functions and operators are constructed from LM_L functions in R_3, such first-order states can be assigned a value of L, being the value of L in R_3 which projects the function. This numerical value, referred to as $J(\Gamma)$, was discussed in Section V. This is a clear example of the symmetry-ascent selection rule, where an R_3 condition is applied to define the existence of an octahedral 3Γ symbol which can, in fact, be used, through the NSH factorization, to define the existence of matrix elements of even lower symmetry.

Note, however, that when defining a numerical value for Γ, to be used in the phase factor preceding the 3Γ symbol, a different procedure is adopted. For the specific application of defining this phase factor, $J(\Gamma)$ is defined as the lowest nonzero value of L which projects Γ (in O_h) from R_3. Thus, for example, $J(T_{2g}) = 2$ (see Table I), irrespective of the actual derivation of the T_{2g} function concerned (52).

Taken together, properties 1 and 2 show that for representations of single groups (ignoring spin) at least n-basic numerical labels are necessary to distinguish all the components of a representation. These component values are connected by a modular arithmetic to modulus n. Property 2 indicates that the chosen number should be symmetric about $n = 0$, that is, chosen in the range $-n/2$ to $n/2$. However, property 3 of finite components indicates that if $m_j = n/2$ and $-n/2$ occur together without the appearance of other m_j values in the linear combination, the conjugation properties of such components is anomalous. Two different types of linear combination are possible: symmetric and antisymmetric. These two types of finite component are distinguished by component labels $\gamma = (n/2)+$ or $(n/2)-$

While some aspects of the behavior of 3Γ symbols are dealt with below, refer to Ref. 52 for a detailed discussion of the subtleties involved. It is eminently clear, however, that these numerical substitutions permit the replacement of the conventional alphabetical algebra of finite-group direct products with a numerical summation algebra, very similar to that used in combining j and m_j values in R_3.

As examples of the use of these numerical values for finite-group representation Γ and component γ labels, the conjugation (denoted by *) properties of finite-group components can be defined (52) as follows:

$$(\Gamma\gamma)^* = (-1)^{J(\Gamma)-\gamma}(\Gamma - \gamma) \qquad [\text{cf.}\,(JM_J)^* = (-1)^{J-M_J}(J - M_J)]$$

except in the case of $\gamma = fn/2$, where

$$(\Gamma\gamma+)^* = (-1)^{J(\Gamma)+\gamma}(\Gamma\gamma+) \qquad [\text{cf.}(JO)^* = (-1)^{J+O}(JO)] \tag{103}$$

and

$$(\Gamma\gamma-)^* = (-1)^{J(\Gamma)+\gamma+1}(\Gamma\gamma-)$$

Also note that

$$|\pm\gamma| = \gamma \qquad |\gamma+| = \gamma \qquad |\gamma-| = \gamma + 1 \tag{104}$$

These phase definitions can now be used to define fully the behavior of 3Γ symbols of any finite-point group.
Thus the permutation behavior of a 3Γ symbol becomes

$$\begin{pmatrix} \Gamma_1 & \Gamma_2 & \Gamma_3 \\ \gamma_1 & \gamma_2 & \gamma_3 \end{pmatrix} = (-1)^{J(\Gamma_1)+J(\Gamma_2)+J(\Gamma_3)} \begin{pmatrix} \Gamma_2 & \Gamma_1 & \Gamma_3 \\ \gamma_2 & \gamma_1 & \gamma_3 \end{pmatrix} \tag{105}$$

which is a direct parallel of Eq. 66. Similarly, the conjugation behavior is

$$\begin{pmatrix} \Gamma_1 & \Gamma_2 & \Gamma_3 \\ \gamma_1 & \gamma_2 & \gamma_3 \end{pmatrix}^* = (-1)^{J(\Gamma_1)+J(\Gamma_2)+J(\Gamma_3)+|\gamma_1^*|+|\gamma_2^*|+|\gamma_3^*|} \begin{pmatrix} \Gamma_1 & \Gamma_2 & \Gamma_3 \\ \gamma_1^* & \gamma_2^* & \gamma_3^* \end{pmatrix} \quad (106)$$

which is parallel to Eq. 65. These phase definitions also quantify the relationship between finite-vector coupling coefficients and 3Γ symbols:

$$<\Gamma_1\gamma_1\Gamma_2\gamma_2|\Gamma_1\Gamma_2\Gamma_T\gamma_T> = (-1)^{J(\Gamma_1)-J(\Gamma_2)+|\gamma_T|}((2\Gamma_T)+1)^{1/2}\cdot\begin{pmatrix} \Gamma_1 & \Gamma_2 & \Gamma_T \\ \gamma_1 & \gamma_2 & (\gamma_T)^* \end{pmatrix} \quad (107)$$

which is a parallel of

$$<J_1M_1J_2M_2|J_1J_2J_TM_T> = (-1)^{J_1-J_2-M_T}[2J_T+1]^{1/2}\begin{pmatrix} J_1 & J_2 & J_T \\ M_1 & M_2 & -M_T \end{pmatrix} \quad (108)$$

The analogous definitions for the application of the Wigner—Eckart theorem to finite-group matrix elements are

$$<\Gamma_1\gamma_1|\Gamma_H\gamma_H|\Gamma_2\gamma_2> = (-1)^{J(\Gamma_1)+|\gamma_1|}\begin{pmatrix} \Gamma_1 & \Gamma_H & \Gamma_2 \\ (\gamma_1^*) & \gamma_H & \gamma_2 \end{pmatrix} <\Gamma_1\|\Gamma_H\|\Gamma_2> \quad (109)$$

which is a parallel of

$$<J_1M_1|C_k^q|J_2M_2> = (-1)^{J_1-M_1}\begin{pmatrix} J_1 & k & J_2 \\ -M_1 & q & M_2 \end{pmatrix} <J_1\|C^q\|J_2> \quad (110)$$

These finite-group relationships reflect conjugation properties of finite-group components which are completely general and independent of both the specific point group or the axis of quantization. However, the rules for existence of specific 3Γ symbols are dependent on the choice of group and axis. Depending on the numerical assignments used for the representation and component labels, different algebras can be devised to demonstrate these existence rules. On all of these quantization axes within any point group, the representation algebra has a common form, which has been previously discussed [triangle rule (Eq. 26), group theoretical requirement (Eq. 28), and component rule (Eq. 29)]. Note that the component rule is modified when an axis contains two collinear rotation elements. Thus on a double axis such as C_4^2 (C_2) both orders are allowed:

$$\gamma_1 + \gamma_2 + \gamma_3 = 0 \quad \text{or} \quad 2 \quad \text{or} \quad 4 \quad (111)$$

In using the component rule with the special labels, $\gamma\pm$, the signs are ignored.

1. The C_4^z Selection Rules

The quantizing axis is of modularity 4, thus the algebra is also to modulus 4, and a 3Γ symbol may exist whenever the sum of the labels is 0 or 4. The rules for permutation are exactly the same as for R_3, and thus no 3Γ symbols with two identical columns can exist unless the top is even under summation of all $J_i(\Gamma_i)$.

2. *The C_3^{xyz} Selection Rules*

The addition is taken to modulus 3. No special labels are necessary except when spin-orbit coupling is considered (52) and the $|\Gamma \pm n/2\rangle$ components are left uncombined by the quantization axis. Consider some examples taken from Table XII, bearing in mind that the operator components are $A_1O + T_2O$. Choosing from A_2 (in C_{3v} matrix), we note that the A_2O states (columns 2 and 3) are connected through the operator to yield a nonzero coefficient of DQ, but zero coefficients of DS and DT. This clearly arises because of the following 3Γ symbols:

$$\begin{pmatrix} A_2 & A_1 & A_2 \\ 0 & 0 & 0 \end{pmatrix} \quad \text{and} \quad \begin{pmatrix} A_2 & T_2 & A_2 \\ 0 & 0 & 0 \end{pmatrix} \tag{112}$$

The first symbol is allowed, but the second contravenes the triangle condition (or the group theoretical condition, $A_2 \times T_2 \times A_2$, does not contain A_1). Similarly, the T_1O and A_2O wave functions are coupled by the operator to yield nonzero values of DS and DT, but the DQ term vanishes. The relevant 3Γ symbols are

$$\begin{pmatrix} T_1 & A_1 & A_2 \\ 0 & 0 & 0 \end{pmatrix} \quad \text{and} \quad \begin{pmatrix} T_1 & T_2 & A_2 \\ 0 & 0 & 0 \end{pmatrix} \tag{113}$$

In this case, the first symbol vanishes because of the triangle condition (or group theoretical condition), whereas the second exists. Now let us consider the $T_1O(P)$ states coupled through the operator. The relevant 3Γ symbols are

$$\begin{pmatrix} T_1 & A_1 & T_1 \\ 0 & 0 & 0 \end{pmatrix} \quad \text{and} \quad \begin{pmatrix} T_1 & T_2 & T_1 \\ 0 & 0 & 0 \end{pmatrix} \tag{114}$$

The first symbol appears to exist since it obeys the group theoretical A_1 requirement. However, it disobeys the triangle requirement since the $J(\Gamma)$ values are 1, 4, and 1, respectively. The DQ component therefore vanishes. The second 3Γ symbol also obeys the group theoretical A_1 requirement. It also obeys the triangle requirement for the second-order operator (1, 2, 1), but not for the fourth-order operator (1, 4, 1). Hence, as seen in Table XII, only the DS coefficient is nonvanishing. Finally, consider the $T_1 \pm 1$ states coupled (separately) to the $T_2\pm$ states. The 3Γ symbols are

$$\begin{pmatrix} T_1 & A_1 & T_2 \\ -1 & 0 & +1 \end{pmatrix} \begin{pmatrix} T_1 & A_1 & T_2 \\ +1 & 0 & -1 \end{pmatrix} \begin{pmatrix} T_1 & T_2 & T_2 \\ -1 & 0 & +1 \end{pmatrix} \quad \text{and} \quad \begin{pmatrix} T_1 & T_2 & T_2 \\ +1 & 0 & -1 \end{pmatrix} \tag{115}$$

First note that because of the complex conjugate relationships (Eq. 104) the opposite signs of the components are mapped in the 3Γ symbol, that is, it is the complex conjugate of $|T_1 + 1>$ which is connected through the operator to the other $|T_1 + 1>$ representation. This did not arise above because the zero components are their own complex conjugates. The first two symbols vanish because of the group theoretical A_1 requirement, or, in general, because of the triangle condition. The second two symbols exist, thus both DS and DT matrix elements appear.

3. *The C_2 Selection Rules*

In a system that is generally adapted to C_2, only two γ_i labels should exist: 0 and 1. The latter is the special half-modulus label carrying either $1+$ or $1-$. However, this system is also adapted to O_h, and thus a division based on some of the O_h properties is necessary.

a. The C_2^z Quantization. The C_2 axis forces all integer representations to be self-conjugate since the C_{2v} tail group contains only unidimensional representations. However, the properties of the coincident C_4^z axis divide these self-conjugate representations into three subclasses containing the following

odd special components	$1\pm\left(=\frac{n}{2}\right)$
even special components	0
even special components	$2\pm$

While $3j$ symbols with an odd phase factor vanish if their m_j values are zero (see Eq. 65), this is not necessarily true of 3Γ symbols. Thus under conjugation, the phase factor in both $3j$ and 3Γ symbols actually contains a term summing the component values. Since this sum is necessarily zero in a $3j$ symbol (see Eq. 27), it is not explicitly indicated in Eq. 65. Similarly, a sum of the moduli of the components appears in the phase factor for a 3Γ symbol (see Eq. 106). This will also generally sum to zero and can then be neglected. However, when one or three $(2-)$ moduli appear, it will sum to an odd number because of the special characteristics of the $(2-)$ label (see Eq. 104). With an odd sum of $J(\Gamma)$ values, the resultant phase factor is even and the 3Γ symbol will not vanish. Thus the following four 3Γ symbols, which yield themselves under conjugation, will exist (52)

$$\begin{pmatrix} T_1 & T_2 & T_2 \\ 1\pm & 1\pm & 2- \end{pmatrix} \quad \text{and} \quad \begin{pmatrix} T_1 & T_2 & A_2 \\ 1\pm & 1\mp & 2- \end{pmatrix} \tag{116}$$

The components must add to 0, modulus 2 or modulus 4, and $\gamma_1 + \gamma_2 \geq \gamma_3$.

b. The C_2^{xy} System. The selection rules are applied in the same way as for C_2^z quantization, but because there is no fourfold axis coincident with

C_2^{xy} there is a redistribution of component labels. The necessary labels are $0\pm$ and $1\pm$, the $0-$ being identified with $2-$ on the C_4^2 axis.

B. Detailed Expansion of a Finite-Group Matrix Element

Once the nonzero matrix elements in a matrix have been determined by use of the selection rules, nonzero elements of the general form shown in Eq. 58 can be evaluated through use of Eqs. 63 and 81. Consider, for example, the following element of matrix Eq. 97 and its expansion:

$$
\begin{aligned}
<d^2F\,T_21(E1)|A_1(A_1)|d^2F\,T_21(E1)> &= <d^2F\,T_21\|A_1O^4|d^2F\,T_21> \\
&= (-1)^{J(T_2)+(1)*}\begin{pmatrix} T_2 & A_1 & T_2 \\ 1 & 0 & -1 \end{pmatrix}\begin{pmatrix} 3 & 4 & 3 \\ T_2 & A_1 & T_2 \end{pmatrix} <d^2\,{}^3F\|C^4\|d^2\,{}^3F> DQ \\
&= (-1)^{2-1} \times \begin{pmatrix} T_2 & A_1 & T_2 \\ 1 & 0 & -1 \end{pmatrix}\begin{pmatrix} 3 & 4 & 3 \\ T_2 & A_1 & T_2 \end{pmatrix}((2\times 2)+1)\begin{pmatrix} 2 & 4 & 2 \\ 0 & 0 & 0 \end{pmatrix} \qquad (117) \\
&\qquad <d^2\,{}^3F\|U^4\|d^2\,{}^3F> DQ
\end{aligned}
$$

By referring to the appropriate tables for the values of the 3Γ symbols (52), partition coefficients (73), $3j$ symbols (18, 55, 56), and reduced unitary matrix elements (40), Eq. 117 becomes

$$
\frac{1}{\sqrt{3}} \times \frac{1}{3\sqrt{22}} \times 5 \times \frac{-\sqrt{11}}{\sqrt{5}} \times \frac{\sqrt{2}}{\sqrt{35}} = \frac{-1}{3\sqrt{21}}\,DQ
$$

in accordance with the appropriate entry on Table XII and in matrix Eq. 97.

The evaluation of finite-group elements involving octahedral representations, but applied to the various subgroups, using Eqs. 63 and 81, therefore provides an extremely easy procedure for the generation of ligand-field matrices. All the relevant 3Γ symbols have been tabulated (52). In addition, one requires tables of finite-group reduced matrix elements or partition coefficients from which the former can be derived. Tables of such partition coefficients exist (73). We have recently published a complete set for finite operator representations A_1, E, T_1 and T_2 up to J values of 7 (76). In conclusion, it is eminently clear that the selection rules and procedures developed here can be used to reduce, very considerably, the effort and computer time necessary to obtain energy matrices.

References

1. J. C. Hempel, J. C. Donini, B. R. Hollebone, and A. B. P. Lever, *J. Am. Chem. Soc.*, *96*, 1693 (1974).

2. J. C. Donini, B. R. Hollebone, G. London, A. B. P. Lever, and J. C. Hempel, *Inorg. Chem., 14*, 455 (1975).
3. F. A. Cotton, *Chemical Applications of Group Theory*, 2nd ed., Wiley, New York, 1971.
4. H. Bethe, *Ann. Phys., 3*, 133 (1929).
5. J. H. Van Vleck, *The Theory of Electric and Magnetic Susceptibilities*, Oxford University Press, Oxford, England 1932.
6. B. N. Figgis, *Introduction to Ligand Fields*, Interscience, New York, 1966.
7. C. J. Ballhausen, *Introduction to Ligand Field Theory"*, McGraw-Hill, New York, 1962.
8. A. B. P. Lever, *Inorganic Electronic Spectroscopy*, Elsevier, Amsterdam, 1968.
9. S. Sugano, Y. Tanabe, and H. Kamimura, *Multiplets of Transition-Metal Ions in Crystals* (*Pure and Applied Physics*, Vol. 33), Academic New York, 1970.
10. J. S. Griffith, *Theory of Transition Metal Ions*, Cambridge University Press, Cambridge, England, 1964.
11. J. S. Griffith, *The Irreducible Tensor Method for Molecular Symmetry Groups*, Prentice Hall, New Jersey, 1962.
12. E. P. Wigner, *Quantum Theory of Angular Momentum*, L. C. Biedenharn and H. Van Dam, Eds., Academic, New York, 1965.
13. E. P. Wigner, *Group Theory*, Academic Press, New York, 1959.
14. G. Racah, *Phys. Rev., 61*, 186 (1942); *62*, 438 (1942).
15. G. Racah, *Phys. Rev. 63*, 367 (1943).
16. G. Racah, *Phys. Rev., 76*, 1352 (1949).
17. U. Fano and G. Racah, *Irreducible Tensorial Sets (Pure and Applied Physics, Vol. 4)*, Academic Press, New York, 1959.
18. B. R. Judd, *Operator Methods in Atomic Spectroscopy*, McGraw-Hill, New York, 1963.
19. B. G. Wybourne, *Symmetry Principles in Atomic Spectroscopy*, Wiley, New York, 1970.
20. G. Maki, *J. Chem. Phys., 28*, 651 (1958).
21. J. C. Donini, B. R. Hollebone, and A. B. P. Lever, *J. Am. Chem. Soc., 93*, 6455 (1971).
22. M. Gerloch and R. C. Slade, *Ligand Field Parameters*, Cambridge University Press, Cambridge, England, 1973.
23. M. Gerloch, J. Kohl, J. Lewis, and W. Urland, *J. Chem. Soc. A*, 3283 (1970) and references, therein.
24. E. König and S. Kremer, *Theoret. Chim. Acta, 32*, 27 (1973).
25. S. E. Harnung and C. E. Schäffer, *Structure and Bonding, 12*, 201 (1972).
26. W. Kauzmann, *Quantum Chemistry*, Academic, New York, 1957.
27. G. F. Koster, J. O. Dimmock, R. G. Wheeler, and H. Statz, *Properties of the Thirty-Two Point Groups*, Massachusetts Institute of Technology Cambridge, Mass., 1963.
28. J. S. Griffith, *Mol. Phys., 8*, 217 (1964).
29. C. E. Schäffer, *Wave Mechanics, The First Fifty Years*, W. C. Price, S. S. Chissick, and T. Ravensdale, Eds., Butterworths, London, 1973, p. 174.
30. H. L. Schläfer and G. Gliemann, *Basic Principles of Ligand Field Theory*, Wiley-Interscience, New York, 1969.
31. D. S. McClure *Advances in the Chemistry of the Coordination Compounds*, S. Kirschner, Ed., Macmillan, New York, 1961, p. 498.
32. H. Yamatera, *Bull. Chem. Soc. Japan, 31*, 95 (1958).
33. E. Larson, and G. N. La Mar, *J. Chem. Educ., 51*, 633 (1974).
34. C. K. Jørgensen, R. Pappalardo, and H. H. Schmidtke, *J. Chem. Phys., 39*, 1422 (1963).

35. C. E. Schäffer and C. K. Jørgensen, *Mol. Phys., 9*, 401 (1965).
36. C. E. Schäffer, *Structure and Bonding, 5*, 68 (1968); *Theoret. Chim. Acta, 4*, 166 (1966); *Pure and Appl. Chem., 24*, 361 (1970).
37. C. E. Schäffer and C. K. Jørgensen, *Mat. -fys. Medd. Selsk., 34*, 13 (1965).
38. P. L. Meredith and R. A. Palmer, *Inorg. Chem., 10*, 1049 (1971).
39. A. L. Companion and M. A. Komarynsky, *J. Chem. Educ., 41*, 257 (1964).
40. C. W. Nielson and G. F. Koster, *Spectroscopic Coefficients for the p^n, d^n and f^n Configurations*, Technology Press, Massachusetts Institute of Technology Cambridge, Mass., 1963.
41. M. Hamermesh, *Group Theory and its Application to Physical Problems*, Addison-Wesley, Reading, Mass., 1962.
42. L. Jansen and M. Boon, *Theory of Finite Groups*, Interscience, New York, 1967.
43. F. A. Matsen and O. R. Plummer, *Group Theory and its Applications*, E. M. Loebl, Ed., Academic, New York, 1968.
44. C. E. Soliverez, *Int. J. Quant. Chem., 7*, 1139 (1973).
45. J. S. Lomont, *Application of Finite Groups*, Academic, New York, 1959.
46. A. Gamba, *J. Math. Phys., 9*, 186 (1968).
47. E. B. Wilson, Jr., J. C. Decius, and P. C. Cross, *Molecular Vibrations*, McGraw-Hill, New York, 1955.
48. P. McWeeny, *Symmetry—An Introduction to Group Theory and its Applications*, Pergamon, New York, 1963.
49. M. L. Ellzey, Jr., *Int. J. Quant. Chem., 7*, 253 (1973).
50. P. J. McCarthy, G. London, and A. B. P. Lever, *Can. J. Chem.*, in press.
51. B. R. Hollebone and J. C. Donini, *Theoret. Chim. Acta, 37*, 233, (1975).
52. J. C. Donini, and B. R. Hollebone, *Theoret. Chim. Acta, 42*, 97 (1976).
53. P. A. M. Dirac, *Quantum Mechanics*, Oxford, England, 1947.
54. E. U. Condon and G. H. Shortley, *The Theory of Atomic Spectra*, Cambridge University Press, Cambridge, England, 1963. (Reprint of 1935 original edition.)
55. D. M. Brink and G. R. Satchler, *Angular Momentum*, Clarendon Press, Oxford, England 1968.
56. M. Rotenberg, R. Bivins, N. Metropolis, and J. K. Wooten, Jr., *The 3-j and 6-j Symbols*, Technology Press, Massachusetts Institute of Technology, Cambridge, Mass., 1959.
57. B. R. Hollebone, A. B. P. Lever, and J. C. Donini, *Mol. Phys., 22*, 155 (1971).
58. C. E. Harnung and C. E. Schäffer, *Struct. and Bonding., 12*, 257 (1972).
59. B. R. Hollebone and J. C. Donini, *Theoret. Chim. Acta, 39*, 33 (1975).
60. Ref. 22, Chapters 5 and 7.
61. C. E. Schäffer, *Theoret. Chim. Acta, 34*, 237 (1974).
62. A. B. P. Lever and B. R. Hollebone, *Inorg. Chem., 11*, 2183 (1972).
63. A. B. P. Lever and B. R. Hollebone, *J. Am. Chem. Soc., 94*, 1816 (1972) and references therein.
64. E. König, P. Gutlich, and R. Link, *Chem. Phys. Lett., 15*, 302 (1972).
65. M. Keeton, B. Fa Chun Chou, and A. B. P. Lever, *Can. J. Chem., 49*, 192 (1971) and references therein. Erratum *Can. J. Chem.,51*, 3690 (1973).
66. A. B. P. Lever, *Coord. Chem. Rev., 3*, 119 (1968).
67. D. A. Rowley, *Inorg. Chem., 10*, 397 (1971).
68. A. F. Schreiner and D. J. Hamm, *Inorg. Chem., 12*, 2037 (1973).
69. R. L. Chiang and R. S. Drago, *Inorg. Chem., 10*, 453 (1971) and references therein.

70. V. Balzani and V. Carassiti, *Photochemistry of Coordination Compounds*, Academic, New York, 1970.
71. E. König and S. Kremer, *Int. J. Quant. Chem., 8*, 347 (1974).
72. J. L. Prather, National Bureau of Standards Technical Notes No. 19 (1955).
73. Tang Au-Chin et al., *Scientia Sinica, 15*, 610 (1966).
74. J. C. Donini (Ph.D. thesis), "Symmetry: Its Application to the Spectra of Complexes," York University, Downsview, Ontario, Canada, 1973.
75. M. Gerloch and D. J. Mackay, *J. Chem. Soc A*, 2605 (1971).
76. J. C. Donini and B. R. Hollebone, *Theoret. Chim. Acta, 42*, 111 (1976).

Chemical Applications of Magnetic Anisotropy Studies on Transition Metal Complexes

by S. MITRA

Chemical Physics Group
Tata Institute of Fundamental Research
Bombay-5, India

I. INTRODUCTION

Although magnetic anisotropy is a very old technique, its application to problems of modern inorganic chemistry is rather new. Earlier studies in this area on paramagnetic complexes were mostly directed toward verifying the basic postulates of ligand field theory (LFT), and outstanding success was achieved in most cases. Later studies provided several interesting applications of the technique to understand the nature of ligand fields (LFs) and thus helped in the formulation and elaboration of the LFT. It is, however, only during the past decade that interest in this area has been generated among chemists, and a large potential scope for the application of this technique to inorganic chemistry has emerged. The hazards of a conventional approach to deduce chemical information from the magnetic measurements on polycrystalline samples have now become abundantly clear, and it is being increasingly realized that the measurements on single crystals provide the most accurate description. The application of this technique to understand and deduce the electronic structure of a variety of transition metal complexes has been the most illuminating one and has demonstrated, in several cases, the superiority of the technique over other existing ones. The sensitive nature of the paramagnetic anisotropy to even small variations in LF and stereochemical arrangements makes it a convenient probe to deduce the chemically important informations. Recent studies highlighting the necessity of paramagnetic anisotropy measurements for an accurate evaluation of the dipolar term in the isotropic proton magnetic resonance investigations have added another dimension to this technique.

Two general review articles on magnetic anisotropy appeared a few years ago. In one article the present author (1) discussed the details of the experimental techniques and surveyed broadly the measurements carried out on a wide range of para- and diamagnetic crystals, including transition metal and rare earth complexes, minerals, semiconductors, aromatic and aliphatic hydrocarbons, high polymers, and so forth. In the other article Hall and Horrocks (2) also discussed the experimental details and, in addition, listed the paramagnetic compounds whose anisotropies have been reported. The nature and scope of these review articles did not allow a clear and specific exposition of the chemical applications of this technique. In view of the increasing interest among chemists in the measurement of magnetic anisotropy, it now appears important that its chemical applications should be highlighted. It appears equally important to stress the theoretical reasoning and strategy employed in obtaining such information from the anisotropy data and to bring out, in particular, the superiority and limitations of the technique in solving these problems. The present article is intended to serve this purpose. It is hoped that the article will also form an integral part of modern magnetochemistry and will supplement the existing body of review articles and books on magnetochemistry, which are strikingly lacking in a treatment of these applications.

To achieve the above objectives, we have chosen specific examples to illustrate the application of the technique and have not tried to make this article an up-to-date reference list. Almost all the major areas of application are covered. The discussion is confined to the complexes of *d*-block elements, and those of the *f*-block elements are excluded. The latter is not excluded because of any unprivileged status of this interesting series of molecules. The theoretical basis for interpretation and the types of information obtained from the anisotropy data on rare earth complexes are quite different from those for transition metal complexes, hence the two merit separate treatments.

II. EXPERIMENTAL ASPECTS

The experimental methods for the measurement of paramagnetic anisotropy have been discussed in detail elsewhere (1, 2), and therefore only a brief mention will be made here of the salient features of the instrument and experimental techniques. The notation and terms to be used later will be explained, and a somewhat elaborate discussion of the calculation of principal molecular anisotropies (or susceptibilities) in monoclinic system will be included at the end of this section. The principal susceptibilities can be determined in some cases by a Faraday-type setup, but direct measurement of magnetic anisotropy is more convenient and accurate.

Among the different methods employed for the measurement of magnetic anisotropy, the "critical-couple" (3) and "null-deflection" (4) methods are the most common. The latter is primarily a slight modification of the former, and, being essentially a null-type, it is more convenient and accurate. The torsion balance in both methods is essentially the same, and the one employed in the null-deflection method is described here. The torsion balance consists of a torsion head (reading accurately to 0.1°) which is attached to a quartz fiber (approximately 10μ thick) about 30-cm long. The lower end of the quartz fiber is attached to a thin pyrex-glass rod of similar length. A hexagonal mirror system is fixed near the upper end of this rod. The specimen crystal with the desired orientation is attached to the lower end of the glass rod with an adhesive cement. The system is then fixed so that the crystal lies in the most homogeneous part of the magnetic field (see Fig. 1). When the magnetic field is applied, the crystal with any (*hkl*) plane horizonal tends to rotate so that the direction of the greater susceptibility in the plane (χ_{max}) lies along the magnetic field. The rotation of the crystal is followed by the mirrors using lamp and scale arrangement. By alternately applying the field and rotating the torsion head, the crystal is brought to the "zero position" or "setting position," as it is generally called. At this position, χ_{max} lies along the magnetic field.

The couple acting on an anisotropic crystal in a plane containing the susceptibilities χ_{max} and χ_{min} is given by

$$C = \frac{m}{2M}[H^2(\chi_{max} - \chi_{min}) \sin 2\phi] \tag{1}$$

where m and M are the mass and molecular weight of the crystal, H is the intensity of the applied magnetic field, and ϕ is the angle between χ_{max} and H. The couple is maximum when $\phi = 45°$. In the null-deflection method, the crystal is set at one of the 45° positions and the field is applied. The crystal rotates toward the setting position, and the torsion head is rotated to bring the crystal back to its original position. If τ is the torsion constant of

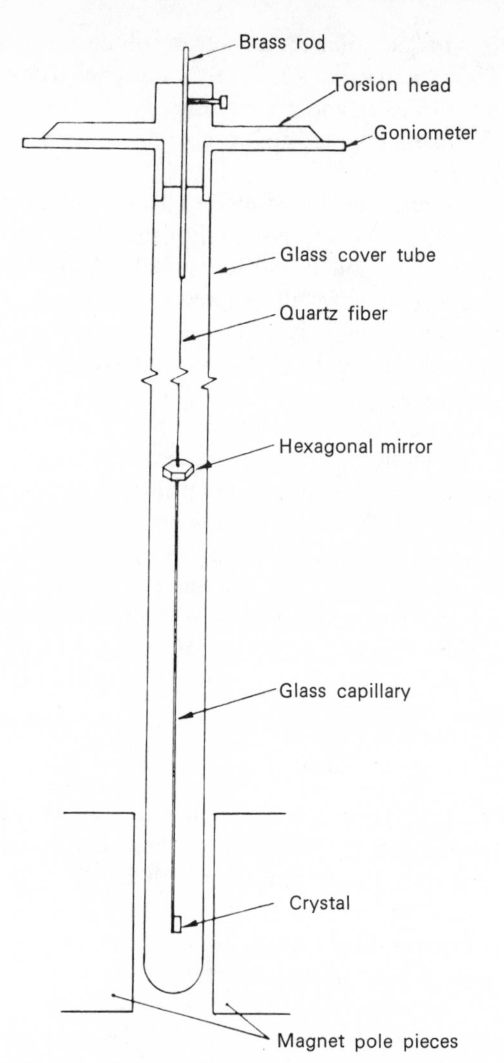

Fig. 1. Schematic diagram of a room-temperature anisotropy apparatus.

the fiber and α is the angle through which the torsion head is rotated from the 45° position, then the couple (C) is given by $\tau\alpha$. We get, therefore,

$$\tau\alpha = \frac{m}{2M}[H^2(\chi_{max} - \chi_{min})] \tag{2}$$

which gives

$$\Delta\chi = (\chi_{max} - \chi_{min}) = \frac{2m}{M} \cdot \frac{\tau}{H^2} \cdot \alpha \tag{3}$$

Equation 3 determines the $\Delta\chi$ once H and τ are known. H can be accurately measured, and the methods to determine τ are described elsewhere (5, 6).

It is often convenient to determine the anisotropy of the unknown crystal with respect to a secondary standard of accurately known anistropy. This obviates the difficulty of measuring H and τ everytime. It has been pointed out (7) that the single crystal of copper sulfate pentahydrate is a very suitable secondary standard for this purpose.

The number of measurements required to define the principal crystalline anisotropies depends on the crystal system. For the crystal belonging to axial systems (e.g., trigonal, tetragonal, and hexagonal), the principal crystalline susceptibilities lie along the crystallographic axes, and $\chi_a = \chi_b = \chi_\perp$ and $\chi_c = \chi_\parallel$. Only one measurement is, therefore, required to determine the $\pm(\chi_\parallel - \chi_\perp)$. In the orthorhombic system, as well, the principal crystalline susceptibilities lie along the crystal axes, but here $\chi_a \neq \chi_b \neq \chi_c$, hence measurements along two axes are required. In both cases, the sign of anisotropies is determined by noting the orientation of the crystal at the setting position. In monoclinic crystals, one of the principal crystal susceptibilities, for example, χ_3, lies along the b- axis of the crystal, whereas the other two crystal susceptibilities, χ_1 and χ_2, lie in the (010) plane making angles Ψ and ϕ with the c and a axes of the crystal, respectively. Assume that χ_1 is greater than χ_2. The expressions connecting the anisotropies measured with a, b, and c axes, vertical with the principal crystalline anisotropies, are given below:

$$(\Delta\chi)_a = \pm[(\chi_1 - \chi_2)\sin^2\phi - (\chi_1 - \chi_3)] \tag{4}$$

$$(\Delta\chi)_b = (\chi_1 - \chi_2) \tag{5}$$

$$(\Delta\chi)_c = \pm[(\chi_1 - \chi_2)\cos^2\Psi - (\chi_1 - \chi_3)] \tag{6}$$

The positive and negative signs in Eqs. 4 and 6 are taken for the b axis lying parallel and perpendicular to the magnetic field at the setting position. Measurements along the two axes, together with a determination of Ψ or ϕ, are therefore needed to obtain $(\chi_1 - \chi_2)$ and $(\chi_1 - \chi_3)$. A knowledge of the average susceptibility (or one of the principal susceptibilities) uniquely determines the values of the principal susceptibilities.

In the triclinic system, there is no relationship between the crystallographic and principal susceptibility axes, and the methods for the determination of the principal crystal anisotropies (or susceptibilities) are tedious. These are discussed in detail elsewhere (1).

Once the principal crystalline anisotropies (or susceptibilities) are known, the corresponding molecular quantities are obtained by a tensor transformation. This transformation is quite straightforward in the case of axial or orthorhombic crystals (1). In the case of monoclinic cyrstals, this is not so unless axial symmetry is assumed when the molecular anisotropies are simply

$$K_\parallel > K_\perp: (K_\parallel - K_\perp) = 2(\chi_1 - \chi_2) - (\chi_1 - \chi_3) \tag{7}$$

$$K_\perp > K_\parallel: (K_\perp - K_\parallel) = (\chi_1 - \chi_2) + (\chi_1 - \chi_3) \tag{8}$$

given by where the parallel and perpendicular subscripts refer to the quantities parallel and perpendicular to the symmetry axis. In the triclinic crystal, the situation remains slightly complicated, even if the assumption of axial symmetry is made. This has been discussed in Ref. 1.

When the axial symmetry is not assumed, the calculation of principal molecular anisotropies (or susceptibilities) in a monoclinic crystal is not straightforward, and difficulties have been encountered in such calculations in the past (8). The general treatment of the tensor transformation of crystalline quantities into molecular ones in a monoclinic system has been given by Lonsdale and Krishnan (9). Below we discuss two simplified versions of this treatment, as they have not been included in the earlier articles (2, 1).

Leela (10) simplified the general equations of Lonsdale and Krishnan and wrote them in the following form:

$$K_z = \frac{-(SX+TY) \pm [R(b-a)^2 \cdot (X^2+Y^2)-(SY-TX)^2]^{1/2}}{X^2+Y^2} \tag{9}$$

$$K_x = \frac{bP-Q}{b-a} + \frac{c-b}{b-a} \cdot K_z \tag{10}$$

$$K_y = \frac{Q-aP}{b-a} - \frac{c-a}{b-a} \cdot K_z, \tag{11}$$

where, $a = \alpha_1^2+\gamma_1^2$,
$b = \alpha_2^2+\gamma_2^2$,
$c = \alpha_3^2+\gamma_3^2$,
$d = \alpha_1^2-\gamma_1^2$,
$e = \alpha_2^2-\gamma_2^2$,
$f = \alpha_3^2-\gamma_3^2$,
$l = 2\alpha_1\gamma_1$,
$m = 2\alpha_2\gamma_2$,
$n = 2\alpha_3\gamma_3$,
$P = (\chi_1+\chi_2+\chi_3)$,
$Q = (\chi_1 + \chi_2)$,
$R = (\chi_1 - \chi_2)^2$.
$S = P(bd - ac) + Q(c - d)$
$T = P(lb - am) + Q(m - l)$,
$X = (f - d)(b - a) - (c - d)(c - a)$
$Y = (n - l)(b - a) - (m - l)(c - a)$

and $(\alpha, \beta, \gamma)_i$ are the direction cosines of principal molecular susceptibilities (K_x, K_y, K_z) relative to the orthogonal set of axes a, b, and c^*:

	a	b	c^*
K_x	α_1	β_1	γ_1
K_y	α_2	β_2	γ_2
K_z	α_3	β_3	γ_3

The calculation of K_x, K_y, and K_z from the above equations appears easy, but it should be noted that a singularity in Eqs. 10 and 11 exists when $b = a$ (i.e., $\beta_1 = \beta_2$). We will discuss its implication in Section IV. E.

In view of this shortcoming, Ganguli et al (11) suggested a method in which the crystal anisotropies are transformed directly into the molecular anisotropies. Assuming that the coordinate system is the same as shown above (a, b, c^*), they deduced from Ref. 9,

$$(\Delta\chi)_b = [\{(\alpha_1^2 - \gamma_1^2)(K_x - K_z) + (\alpha_2^2 - \gamma_2^2)(K_y - K_z)\}^2 + 4\{\alpha_1\gamma_1(K_x - K_z) + \alpha_2\gamma_2(K_y - K_z)\}^2]^{1/2} \quad (12)$$

$$(\Delta\chi)_a = \pm[(\beta_1^2 - \gamma_1^2)(K_x - K_z) + (\beta_2^2 - \gamma_2^2)(K_y - K_z)] \quad (13)$$

Values of $(K_x - K_z)$ and $(K_y - K_z)$ can be obtained from Eqs. 12 and 13 by solving the quadratic equations. In general, two solutions are obtained, and the correct one is decided by comparing the experimentally determined values of ϕ with those calculated by the following equations:

$$\cos 2(90 - \phi) = \frac{(K_x - K_z)(\alpha_1^2 - \gamma_1^2) + (K_y - K_z)(\alpha_2^2 - \gamma_2^2)}{(\Delta\chi)_b} \quad (14)$$

$$\sin 2(90 - \phi) = \frac{2\alpha_1\gamma_1(K_x - K_z) + 2\alpha_2\gamma_2(K_y - K_z)}{(\Delta\chi)_b} \quad (15)$$

This method has been applied (11, 12) to a number of monoclinic crystals, and values of K_i have been obtained. Furthermore, the method does not suffer from the limitation commented upon earlier. It should be emphasised that it is always desirable to calculate K_x, K_y, and K_z when the structural data are available. The advantage of Eqs. 7 and 8 is that no such structural data are required, but assumption of axial symmetry must be made, which may not be valid. In the case of crystals belonging to axial, orthorhombic and triclinic systems, the structural data are required even when the axial symmetry is assumed.

Before we close this section, mention should also be made to various corrections which are sometimes important but are nevertheless overlooked. Shape and diamagnetic anisotropies are two such factors. The former arises because of an asymmetric shape of the crystal, which produces an anisotropic magnetization. It can also arise if the magnetic field is inhomogeneous. To take account of the shape anisotropy, the crystals should be, if possible, ground into cylindrical or spherical shapes, and the applied magnetic field should be as homogeneous as possible. Fortunately, the shape anisotropy is generally very small (except for S-state ions) and can be ignored. For a paramagnetic crystal, there is also a couple that results from the anisotropy in the diamagnetic susceptibility, and this should be allowed for when determining the true paramagnetic anisotropy. To correct for this factor, measurement should be made on an isomorphous diamagnetic compound and correction effected (3). This correction is again usually small, but it becomes significant when there is an extensive delocalized pi system in the molecule (as in the metal phthalocyanines) and/or the anisotropies are very small (*S*-state ions). A detailed discussion of this correction is available in Ref. 1.

Finally, we usually refer to the experimental data in terms of the principal magnetic moments, given as $\mu_i^2 = 7.998\, K_i \cdot T$. Therefore $(\mu_\perp^2 - \mu_\parallel^2) = \Delta\mu = 7.998\,(K_\perp - K_\parallel)\cdot T$. The susceptibilities will be expressed throughout this article in units of $[10^{-6}\text{cm}^3/\text{mole}]$.

III. THEORETICAL CONSIDERATIONS

The anisotropy in the magnetic susceptibility of a paramagnetic ion arises, in general, from the noncubic nature of the LF acting on the metal ion. Since the spin contribution to the susceptibility is spatially isotropic, the paramagnetic anisotropy usually arises from the anisotropic nature of the orbital contribution; thus it can be used as a sensitive probe to understand those properties which depend on the orbital contribution, for example, electronic structure of the metal ion, ligand fields, and other allied chemical features. The situation can be contrasted to the average magnetic moment ($\bar{\mu}$), where most of the contribution comes from the spin part which only characterizes the number of unpaired electrons. It follows, therefore, that the average magnetic moment would be, in general, insensitive to the LF and electronic structure of the paramagnetic ion.

If the ground state of a paramagnetic ion is an orbital singlet, it has no orbital contribution in the ground state, hence no anisotropy is expected. A finite anisotropy is, however, produced by mixing in of the ground state with some excited states through spin-orbit (S–O) coupling. Evidently the anisotropy in this case would depend on the strength of the S–O coupling parameter and the energy separation between the ground and excited states. When the ground state of a paramagnetic ion is "formally" an orbital triplet, the susceptibility is expected to be highly anisotropic. These are, however, very qualitative and simple generalizations, and they are not always strictly obeyed. Nevertheless, they reflect the general features of the experimental observations, especially on simple systems. In Table I we list some such conclusions for octahedrally* coordinated first-row transition metal complexes. Those for the tetrahedrally* coordinated complexes will follow similarly.

*Throughout this article, the terms "octahedrally," 'tetrahedrally," "octahedral," or "tetrahedral" mean, unless otherwise mentioned, that the octahedron (or tetrahedron) is distorted.

Table I shows that the anisotropies of Cr(III) and Fe(III) ions (with orbital singlet ground state) are extremely small, whereas those for the iron(II) and cobalt(II) ions with orbital triplet ground state are very high. However, the anisotropy of the V^{3+} ion is quite small, contrary to the expected high value because of the orbital triplet ground state. The observed anisotropy of the vanadium(III) complexes is small since the splitting of the ground triplet in axial field is very large with effectively an orbital singlet lowest. This shows how sensitive the paramagnetic anisotropy is to the splitting of the ground state, a result which will be discussed in detail in the next section. For the octahedral and tetrahedral transition metal complexes, the LF acting at the metal ion can be viewed as composed of a predominant symmetric part (cubic field) with a small axial (or rhombic) part superimposed over it. This simplification is convenient, and it helps us to categorize the octahedral and tetrahedral complexes according to their ground states in the cubic field, as is shown in the next section.

For the metal ions with orbital singlet ground state and $S > \frac{1}{2}$, the S–O interaction partially lifts the spin degeneracy of the ground state and leaves a set of spin states. This is called zero-field splitting (ZFS). For the high-spin iron(III) ion, for example, the sixfold spin degeneracy of the ground $^6A_1$ state is partly removed into a set of three Kramer's doublets; the separation between which is characterized by a parameter, called the ZFS parameter. We shall see in the next section that this parameter, which is very sensitive to the paramagnetic anisotropy, provides an easy description of the nature of LF.

For complexes of lower symmetry (e.g., square planar, square pyramidal, etc.), the situation is fairly complicated. These complexes generally represent the limiting situation of an octahedral arrangement with, for example, a very large tetragonal distortion. The original t_{2g} (d_{xy}, d_{xz}, d_{yz}) and e_g $(d_{z^2}, d_{x^2-y^2})$ orbitals in pure octahedral symmetry are split in the tetragonal field into a singlet (d_{xy}) and a doublet $(d_{xz,yz})$ and into two singlets $(d_{z^2}, d_{x^2-y^2})$, respectively. As the tetragonal field is increased, the antibonding $d_{x^2-y^2}$ orbital is destabilized and lies at much higher energy than the remaining ones, which come much closer to each other. The ordering of these close lying orbitals, therefore, becomes uncertain and can best be decided by experiment. Since the magnetic anisotropy is extremely sensitive to the orbital contribution, its measurement is expected to provide this information.

When discussing the experimental results in the next section, we describe the theoretical aspects mostly in an illustrative form. Derivation of the expressions and their complications are avoided as they are avilable in standard books on LFT. We use the theory in a form, often simplified, which helps us to understand the salient features of the experimental results. Since

TABLE I

Cubic-Field Ground States, Spin-Orbit Coupling Parameter, and Expected and Observed Anisotropies for $3d^n$-Transition Metal Ions in Octahedrally Coordinated Complexes

Ions →	Ti^{3+} (d^1)	V^{3+} (d^2)	Cr^{3+} (d^3)	Mn^{3+} (d^4)	Fe^{3+}, Mn^{2+} (d^5)	Fe^{2+} (d^6)	Co^{2+} (d^7)	Ni^{2+} (d^8)	Cu^{2+} (d^9)
λ Free ion, cm^{-1}	+155	+105	+92	+89	92	−100	−172	−315	−830
Cubic-field ground states	$^2T_{2g}$	$^3T_{1g}$	$^4A_{1g}$	^5E_g	$^6A_{1g}$	$^5T_{2g}$	$^4T_{1g}$	$^3A_2$	^2E
Expected anisotropy	Large	Large	Very small	Moderate	Very small	Large	Large	Small	Large
Observed anisotropy	5%	2%	<1%	?	<1%	25–30%	25–30%	6–8%	20–30%

Large: >20%; moderate: 10–20%; small: 1–10%; very small: <1%.

One-electron spin-orbit coupling parameter (ζ) is related to λ by $\lambda = \pm\zeta/2S$.

the measurement of paramagnetic anisotropy is still mostly limited to the liquid nitrogen temperature range, the theoretical expressions can often be simplified.

IV. APPLICATIONS TO METAL COMPLEXES

A. Nature of Ligand Fields and Electronic Structure of Octahedral and Tetrahedral Complexes with Cubic-Field Orbital Singlet and Doublet Ground States

1. Orbital Singlet

As mentioned in the previous section, the S–O coupling mixing of the ground orbital singlet with the excited states produces the paramagnetic anisotropy. This mixing causes a g-anisotropy as well as the ZFS of the ground state for ions with $S \geq 1$. Measurement of paramagnetic anisotropy is, in general, helpful in determining these quantities, which in turn give information about the excited states.

We consider here examples of $^4A_2$ and $^3A_2$ ground states. The $^6A_1$ ground state will be discussed separately in Section IV.E.

a. $^4A_2$ Ground State. We now consider specific examples of octahedrally coordinated chromium(III) and tetrahedrally coordinated cobalt(II) complexes, both of which have the $^4A_2$ ground state in the LF (Fig. 2). In the calculation only the first excited state, $^4T_2$, is considered as the next

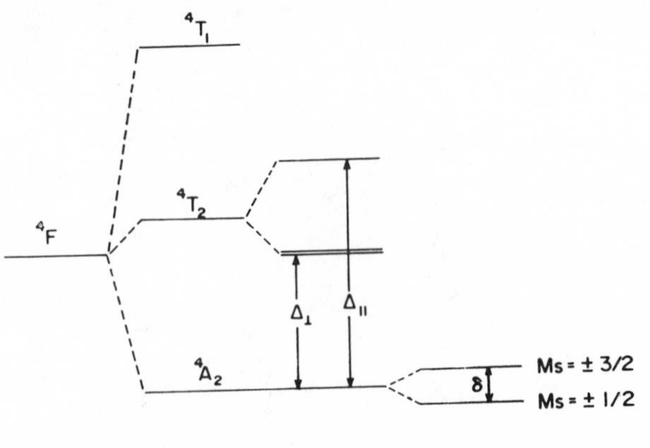

Fig. 2. Energy level diagram for a tetrahedral cobaltous (or octahedral Cr^{3+}) ion: (*a*) free ion; (*b*) cubic field; (*c*) axial field; (*d*) spin-orbit coupling.

excited state, $^4T_1$, does not directly mix with $^4A_2$. The principal molecular susceptibilities are therefore given by the following expressions in axial symmetry:

$$K_{\parallel} = \frac{N\beta^2}{3kT}\left[\frac{15}{4} g_{\parallel}^2\left(1 - \frac{2\delta}{5kT}\right)\right] + \frac{8N\beta^2\kappa^2}{\Delta_{\parallel}} \tag{16}$$

$$K_{\perp} = \frac{N\beta^2}{3kT}\left[\frac{15}{4} g_{\perp}^2\left(1 + \frac{\delta}{5kT}\right)\right] + \frac{8N\beta^2\kappa^2}{\Delta_{\perp}} \tag{17}$$

where $\quad g_{\parallel} = 2(1 - 4\kappa\lambda/\Delta_{\parallel}),$

$$g_{\perp} = 2(1 - 4\kappa\lambda/\Delta_{\perp}),$$

$$\delta = 8\lambda^2(1/\Delta_{\parallel} - 1/\Delta_{\perp}) \tag{18}$$

Here κ and λ are the orbital-reduction and spin-orbit coupling parameters, respectively; $\Delta_{\parallel}$ and $\Delta_{\perp}$ are as shown in Fig. 2. δ is the ZFS parameter. The expressions show that the principal susceptibilities (and anisotropy) depend on three terms: the term on the g values which varies as $1/T$; the term in-

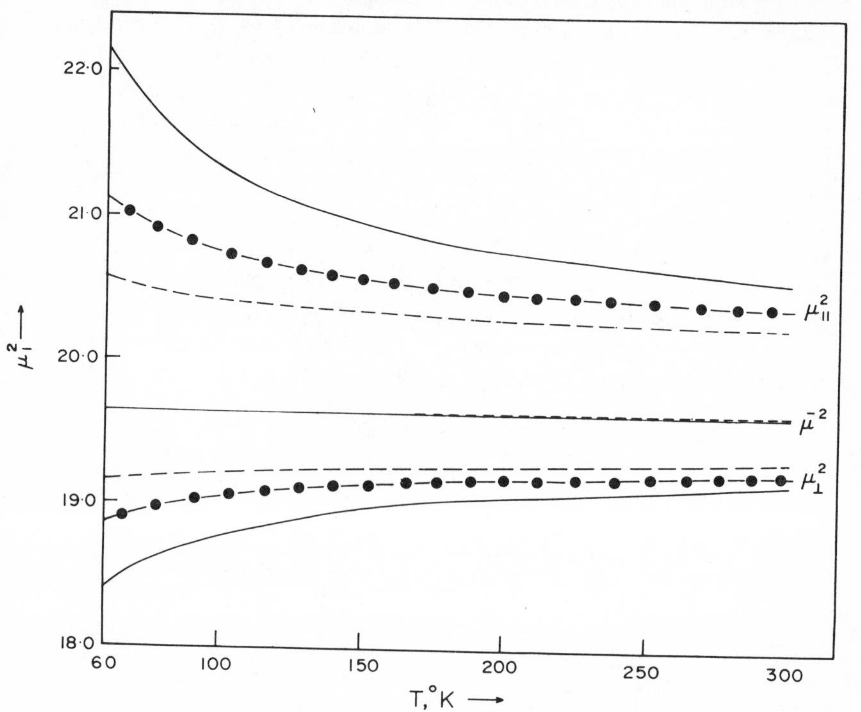

Fig. 3. The variation of μ_i^2 vs. T for the $^4A_2$ term (corresponding to octahedral d^3 or tetrahedral d^7 configuration). The TIP term is neglected, and $g_{\parallel} = 2.32$ and $g_{\perp} = 2.27$: ———, $\delta = -10\ \text{cm}^{-1}$; —·—·—, $\delta = -5\ \text{cm}^{-1}$; — — — —, $\delta = -2\ \text{cm}^{-1}$.

volving ZFS which varies as $1/T^2$; and, of course, the small temperature independent term (TIP). Neglecting the TIP terms, Eqs. 16 and 17 can be conveniently plotted in terms of principal moments for different values of δ. This is shown in Fig. 3. Values of $g_{\parallel}$, $g_{\perp}$, and λ are typical for tetrahedral cobaltous ions. The figure shows the marked effect of variation in δ on the principal magnetic moments (hence on the anisotropy) and the insensitive nature of the average magnetic moment to δ. The study of single crystals, even in the liquid nitrogen temperature range, provides, therefore, a sensitive method to determine the ZFS and other LF parameters. Equations 16 and 17 show that the paramagnetic anisotropy of octahedral Cr(III) complexes will, in general, be very small since here $\Delta_{\parallel}$ and $\Delta_{\perp}$ are of the order of 15,000 cm^{-1}; λ is also very small. The g values will be almost isotropic and δ will be very small. The anisotropy will then vary approximately as T^{-2}. For the tetrahedral cobaltous complexes, the situation is shown in Fig. 3. We now discuss the experimental results on these complexes in the light of the above theoretical approach.

(1). Octahedral Chromium(III) Complexes. Measurements on a few octahedrally coordinated chromium(III) complexes have been reported (13, 14). As expected, the anisotropies are extremely small, hence the corrections for diamagnetic and shape anisotropies become important (see Table II). For example, the paramagnetic anisotropy of potassium and ammonium tris(oxalato)chromium(III) is only about 0.25% of its average susceptibility. Furthermore, the anisotropies obey very nearly, as expected, a $1/T^2$ variation at low temperatures. In view of the extremely low anisotropies, it is difficult to determine reliable values of δ and other LF parameters from the anisotropy measurements in the 80 to 300°K temperature range, although such attempts have been made (14). Measurements at lower temperatures, especially in the liquid helium temperature range, are required to deduce accurate values of these parameters. ZFS in chromium(III) complexes is generally of the order of 0.3 cm^{-1}.

TABLE II
Room-Temperature Magnetic Anisotropy Data (in 10^6 cm^3 $mole^{-1}$) on Trivalent Chromium(III) Compounds

Compound	$\lvert\chi_1 - \chi_2\rvert$	$\lvert\chi_1 - \chi_3\rvert$
$K_3Cr(_2CO_4)_3 \cdot 3H_2O$	20	19
$(NH_4)_3Cr(C_2O_4)_3 \cdot 3H_2O$	4.3	10.3
$K_3Al(C_2O_4)_3 \cdot 3H_2O$	6.9	13.1

(2). Tetrahedral Cobaltous Complexes. Measurements on a number of tetrahedrally coordinated cobalt(II) complexes have been reported over a

wide range of temperature. Cs_3CoCl_5 has been most extensively studied, perhaps because of its convenient and simple crystallographic features (15). The anisotropy has been directly measured between 80° to 300°K (16–18), the principal susceptibilities being measured down to 4°K. ESR, single-crystal polarized spectra, and Zeeman field studies have also been reported. The anisotropies of Cs_3CoCl_5 together with two other tetrahedral cobaltous compounds are included in Table III. The anisotropies are small (ca. 6–7% of the average susceptibility), but are much larger than the octahedral chromium (III) complexes. This implies, of course, that $\Delta_\parallel$ and $\Delta_\perp$ are much smaller (characteristic of tetrahedral LF); hence δ is much larger than those in the octahedrally coordinated Cr(III) complexes.

TABLE III
Room-Temperature Anisotropies (in 10^6 cm^3 $mole^{-1}$) for Some Tetrahedrally Coordinated Cobaltous Compounds

Compounds	$(K_\parallel - K_\perp)$	Refs.
Cs_3CoCl_5	652	16
Cs_2CoCl_4	646	20
$K_2Co(NCS)_4 \cdot 4H_2O$	118	18

There are four adjustable parameters in Eqs. 16 and 17: $\Delta_\parallel$, $\Delta_\perp$, λ, and κ. The temperature dependence of molecular anisotropy can determine only two of the parameters. Normally, some reasonable values of λ and κ are assumed. If the anisotropy in the TIP terms is neglected, then Eqs. 16 and 17 can be expressed in the following form:

$$(\mu_\parallel^2 - \mu_\perp^2) = \frac{15}{4}(g_\parallel^2 - g_\perp^2) - \frac{3}{4} \cdot \frac{\delta}{T}(2g_\parallel^2 + g_\perp^2) \qquad (19)$$

Equation 19 shows that a plot of $\Delta\mu^2$ versus $1/T$ should be a straight line, the slope of which determines, even by visual inspection, the sign of δ. A similar plot of the experimental anisotropy data in the 80° to 300°K temperature range for Cs_3CoCl_5 is shown in Fig. 4. It is interesting that the experimental data lie exactly on a straight line, suggesting that the anisotropy in the TIP term is negligible. The slope of the line establishes that δ is negative ($Ms = \pm 3/2$ lying lowest). A detailed fitting of the data (17) gives $\delta = -9$ cm^{-1}. This leads to $\Delta_\parallel > \Delta_\perp$, which is in agreement with the optical and Zeeman field studies. This ordering disagrees, however, with the reported analysis (18) of the anisotropy data based on a point charge model. We will come back to this point in Section IV.E.

The above analysis of the paramagnetic anisotropy data, even in the liquid nitrogen temperature range, affords a convenient and accurate determination of the ZFS and other parameters. Figure 4 further shows that the

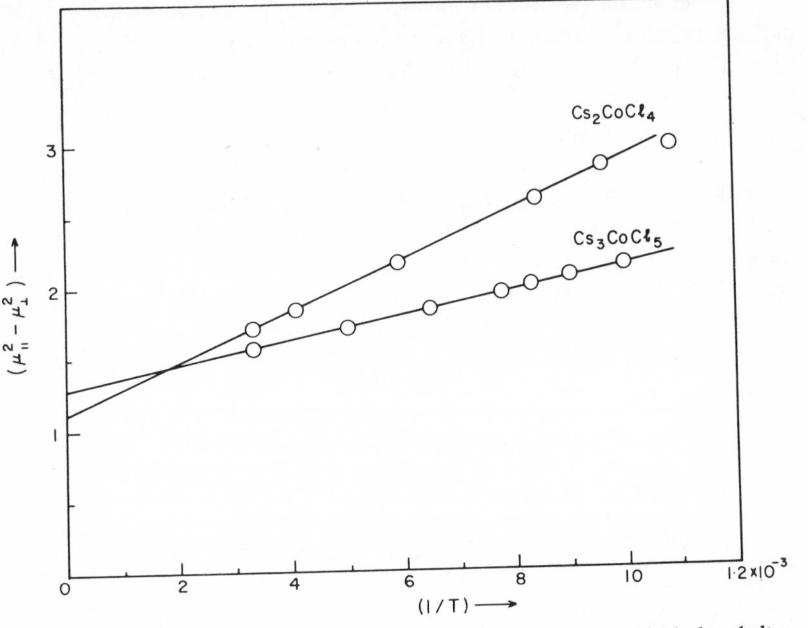

Fig. 4. Temperature dependence of $(\mu_{\parallel}^2 - \mu_{\perp}^2)$ for two typical tetrahedral cobaltous complexes. The circles are the experimental data taken from Refs. 16 and 18. See text for the data on Cs_2CoCl_4.

effect of exchange interaction is negligible in the 80° to 300°K range. This is an advantage as the exchange interaction tends to complicate the interpretation of the Zeeman field and ESR studies usually carried out at very low temperatures.

The magnetic anisotropy of Cs_2CoCl_4 is observed to be very similar to that of Cs_3CoCl_5 (see Table III). There was some confusion about the sign of molecular anisotropy of this crystal. However, a careful remeasurement established (20) that $K_{\parallel} > K_{\perp}$. The temperature dependence of its molecular anisotropy (as recalculated by the present writer from Ref. 18 after the necessary correction) is given in Fig. 4. δ is again negative. A remeasurement over the temperature range is required before accurate values of LF parameters can be deduced.

The lower anisotropy of $K_2Co(NCS)_4$ is interesting and implies that the excited ${}^4T_2$ state lies much higher in energy than in the chlorides because of a stronger LF in the $[Co(NCS)_4]^{2-}$ anion.

b. ${}^3A_2$ Ground State. We now discuss the magnetic behavior of a few octahedrally coordinated nickel(II) complexes which have a ${}^3A_2$ ground state in the cubic field. The energy level diagram is similar to that shown in Fig. 2, with the first excited state being ${}^3T_2$. The ${}^3A_2$ ground state is zero-field split into $Ms = 0$ and $Ms = \pm 1$, with δ being the ZFS parameter. The principal magnetic moments are given by

$$\mu_{\parallel}^2 = 2g_{\parallel}^2 \cdot \frac{3e^{-x}}{1 + 2e^{-x}} + \frac{8N\beta^2\kappa^2}{\Delta_{\parallel}}$$
$$\mu_{\perp}^2 = 2g_{\perp}^2 \cdot \frac{3(1 - e^{-x})}{x(1 + e^{-x})} + \frac{8N\beta^2\kappa^2}{\Delta_{\perp}} \qquad (20)$$

where $x = \delta/kT$ and $\Delta_{\parallel}$ and $\Delta_{\perp}$ are quantities as shown in Fig. 2. Neglecting terms with higher powers in x in the expansion of the Eq., we get

$$(\mu_{\perp}^2 - \mu_{\parallel}^2) = 2(g_{\perp}^2 - g_{\parallel}^2) + \frac{\delta}{3kT}(2g_{\parallel}^2 + g_{\perp}^2) + 24kT\left(\frac{1}{\Delta_{\perp}} - \frac{1}{\Delta_{\parallel}}\right)\kappa^2 \quad (21)$$

For the octahedral nickel(II) complex $g_{\parallel} \simeq g_{\perp}$. Therefore, Eq. 21 can be further simplified,

$$(\mu_{\perp}^2 - \mu_{\parallel}^2) = \frac{\delta}{kT} \cdot \bar{g}^2 + 24kT\left(\frac{1}{\Delta_{\perp}} - \frac{1}{\Delta_{\parallel}}\right)\kappa^2 \qquad (22)$$

If the anisotropy in the *TIP* contribution is neglected (at very low temperatures) then $\Delta\mu^2$ is expected to vary as T^{-1} (or ΔK as T^{-2}).

Measurements on a series of nickel(II) Tutton salts, sulphates, and

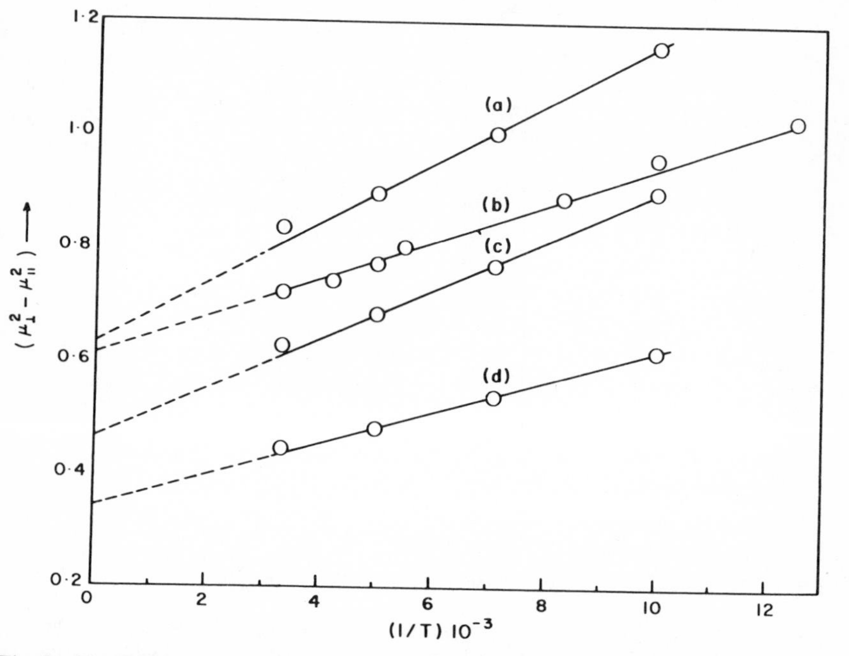

Fig. 5. Temperature dependence of $(\mu_{\perp}^2 - \mu_{\parallel}^2)$ for a few octahedral nickel (II) complexes: (*a*) $NiSO_4\ 6H_2O$; (*b*) Ni(tu) Cl_2; (*c*) $K_2Ni(SO_4)_2\ 6H_2O$; (*d*) $(NH_4)_2Ni(SO_4)_2\ 6H_2O$. The experimental data (shown by the circles) are taken from Refs. 21 and 22.

fluosilicates have been reported (21). The molecular anisotropy of these similarly constituted Tutton salts varies quite significantly among themselves, which has been ascribed to a long-range effect. A plot of the experimental data as $\Delta\mu^2$ versus $1/T$ gives, as expected, a straight line (see Fig. 5). Here, δ is generally small (~ 3 cm^{-1}) and positive, that is, $Ms = 0$ lies below $Ms = \pm 1$. There is some controversy about the nature of LF, even in these simple compounds, and it has been suggested that the LF parameters are temperature dependent (21). This conclusion based on axial symmetry is, however, doubtful as recent calculations (12) show a very large in-plane anisotropy in the nickel ammonium Tutton salt, thus invalidating the previous assumption of axial symmetry. The study on octahedrally coordinated dichlorotetrakis-(thiourea)nickel(II) is interesting as it shows a much larger positive value for δ ($\simeq +8$ cm^{-1}) (22).

2. *Orbital Doublet*

Octahedrally coordinated copper(II) complexes provide a good example of compounds which have an orbital doublet (^2E_g) lying lowest in the cubic LF. The effect of a tetragonal distortion is to split the lowest ^2E into singlets

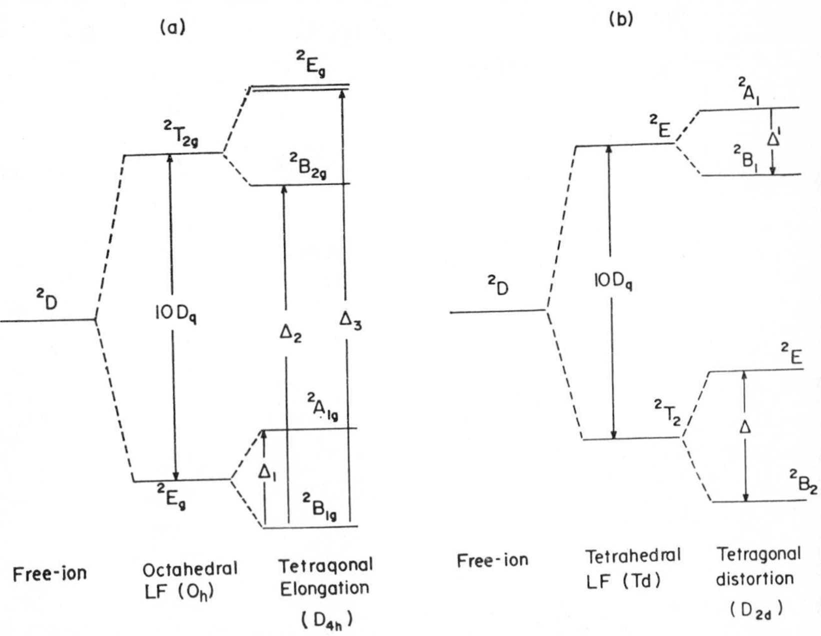

Fig. 6. Energy level diagrams for octahedrally (*a*) and tetrahearally (*b*) coordinated Cu(II) complexes. Fig. 6*b* is also applicable to low-spin octahedral d^5 complexes with some change.

and the excited $^2T_2$ into a singlet and a doublet (see Fig. 6*a*). The separation between ^2E_g and $^2T_{2g}$ is 10 D_q, where D_q is the cubic-field splitting parameter. Figure 6*a* depicts the situation for a tetragonal elogation when $^2B_{1g}$ lies below $^2A_{1g}$. The principal susceptibilities are then given by

$$\begin{aligned} K_{\parallel} &= \left(\frac{N\beta^2}{4kT}\right)g_{\parallel}^2 + \frac{8N\beta^2\kappa^2}{\Delta_2} \\ K_{\perp} &= \left(\frac{N\beta^2}{4kT}\right)g_{\perp}^2 + \frac{2N\beta^2\kappa^2}{\Delta_3} \\ g_{\parallel} &= 2\left(1 - \frac{4\kappa\lambda}{\Delta_2}\right) \\ g_{\perp} &= 2\left(1 - \frac{\kappa\lambda}{\Delta_3}\right) \end{aligned} \tag{23}$$

Δ_2 and Δ_3 are as shown in Fig. 6*a*. If the distortion is a tetragonal compression, $^2A_{1g}$ usually lies below $^2B_{1g}$. The principal susceptibilities are then given by

$$\begin{aligned} K_{\parallel} &= \left(\frac{N\beta^2}{4kT}\right)g_{\parallel}^2 \\ K_{\perp} &= \left(\frac{N\beta^2}{4kT}\right)g_{\perp}^2 + \frac{6N\beta^2\kappa^2}{\Delta_3} \end{aligned} \tag{24}$$

where $g_{\parallel} = 2$ and $g_{\perp} = 2\,(1 - 3\kappa\lambda/\Delta_3)$. Δ_3 is now the separation between $^2A_{1g}$ and ^2E_g. Equations 23 and 24 show that, in the case of tetragonal elongation, $K_{\parallel} > K_{\perp}$, whereas for tetragonal compression $K_{\perp} > K_{\parallel}$. Thus a measurement of paramagnetic anisotropy can help to distinguish between these two possibilities.

An extensive measurement exists on a series of copper(II) Tutton salts (23), all of which have $K_{\parallel} > K_{\perp}$, in agreement with the available x-ray structural data. As in the nickel(II) Tutton salts, here again large variation in the paramagnetic anisotropies is observed among the different members of these similarly constituted salts. Equation 23 shows that the $(K_{\parallel} - K_{\perp})$ should obey an $(A + B/T)$-type temperature variation. Difficulties have, however, been encountered in such a fitting and a large $1/T^2$ term was observed (23). This has been interpreted to imply that the LF acting on the copper(II) ion in these salts is perhaps temperature dependent. Recent ESR measurements on copper(II) ammonium sulfate hexahydrate also show that the *g* values are temperature dependent. It should be pointed out that the above analysis of paramagnetic anisotropy data is based on the axial symmetry. Recent calculations of K_x, K_y, and K_z using x-ray data bring out, however, a strong rhombic character of the principal susceptibilities (12, 24). Hence a reinterpretation of the data has been attempted (24).

B. Nature of Ligand Fields and Electronic Structure of Octahedral and Tetrahedral Complexes with Cubic-Field Orbital Triplet Ground States

We discuss here the magnetic properties of the transition metal ions with a formally orbital triplet lying lowest in the cubic part of the LF. The lower symmetry field splits the ground state into an orbital doublet and a singlet (in the axial field) or into three orbital singlets (in the rhombic field), with an overall separation generally of the order of the thermal energy at room temperature. The magnetic properties in such cases give useful information about the molecular distortion, nature of bonding, and so forth. A very extensive study of the magnetic properties of complexes with orbital triplet ground term has therefore been made. Usually axial symmetry is assumed (although it may not be true in many cases), which helps to rationalize the magnetic properties in terms of three disposable parameters: Δ (the axial field splitting), λ, and κ. These parameters are useful chemical quantities; Δ, for example, can be correlated to the degree of geometric distortion in the basic octahedron or tetrahedron. Significance of the orbital reductor factor (κ) has been discussed in detail elsewhere (25). It is therefore important to ascertain the accuracy of the values of these parameters as deduced from the experimental results.

One of the common methods for deducing these parameters has been to analyze the average magnetic susceptibility data over a range of temperature (usually 77°–300°K). Evidently a unique choice of these three parameters would be impossible from the temperature-dependent values of the average magnetic moment alone. To avoid this difficulty, a composite parameter, $\nu\,(= \Delta / \lambda)$, has been used by Figgis (26). It has, however, been observed that the average magnetic moment is quite insensitive to these LF parameters and often leads to large errors in the values deduced from the consideration of this single source of measurement. In this section we highlight these aspects and demonstrate how the measurement of paramagnetic anisotropy provides a very sensitive and accurate tool to determine these LF parameters. The discussion is limited to only a few specific examples with $^2T_2$, $^3T_1$, and $^5T_2$ ground states, but the conclusions are fairly general and apply to other triplet ground terms as well. Since the majority of the studies so far reported are based on axial symmetry, the discussion here is also limited to this assumption, although we mention of the rhombic LF wherever possible.

1. $^2T_2$ *Ground State*

Octahedral d^1 and low-spin d^5 ions, as well as the tetrahedral d^9 ions, have a formal $^2T_2$ cubic-field ground term. In the octahedral complexes, the excited terms lie much higher in energy, hence their effect is generally neglected. This is, however, not the case for tetrahedral d^9 complexes [e.g., Cu (II)] because of much smaller LF splitting and higher spin-orbit coupling. In view of this difference, we will consider these cases separately.

a. Octahedral Complexes. The splitting of the $^2T_2$ term under the action of axial LF is shown in Fig. 6*b*. We assume the convention that Δ is positive when $^2B_2$ lies below ^2E (^2T). Explicit expressions for the principal molecular susceptibilities of the $^2T_2$ term are available (27–30); Ref. 29 and 30 give the derivation in detail. We discuss below the results of these calculations of the principal and average magnetic moments and in particular, their variations with Δ and κ.

A plot of the temperature variation of the principal and average magnetic moments for different values of Δ, keeping $\kappa = 1.0$, is shown in Fig. 7 (31). The effect of varying Δ is to separate out $\mu_{\parallel}$ and $\mu_{\perp}$ in such a way as to keep the $\bar{\mu}$ almost constant. Thus variation in Δ, although dramatically affecting the anisotropy, has very little effect on the $\bar{\mu}$. For $\lambda = -1000\ cm^{-1}$ (a value relevant for low-spin d^5 ions such as Ru^{3+}), the change in $(\mu_{\parallel} - \mu_{\perp})$ at room temperature in going from $\Delta = -500\ cm^{-1}$ to $-5000\ cm^{-1}$ is about 200%, whereas the corresponding change in $\bar{\mu}$ is only about 3%. The situation is similar when Δ is positive. Here, of course, an orbital singlet lies lowest

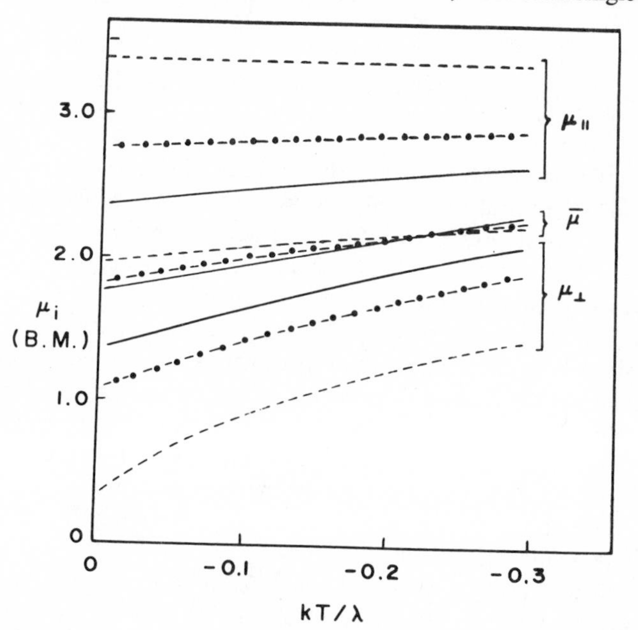

Fig. 7. Temperature dependence of principal and average magnetic moments of the $^2T_{2g}$ term with $\kappa^2 = 1$: ————, $\Delta = -5000\ cm^{-1}$; —·—·—, $\Delta = -1000\ cm^{-1}$; ———, $\Delta = -500\ cm^{-1}$ (31).

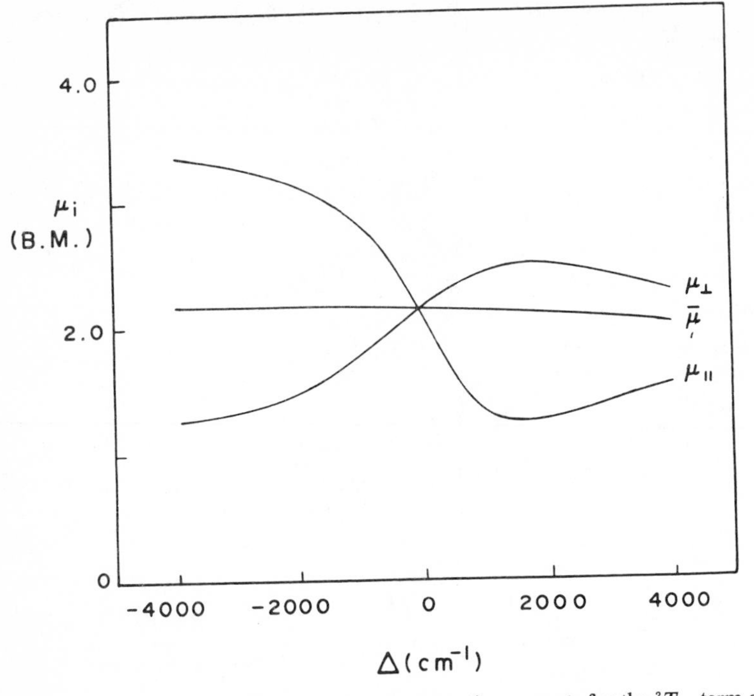

Fig. 8. Variation of principal and average magnetic moments for the $^2T_{2g}$ term at 300°K with $\kappa^2 = 1.0$ (45).

and $\mu_\perp > \mu_\parallel$. The situation is further illustrated in Fig. 8, where the variation of $\mu_\parallel$, $\mu_\perp$, and $\bar{\mu}$ with Δ at room temperature is shown for constant values of λ and κ. These results explain why the average magnetic moment $\bar{\mu}$ is so insensitive to changes in the sign and magnitude of Δ, hence the large errors in the values of Δ deduced from the average magnetic moments that could be incurred (31). An important point is the unambiguous choice of the sign of Δ once the sign of the paramagnetic anisotropy is experimentally determined.

The effect on the principal moments of the variation in κ follows a similar pattern (31) (Fig. 9). For all values of kT/λ and a fixed negative value of Δ, $\mu_\perp$ is very little affected, whereas $\mu_\parallel$ decreases considerably when κ^2 is reduced from 1.0 to 0.5, making the magnetic anisotropy ($\mu_\parallel - \mu_\perp$) quite sensitive to the variations in κ. The reduction in $\bar{\mu}$ is mainly due to changes in $\mu_\parallel$, hence $\bar{\mu}$ remains less sensitive.

An excellent example where all these features have been beautifully illustrated was provided by Gregson and Mitra (31) on tris(acetylacetonato)-ruthenium(III), $Ru(acac)_3$. The compound is low spin with one unpaired electron in the t_{2g}^5 configuration. The average magnetic susceptibility has

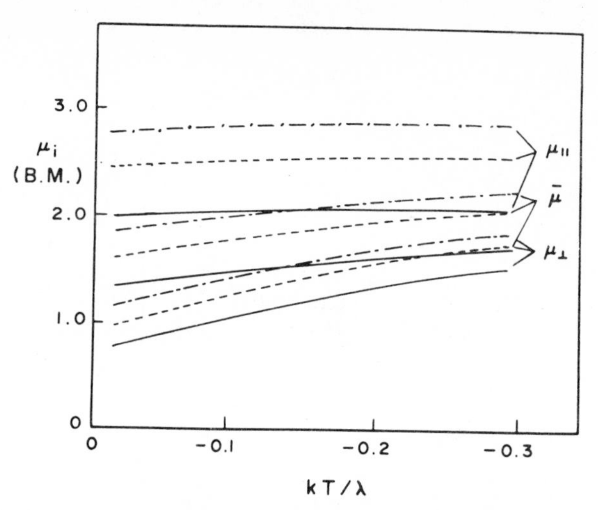

Fig. 9. Temperature dependence of principal and average magnetic moments for the $^2T_{2g}$ term with $\Delta = -1000\ cm^{-1}$: ———, $\kappa^2 = 0.5$; – – – –, $\kappa^2 = 0.7$; —·—·—, $\kappa^2 = 1.0$ (31).

been reported between 77° to 300°K and the results interpreted by fitting the data to the $^2T_2$ model (32). The following set of LF parameters was deduced:

$$\Delta = -5000\ cm^{-1} \qquad \kappa^2 = 0.9 \qquad \text{and} \qquad \lambda = -1200\ cm^{-1}$$

Gregson and Mitra (31) reported the paramagnetic anisotropy of the single crystals of this compound in the same temperature range; their experimental results are shown in Fig. 10. By fitting the temperature dependence of $\mu_{\parallel}$, $\mu_{\perp}$, and $\bar{\mu}$, the following set of LF parameters was deduced which gave the best fit (solid line in Fig. 10):

$$\Delta = -500\ cm^{-1} \qquad \kappa^2 = 0.9 \qquad \text{and} \qquad \lambda = -1200\ cm^{-1}$$

The value of Δ is thus one-tenth of that deduced earlier by Figgis et al (32) from the average magnetic moment. Figure 10 also includes the theoretical curves (dashed curves) using the parameters deduced by Figgis et al. It is observed that while these two widely different values of Δ give good fit with the average magnetic moment, the one deduced from the consideration of the $\bar{\mu}$ alone predicts values of $\mu_{\parallel}$ and $\mu_{\perp}$ which are vastly different from the experimentally observed ones.

The above experimental observation highlights the usefulness of paramagnetic anisotropy measurement in deducing accurate values of the LF parameters. It also illustrates the danger inherent in using the average

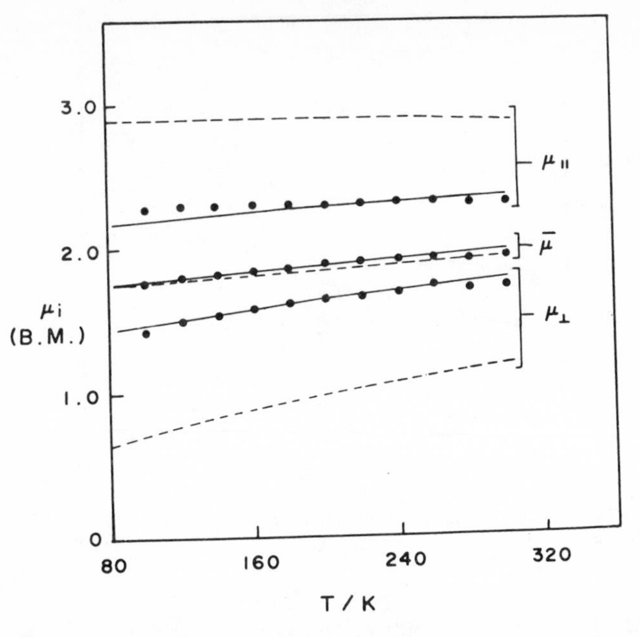

Fig. 10. Molecular moments of $Ru(acac)_3$: ●, experimental results (31). The solid curves are the theoretical ones deduced by Gregson and Mitra (31), and the dashed curves correspond to those deduced from the consideration of average magnetic moments alone by Figgis et al. (32).

magnetic susceptibility alone in deducing these parameters. It is of interest that a single-crystal ESR study (33) on $Ru(acac)_3$ doped in $Al(acac)_3$ gives $\Delta \simeq -700 cm^{-1}$.

$K_3Fe(CN)_6$, another well-known example of $^2T_2$ ground state, provides support to the above conclusion. It had been noticed earlier (26) that the average magnetic susceptibility data on $K_3Fe(CN)_6$ were incapable of deciding unambiguously the sign and magnitude of Δ. A detailed magnetic anisotropy study later (34) elaborated this point and established the sign of Δ. A subsequent calculation (35) based on rhombic LF (as is indeed pointed out by the anisotropy measurement) attempted to correlate the magnetic data with the quadrupole splitting measurements and deduced $\Delta_1 = 85$ cm^{-1} and $\Delta_2 = 150$ cm^{-1}. It is interesting to note that $(\chi_1 - \chi_3)$ shows a maximum at ca. 130°K. It has been predicted that some sort of phase transition occurs at this temperature (35).

It should be appreciated that when λ is small (e.g., Ti^{3+}), the average magnetic moment need not always be insensitive to Δ, and in some favorable cases, it may be possible to deduce reliable values of LF parameters from the average susceptibility data alone. Nevertheless, a single-crystal study, even in such cases, would be desirable.

b. Tetrahedral Complexes. The ground state of the tetrahedrally coordinated d^9 ions (e.g., Cu^{2+}) is ${}^2T_2$, but the excited, ${}^2E({}^2D)$ state lies much closer because of the smaller value of D_q (cf. Fig. 6*b*). The effect of this ${}^2E$ state can not, therefore, be neglected and the magnetic properties of tetrahedral Cu(II) complexes should be analyzed in terms of a complete ${}^2D$ model.

We have described earlier, based on the ${}^2T_2$ model, the variation of principal magnetic moments with temperature and Δ. The results show that the sign of Δ is uniquely and directly determined by the sign of the magnetic anisotropy (cf. Fig. 7 and 8). In Fig. 11, the variation of $(\mu_{\parallel} - \mu_{\perp})$ with Δ/λ is shown for the complete ${}^2D$ term, including the effect of the excited ${}^2E$ state (36). The result is quite interesting as it shows that the sign of the magnetic anisotropy is now not directly related to the sign of Δ, as it was in the ${}^2T_2$ model (cf. Fig. 8). Over a considerable range of both positive and negative vaules of Δ, $\mu_{\parallel}$ is found to be larger than $\mu_{\perp}$. The range over which $u_{\perp} > \mu_{\parallel}$ is extremely limited, suggesting it to be an unlikely possibility. The existing experimental data on tetrahedral Cu(II) complexes show that $\mu_{\parallel} > \mu_{\perp}$, in all cases, thus supporting the above observation.

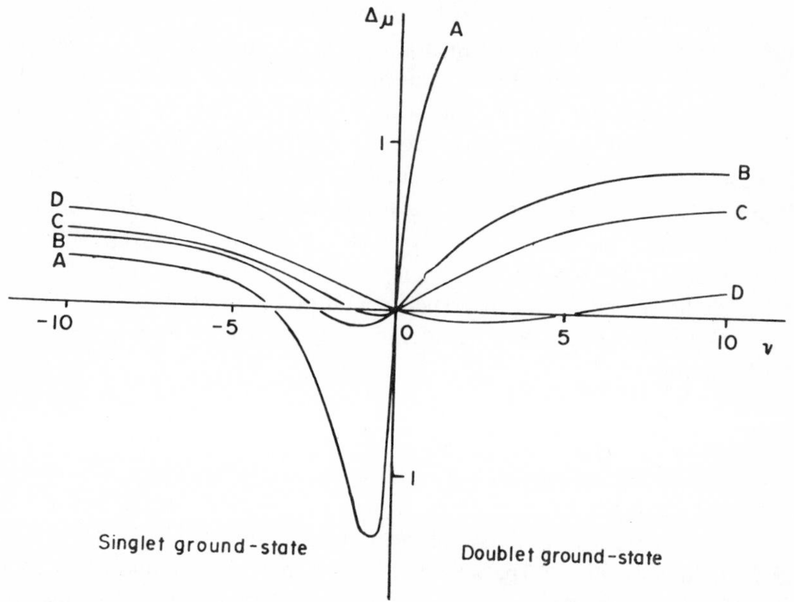

Fig. 11. The variation of the anisotropy, $\Delta\mu = (\mu_{\parallel} - \mu_{\perp})$, with $\nu\,(= \Delta/\lambda)$ for the axially distorted ${}^2D$ term when $\kappa = 1.0$ (36): kT/λ is (A), -0.1; (B), -0.5; (C), -1.0; (D), -2.0.

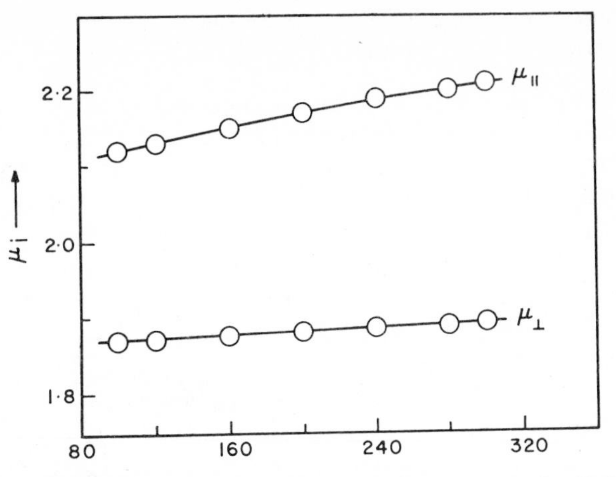

Fig. 12. Temperature dependence of the principal magnetic moments for Cs_2CuCl_4. The circles are the experimental data (16), and the solid lines are theoretical ones based on a ^2D model.

Some of these features can be illustrated by taking the example of the well-known tetrahedral copper(II) complex, Cs_2CuCl_4. The magnetic anisotropy has been reported (37–39) between 300° to 90°K, and the results are summarized in Fig. 12. Experimentally, $\mu_{\parallel} > \mu_{\perp}$ in accord with the above expectation. The temperature variation of anisotropy clearly shows that Δ/λ must be negative, that is, the orbital siglet $^2B_2$ lying below $^2E(^2T)$ (see Fig. 11). A fitting (38, 39) of the experimental data to the ^2D model gives $\Delta = +5000\ \text{cm}^{-1}$ and $\Delta' \simeq 100\ \text{cm}^{-1}$ (cf. Fig. 6*b*). These values are in agreement with the single-crystal polarized spectral studies (40). A point of interest is that the fitting of temperature-variation of average-magnetic-moment data of Cs_2CuCl_4 to the $^2T_2$ model would give (39) $\Delta = +5000\ \text{cm}^{-1}$, but it predicts, contrary to the experimental observation, $\mu_{\perp} > \mu_{\parallel}$. Another interesting point is the large reduction in κ, (even anisotropic) deduced to fit the data.

2. $^5T_2$ *Ground State*

Octahedrally coordinated iron(II) complexes are perhaps the best documented examples of complexes with a $^5T_2$ ground state. They provide another good example to illustrate the usefulness of paramagnetic anisotropy study to deduce the LF parameters. This is particularly important in the case of iron(II) complexes in view of the virtual absence of ESR studies.

The free-ion ground term for an iron(II) ion is ^5D, which splits in the cubic part of the LF into $^5T_2$ and ^5E, the former lying about $10{,}000\ \text{cm}^{-1}$ below the latter. The energy level diagram is thus similar to Fig. 6*b*. Because

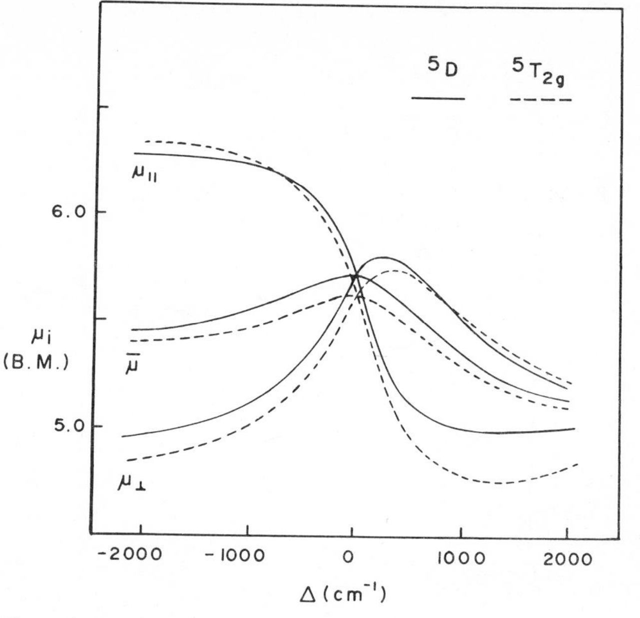

Fig. 13. The variation of principal and average magnetic moments at 300°K on $^5T_{2g}$ and ^5D models (47).

of the large separation and a weak spin-orbit coupling parameter of the iron-(II) ion ($\lambda \simeq -100$ cm^{-1}), the effect of the ^5E term is small and usually neglected. The magnetic properties are thus relationalized on the basis of the $^5T_2$ model with axial nature of LF. The convention here for the sign of Δ remains the same as that in $^2T_2$.

The theoretical behavior of the principal molecular susceptibilities on the $^5T_2$ model has been very extensively investigated (41–44). A more complete calculation on the ^5D model is also now available (45–47). In Fig. 13 we have plotted the variation of $\mu_\parallel$, $\mu_\perp$, and $\bar{\mu}$ at room temperature, with Δ based on both the $^5T_2$ and ^5D models. The figure shows, as in the $^2T_2$ model but unlike the ^2D one, that once the sign of molecular anisotropy is known from the experiment, the sign of Δ is fixed. On the other hand, the average magnetic moment remains insensitive to Δ and cannot be used to decide its sign. It is also observed that the effect of including ^5E (^5D) term is, as expected, small, except on $\mu_\parallel$ when Δ is large and positive. Our further discussion is therefore confined to the $^5T_2$ model.

We consider next the temperature variation of $\mu_\parallel$, $\mu_\perp$, and $\bar{\mu}$ for both the positive and negative values of Δ. This is shown in Figs. 14 and 15 (44, 45). The trend is similar to that observed in the case of $^2T_2$, the paramagnetic anisotropy being very sensitive to even small changes in Δ and $\bar{\mu}$ being

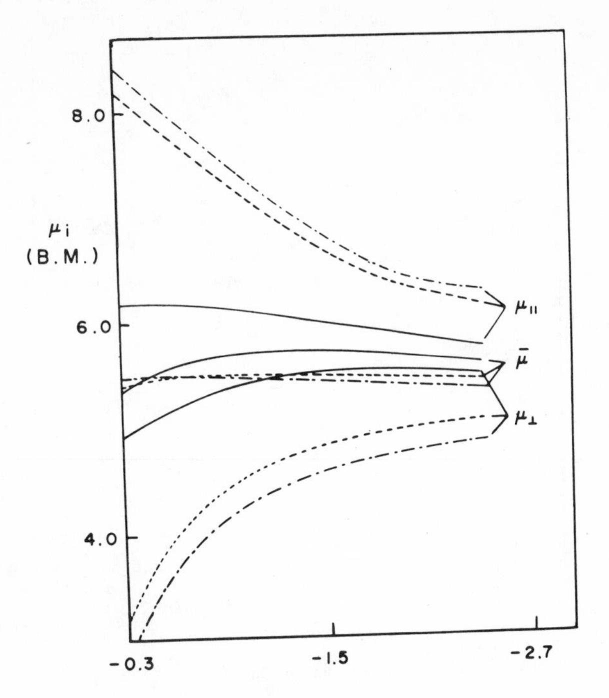

Fig. 14. Theoretical magnetic moments for the $^5T_{2g}$ term with $\kappa^2 = 1.0$: ———, $\Delta = -100$ cm^{-1}; — — —, $\Delta = -1000$ cm^{-1}; —·—·—, $\Delta = -3000$ cm^{-1} (44).

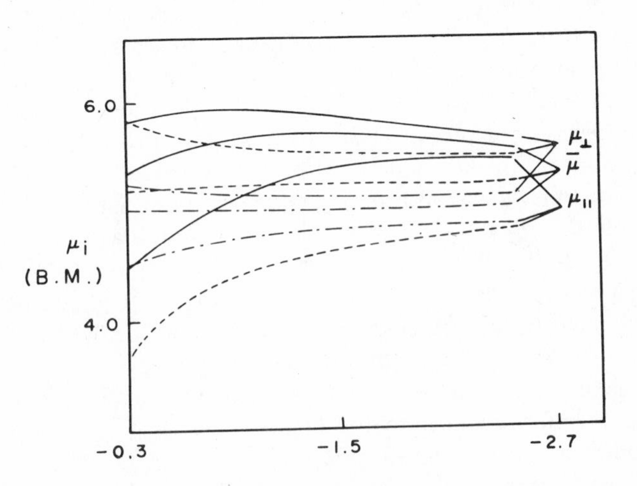

Fig. 15. Theoretical magnetic moments for the $^5T_{2g}$ term with $\kappa^2 = 1.0$ for positive values values of Δ: ———, $\Delta = 100$ cm^{-1}; — — —, $\Delta = 1000$ cm^{-1}; —·—·—, $\Delta = 3000$ cm^{-1} (44, 45).

rather insensitive. For example, the change in $\bar{\mu}$ at room temperature with $\lambda = -100\ cm^{-1}$ in going from $\Delta = -100\ cm^{-1}$ to $\Delta = -1000\ cm^{-1}$ is only about 3%, whereas the corresponding change in $\Delta\mu$ is about 100%. Reference to Fig. 15 for positive values of Δ shows that the situation is similar in relation to $\Delta\mu$, but in this case the temperature variation of $\bar{\mu}$ is not so insensitiv to Δ, even in the 300° to 80°K temperature range. The effect of variation in the orbital reduction parameter has been shown to be similar (44). The $\bar{\mu}$ seems to be more insensitive to κ than was observed in the $^2T_2$ ground state.

We conclude, as in the case of $^2T_2$ state, that the paramagnetic anisotropy measurement provides an accurate method to deduce the LF parameters, while an attempt to deduce these parameters from the average magnetic moments could be quite misleading and erroneous. We describe below two representative examples and discuss the experimental features.

a. Ferrous Fluosilicate Hexahydrate. The principal molecular susceptibility of this uniaxial crystal has been reported between 1.5° to 300°K (48, 49), and direct measurement of paramagnetic anisotropy was between 80° to 300° K(45, 46, 50). The results of these measurements are summarized in Fig. 16. The experimental data show that $\mu_\perp > \mu_\parallel$. This determines the sign of Δ to be positive. There is, however, no agreement on the exact values of Δ, κ, and λ deduced. Values of Δ between 750 to 1200 cm^{-1}, λ values between -80 to $-100\ cm^{-1}$, and κ^2 between 0.7 to 1.0 have been arrived at

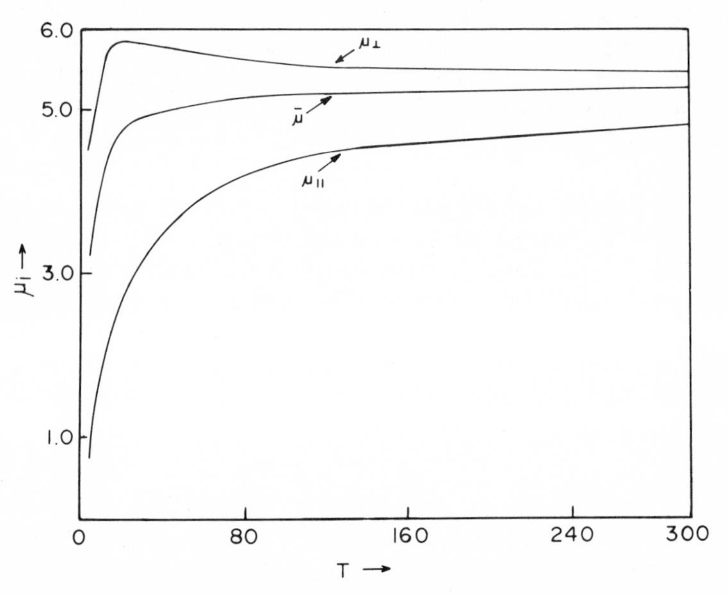

Fig. 16. Temperature dependence of principal and average magnetic moments for $FeSiF_6 \cdot 6H_2O$

by considering the average and principal moments. The temperature dependence of quadrupole splitting has been fitted to $\Delta = 760\ \mathrm{cm}^{-1}$, $\lambda = -80\ \mathrm{cm}^{-1}$, and $\kappa^2 = 0.8$ (51). It has also been deduced that Δ is temperature dependent. It should be appreciated that the paramagnetic anisotropy data at different temperatures can only determine two parameters uniquely. Since the $\bar{\mu}$ is rather insensitive to these parameters, a unique choice of all three parameters becomes difficult. However, considering the temperature dependence of quadrupole splitting and the principal and average magnetic moments of $FeSiF_6 \cdot 6H_2O$, Gregson (45) has deduced the following: $\Delta = 1000\ \mathrm{cm}^{-1}$, $\lambda = -95\ \mathrm{cm}^{-1}$, and $\kappa^2 = 0.95$, based on $^5T_2$ and the more complete ^5D model. It should be noted that the excited ^5E remains unsplit as the symmetry of the LF in $FeSiF_6 \cdot 6H_2O$ is trigonal.

b. Ferrous Ammonium Sulfate Hexahydrate. The magnetic properties of this well-known member of the ferrous Tutton salts (abbreviated here to FASH) have been again very extensively studied. Measurements of average susceptibility extend down to about 1°K. Paramagnetic anisotropies were reported by a large number of workers down to 80°K (42, 52, 53). The first theoretical study by Bose et al. (54) assumed $K_{\parallel} > K_{\perp}$ (i.e., negative Δ); they deduced that Δ was temperature dependent. Later studies by Konig and Chakravarty (42) also assumed $K_{\parallel} > K_{\perp}$ for a theoretical interpretation of the data. Average susceptibility data were, however, interpreted by Figgis et al. (54*a*) in terms of a positive Δ, which implied that $K_{\perp} > K_{\parallel}$. The discrepancy was removed later by Gregson and Mitra (47) who determined the sign of moecular anisotropy of FASH with the help of then available x-ray structural data (55). They established $K_{\perp} > K_{\parallel}$ in this crystal so that Δ was positive (cf. Fig. 13). The temperature dependence of the principal moments can be explained (47) with $\Delta = 300\ \mathrm{cm}^{-1}$, $\lambda = -90\ \mathrm{cm}^{-1}$, and $\kappa = 0.9$.

In all the above analyses an axial symmetry was assumed. It was noted, however, that the temperature dependence of quadrupole splitting cannot be explained on the basis of this assumption and a rhombic symmetry must be considered (56). A calculation of K_x, K_y and K_z, with the help of available structural data, shows (12, 56) a large rhombic component, although a reasonable explanation of the magnetic data is possible, even on the assumption of axial symmetry (47).

3. $^3T_1$ Ground State

Both octahedral trivalent vandium and tetrahedral divalent nickel ions have a $^3T_1$ ground state in the cubic LF. The energy level diagram appropriate to both these situations is shown in Fig. 17. An interesting point is that S—O interaction leaves a nonmagnetic level lowest with a spin doublet

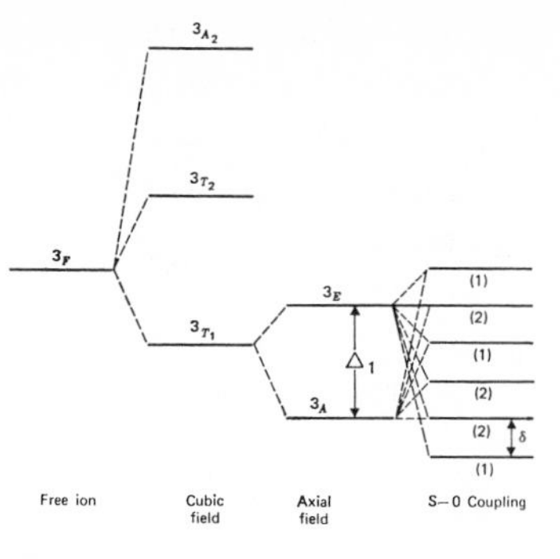

Fig. 17. Energy level diagram for octahedrally coordinated vanadium(III). The values in parantheses represent the total degeneracy.

above it. There has been interest, in particular, in the determination of this splitting (δ).

Results of magnetic anisotropy measurements on the octahedrally coordinated potassium tris(oxalato)vanadate trihydrate have been reported (57) and explained on the basis of the energy level diagram of Fig. 17. The anisotropy at room temperature is very small (1.5%) and increases rapidly at lower temperatures. The feeble anisotropy implies that Δ_1 is positive and large. A value of $\Delta_1 \simeq 900$ cm^{-1} has been deduced (57). δ is found to be about 8 cm^{-1}.

A single-crystal magnetic susceptibility study by McElearney et al. (58) on guanidinium vanadium(III) sulfate hexahydrate highlighted the advantage of single-crystal measurement over the average data, even in the liquid helium temperature range. The single-crystal susceptibility data in the temperature range 1.5° to 20°K gave $\delta = 3.8$ cm^{-1}, which is about half the value deduced from the average susceptibility data in the same temperature range. (59). The value of δ obtained from the ESR studies matches with that deduced from the single-crystal susceptibility data.

A very detailed analysis of the magnetic properties of tetrahedral nickel(II) complexes has been reported by Gerloch and Slade (60). Some limited experimental data are also reported elsewhere (61). In all cases, the anisotropies are much larger than the corresponding vandium(III) complexes, suggesting small values of Δ_1. For tetraethylammonium tetrachloronickelate (II) and bis(*N*-isopropylsalicylaldiminato)nickel(II) complexes, Δ_1 is deduced to be 160 and 380 cm^{-1}. The former complex shows a very interesting phase transition in the anisotropy around 220°K.

The foregoing analysis of the magnetic properties of these orbital triplet ground-state complexes has demonstrated the advantages and superiority of the single-crystal magnetic studies. We have shown the inadequacy of the average susceptibility data for deducing accurate values of the LF parameters and the reliability of the single-crystal magnetic studies for this purpose. We have also pointed out the difficulties in the analysis of the magnetic anisotropy data which often arise in cases where the convenient assumption of axial symmetry may not be valid.

C. Electronic Structure of Square Planar and Square Pyramidal Metal Complexes

As mentioned in Section III, the electronic ground state of a transition metal ion in a perfect octahedral or tetrahedral LF is known *a priori*. This is not, however, the case with square planar and square pyramidal complexes. These complexes can be viewed as special cases of an octahedral arrangement with very large tetragonal distortion. In such cases, the relative ordering of the *d* orbitals becomes uncertain. When the *d* electrons are fed into these orbitals, depending on the relative strength of interelectronic repulsion and the LF, different spin situations and electronic ground states are possible. Even for the same electronic $3d^n$ configuration and similar molecular geometry, different spin situations occur. An example is provided by the ferric ion in the square pyramidal iron(III) hemins and halobis(dithiocarbamato)iron(III), where the former has a spin state of $S = 5/2$ and the latter has $S = 3/2$. For a particular spin state there are again different possibilities for the ground and excited states which have to be determined by experimentation.

Measurement of average susceptibility determines the spin state of the metal ion, but it does not, in general, give any information regarding the nature and position of ground and excited states. Optical spectroscopy has been very extensively applied for this purpose and has been successful in several cases. However, the weak *d-d* bands are often masked by strong charge-transfer bands, thus identification and assignment of these weak *d-d* transitions become difficult. ESR and several other techniques have now been applied to understand the electronic structure of the metal ions, and each method has its merits and limitations.

Recently, paramagnetic anisotropy data was utilized to deduce the electronic structure of a series of square planar and square pyramidal complexes. In general, the success achieved is quite encouraging; in some cases it is outstanding. An important point, which will become evident later in this section, is the superiority of this technique over the other existing ones in understanding the electronic structure of this class of low-symmetry complexes.

1. Square Planar Complexes

Square planar geometry is not very common for transition metal ions, in general, and only a few examples are known which are paramagnetic. Tran-

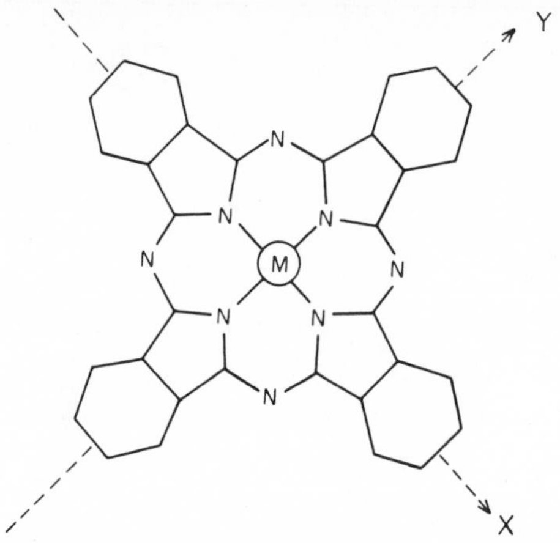

Fig. 18. Molecular geometry of the planar divalent metal phthalocyanines. The unlabeled atoms are the carbon atoms. The hydrogen atoms are not shown.

sition metal phthalocyanines (MPc's) provide the best known example in which a number of metal ions achieve almost a perfect square planar geometry. In Fig. 18 the molecular geometry of the planar metal phthalocyanine is shown. A detailed measurement of paramagnetic anisotropy on a number of MPc's with d^5, d^6, d^7, and d^9 electronic configurations has recently been reported. Measurements have also been reported on other square planar, d^7 and d^9 complexes. We discuss below examples of each electron configuration separately.

a. d^5 (Mn^{2+}). Manganese(II) phthalocyanine provides the only established example of a square planar d^5 ion. Measurements of paramagnetic anisotropy have been reported between 300° to 90°K and the average suscep tibility between 300° to 1°K(62). The average magnetic moment of 4.3 B.M. and its temperature dependence in the 300° to 90°K range establish that the Mn(II) ion in MnPc has an intermediate spin state of $S = 3/2$. There are several possibilities in which the five d electrons can be arranged among the d orbitals to give $S = 3/2$. The average magnetic data does not, however, help in deciding between these possibilities and the ground state.

Measurement of paramagnetic anisotropy shows that MnPc is highly anisotropic and $K_\perp > K_\parallel$. These results have been interpreted (62) to indicate an electron configuration $(d_{xy})^2\,(d_{xz}, d_{yz})^2\,(d_{z^2})^1\,(d_{x^2-y^2})^0$, giving $^4A_2$ as the ground state. An alternative possibility $(d_{xy})^1\,(d_{xz}, d_{yz})^3\,(d_{z^2})^1\,(d_{x^2-y^2})^0$, giving ^4E as the ground state, can be discarded as it predicts $K_\parallel > K_\perp$. The temperature dependence of $\mu_\parallel$, $\mu_\perp$, and $\bar{\mu}$ is shown in Fig. 19.

We now describe an LF interpretation (63) of the experimental data given in Fig. 19 and discuss how the magnetic anisotropy measurements on

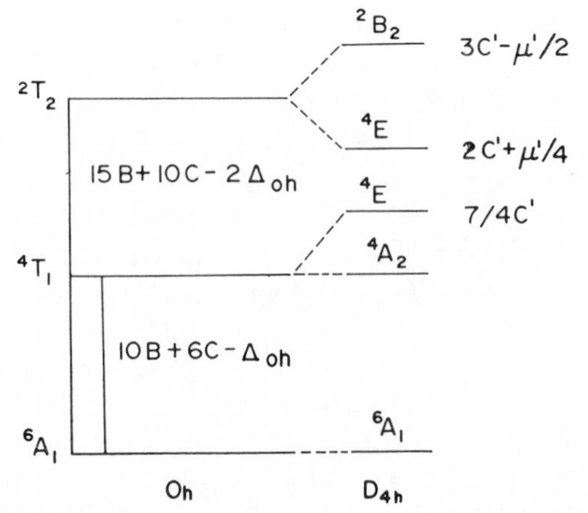

Fig 19. Temperature dependence of the principal and average magnetic moments for manganese(II) phthalocyanine. Circles are the experimental data taken from Ref. 62. The solid lines are theoretical curves taken from Ref. 63.

this molecule help us in understanding the details of its electronic structure. A d^5 ion such as Mn(II) possesses a total of 43 high-, low-, and intermediate-spin multiplets in a crystal field of cubic symmetry. However, Harris (64) has shown that only the lowest 24 states ${}^2T_{2g}$ (t_{2g}^5) and the ${}^4T_{1g}$ (t_{2g}^4 e_g^1) and

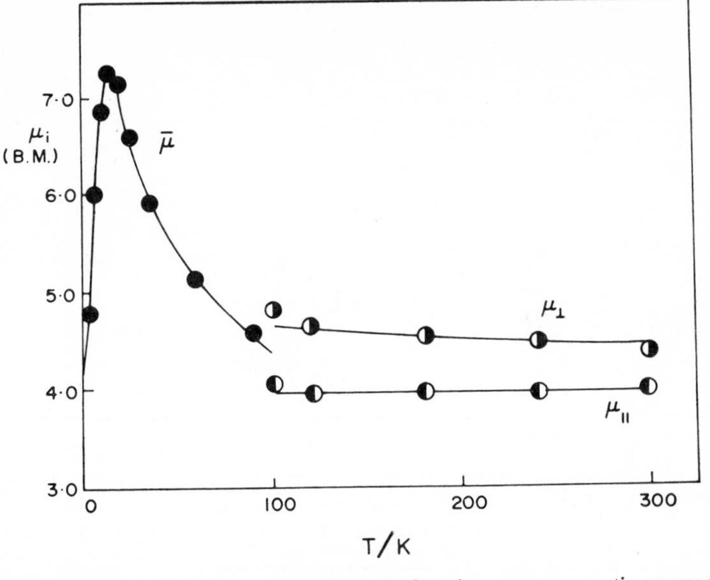

Fig. 20. The first-order electrostatic and crystal-field interactions for the sextet, quartet, and doublet states of the d^5 configuration.

$^6A_1(t_{2g}^3\ e_g^2)$ strong field terms mix significantly via S–O coupling. In an LF of strong tetragonal symmetry, the cubic field $^2T_{2g}$ term splits into $^2B_2$ and ^2E, whereas the $^4T_1$ term splits into ^4E and $^4A_2$, the $^6A_1$ term being orbitally nondegenerate. The energies of the resulting five electronic states are given in Fig. 20. Here, Δ_{oh} is the cubic crystal field parameter, and C' and μ' relate to the differences between axial and in-plane field strengths (64). In Fig. 21, regions of parameter space for an Mn(II) ion have been delineated. It is observed that a quartet ground state occurs for $\Delta_{oh} > 28{,}000\ \text{cm}^{-1}$ and $C' > 7500\ \text{cm}^{-1}$. Furthermore, for lower values of Δ_{oh} and larger value of C', the zero-field split $^4A_2(\pm\ 1/2)$ becomes the ground state. Taking the values of C' and Δ_{oh} on which $^4A_2$ lies lowest, the results of the calculation of

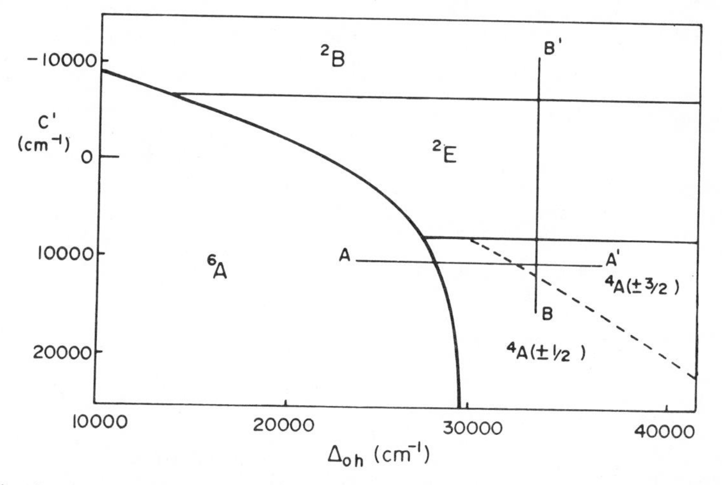

Fig. 21. Regions of parameter space defining ground states for the d^5 manganous ion, calculated with the free-ion values for B, C, and ζ (63).

principal magnetic moments are shown in Figs. 22 and 23. The figures show that the principal moments depend very markedly on the ground state, and even the sign of anisotropy in the magnetic moment $(\mu_\perp - \mu_\parallel)$ changes depending on the sign of the ZFS of $^4A_2$. Since in MnPc $\mu_\perp > \mu_\parallel$, $^4A_2(\pm\ 1/2)$ must lie lowest. A detailed fitting (63) showed that $^4A_2(\pm\ 3/2)$ lies about $40\ \text{cm}^{-1}$ above $^4A_2\ (\pm\ 1/2)$. The next excited state was deduced to be $^6A_1$ $(\pm\ 5/2)$ at about $600\ \text{cm}^{-1}$. ^4E lies much higher in energy. The paramagnetic anisotropy studies on MnPc have therefore been most useful in deciding the sign and magnitude of the ground state ZFS, and in determining the nature and position of the excited $^6A_1$ and ^4E states and even the sign of the ZFS of the excited $^6A_1$ state. MnPc shows an interesting ferromagnetic interaction at lower temperatures which will be discussed later.

Marathe and Mitra (65) have pointed out that the above model of Har-

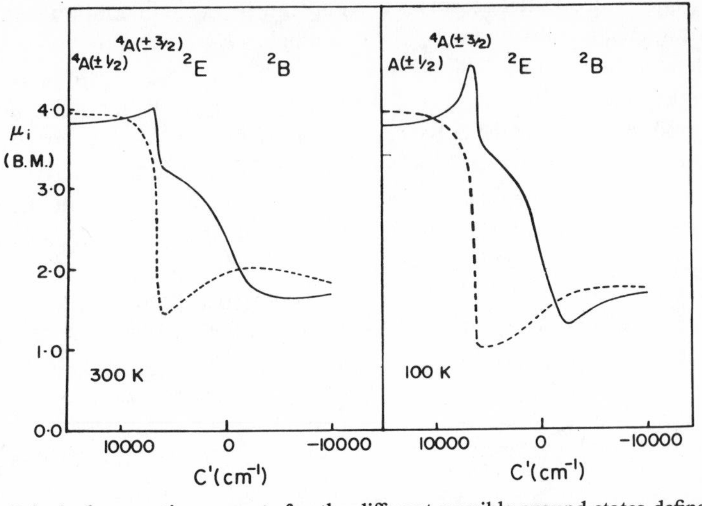

Fig. 22. Principal magnetic moments for the different possible ground states defined by section BB' of Fig. 21: ———, $\mu_{\parallel}$; — — — —, $\mu_{\perp}$ (63).

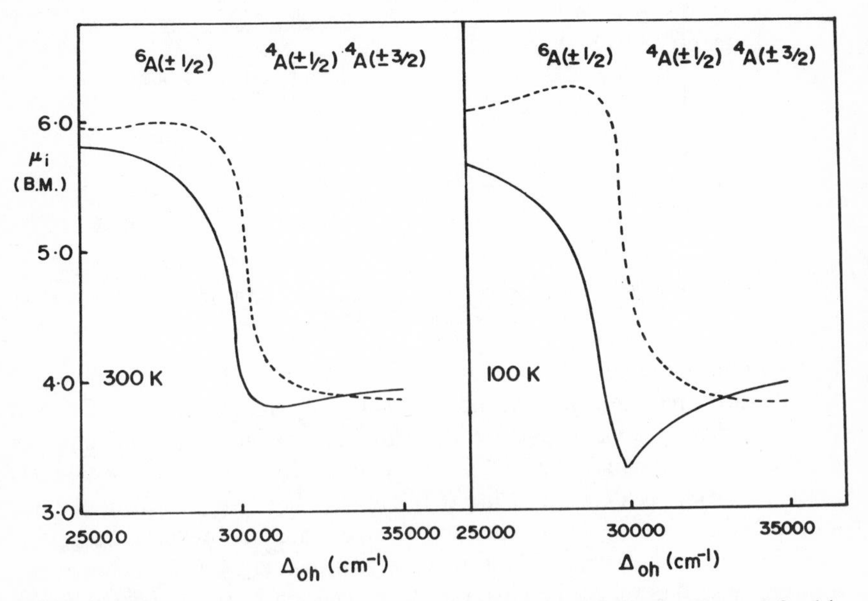

Fig. 23. Principal magnetic moments for the different possible ground states defined by AA' of Fig. 21: ———, $\mu_{\parallel}$; — — — —, $\mu_{\perp}$ (63).

ris suffers from a serious drawback. While the calculation in Harris's model is done in D_{4h} symmetry, the wave functions used for the calculation are those of cubic field. The effect of this approximation is particularly significant in choosing the wave function of ^4E. Using the correct wave functions, Marathe and Mitra obtained the following expressions of ZFS (δ) and $g_\perp$:

$$\delta = \zeta^2\left[\frac{1}{9E_1} + \frac{1}{5E_2} - \frac{1}{2E_3}\right]$$

$$g_\perp = 2\left(1 + \frac{\kappa\zeta}{3E_1}\right) \tag{25}$$

which can be compared with Harris's expression

$$\delta = \zeta^2\left[\frac{1}{36E_1} + \frac{1}{5E_2} - \frac{1}{2E_3}\right]$$

$$g_\perp = 2\left(1 + \frac{\kappa\zeta}{12E_1}\right) \tag{26}$$

E_1, E_2, and E_3 are the energies of excited ^4E, ^6A, and ^2E states. The above model does not significantly alter the LF parameters in MnPc since the contribution to ZFS in this molecule comes mainly from $^6A_1$, which contributes equally in both the models. However, we will see the large effect of the latter correction in some other spin $S = 3/2$ cases discussed later in this section.

b. (d^6 Fe^{2+}). Iron(II) phthalocyanine provides a unique example of a paramagnetic d^6 ion in a square planar environment. The electronic structure of this molecule has been the subject of discussion for a long time, as its room temperature, average magnetic moment of 3.96 B.M. lies between the values expected for the $S = 1$ and $S = 2$ spin states. Recently, detailed measurements of average susceptibility between 300° to 1°K, paramagnetic anisotropy between 300° to 90°K, and magnetization between 100 Oe to 15000 Oe at 1.4°K have been reported (66, 67). Extensive Mössbauer and other measurements are also available. The results of these magnetic measurements, summarized in Fig. 24, establish that the spin state of the ferrous ion in FePc is $S = 1$. The large deviation in the $\bar{\mu}$ at room temperature from the spin-only value of 2.83 B.M. can be thought to be qualitatively due to a large orbital contribution. Furthermore, the rapid decrease in $\bar{\mu}$ at lower temperatures suggests a large ZFS of the $S = 1$ state, with probably Ms = 0 lying below Ms = ± 1 level.

The paramagnetic anisotropy data on FePc suggest an electron configuration $(d_{xy})^2\ (d_{xz}, d_{yz})^3\ (d_{z^2})^1\ (d_{x^2-y^2})^0$. When electron repulsion terms are taken into account, this configuration gives $^3A_2$ as the ground state (the previous deduction of $^3B_2$ ground state is in error, although anisotropy data cannot distinguish between these two possibilities). Spin-oribt interaction with some excited states splits the ground term into Ms = 0 and Ms = ± 1, with the ZFS separation given by D. Neglecting the TIP contributions (66, 67), we get for the ZF split $^3A_2$ state

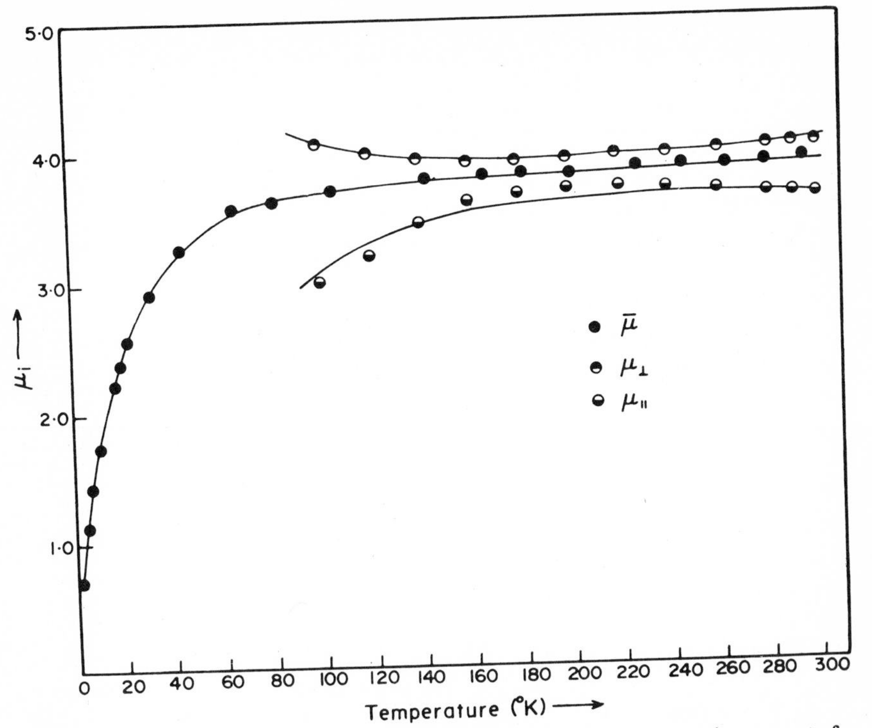

Fig. 24. Temperature dependence of the principal and average magnetic moments for ion(II) phthalocyanines. Circles are the experimental data (66), and solid curves are the theoretical ones calculated with $g_\parallel = g_\perp = 2.74$ and $D = 64$ cm^{-1} (see text).

$$K_\parallel = \left(\frac{2N\beta^2}{kT}\right) g_\parallel^2 (e^d + 2)^{-1} \tag{27}$$

$$K_\perp = \left(\frac{2N\beta^2}{D}\right) g_\perp^2 \left[\frac{(e^d - 1)}{(e^d + 2)}\right] \tag{28}$$

where $d = D/kT$.
At $T \to 0$, $K_\parallel \to 0$, and Eq. 28 reduces to

$$K_\perp = \left(\frac{2N\beta^2}{D}\right) g_\perp^2 \tag{29}$$

Hence,

$$\bar{\chi}_{T=0} = \left(\frac{4N\beta^2}{3D}\right) g_\perp^2 \tag{30}$$

Equation 30 requires that $\bar{\chi}$ be temperature independent at very low temperatures. This is indeed observed below 10°K at $\bar{\chi} = 0.040$ cm^3 mole^{-1}. Substituting the value of $\bar{\chi}$ in Eq. 30 we get

$$g_{\perp}^2 = 0.1173\,D \tag{31}$$

Substituting the value of $g_{\perp}$ in Eq. 28, we get

$$K_{\perp} = 2 \times 0.1173N\,\beta^2 \left[\frac{(e^d - 1)}{(e^d + 2)}\right] \tag{32}$$

Equation 32 can be tested against the experimental results of $K_{\perp}$ between 300° to 100°K. A constant value of $D = 64\ \text{cm}^{-1}$ is obtained, which from Eq. 31 gives $g_{\perp} = 2.74$. Substituting the value of D and $K_{\parallel}$ in Eq. 27, we obtain $g_{\parallel} = 2.74$. In Fig. 24, the theoretical curves are shown for $g_{\parallel} = g_{\perp} = 2.74$ and $D = 64\ \text{cm}^{-1}$.

The above analysis of paramagnetic anisotropy data apparently provides a very unique and simple description of the ligand field and electronic structure of FePc. However, the situation is far from satisfactory. The origin of the large ZFS and high g values remains unknown. There must be several low-lying states (some even within kT) which mix with the ground state to produce the large values of D and g_i, but no account has been taken in the above model of these states. It is interesting to note here that, unlike in other complexes, even the analysis of low-temperature, average magnetic moment data on FePc was adequate to give a reasonable account of its electronic structure (67). This is because of the very large component of orbital contribution in $\bar{\mu}$.

c. d^7 (Co^{2+}). Square planar geometry for cobaltous complexes is not very uncommon, and measurement of paramagnetic anisotropy has been reported on a few of them. The average susceptibility data show that they are all low spin ($S = \frac{1}{2}$), the magnetic moment lying between 2.2 to 2.6 B.M., with very little dependence on temperature. The average data are, however, of no avail in understanding the electronic structure of these interesting molecules beyond establishing the spin state. The measurement of paramagnetic anisotropy on the planar cobaltous complexes has, however, been most successful in revealing the finer details of their electronic structures, and it has led to some very important results of general consequence. We now discuss the magnetic properties of these complexes.

(1) Cobalt(II) Phthalocyanine. Paramagnetic anisotropy of CoPc has been reported by Martin and Mitra (68) between 300° to 90°K, and the temperature dependence of $\mu_{\parallel}$ and $\mu_{\perp}$ is shown in Fig. 25. The figure shows that $\mu_{\perp} > \mu_{\parallel}$, and while $\mu_{\parallel}$ is independent of temperature, $\mu_{\perp}$ decreases with it. The d^7 electron configuration of the Co(II) ion in CoPc can be assumed as three holes in the d^{10} configuration, with the $(d_{x^2-y^2})$ being doubly occupied. The unpaired hole can reside in any of the remaining orbitals: (d_{xz}, d_{yz}), d_{xy}, and d_{z^2}. If the hole lies in the (d_{xz}, d_{yz}) orbital, we get $\mu_{\parallel} > \mu_{\perp}$, which is contrary to experimental results. If the hole is in the (d_{xy}) orbital, then

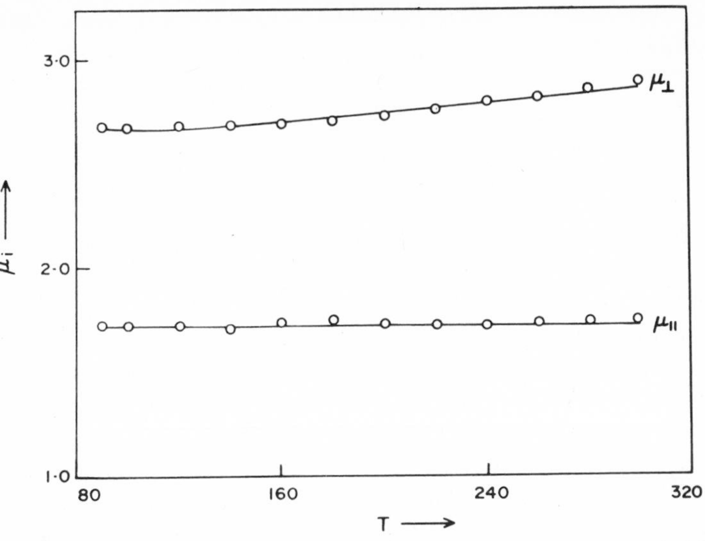

Fig. 25. Temperature dependence of the principal magnetic moments for cobalt(II) phthalocyanine. Circles are experimental points, and solid curves are the theoretical plots of Eq. 33, using $\Delta E = 2400\ \text{cm}^{-1}$ and $\zeta = 400\ \text{cm}^{-1}$ (68).

$\mu_\parallel$ and $\mu_\perp$ are expected to vary appreciably with temperature, which again is not experimentally observed. If the hole lies in the (d_{z^2}) orbital, then by a first-order calculation (68) in D_{4h} symmetry,

$$K_\parallel = \frac{N\beta^2}{kT}$$

$$K_\perp = \frac{N\beta^2}{4kT}\left[2 + \frac{6\zeta}{\Delta E}\right]^2 + \frac{3N\beta^2}{\Delta E - \zeta/2} + \frac{3N\beta^2}{\Delta E + \zeta/2} \tag{33}$$

where ΔE is the energy separation between (d_{z^2}) and (d_{xz}, d_{yz}) orbitals. Equation 33 shows that $\mu_\perp > \mu_\parallel$, and while $\mu_\parallel$ is independent of temperature, $\mu_\perp$ should decrease slightly with the decrease in temperature. This is exactly what is experimentally observed. The experimental data were fitted to $\Delta E = 2400\ \text{cm}^{-1}$ and $\zeta = 400\ \text{cm}^{-1}$ (solid line in Fig. 25). Thus the paramagnetic anisotropy data enable us to decide the ground state of the CoPc and the position of the excited state. It is interesting that the analysis of the ESR data also led to the same conclusion about the CoPc.

(2) Bis(dithioacetylacetonato)cobalt(II) and Trans-dimethyl-bis(diethylphenylphosphine)cobalt(II). The magnetic behaviors of the single crystals of these two planar cobaltous compounds are very similar. We therefore confine our discussion mostly to the former one, abbreviated to Co $(SacSac)_2$, and quote the results on the latter.

The molecular geometry of $Co(SacSac)_2$ is shown in Fig. 26. The molecule is strictly planar, the Co–S bond distances are all equal, and S–Co–S angles deviate only by 2° from the ideal 90°. The x and y axes in both molecules lie in the plane of the molecule as shown, and the Z axis is taken perpendicular to the molecular plane. In $CoM_2(PEt_2Ph)_2$, the cobalt atom is bonded to two carbon and two phosphorus atoms in a planar geometry.

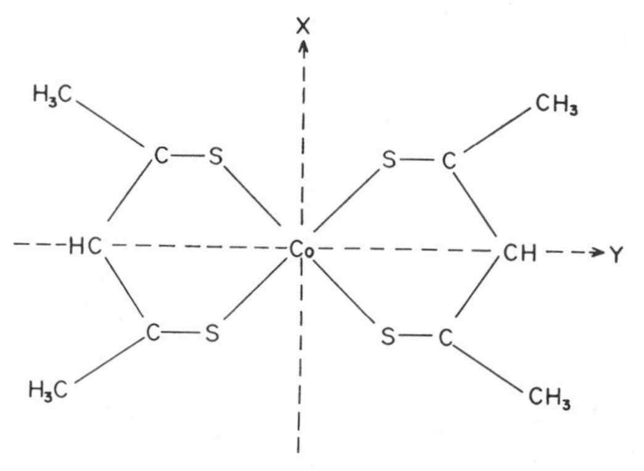

Fig. 26. Molecular geometry of dithioacetylacetonato cobalt(II), $Co(SacSac)_2$. The X and Y axes lie in the plane of the molecule, with the Z axis perpendicular to it.

Measurement of paramagnetic anisotropy on both these molecules has been reported (69–71) between 300° to 90°K, and the results at two extreme end temperatures are summaried in Table IV. The table illustrates the outstanding point that the magnetic properties of these two planar coabltous pounds are quite different from those of CoPc. While CoPc shows no in-plane anisotropy, these two compounds show an abnormally large in-plane anisotropy. This alone suggests that the detailed electronic structure of these planar cobalt(II) compounds must be very different from that of CoPc. A noteworthy feature is that almost the entire molecular anisotropy of $Co(SacSac)_2$ and $CoM_2(PEt_2Ph)_2$, especially in the former one, is reflected in the plane of the molecule, which is geometrically fairly symmetric. This indicates that there is perhaps no simple correspondence between the symmetric nature of the molecular geometry and magnetic properties. The temperature dependence of the principal moments of $Co(SacSac)_2$ (69, 70) is shown in Fig. 27.

An interpretation of the magnetic properties of these two molecules has been attempted on the crystal-field model. Following the arguments on CoPc in D_{2h} symmetry, the unpaired hole can reside in any of the four orbitals: $(d_{x^2-y^2})$, (d_{xz}), (d_{yz}), and (d_{z^2}). Choosing each alternative in turn and comparing the theoretical predictions with the experimental results, Gregson et al. (69, 70) came to the conclusion that the experimental data of Co(Sac-

TABLE IV

Principal Molecular Moments (μ_i) and ESR Data on $Co(SacSac)_2$ and Other Planar Cobalt(II) Compounds[a]

$Co(SacSac)_2$					β-CoPc		$Co(MNT)_2$	
ESR		Principal molecular moments			ESR		ESR	
g_i	$\|A\|$	μ_i	300°K	90°K	g_i	$\|A_i\|$	g_i	$\|A_i\|$
$g_x = 3.280$	$A_x = 105$	μ_x	2.93	2.80	$g_x = 2.42$	$A_x = 66$	$g_x = 2.798$	$A_x = 50$
$g_y = 1.904$	$A_y = 35$	μ_y	1.89	1.92	$g_y = 2.42$	$A_y = 66$	$g_y = 2.025$	$A_y = 28$
$g_z = 1.899$	$A_z = 35$	μ_z	1.77	1.52	$g_z = 2.01$	$A_z = 116$	$g_z = 1.977$	$A_z = 23$

[a] The magnetic moments are expressed in Bohr magnetons and the A_i values in 10^4 cm^{-1} units.

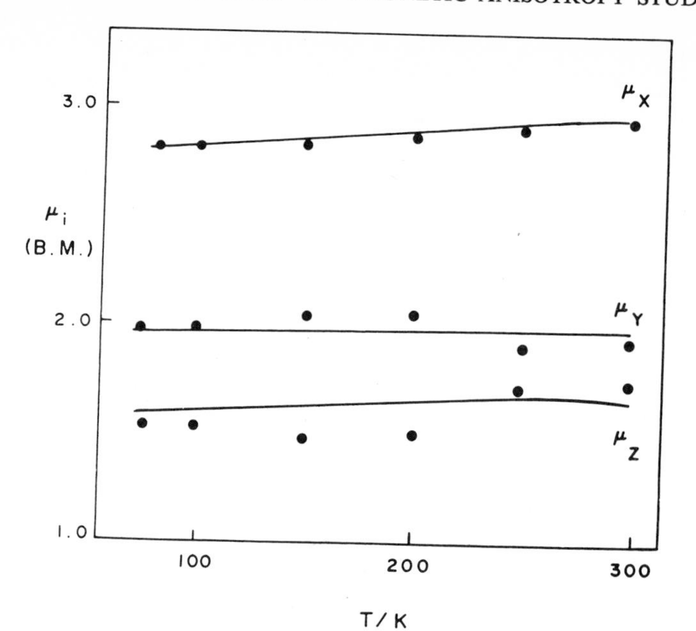

Fig. 27. Temperature dependence of the principal magnetic moments for Co(SacSac)$_2$. Circles are experimental results, and solid lines are the theoretical plots (45, 77).

Sac)$_2$ agree best with the unpaired hole being in the (d_{z^2}) orbital. Taking $\zeta = 400\ \text{cm}^{-1}$, good agreement with the experimental data was obtained for excitation energies: $(d_{yz}) \leftarrow (d_{z^2}) = 2000\ \text{cm}^{-1}$ and $(d_{xz}) \leftarrow (d_{z^2}) = 10{,}000\ \text{cm}^{-1}$ (see Fig. 27). Similar results were also obtained for $CoM_2(PEt_2Ph)_2$.

The large in-plane anisotropy observed in such planar compounds deserves further comment. Other things being equal, the mixing of $(d_{x^2-y^2})$ into the ground state (d_{z^2}) introduces an anisotropy into the plane of the molecule, which in these cases is obviously very important. With no mixing, the total anisotropy must arise from the splitting of the normally degenerate (d_{xz}, d_{yz}) pair and abnormally large splitting would result. It seems more likely that part of the in-plane anisotropy arises from this mixing. In the fitting of the data on Co(SacSac)$_2$ this mixing has been considered. Admixtures of this kind have little effect on the alternative ground-state configurations, where the total anisotropy must arise entirely from the energy difference between (d_{xz}) and (d_{yz}), thus offering further support for a (d_{z^2}) ground state. Proton magnetic resonance studies are in agreement with this choice of ground state (72).

It is of interest that ESR studies on these planar cobaltous compounds show similar in-plane anisotropies and lead to the same conclusion about the ground state (69, 71).

d. d^9 (Cu^{2+}). Measurements of paramagnetic anisotropy on several square planar Cu(II) complexes have been reported. Below we discuss representative measurements to illustrate the information that can be obtained. The electronic configuration of a Cu(II) ion, with respect to the coordinate system of Fig. 18, is $(d_{xy})^2\ (d_{xz}, d_{yz})^4\ (d_{z^2})^2\ (d_{x^2-y^2})^1$, which gives $^2B_1$ as the ground state. The principal susceptibilities are given by

$$\begin{aligned} K_{\parallel} &= \frac{N\beta^2}{4kT}\left[2 + \frac{8\zeta}{\Delta E_{\parallel}}\right]^2 + \frac{8N\beta^2\kappa^2}{\Delta E_{\parallel}} \\ K_{\perp} &= \frac{N\beta^2}{4kT}\left[2 + \frac{2\zeta}{\Delta E_{\perp}}\right]^2 + \frac{2N\beta^2\kappa^2}{\Delta E_{\perp}} \end{aligned} \tag{34}$$

where $\Delta E_{\parallel}$ and $\Delta E_{\perp}$ are energies corresponding to $(d_{xy}) \leftarrow (d_{x^2-y^2})$ and $(d_{xz}, d_{yz}) \leftarrow (d_{x^2-y^2})$ electronic transitions, respectively. Thus from a measurement of paramagnetic anisotropy of the planar copper(II) compound, it is possible, at least in principle, to decide the positions of the excited states, provided some estimate of ζ and κ is known.

This was successfully demonstrated (73) in CuPc where two sets of ESR studies employed very differing values of $\Delta E_{\parallel}$ and $\Delta E_{\perp}$. For example, Kivelson and Nieman (74) used $\Delta E_{\parallel} = 17{,}000\ cm^{-1}$ and $\Delta E_{\perp} = 25{,}000\ cm^{-1}$, whereas Harrison and Assour (75) suggested that $\Delta E_{\parallel} = 27{,}000\ cm^{-1}$ and $\Delta E_{\perp} = 17{,}000\ cm^{-1}$. In Fig. 28, the temperature variation of the experi-

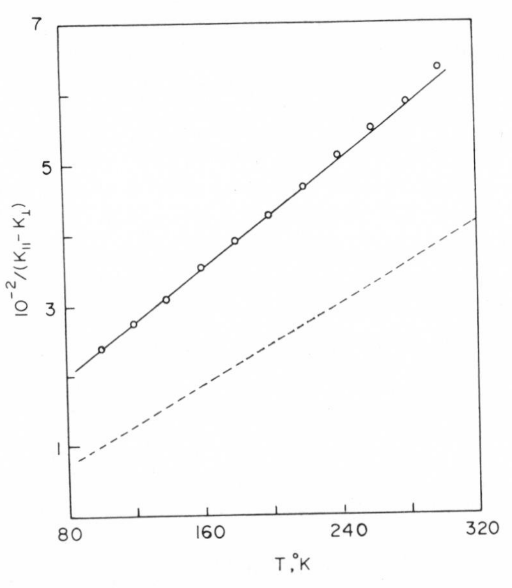

Fig. 28. Variation of molecular magnetic anisotropy with temperature for β-copper(II) phthalocyanine. Open circles are the experimental data (73). The solid and dashed lines are the theoretically calculated ones, using energy separations as suggested by Harrison and Assour (75) and Kivelson and Nieman (74).

mental paramagnetic anisotropy data is shown, together with the plot of Eq. 34, with the energy separations suggested by Kivelson and Nieman (dashed line) and Harrison and Assour (solid line). Here, $\zeta = 500\ cm^{-1}$ and $\kappa^2 = 0.7$ have been taken. The close agreement of the experimental data with the latter assignment appears encouraging. It has been pointed out (73) that variation in the values of ζ and κ will not alter the conclusion.

The situation in copper(II) acetylacetonate is rather complicated. Extensive single-crystal polarized spectral studies have been reported on this planar molecule, and there is no complete agreement between the various proposed assignments. It is, however, agreed that the band at 15,600 cm^{-1} is due to $(d_{xy}) \leftarrow (d_{x^2-y^2})$ transition $(\Delta E_{\parallel})$. Bands at 14,500 and 18,000 cm^{-1} can be assigned to $(d_{xy}, d_{yz}) \leftarrow (d_{x^2-y^2})$ transition $(\Delta E_{\perp})$. Paramagnetic anisotropies of $Cu(acac)_2$ have been measured (76, 77) in the 300° to 90°K temperature range and compared with the above values of $\Delta E_{\parallel}$ and $\Delta E_{\perp}$, as in CuPc. Unfortunately it is rather difficult to decide between the two values of $\Delta E_{\perp}$ since even a large change in its value has little effect on $K_{\perp}$, thus the magnetic anisotropy of $Cu(acac)_2$ does not offer much propsect of deciding between alternative assignments of its electronic spectrum. In CuPc, the paramagnetic anisotropy was very sensitive to ΔEs since the two alternative assignments involved large change in $\Delta E_{\parallel}$ as well as that in $\Delta E_{\perp}$.

It is interesting to compare the anisotropy of planar Cu(II) ions bonded to O, N, and S as ligands. The data for the first two types have already been discussed here. For S-bonded compounds the data from the measurements on Cu(II) dithiocarbamate can be taken (7, 77). In Table V, the percentage anisotropy of three Cu(II) systems at room temperature is compared. The trend in the variation of the magnetic anisotropy is in accord with the covalent character of Cu(II)-X bonds; namely, S > N > O.

TABLE V
Comparison of Percentage Molecular Anisotropy at Room Temperatre for the Cu(II) Ion Bonded to Oxygen, Nitrogen, and Sulfur (7, 73, 77)

Compound	Metal-ligand bond	$(\Delta K/K)$
$Cu(acac)_2$	Cu—0	20
CuPc	Cu—N	11%
$Cu(dtc)_2$	Cu—S	6.6%

A point regarding the magnetic anisotropy of CuPc deserves mention. In Fig. 29, the temperature variation of the observed crystal anisotropy is shown. The observed crystal anisotropy is found to *decrease* with the decreasing temperature. This anomaly, however, disappears when the observed anisotropy is corrected for the diamagnetic anisotropy of the CuPc molecule, and normal behavior is observed for the corrected anisotropies. CuPc has an inner macroring consisting of a delocalized pi-system and four nearly isolated

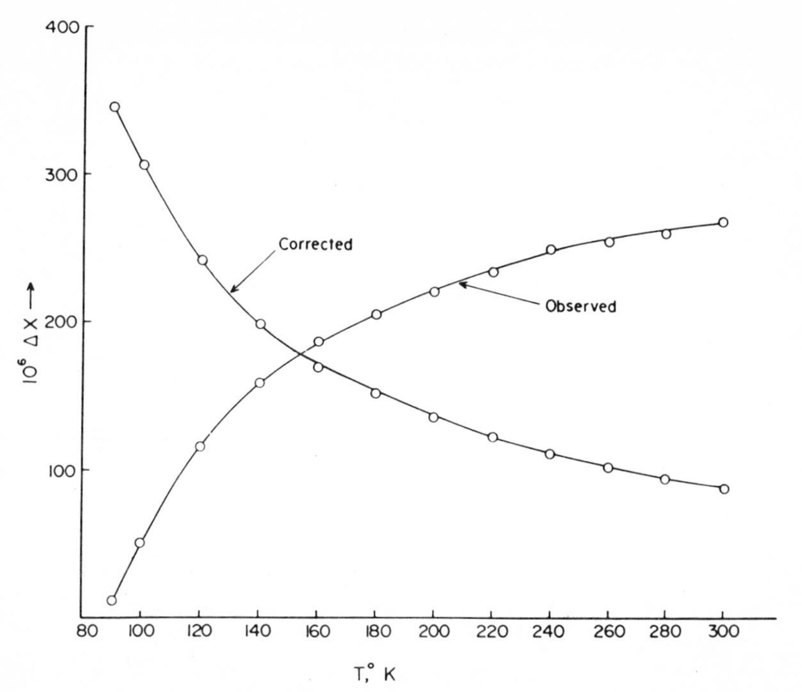

Fig. 29. The temperature dependence of the observed and corrected (for diamagnetic anisotropy) anisotropies for copper(II) phthalocyanine with "*b*" axis vertical (77).

benzene rings (cf. Fig 18) which produce a large diamagnetic anisotropy. In fact, the diamagnetic anisotropy of CuPc (assumed to be the same as that of ZnPc) is larger than its paramagnetic anisotropy at room temperature. Since it is temperature independent and opposite in sign, the observed anisotropy decreases as the temperature is lowered. No such anomaly is observed for manganese (II), iron(II), and cobalt(II) phthalocyanines since the paramagnetic anisotropies in these compounds were much larger than the corresponding diamagnetic anisotropy, although correction for diamagnetic anisotropy in all cases is important. CuPc is a striking example which illuminates the importance of correcting for diamagnetic anisotropy, particularly in cases where extensive delocalized pi-systems exist.

2. *Square Pyramidal Complexes*

Paramagnetic anisotropies of square pyramidal, transition metal complexes of vanadium(IV), iron(III), cobalt(II), nickel(II), and copper(II) have been examined, and the studies have been very useful in understanding the electronic structure of these ions. The molecular geometry of these complexes is generally distorted square pyramidal, and we take this into account in our analysis of magnetic data wherever desirable.

a. Vanadyl Complexes (d^1). Measurements of paramagnetic anisotropy on vanadyl complexes are confined to $VO(acac)_2$ alone, which forms stable noncubic single crystals. The electronic spectrum of $VO(acac)_2$ has been studied by several workers. Since the optical spectrum of $VO(acac)_2$ arises from the d^1 configuration, it is expected to be easy to deal with, theoretically. However, the assignment of various bands is still uncertain and being debated. In the Ballhausen and Gray (B&G) scheme, band I (10,500–16,000 cm^{-1}) is assigned to the $(d_{xy}) \rightarrow (d_{xz}, d_{yz})$ transition and band II (16,000–20,000 cm^{-1}) is assigned to the $(d_{xy}) \rightarrow (d_{x^2-y^2})$ transition (78). Band III is variously assigned to a d–d charge transfer, and/or spin-forbidden $n \rightarrow \pi^*$ or $\pi \rightarrow \pi^*$ transitions. In the cluster scheme of Selbin et al. (79), band II is assigned to the $(d_{xy}) \rightarrow (d_{z^2})$ transition.

Gregson and Mitra (80) have reported the paramagnetic anisotropy of $VO(acac)_2$ between 300° to 90°K and have attempted to rationalize the optical spectrum. They found that $K_\perp > K_\parallel$. Furthermore, ESR measurement shows that $g_\perp > g_\parallel$. The ground state of the vanadyl ion is an orbital singlet ($^2B_2$), thus Eq. 23 is applicable here. Using the experimental g_i values, it was observed that only the B & G scheme gives the correct sign of anisotropy ($K_\perp > K_\parallel$); the other scheme leads to the opposite sign ($K_\perp < K_\parallel$). The temperature dependence of the principal susceptibilities is also reproduced well by the B & G scheme.

There is another interesting feature in $VO(acac)_2$ which deserves attention (81). Assuming that the orbital reduction factor is anisotropic ($\kappa_\parallel$, $\kappa_\perp$), Eq. 23 can be written as

$$(\mu_\perp^2 - \mu_\parallel^2) = \Delta\mu^2 = \tfrac{3}{4}(g_\perp^2 - g_\parallel^2) + 3kT\left(\frac{2\kappa_\perp^2}{\Delta E_\perp} - \frac{8k_\parallel^2}{\Delta E_\parallel}\right) \tag{33}$$

where $\Delta E_\parallel$ and $\Delta E_\perp$ refer to the $(d_{xy}) \rightarrow (d_{x^2-y^2})$ and $(d_{xy}) \rightarrow (d_{xz}, d_{yz})$ transitions, respectively. A plot of the experimental data on $VO(acac)_2$ as $\Delta\mu^2$ versus T shows that the experimental data lie, as expected, on a straight line with a positive slope (Fig. 30). This suggests that $(2\kappa_\perp^2/\Delta E_\perp - 8\kappa_\parallel^2/\Delta E_\parallel)$ must be positive, that is, $\Delta E_\perp$ must be much smaller than $\Delta E_\parallel$ (which is consistent with the B & G scheme) and $k_\parallel \ll k_\perp$, indicating that the V = O bond is more covalent than the in-plane bonding. It can be easily seen that the latter requirement is essential, otherwise $\Delta E_\parallel > 4\Delta E_\perp$, which is highly unlikely. This analysis highlights the chemical information that can be deduced from the paramagnetic anisotropy data and the importance of a proper method of analysis of the data. We will discuss this point later.

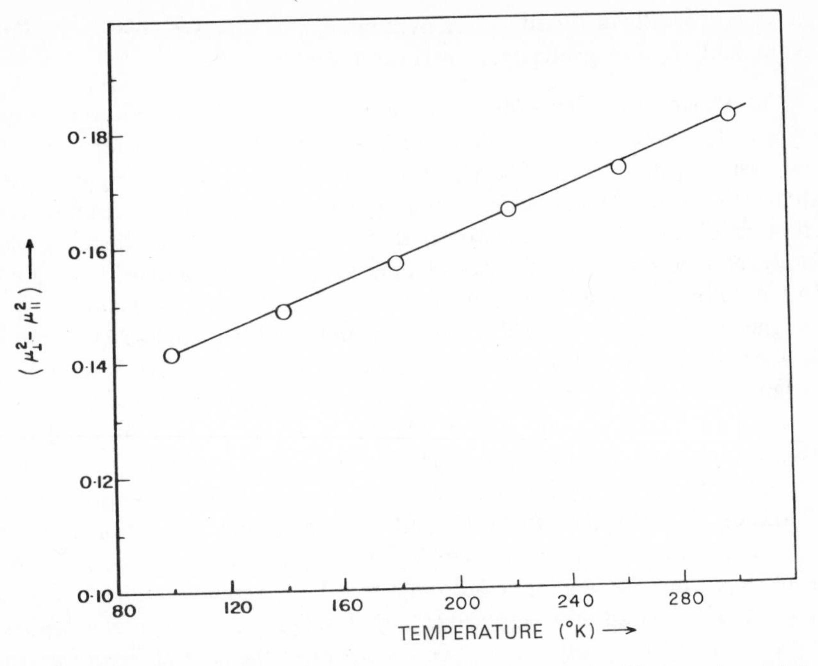

Fig. 30. Temperature dependence of ($\mu_\perp^2 - \mu_\parallel^2$) for $VO(acac)_2$. The experimental data (open circles) are taken from Ref. 80.

b. Iron(III) Complexes (d^5). Halobis(diethyldithiocarbamato)iorn(III) $Fe(dtc)_2X$ provides a series of molecules in which the ferric ion achieves the rare intermediate spin state $S = 3/2$, the usual spin state of the iron(III) ion being $S = 1/2$ or $5/2$. The molecular geometry around the iron atom is grossly square pyramidal, with the halogen atom being at the apex of the pyramid and the iron atom at the center, about 0.6 Å above the mean rectangular-based plane of the four sulfur atoms belonging to the two dithiocarbamate ligands (Fig. 31). The local symmetry at the iron site is close to C_2. Because of the uncommon spin state, the magnetic and other properties of the $Fe(dtc)_2X$ complexes have been very well studied. Extensive low-temperature, average magnetic susceptibility and Mössbauer and some ESR studies have been reported (82), which establish that the ground state of the ferric ion in these molecules is an orbital singlet ($^4A_2$); however, they do not lead to any further understanding about the electronic structure of these molecules. A very accurate and direct estimate of the ZFS of the ground $^4A_2$ state was also provided by measurements in the far infrared region (83). In the interpretation of almost all these studies, axial symmetry was assumed.

Recently, the paramagnetic anisotropies of $Fe(dtc)_2X$ (X = Cl, Br, I) have been reported and successfully applied by Ganguli et al. (84,85) to deduce the ground- and excited-state properties of the ferric ion in these molecules.

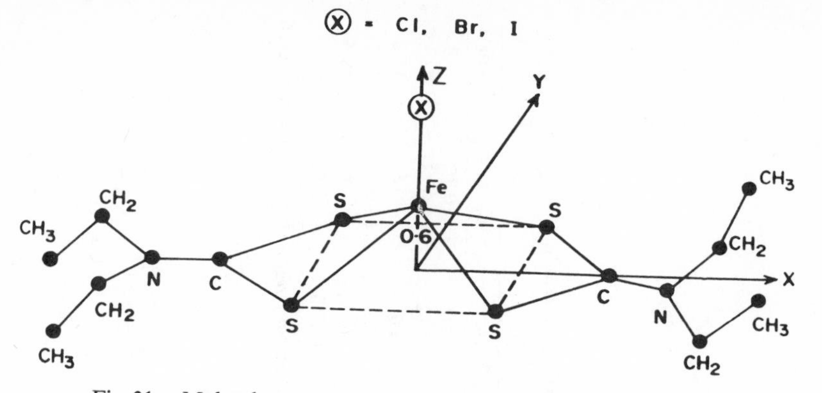

Fig. 31. Molecular geometry of halobis ferric diethyldithiocarbamate.

Table VI includes the room-temperature, paramagnetic anisotropy data on these derivatives. Here, the x and y axes lie in the mean plane of the four sulfur atoms, and the z axis is perpendicular to this plane along the Fe-X direction. The table illustrates the interesting point that the susceptibilities are strongly rhombic. Also, note the variation of the in-plane anisotropy $(K_x - K_y)$ along the series from the chloro to the iodo derivative. In Fig. 32, the temperature dependence of the experimental data is summarized. $Fe(dtc)_2Br$ shows an interesting phase transition near 220°K, and the crystal disintegrates below this temperature.

On the basis of the sign of the anisotropy $(K_x > K_y > K_z)$, Ganguli et al. propose that the ground state configuration for the ferric ion in this series is $(d_{x^2-y^2})^2$ $(d_{yz})^1$ $(d_{xz})^1$ $(d_{z^2})^1$, giving ${}^4A_2$ as the ground state; other alternatives would give $K_z > K_x$ or K_y, which is not in agreement with the experimental result (see Table VI). Using the appropriate spin Hamiltonian,

$$\mathscr{H} = D\left[S_z^2 - \tfrac{1}{3}S(S+1)\right] + E(S_x^2 - S_y^2) + \beta[g_xH_xS_x + g_yH_yS_y + g_zH_zS_z] \tag{36}$$

the following expression for the principal magnetic moments has been deduced:

$$\mu_i^2 = 3g_i^2\left(\frac{4A_i^2}{\delta^2} + \frac{1}{4}\right) + \left(\frac{24B_i^2}{X\delta^2} - \frac{2A_i}{\delta}\right)\left(\frac{1-e^{-X}}{1+e^{-X}}\right) \tag{37}$$

where

$$i = x, y, z$$

$$A_x = \tfrac{1}{2}(3E - D) \qquad B_x = -\tfrac{1}{2}\,(D + E)$$
$$A_y = -\tfrac{1}{2}\,(3E + D) \qquad B_y = \tfrac{1}{2}\,(D - E)$$
$$A_Z = D \qquad B_Z = E$$

$$X = \frac{\delta}{kT}$$

TABLE VI

Room-Temperature Molecular Anisotropies (in 10^6 cm^3 $mole^{-1}$) and the Calculated Spin Hamiltonian Parameters for the $Fe(dtc)_2X$ System (85)[a]

Compound	$(Kx - Ky)$	$(Ky - Kz)$	g_x	g_y	D, cm^{-1}	E, cm^{-1}	δ, cm^{-1}
$Fe(dtc)_2Cl$	185	269	2.059	2.039	1.09	−0.74	3.4
			(2.057)	(2.039)	(1.12)	(−0.78)	(3.85)
$Fe(dtc)_2$ Br	213	532	2.065	2.040	8.0	−0.6	16.0
			(2.070)	(2.046)	(6.9)	(−0.6)	(15.1)
$Fe(dtc)_2I$	417	598	2.089	2.047	9.34	−1.64	19.5
			(2.082)	(2.047)	(9.37)	(−1.90)	

[a] The values in parantheses for g_x, g_y, D, and E are obtained by the graphical method, whereas those in parentheses for δ are measured directly by far infrared spectroscopy.

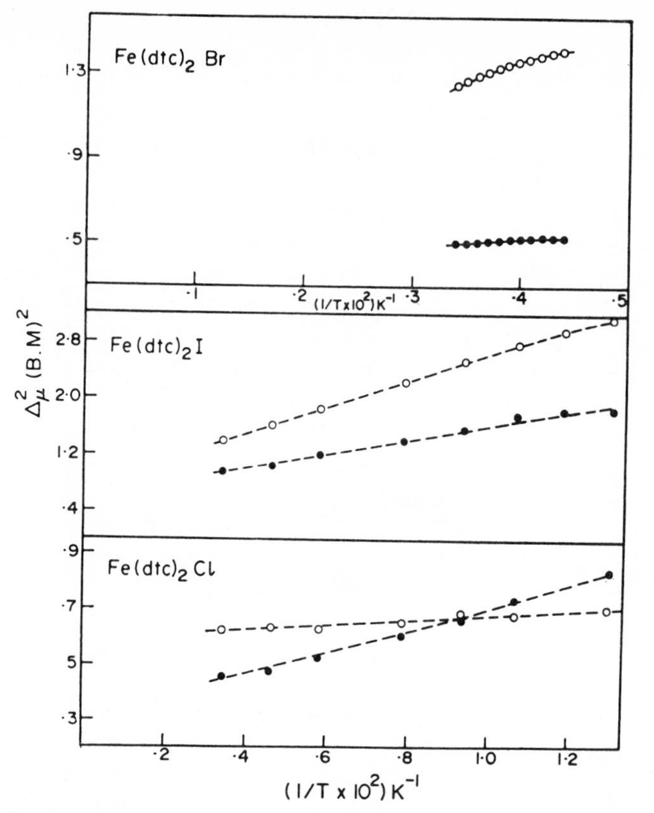

Fig. 32. Temperature dependence of $\Delta\mu^2$ for $Fe(dtc)_2X$. The circles are the experimental data: 0, $(\mu_y^2 - \mu_z^2)$; •, $(\mu_x^2 - \mu_y^2)$ (85).

Here, D and E are the axial and rhombic ZFS parameters. δ is the total ZFS (separation between the two Kramer's doublets, $Ms = \pm\frac{1}{2}$ and $\pm 3/2$) and is given by

$$\delta = 2[D^2 + 3E^2]^{\frac{1}{2}} \tag{38}$$

Equation 37 can be simplified if terms with second and higher powers of $(1/T)$ are neglected, which may be valid in the 80° to 300°K temperature range for small δ. The simplified form of the equation can be written as

$$\begin{aligned} \mu_x^2 - \mu_z^2 &= \frac{15}{4}(g_x^2 - 4) + \left(\frac{3}{kT}\right)[4D + \tfrac{1}{2}(D - 3E)g_x^2] \\ \mu_x^2 - \mu_y^2 &= \frac{15}{4}(g_x^2 - g_y^2) + \left(\frac{3}{2kT}\right)[(D - 3E)g_x^2 - (D + 3E)g_y^2] \end{aligned} \tag{39}$$

Here, g_z has been taken to equal 2, which is characteristic of the $^4A_2$ ground state mixed with the ^4E excited state.

A plot of Eq. 39 as $\Delta\mu^2$ versus $1/T$ would give straight lines, the slope and intercepts of which uniquely determine D, E, g_x, and g_y. The experimental data plotted in Fig. 32 as $\Delta\mu^2$ versus $1/T$ lie fairly well on straight lines, from the slopes and intercepts of which the values of these parameters were uniquely deduced, which agree well with those obtained by least-square fit of the experimental data to Eq. 37. The values obtained by least-square fit are included in Table VI. The significance and convenience of graphically deducing the LF parameters from a suitable plot of the magnetic anisotropy data have been stressed (84). The values of δ deduced from the anisotropy data agree very well with the values obtained from far infrared measurements. It should also be appreciated that the paramagnetic anisotropy measurements on the Fe(dtc)$_2$X system were independently able to give very accurate values of D and E, which could not otherwise be obtained, even from far infrared, Mössbauer, and (to some extent) ESR studies.

Equation 39 can be further simplified in the case of axial symmetry. We then get ($\mu_x = \mu_y = \mu_\perp, \mu_z = \mu_\parallel$):

$$\mu_\perp^2 - \mu_\parallel^2 = \frac{15}{4}(g_\perp^2 - 4) + \frac{3D}{2kT}\left(g_\perp^2 + 8\right) \tag{40}$$

A plot of Eq. 40 as $\Delta\mu^2$ versus $1/T$ gives a straight line with positive or negative slope, depending on the sign of D. Hence the sign of D can be immediately determined once the slope is known.

An LF calculation to explain the paramagnetic anisotropy data on the

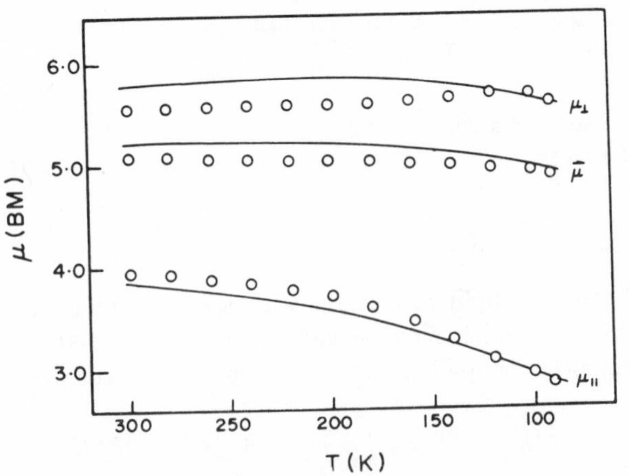

Fig. 33. Temperature dependence of the principal and average magnetic moments for the square pyramidal [(Ph$_2$MeAsO)$_4$CoNo$_3$)$^+$No$_3^-$ (88). The results for the analogous nickel(II) compound are similar. The solid curves are the calculated ones (88).

$Fe(dtc)_2X$ system has recently been done (86). The calculation is basically an extension to rhombic symmetry of that reported in axial symmetry on MnPc, with the modification mentioned there. The calculation shows that the first excited states (above the ground $^4A_2$ state) are the components of ^4E, with a separation between 6000 to 7000 cm^{-1}. Components of $^6A_1$ lie much higher in energy. This scheme can be constrasted to MnPc, where the $^6A_1$ lies very close to the ground $^4A_2$ state. The present energy level scheme deduced from the paramagnetic anisotropy data is, however, in complete disagreement with

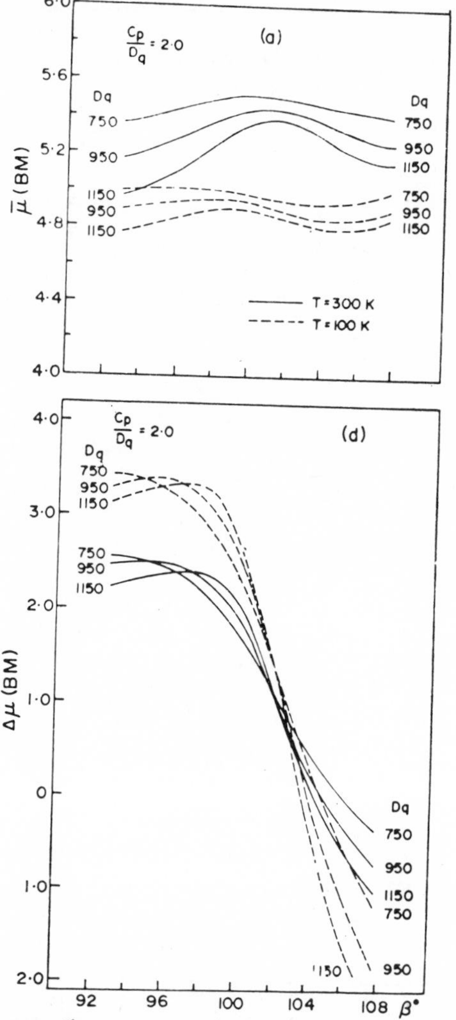

Fig. 34. Average magnetic moments and anisotropies of d^7, ^4F, and 4_p terms in a crystal field of C_{4v} symmetry: $\kappa = 1.0$, [D_q (apical)/D_q(basal)] = 1.0; $B = 850$ cm^{-1} (88).

that deduced earlier (87) from the consideration of average susceptibility and Mössbauer-effect data, in which the first excited state was proposed to be $^6A_1$, with ^4E being very far away. This latter scheme will not, however, produce the observed large in-plane anisotropy and the temperature dependence of the principal moments, hence it can be easily discarded.

c. Nickel(II) and Cobalt(II) Complexes. Measurements of paramagnetic anisotropy on some square pyramidal nickel(II) and cobalt(II) complexes have been reported by Gerloch et al.(88), and a very informative analysis of

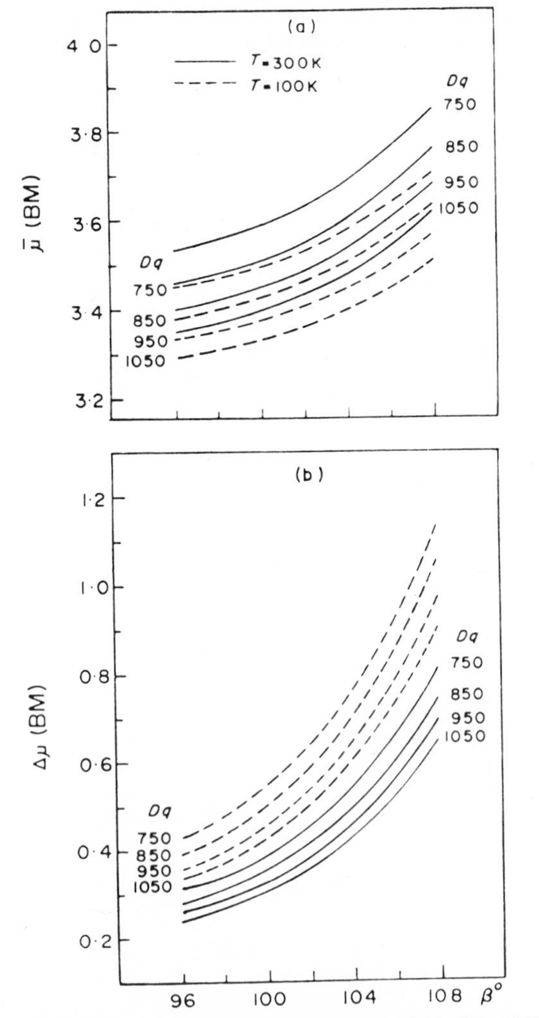

Fig. 35. Average magnetic moments and anisotropies of d^8, ^3F and ^3P terms in a crystal field of C_{4v} symmetry: $\kappa = 1.0$; $B = 800\ \text{cm}^{-1}$; $C_p/D_q = 3.0$ (88).

the data has been presented. Because of the similarity of the theoretical analyses, we present the results on both these ions together.

Both the nickel(II) and cobalt(II) complexes of ethyldiphenylarsine oxide are highly anisotropic, and the temperature dependence of the principal magnetic moments are similar (Fig. 33). The interpretation of the data given by Gerloch et al. is based on a point-charge model in terms of five disposable parameters: β, the effective angle between the axial and basal bonds; D_q (apical) and D_q (basal), the fourth-order radial integrals for apical and basal ligands, respectively; and C_p (apical) and C_p (basal), the second-order radial integrals for the same ligands. They have examined the effect on the average susceptibility and the paramagnetic anisotropy of the variation in all these parameters. Some representative types are shown in Figs. 34 and 35. The effect of the change in β on the $\Delta\mu$, in the case of the Co(II) complex, is quite dramatic, hence the anisotropy data could afford to give information on these kinds of structural parameters. The fitting of the experimental data in Fig. 33 showed that $[D_q \text{ (apical)}/D_q \text{ (basal)}]$ is nearly equal to unity, although a unique choice of the parameters was found to be difficult. For fitting the magnetic data, Gerloch et al. described a mapping procedure which is quite useful.

Recently, the paramagnetic anisotropy of low-spin Co(salen)(py) has been reported (89). The results show the interesting feature that the principal moments decrease with decreasing temperature (see Fig. 36). Note the large in-plane anisotropy. This unusual temperature dependence of the principal moments cannot derive only from a spin-doublet ground state, and the influence of a low-lying, excited quartet term has been suggested to dominate at

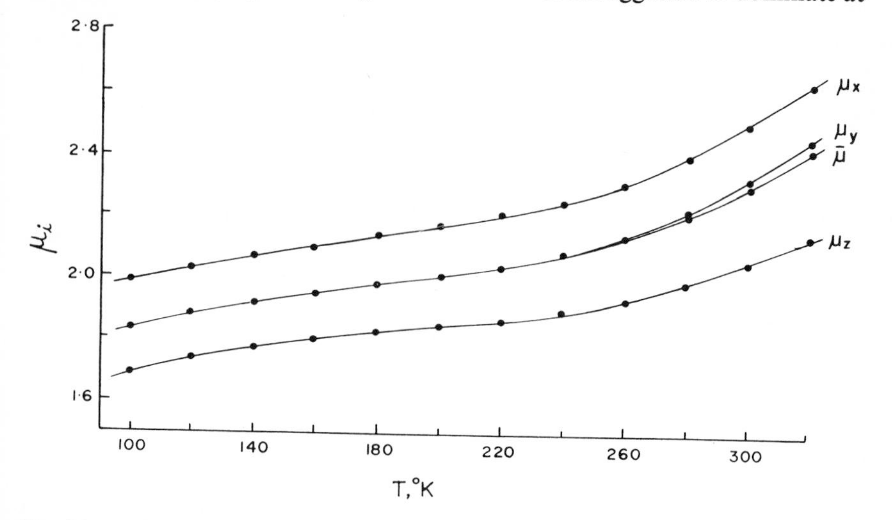

Fig. 36. Temperature dependence of the principal and average magnetic moments for square pyramidal Co(salen)py (89).

higher temperatures. A quantitative analysis of the data has not as yet been reported.

In conclusion, we have shown, in this subsection, the great usefulness of the paramagnetic anisotropy data in understanding the ground-and excited-electronic-state properties of the square planar and square pyramidal complexes. We have also highlighted its superiority over other methods, especially in $Fe(dtc)_2X$.

D. Zero-Field Splitting and Electronic Structure of High-Spin d^5 Ions

High-spin d^5 ions (e.g., Fe^{3+} and Mn^{2+}) have ${}^6A_1$ as the electronic ground state, hence they are generally expected to be very feebly anisotropic. As mentioned in Section III, the paramagnetic anisotropy arises entirely from the ZFS of the ground ${}^6A_1$ state and therefore serves as a very sensitive method for its determination. There is considerable interest in the determination of the ZFS of the d^5 ions as it is an important physical quantity in the ESR, specific heat, and adiabatic demagnetization experiments. Usually this splitting is very small (~ 1 cm^{-1}) for most of the simple complexes. For systems with nearly cubic symmetry, the splitting is of the order of 0.001 cm^{-1}. However, in some low-symmetry system (e.g., proteins), the ZFS of the ground ${}^6A_1$ state is very large and of the order of 10 cm^{-1}. Recently, interest in the ZFS in iron(III) complexes has been directed toward an understanding of the physical properties of hemoglobin and related molecules (90).

The origin of ZFS in cubic (or nearly cubic) d^5 systems is not very clear. It has been, for example, explained on the basis of high-order perturbation due to the cubic field and spin-orbit coupling (91, 92). On the other hand, it is believed by some workers to arise predominantly from spin-spin coupling (93–95). In complexes with axial (or rhombic) symmetry, the origin of the ZFS is rather easy to understand. Griffith (96) has ascribed it to arise from the second order effect of spin-orbit coupling between ${}^6A_1$ and the lowest ${}^4T_1$ state which is grossly split by axial crystal field. This theory of ZFS in axial crystal fields explains well the ESR data in various hemoglobins and was able to predict the existence of large ZFS in these proteins.

Since excited ${}^4T_1$ state usually lies about 10,000 cm^{-1} above the ${}^6A_1$, the magnetic properties of the high-spin d^5 ions are generally explained on the basis of spin-Hamiltonian formalism. We consider here the situation of axially symmetric complexes. The simple form of the spin Hamiltonian which has been commonly used is (90, 97)

$$\mathscr{H} = D \cdot S_z^2 + 2\beta\bar{S} \cdot H \tag{41}$$

The splitting of the ${}^6A_1$ term on the basis of Eq. 41 is shown in Fig. 37. The

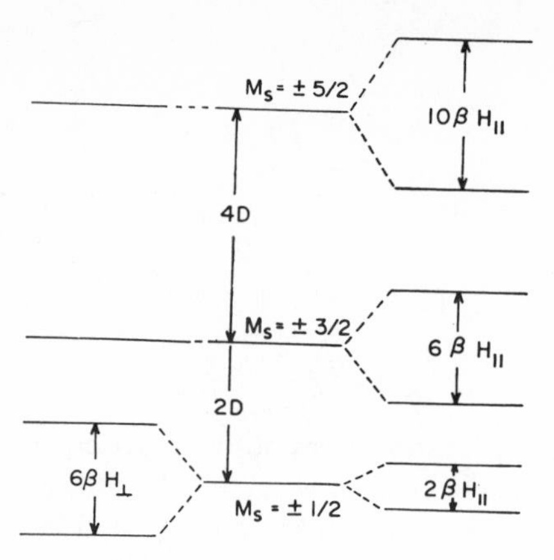

Fig. 37. Splitting of $^6A_1$ ground term in (*a*) zero, (*b*) perpendicular, and (*c*) parallel magnetic fields due to the Hamiltonian given by Eq. 41 (see text).

three Kramer's doublets have a total energy separation of $6D$, with the convention that when D is positive, $\pm 1/2$ lies lowest. The energies of the components of $^6A_1$ for magnetic fields parallel and perpendicular to the symmetry axes are given by the following:

For the magnetic field parallel to the symmetry axis:

$$E = \begin{cases} 0 \pm \beta H_{\parallel} \\ 2D \pm 3\beta H_{\parallel} \\ 6D \pm 5\beta H_{\parallel} \end{cases}$$

For the magnetic field perpendicular to the symmetry axis:

$$E = \begin{cases} 0 \pm 3\beta H - \dfrac{4\beta^2 H_{\perp}^2}{D} \\ 2D + \dfrac{11}{4} \cdot \dfrac{\beta^2 H_{\perp}^2}{D} \\ 6D + \dfrac{5}{4} \cdot \dfrac{\beta^2 H_{\perp}^2}{D} \end{cases}$$

Substituting these energies in Van Vleck's equation, the expressions for principal magnetic moments are given by

$$\mu_{\parallel}^2 = \frac{3}{1 + e^{-2X} + e^{-6X}} (1 + 9e^{-2X} + 25e^{-6X})$$

$$\mu_{\perp}^2 = \frac{3}{1 + e^{-2X} + e^{-6X}} \left[\left(9 + \frac{8}{X}\right) - \left(\frac{11}{2X}\right)e^{-2X} - \left(\frac{5}{2X}\right)e^{-6X}\right] \tag{42}$$

where

$$X = \frac{D}{kT}$$

Equation 42 can be written as

$$\mu_\perp^2 - \mu_\parallel^2 = \frac{3}{(1 + e^{-2X}e + e^{-6X})}\left[8\left(1 + \frac{1}{X}\right) - \left(\frac{11}{2X} + 9\right)e^{-2X} - \left(\frac{5}{2X} + 25\right)e^{-6X}\right] \quad (43)$$

Let us first consider some limiting cases of the above equations. When $X \rightarrow 0$ (i.e., at very high temperatures), $\mu_\parallel = \mu_\perp = 5.92$ B.M. As $X \rightarrow \infty$(i.e., at very low temperatures), $\mu_\parallel = 1.73$ B.M. and $\mu_\perp = 5.19$ B.M. Thus large anisotropy is expected at low temperatures, even for moderate values of D. Equation 43 can be simplified further if we retain only the terms up to the first power in X in the expansion. We then get

$$(\mu_\perp^2 - \mu_\parallel^2) = 112\frac{D}{kT} \quad (44)$$

Thus $\Delta\mu^2$ is expected to vary as $1/T$, and $(K_\perp - K_\parallel)$ is expected to vary as $1/T^2$. It should be noted that Eq. 44 is a very good approximation for most complexes in the liquid nitrogen temperature range.

In Fig. 38, the variation of the principal magnetic moments with the ZFS parameter is shown for three fixed temperatures. The figure shows that when D is positive, $\mu_\perp > \mu_\parallel$, whereas for negative values of D, $\mu_\parallel > \mu_\perp$. Thus the sign of D is uniquely determined from the sign of paramagnetic

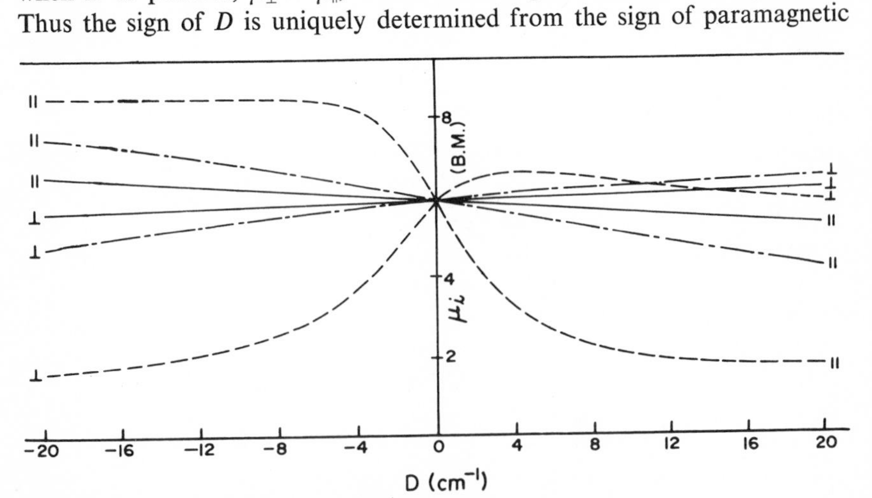

Fig. 38. Variation of principal magnetic moments with the ZFS parameter (D): ———, 300°K; —·—·—, 100°K; — — — —, 10°K (98).

anisotropy. Furthermore, for the same value of D, the anisotropy is larger for negative D. This is because when D is positive, $Ms = \pm\frac{1}{2}$ lies lowest which gives $g_{\parallel} = 2$, $g_{\perp} = 6$. For negative D, that is, $Ms = \pm 5/2$ lowest, $g_{\parallel} = 10$, $g_{\perp} = 0$. Thus the anisotropy is much larger when D is negative. In addition, the figure shows that as D increases, the anisotropy also increases, the increase being quite marked for negative D. We further observe that the anisotropy increases quite fast as the temperature is lowered, assuming very large values at low temperatures. All these features of Fig. 38 demonstrate the sensitive nature of the paramagnetic anisotropy to the ZFS of the $^6A_1$ state and clearly make it a very accurate method of its determination. A point of interest is, however, the insensitive nature of the anisotropy to the variation in D at very low temperatures, especially for $|D| > 8\ \text{cm}^{-1}$.

In the derivation of Eqs. 42 and 43 it has been assumed that $\beta H/kT \ll 1$ and $D/kT \ll 1$. However, both these assumptions are not valid when measurements are done at very high magnetic fields and/or at very low temperatures. In such limits, the magnetic susceptibility shows saturation effect, and the use of Van Vleck's equation for the calculation of susceptivility is not correct. In such cases, the susceptibility should be calculated using the thermodynamic expression for susceptibility:

$$\chi = -\frac{N}{H}\left(\frac{\Sigma \frac{\partial E}{\partial H} \cdot \exp(-E/kT)}{\Sigma \exp(-E/kT)}\right) \tag{45}$$

This has been pointed out by Gerloch et al (98). and Marathe and Mitra (99). Gerloch et al. have calculated the principal susceptibilities while taking this factor into account. Figure 38 includes this correction, which is singnificant only for the 10°K curve.

Let us now discuss the significance of the ZFS parameter (D) in terms of the crystal-field approach. The $^6A_1$ ($t_2^3 e^2$) ground state has a nonzero matrix element of spin-orbit coupling with the $^4T_1$ (t^4e) excited state; all other matrix elements with the quartet states are zero. When the symmetry of the crystal field is lower than cubic, the $^4T_1$ state splits into two or three components. Griffith (96) assumed a large axial splitting and considered the $^4A_2$ component only, this lying at a lower energy than the ^4E component. Kotani (97) has considered the effect of a rhombic crystal field when $^4T_1$ splits into three components. The ZFS parameter D is then given by where

$$D = \frac{\zeta^2}{10}\left[\frac{2}{E_z} - \frac{1}{E_x} - \frac{1}{E_y}\right] \tag{46}$$

E_x, E_y, and E_z are the energies of the three components of $^4T_1$ relative to the $^6A_1$ ground state. In an axial crystal field, $E_z = E_{\parallel}$ and $E_x = E_y = E_{\perp}$, and thus Eq. 46 reduces to

$$D = \frac{\zeta^2}{5}\left[\frac{1}{E_\parallel} - \frac{1}{E_\perp}\right] \simeq \frac{\zeta^2}{5}\left[\frac{\Delta E}{E^2}\right] \tag{47}$$

where E is the mean energy of the components of ${}^4T_1$ relative to the ${}^6A_1$ state. Thus the splitting (or position) of the excited ${}^4T_1$ state can be determined once D is known.

We describe below some experimental results on d^5 ions and discuss them on the basis of the above theory.

1. *Manganese(II) Tutton Salts*

Measurements of paramagnetic anisotropy of high-spin manganese(II) Tutton salts have been reported (3) as early as 1935, and a detectable anisotropy of the ${}^6A_1$ ground state was measured. In Table VII, some typical values

TABLE VII
Room-Temperature Anisotropies (in 10^6 cm^3 mole^{-1}) of Some High-Spin Manganese(II) and Iron(III) Compounds

Compound	$\mid \chi_1 - \chi_2 \mid$	$\mid \chi_1 - \chi_3 \mid$
$(NH_4)_2Mn(SO_4)_2 \cdot 6H_2O$	8.7	6.0
$Rb_2Mn(SO_4)_2 \cdot 6H_2O$	8.9	6.2
$(NH_4)_2Zn(SO_4)_2 \cdot 6H_2O$	1.2	0.8
$Fe(acac)_3$	$\chi_b - \chi_a = 51$	$\chi_b - \chi_c = 27$
$K_3Fe(oxalate)_3 \cdot 3H_2O$	$\chi_e - \chi_b = 58$	$\chi_{a'} - \chi_b = 70$

at room temperature are given. The anisotropies are extremely small, indicating that D is very small. In fact, D values for most of the Mn(II) ions are usually less than 0.1 cm^{-1}. A point worth noting is that since the observed anisotropies at room temperature are extremely small, the correction for diamagnetic and shape anisotropy becomes very important. In Ref. 3 this consideration was very elegantly taken into account. A complete analysis of the data on Mn(II) Tutton salts has not been reported, although an attempt was made (3) to deduce D from the room-temperature data alone. The analysis was based on several assumptions (which were not correct), but predicted values of D were not very different from those obtained from adiabatic demagnetization and specific heat measurements.

2. *Tris(acetylacetonato)iron(III)and Potassium Tris(oxalato)iron(III) Trihydrate*

The paramagnetic anisotropy measurements on these two high-spin iron(III) complexes have been reported by Gerloch et al (98). between 90° to 300°K, and the data were analyzed in terms of Eq. 42. In both complexes $\mu_{\parallel} > \mu_{\perp}$; therefore, D is negative. Anisotropies are very small at room temperature (see Table VII), but the corrections for the diamagnetic and shape anisotropies do not appear to have been effected. By fitting the temperature dependence of the principal moments to Eq. 42, the following values of D have been deduced (98):

Complex	D, cm^{-1}	Calculated excited-state splitting, E, cm^{-1}
Fe(ac ac)$_3$	-0.11 ± 0.04	-350 ± 125
$K_3Fe(oxalate)_3 \cdot 3H_2O$	-0.55 ± 0.07	-1925 ± 230

Using these values of D, ΔE can be calculated from Eq. 47. E was obtained from the relation $E = 10B + 6C - 10D_q = 10{,}568$ cm^{-1}, with $B = 609$ cm^{-1}, $C = 3284$ cm^{-1} and $Dq = 1522$ cm^{-1}. ζ was taken at 400 cm^{-1}.

Note that although the paramagnetic anisotropy data for Fe(acac)$_3$ and $K_3Fe(oxalate)_3 \cdot 3H_2O$ were explained by Gerloch et al (98). on the basis of Eq. 42, the molecular anisotropies of these complexes do not strictly vary as $1/T^2$ as demanded by the above equation (cf. Eq. 44). The deviation from $1/T^2$ variation may not be fully attributed to the neglect of the diamagnetic and shape anisotropy corrections since this deviation is observed even in $K_3Fe(oxalate)_3 \cdot 3H_2O$ below 150°K, where the anisotropies are relatively much larger and the above corrections may not be very important. We shall examine the reason for this discrepancy later.

3. *Iron(III) Heme Proteins*

As mentioned earlier, ferric heme proteins provided the first example for large ZFS. It was originally measured by the ESR technique, but only an upper limit of this parameter could be fixed. For an accurate determination of D, it is necessary to perform the ESR measurement at different frequencies, preferably at higher frequencies when a deviation in $g_{\perp}$ from its expected value of 6 is observed. D can then be calculated from the relation (101)

$$g_{\perp}^{eff} = 3g_{\perp}\left[1 - 2\left(\frac{g_{\perp}\beta H}{2D}\right)^2\right] \tag{48}$$

Such measurements have recently been reported up to 4 mm, and accurate values of D were obtained for ferric myoglobin fluoride (101).

Measurements of magnetic anisotropy of ferric myoglobin flouride, Fe(Mb)F, and ferric met myoglobin, Fe(Mb)H_2O, have been done. The first measurement was reported on Fe(Mb)H_2O single crystals between 77° to 300°K, and a value of $D = 2.5$ cm^{-1} was deduced (102). However, the results were not very reliable since the observed anisotropies were small, and no attempt was made to correct for the huge diamagnetic anisotropy of the $Fe(Mb)H_2O$, which may be comparable to, if not larger than, its paramagnetic component in the liquid nitrogen temperature range.

A detailed measurement of magnetic anisotropy was subsequently reported on the single crystals of $Fe(Mb)H_2O$ and Fe(Mb)F (103). Measurements were made at liquid nitrogen, hydrogen, and helium temperatures, and the experimental data are summarized in Fig. 39. As expected, the anisotropies are large at lower temperatures, with $\mu_\perp > \mu_\parallel$. For example, $\Delta\mu^2 = 8$ B.M. for Fe(Mb)F at 77°K, which can be compared to $\Delta\mu^2 = 1.12$ B.M. for $K_3Fe(OX)_3 \cdot 3H_2O$ at 90°K. The experimental data in Fig. 39 were fitted to Eq. 43, which gave $D = 10.5\ cm^{-1}$ for $Fe(Mb)H_2O$ and $D = 6.5\ cm^{-1}$ for Fe(Mb)F. These values agree very well with the later far infrared measurements of the ZFS (83). It is interesting that a value of $D = 28\ cm^{-1}$ was deduced from the average susceptibility data on the ferric myoglobin fluoride (104).

In spite of the success of the above work on heme proteins in giving an accurate value of D, there are some discrepancies in this study. It has been pointed out (103) that there was a calibration error in the measurement of paramagnetic anisotropy, which was corrected by a scaling factor. The scaling factor was determined by taking the theoretically calculated $(\mu_\perp^2 - \mu_\parallel^2)$ values at 4.2°K with Eq. 43 for the D values quoted above, as the true experimental values of the anisotropy. The same scaling factor was used for fitting the data at liquid hydrogen and nitrogen temperatures. While the agreement over the entire temperature range is good (cf. Fig. 39), this is by no means satisfactory. Furthermore, no account of the dimagnetic anisotropy of these molecules has

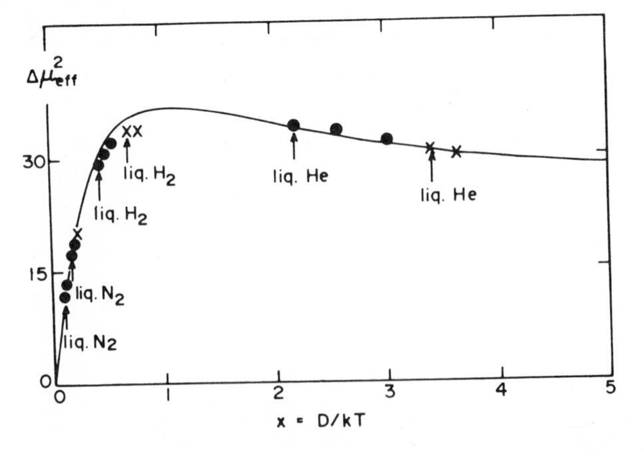

Fig. 39. Temperature dependence of $\Delta\mu^2$ $(\mu_\perp^2 - \mu_\parallel^2)$ for Fe(Mb)F (●) and $Fe(Mb)H_2O$ (X). The solid line was calculated using the Hamiltonian (Eq. 41), with values of D given in the text (103).

been taken. Also, the measurements were made at high magnetic fields and the saturation effects may be important at liquid helium temperatures. This factor, however, was not considered

It has recently been pointed out by Marathe and Mitra (99, 100) that the spin Hamiltonian given by Eq. 41 is inadequate to account accurately for the magnetic properties of high-spin d^5 systems even after considering the saturation effect. Analyzing the average magnetic susceptibility data on $S = 5/2$ hemin and ferric tris(pyrrolidine)dithiocarbamate, they showed that the above spin Hamiltonian gave an incorrect magnitude of the ZFS parameter. In the case of hemin, Eq. 41 gave a value of D which was about one and a half times larger than the value directly measured by the far infrared technique (83). For the ferric dithiocarbamate, the above Hamiltonian could not even reproduce the temperature dependence of the average magnetic susceptibility at low temperatures. It has been suggested (99, 100) that the inclusion of fourth-order crystal-field terms is very important and can indeed explain the above discrepancies quite elegantly. The spin Hamiltonian appropriate to the tetragonal and trigonal symmetry, including the fourth-order terms, can be written as below:

Tetragonal:

$$\mathscr{H} = 2\beta H \cdot S + \frac{1}{3}(B_2^0\, O_2^0) + \frac{1}{60}(B_4^0\, O_4^0) + \frac{1}{12}(B_4^4\, O_4^4) \tag{49}$$

Trigonal:

$$\mathscr{H} = 2\beta H \cdot S + \frac{1}{3}(B_2^0\, O_2^0) + \frac{1}{60}(B_4^0\, O_4^0) - \frac{1}{3}\sqrt{2}\,(B_4^3\, O_4^3) \tag{50}$$

Here, B_2^0 becomes D as defined in Eq. 41 when the fourth-order terms are neglected. The effect of the fourth-order terms is to induce mixing between $|\pm 5/2>$ and $|\pm 3/2>$ in the tetragonal symmetry and between $|\pm 5/2>$ and $|\pm 1/2>$ in the trigonal symmetry. The magnetic properties are therefore expected to be altered significantly depending upon the magnitude of this mixing (i.e., the strength of the fourth-order terms). For $H = O$, the energies of the three Kramers doublets can be given in both symmetries as shown below:

Tetragonal:

$$E\left(\cos\theta\,|\pm\frac{5}{2}> + \sin\theta\,|\mp\frac{3}{2}>\right)$$

$$= \frac{4}{3}B_2^0 - B_4^0 + [(2B_2^0 + 2B_4^0)^2 + 5(B_4^4)^2]^{1/2}$$

$$E\left(\cos\theta\,|\mp\frac{3}{2}> - \sin\theta\,|\pm\frac{5}{2}>\right) \tag{51}$$

$$= \frac{4}{3} B_2^0 - B_4^0 - [(2B_2^0 + 2B_4^0)^2 + 5(B_4^4)^2]^{1/2}$$

$$E(| \pm \tfrac{1}{2} >) = -\frac{8}{3} B_2^0 + 2B_4^0$$

where

$$\tan 2\theta = 5 \frac{B_4^4}{(2B_2^0 + 2B_4^0)}$$

Trigonal:

$$E\left(\cos\theta \,|\, \pm \frac{5}{2} > \mp \sin\theta \,|\, \mp \frac{1}{2} >\right)$$

$$= \frac{1}{3} B_2^0 + \frac{3}{2} B_4^0 + \frac{1}{2} [(6B_2^0 - B_4^0)^2 + 80(B_4^3)^2]^{1/2}$$

$$E\left(| \pm 3/2 >\right) = -\frac{2}{3} B_2^0 - 3B_4^0 \tag{52}$$

$$E\left(\cos\theta \,|\, \mp \frac{1}{2} > \pm \sin\theta \,|\, \pm \frac{5}{2} >\right)$$

$$= -\frac{1}{3} B_2^0 + \frac{3}{2} B_4^0 - \frac{1}{2} [(6B_2^0 + B_4^0)^2 + 80(B_4^3)^2]^{1/2}$$

where

$$\tan 2\theta = 4\sqrt{5} \frac{B_4^3}{(6B_2^0 - B_4^0)}$$

A point of interest in Eq. 52 is that the $| \pm 5/2 >$ and $| \pm 1/2 >$ states mix in opposite sense, hence magnetic moments intermediate to the extreme values are observed (100).

Let us now consider the effect of the fourth-order term on the magnetic anisotropy. This is illustrated in Fig. 40, in the case of trigonal symmetry, for various values of $\sin^2 \theta$, the mixing parameter. The figure shows the dramatic effect of the inclusion of fourth-order terms. For a particular value of the mixing coefficient, $\sin^2 \theta = 0.48$, the anisotropy becomes zero and then changes sign as $\sin^2 \theta$ increases. The figure also shows the sensitive nature of the paramagnetic anisotropy to this mixing coefficient and suggests that the anisotropy measurement, even between 77° to 120°K, should be able to support the above prediction regarding the existence of the fourth-order crystal-field terms.

An experimental verification of the inadequacy of Eq. 41 and the necessity of the fourth-order ligand-field terms has been provided by Ganguli et al. (105). They have measured the paramagnetic anisotropy of the high-spin ferric tris(pyrrolidyl) dithiocarbamate, $Fe(Pydtc)_3$, in which the

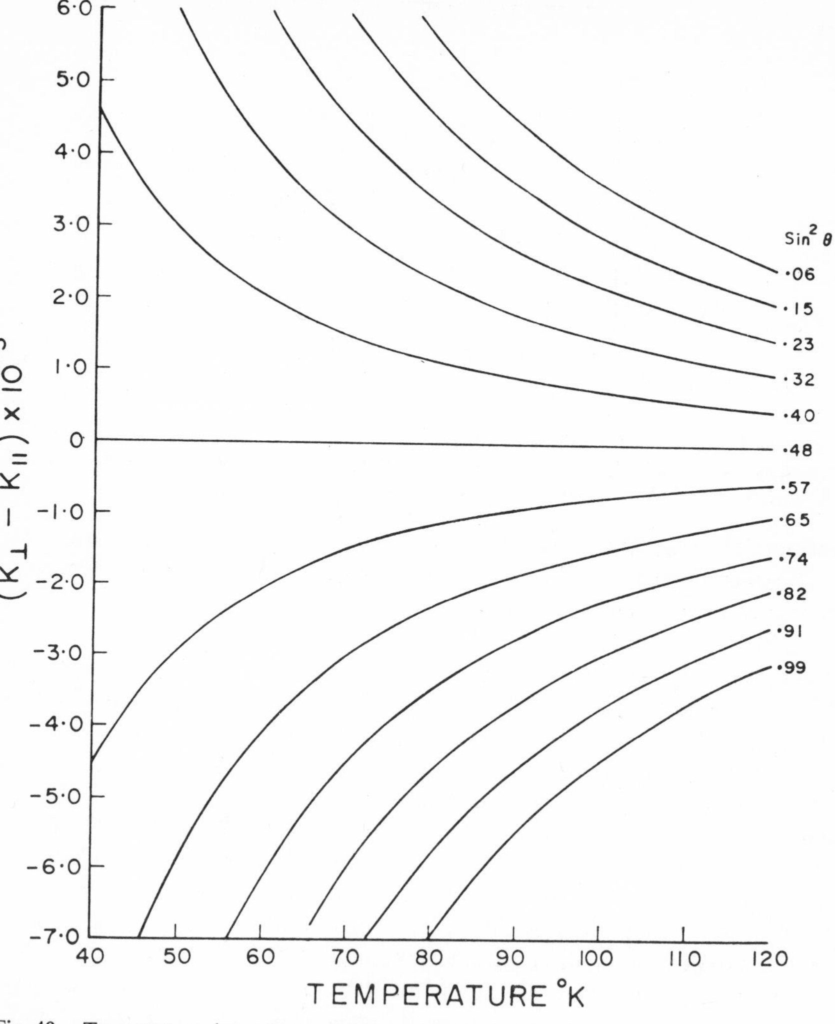

Fig. 40. Temperature dependence of the anisotropy for various values of $\sin^2 \theta$ in the trigonal symmetry, using the Hamiltonian given by Eq. 50.

ZFS has been measured by the far infrared technique (83), and it was deduced that $\pm$ 5/2 lies lowest and $D = -2.14$ cm^{-1} was obtained through Eq. 41. In Fig. 41, the experimental paramagnetic anisotropy data are shown with the theoretical plot of Eq. 43 for $D = -2.14$ cm^{-1}. The discrepancy is clear. Also included in the figure is the theoretical plot of Eq. 52, with $\sin^2 \theta$ = 0.70, which reproduces the data extremely well and is also consistent with

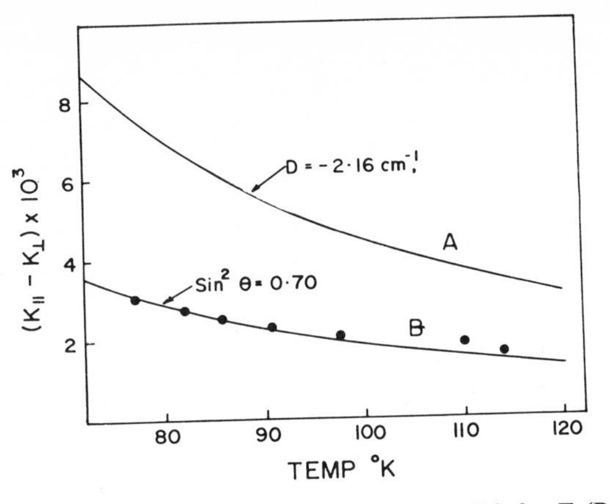

Fig. 41. Temperature dependence of the anisotropy ($K_{\parallel} - K_{\perp}$) for Fe(Pydtc)$_3$. The solid line A was caluclated using Eq. 43, with no fourth-order terms considered, whereas line B includes the fourth-order terms. The best fit was observed when $\sin^2 \theta = 0.70$ (105).

the far infrared data. It can be mentioned that Marathe and Mitra (100) had earlier predicted $\sin^2 \theta = 0.66$ from the consideration of low-temperature average magnetic moment.

We now come back to the discrepancy commented upon in the case of $K_3Fe(oxalate)_3 \cdot 3H_2O$. We have seen that D is small for this complex, thus Eq. 44 is valid in this case above 77°K. Equation 44 requires that the plot of $(\mu_{\parallel}^2 - \mu_{\perp}^2) \cdot T$ versus T be a straight line parallel to the temperature axis. The experimental data between 77° to 150°K are plotted similarly (Fig. 42). The results clearly show the inadequacy of the Hamiltonian given by Eq. 41 and point out the necessity of the inclusion of fourth-order crystal-field terms. Equations 51 and 52 involve three disposable parameters and cannot be uniquely determined from the anisotropy data alone. Average susceptibility measurements, preferably at very low temperatures, or far infrared data are required to deduce the three parameters uniquely. For $K_3Fe(oxalate)_3 \cdot 3H_2O$ this analysis was not possible due to the lack of such data. It is, however, clear now that the inclusion of these higher order terms is important to rationalize the magnetic anisotropy data on d^5 ions. Furthermore, we observe that the anisotropy data gives very accurately the values of the mixing coefficient, $\sin^2 \theta$, which determines the composition of the ground state wave function. Such accurate and detailed information of the ground state is, in fact, difficult to obtain from other measurements.

We comment now on the sign of D in these cases. Following the arguments given by Gerloch et al. (98), we note that, in the tetragonal geometry, a crystal-field point-charge approach would predict that elongation would stabilize the d_{z^2} orbital relative to the $d_{x^2-y^2}$ and d_{xz}, d_{yz} pair with respect to

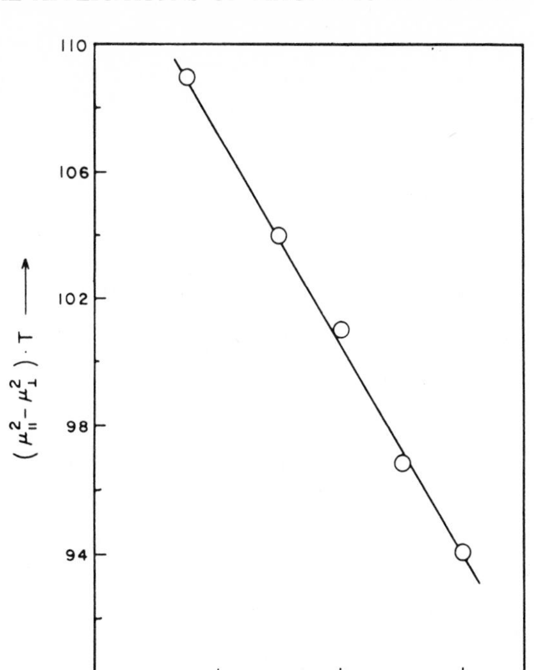

Fig. 42. Temperature dependence of $(\mu_{\parallel}^2 - \mu_{\perp}^2)T$ for $K_3Fe(oxalate)_3 3H_2O$ (105). Notice the large slope of the experimental results. In the absence of the fourth-order crystal-field terms, the slope should have been zero.

d_{xy}. The spin quartet state, corresponding to the first excited state of the molecule, should then correlate with $(d_{xz}, d_{yz})^3 (d_{xy})^1 (d_{z^2})^1$ arising from the $t_4^2 e^1$ $(^4T_1)$ configuration. This level correlates with the ^4E term arising from the octahedral $^4T_{1g}$ (^4G), and the sign of D is predicted to be negative. A reversed ordering of the orbitals leading to the configuration $(d_{xy})^2 (d_{xz}, d_{yz})^2 (d_{z^2})^1$, that is $^4A_2$, is then required to reproduce the experimentally observed positive sign of D in the tetragonal complexes, such as iron(III) hemoglobins. Griffith suggested that pi-bonding with the hemoglobin group involving electron donation from the filled-ligand pi-orbitals could cause this reversal. A similar argument has been given for the negative D in trigonal symmetry by Gerloch et al. on the point-charge model.

In conclusion, the measurement of paramagnetic anisotropy provides a very accurate method for the determination of the ZFS and often gives detailed information about the ground-state wave function. However, a proper analysis of the data appears to be important.

E. Molecular Geometry and Structural Information

The paramagnetic anisotropy of a transition metal complex is very intimately related to the detailed electronic structure of the metal ion, as has been shown. In simple cases, especially in complexes belonging to higher symmetry, the electronic structure of the metal ion is often, to a large extent, governed by the molecular geometry, hence a correlation between the stereochemical arrangement of the ligands and paramagnetic anisotropy would appear possible. Arguments leading to such correlations are often based on simple considerations and are not always valid.

One of the most classic examples of a correlation between paramagnetic anisotropy and stereochemistry is provided by the large difference in the paramagnetic anisotropy of octahedral and tetrahedral cobaltous complexes. The octahedral cobaltous complexes have an orbital triplet $^4T_1$ as the cubic field ground state as compared to the orbital singlet $^4A_2$ in the case of the tetrahedral geometry. It is therefore expected that the magnetic anisotropy of the former would be much larger than that of the latter and can be used to differentiate between these geometries. In Table VIII some typical data at room temperature are listed for the octahedral and tetrahedral cobaltous complexes; the difference between the anisotropies in the two cases is quite marked. It should be noted that the difference between the average magnetic moment, which is often used to differentiate between these geometries, is not so marked and a consideration of this data alone could be misleading.

A similar situation occurs in the octahedral and tetrahedral nickel(II) complexes since the cubic-field ground state in the former is an orbital singlet, whereas that in the latter is an orbital triplet. The magnetic anisotropies are therefore expected to be as marked as in the above example of cobaltous complexes. In Table VIII, some representative data at room temperature are included for the octahedral and tetrahedral nickel(II) complexes. Although

TABLE VIII
Room-Temperature Anisotropies of Cobalt (II) and Nickel (II) Complexes in Different Symmetry

Compound	Symmetry	$(\Delta K/K)$	$\bar{\mu}$(B.M)
$(NH_4)_2\ Co(SO_4)_2 6H_2O$	Octahedral	30%	5.07
Cs_3CoCl_5	Tetrahedral	7%	4.62
$[(Ph_2MeAsO)_4CoNO_3]^+NO_3^-$	Square pyramidal	60%	5.07
$NiSO_4 \cdot 6H_3O$	Octahedral	8%	3.2
$(Et_4N)_2NiCl_4$	Tetrahedral	16%	3.89
(isopropylsal) $NiCl_4$	Tetrahedral	27%	3.24
$[Ph_2MeAsO)_4NiNO_3]^+NO_3^-$	Square pyramidal	25%	3.11

the anisotropies show a similar trend, the difference is not always so large as, for example, the difference between the octahedral $Ni(SO_4)_2 \cdot 6H_2O$ and the tetrahedral tetrachloronickelate(II) complex. The large difference in the percentage anisotropy between the two tetrahedral complexes shows clearly how intimately the anisotropy is related to the detailed electronic structure of the metal ion, even in such complexes of relatively higher symmetry. We discussed the implications of this difference in Section IV. B.

As an extension of this correlation, it has been suggested (106) that the measurement of paramagnetic anisotropy, if applied with care, can differentiate between 5- and 6- (or even 4) coordinated metal complexes. In Table VIII we include the percentage molecular anisotropy data for the square pyramidal cobalt(II) and nickel(II) complexes. The square pyramidal cobaltous complex shows a markedly large anisotropy as compared to the octahedral and tetrahedral complexes, hence the measurement of paramagnetic anisotropy could be used to decide the molecular geometry (106). However, Table VIII shows that the same is not true for the nickel(II) complexes where the anisotropies of the tetrahedral and square pyramidal complexes are very similar and the anisotropies of the octahedral complexes are significantly different. This suggests the possibility of using the paramagnetic anisotropy data to differentiate between the 5- and 6-coordinated nickel(II) complexes, but not between the 5- and 4-coordinated complexes. In the case of ferric complexes this correlation cannot be extended very far. For example, the square pyramidal $Fe(dtc)_2X$ system (see Section IV. C) shows anisotropies at room temperature varying between 3 to 10% of the average susceptibility. Even 6- (or 4-) coordinated iron(III) complexes with large ZFS would show anisotropies of this order (cf. Section IV. D). In this connection it should be mentioned that the high-spin square pyramidal Fe(salen)Cl is quoted to be very highly anisotropic (106), although details of the results are not available. The large anisotropy quoted for this compound is rather surprising as it would require the ZFS to be unusually large, although Mossbauer-effect measurements show the ZFS to be, as expected, very small.

In our above discussion of the correlation between paramagnetic anisotropy and molecular geometry, we referred to the comparison of the molecular anisotropies, although it would indeed be desirable to compare the crystal anisotropies. The comparison of molecular anisotropies, however, presents a better background for our understanding and can be directly extended to the crystal anisotropies as well, However care should be taken to do measurements in different arbitrary directions and to determine the maximum crystal anisotropy.

A more detailed correlation between molecular geometry and paramagnetic anisotropy (hence LF parameters) has recently been attempted. Within the framework of the point-charge model, the ligands can be assumed as point dipoles, and any distortion in the metal-ligand system would affect the crystal field acting on the metal ion and therefore the paramagnetic anisotropy. A correlation between paramagnetic anisotropy and the geometric distortion of the molecule is, therefore, possible on the basis of a point-charge model. A quantitative illustration of how the paramagnetic anisotropy depends on the geometric angular distortion was already shown in Fig. 34 and 35 for the square pyramidal cobalt(II) and nickel(II) complexes.

A good example to demonstrate the success and limitations of such a correlation is provided by the anisotropy studies on tetrahedral cobaltous complexes (18). Figure 43*a* represents a regular tetrahedron constructed from the alternate corners of a cube, through the centers of whose faces pass the Cartesian axes x, y and z. If the tetrahedron is distorted so as to bring the

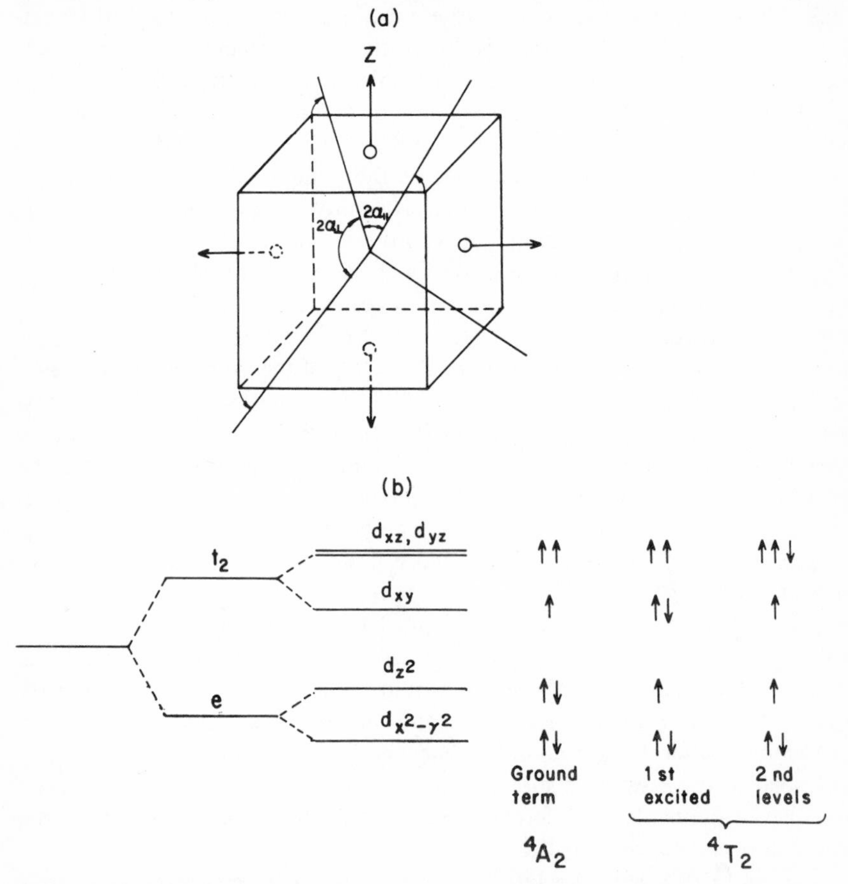

Fig. 43. (*a*) Angular distortion of a regular tetrahedron. (*b*) Splitting of *d*-orbitals and distribution of the electrons in a distorted tetrahedral Co(II) complex (18).

tetrahedral vertices closer to the Z axis, $\alpha_{\parallel} < \alpha_{\perp}$. This distortion decreases the repulsion between the ligand point charges at the tetrahedral positions and an electron in the d_{xy} orbital, whereas that between the charges and electrons in the d_{xz} and d_{yz} orbitals is increased. The splitting of d- orbitals in a distorted tetrahedral crystal field is shown in Fig. 43b. The distortion of the crystal field does not affect the $^4A_2$ ground term which corresponds to the configuration $(d_{x^2-y^2})^2\ (d_{z^2})^2\ (d_{xy})^1\ (d_{xz}\ d_{yz})^2$. The lowest quartet excited configuration that can be written is $(d_{x^2-y^2})^2\ (d_{z^2})^1\ (d_{xy})^2\ (d_{xz}, d_{yz})^2$ and is orbitally nondegenerate. The next higher quartet configuration is $(d_{x^2-y^2})^2$ $(d_{z^2})^1\ (d_{xy})^1\ (d_{xz}, d_{yz})^3$ and is a doublet. Following these arguments based on a point-charge model, Figgis et al (18). showed that for a distortion of the type shown in Fig. 43a, the $^4T_2$ term will split according to the scheme shown in Fig. 2. These simple arguments, therefore, led them to conclude that

$$\alpha_{\parallel} < \alpha_{\perp}, \qquad \Delta_{\parallel} < \Delta_{\perp}$$

$$\alpha_{\parallel} > \alpha_{\perp}, \qquad \Delta_{\parallel} > \Delta_{\perp}$$

where $\Delta_{\parallel}$ and $\Delta_{\perp}$ are as shown in Fig. 2. The values of $\Delta_{\parallel}$ and $\Delta_{\perp}$ were deduced from their measurements of paramagnetic anisotropy on three tetrahedral cobaltous complexes, and the results are given in Table IX. The relative values of $\alpha_{\parallel}$ and $\alpha_{\perp}$, as deduced from the x-ray structural data on Cs_2CoCl_4 and $K_2Co(NCS)_4 \cdot 4H_2O$, were indeed found to be in agreement with the predictions based on the paramagnetic anisotropy (cf. Table IX). In Cs_3CoCl_5, the x-ray structural data of Powell and Wells (15), however, showed $\alpha_{\parallel} > \alpha_{\perp}$, contrary to the expectation from the anisotropy data. This led Figgis et al. to redetermine the crystal structure of Cs_3CoCl_5 more accurately. Their results confirmed the expectation, that $\alpha_{\parallel} < \alpha_{\perp}$, based on the paramagnetic anisotropy data.

The above agreement appears quite impressive and tends to suggest the possibility of employing the magnetic anisotropy data to obtain such structural information (or at least to show such correlation between the geometric parameters and the anisotropy). Furthermore, the above agreement offers strong validity in favor of a point-charge model. Subsequent measure-

TABLE IX
Relationship between the Molecular Susceptibilities, Ligand Field Parameters, and Molecular Distortion (18)

Compounds	Molecular susceptibilities	Δ	α
Cs_3CoCl_5	$K_{\parallel} > K_{\perp}$	$\Delta_{\parallel} < \Delta_{\perp}$	$\alpha_{\parallel} < \alpha_{\perp}$
Cs_2CoCl_4	$K_{\parallel} < K_{\perp}$	$\Delta_{\parallel} > \Delta_{\perp}$	$\alpha_{\parallel} > \alpha_{\perp}$
$K_2Co(NCS)_4 \cdot 4H_2O$	$K_{\parallel} > K_{\perp}$	$\Delta_{\parallel} < \Delta_{\perp}$	$\alpha_{\parallel} < \alpha_{\perp}$

ments have, however, shown that the above agreement may be perhaps fortuitous. Single-crystal polarized spectral and ESR studies (19) on Cs_3CoCl_5 show that, contrary to the prediction of Table IX, $\Delta_{\parallel} > \Delta_{\perp}$. Zeeman field and ESR studies have further shown that D is negative in Cs_3CoCl_5, which immediately yields (cf. Eq. 18) $\Delta_{\parallel} > \Delta_{\perp}$. A more serious discrepancy was later pointed out on Cs_2CoCl_4. It was shown (20) that the earlier experimental paramagnetic anisotropy data on Cs_2CoCl_4 were grossly in error. The correct measurement established that the sign of the molecular anisotropy in Cs_2CoCl_4 is $K_{\parallel} > K_{\perp}$, and not $K_{\perp} > K_{\parallel}$ as deduced by Figgis et al. (cf. Table IX). When these corrections are taken into account, it is clear that the experimental observations are opposite to the predictions contained in Table IX on the basis of a point-charge model. Thus it appears that there is perhaps no direct and simple relationship between the molecular anisotropies and the stereochemical disposition of the ligands in such systems, and any attempt to formulate such a correlation on the basis of a point-charge model appears to be an oversimplification of the true situation.

A slightly different approach, based again on the point-charge model, correlating the paramagnetic anisotropy with the molecular geometric distortion was pursued recently. Here, the crystal field acting on the paramagnetic ion has been parametrized in terms of a number of parameters which include the geometric distortion as one of them. This approach has been applied to a large number of complexes, including some very covalent ones. A brief discussion of this approach was presented in Section IV. C. in the case of the square pyramidal cobalt(II) and nickel(II) complexes. We refrain from giving a detailed account of this correlation as a very elaborate description has recently appeared in a monograph (107). One difficulty of the approach is the large number of parameters involved. It was found that they could not always be uniquely determined from the anisotropy data. Furthermore, the application of a point-charge model to a polyatomic ligand system bonded to the metal ion is always an enigma.

In a crystal belonging to a triclinic or monoclinic system, it is possible, in principle, to deduce the orientation of the molecular symmetry axes from the measurements of paramagnetic anisotropy. This was illustrated very elegantly by Krishnan and Mookherji (107a) in their studies on the triclinic copper sulphate pentahydrate. By doing measurements in a large number of crystallographic planes, they were able to deduce, only from the anisotropy data, the direction cosines of the symmetry axis of the molecule, which matched well with those deduced from the x-ray structural studies. The application of anisotropy studies to obtain the orientation of the molecular planes of aromatic hydrocarbons is, of course, well-known and is surveyed elsewhere (1, 9).

Recently, an attempt was made by Garloch and Quested (8) to deduce the orientation of the LF (i.e., K_i) axes from the measurement of paramagnetic anisotropies on the monoclinic $(NH_4)_2Co(SO_4)_2 \cdot 6H_2O$. Using the direction cosines of the K_i axes, as defined by the molecular geometry of the $[Co^{2+}, 6H_2O]$ distorted octahedron, and Eqs. 9 to 11 for the calculation of K_i, they observed that K_x and K_y at 100°K were negative, which was absurd. This led them to the conclusion that the LF axes do not coincide with the geometric axes, although this coincidence is an implicit assumption in all the calculations of the principal molecular susceptibilities. The x-ray structural study on this crystal was reported at room temperature. It is observed that a calculation at room temperature using Eqs. 12 and 13 gives positive and sensible values for K_x, K_y and K_z, with large in-plane anisotropy (12). This suggests that the K_i-axes coincide at room temperature with the bond axes in $(NH_4)_2Co(SO_4)_2 \cdot 6H_2O$, and that the above conclusion of Gerloch and Quested may not be valid. A determination of the crystal structure at low temperature seems desirable. It should be noted that in this crystal $\beta_1 \simeq \beta_2$ (cf. Section II), and thus the use of Eqs. 9–11 may lead to errors.

F. Isotropic Nuclear Magnetic Resonance and Evaluation of Pseudocontact Shifts

Isotropic NMR shifts in paramagnetic transition metal complexes consist of two contributions: the Fermi contact shift and the pseudocontact (or dipolar) shift. The expression commonly used to obtain the Fermi contact term is of the following form:

$$\left(\frac{\Delta H}{H_0}\right)_{cs} = -\frac{\bar{g}\,\beta S(S+1)}{(\gamma_N/2\pi)3kT}A_s \tag{53}$$

where $\bar{g}$ is the rotationally averaged g value for the complex, γ_N is the nuclear magnetogyric ratio, and ΔH is the shift at an applied magnetic field, H_o. A_s is the hyperfine coupling constant and is proportional to the density of unpaired electron spin occupying the S orbitals centered at the nucleus. Thus the spin-density distribution pattern on the ligand can easily be derived from the knowledge of the contact shift. The above equation shows that the contact shift should obey a Curie law of temperature dependence. The pseudocontact shift arises because of a dipolar interaction between the electronic magnetic moment and the nuclear spin. For a magnetically anisotropic system this contribution is nonzero. McConnell and Robertson (108) deduced a simple expression for the dipolar shift in terms of the g tensor, which is given below:

$$\left(\frac{\Delta H}{H_o}\right)_{pcs} = -\frac{\beta^2 S(S+1)}{3kT} \cdot \frac{(3\cos^2\theta - 1)}{r^3} F(g) \tag{54}$$

where θ is the angle between the symmetry axis and the vector connecting the metal ion center and the nucleus. $F(g)$ is an algebraic function of the g-tensor values, and r the distance between the metal ion and the nucleus.

In the derivation of Eqs. 53 and 54 several assumptions were made, which drastically restrict their application. First, it was assumed that the ground state of the paramagnetic ion is an orbital singlet with no thermally accessible excited state. Second, it was implicitly assumed that for $S \geq 1$, the ZFS of the ground state (an orbital singlet) could be ignored. Third, it was also assumed that the orbital contribution to the isotropic PMR shift could be taken into account indirectly through the use of g-tensor components in Eqs. 53 and 54. All these assumptions are, however, not valid for a large number of metal complexes. For example, the isotropic proton shift in complexes with the orbital triplet ground term shows large deviations from the $1/T$ temperature dependence expected by the above equations. Similar deviations are observed in several high-spin iron(III) complexes where the ZFS of the $^6A_1$ ground term is fairly large. It has also now become evident that the orbital contribution to the isotropic proton shift may not always be taken through g values. In view of these shortcomings, Kurland and McGarvey (109) and Bleaney (110) have deduced general expressions of much wider applicability and emphasized that it is more accurate to express the contact and pseudocontact shifts in terms of the principal susceptibilities. These general expressions in rhombic symmetry are given as follows:

$$\left(\frac{\Delta H}{H_o}\right)_{cs} = -\frac{A_s}{3(\gamma_N/2\pi)\beta}\left(\frac{K_x}{g_x} + \frac{K_y}{g_y} + \frac{K_z}{g_z}\right) \tag{55}$$

$$\left(\frac{\Delta H}{H_o}\right)_{pcs} = \frac{1}{3N}\left[\left\{\frac{1}{2}(K_x + K_y) - K_z\right\}\left\langle\frac{3\cos^2\theta - 1}{r^3}\right\rangle + \frac{1}{2N}(K_y - K_x)\left\langle\frac{\sin^2\theta\cos 2\phi}{r^3}\right\rangle\right. \tag{56}$$

where r, θ, and ϕ are the polar coordinates of the resonating proton. Clearly, the knowledge of the principal susceptibilities (or anisotropies) would be most useful not only for calculating the pseudocontact shift, but also for analyzing the contact shift data. Several features of Eqs. 55 and 56 are noteworthy. First, in highly anisotropic systems, the hyperfine coupling constant (A_s) could be quite different from the value obtained from Eq. 53. This is quite important, and a calculation on this line would be desirable. Second, the contact and pseudocontact shifts need not vary as T^{-1}, as demanded in earlier equations; their temperature dependence would follow that of the principal susceptibilities (or anisotropies). As we have seen during our discussion of the magnetic properties of the metal complexes, the principal susceptibilities and anisotropies of complexes with an orbital triplet ground term, and those having an orbital singlet ground term with large ZFS, do not obey a T^{-1} variation. Evidently, the contact and pseudocontact shifts in such cases show deviations from T^{-1} behavior. Equation 56 further shows that the in-plane anisotropy contributes significantly to the pseudocontact shift. It emphasizes that in cases where the susceptibilities are even slightly rhombic, a customary calculation of this term on the assumption of axial symmetry could seriously underestimate it. In axial symmetry, $K_x = K_y = K_\perp$ and $K_z = K_\parallel$, and Eq. 56 simplifies to

$$\left(\frac{\Delta H}{H_o}\right)_{pcs} = \frac{1}{3N}(K_\perp - K_\parallel)\left\langle\frac{3\cos^2\theta - 1}{r^3}\right\rangle \tag{57}$$

Equations 56 and 57 show how the psuedocontact shift depends directly on the magnetic anisotropy of the system. Evidently, this contribution is expected to be large and predominant in all those cases where the paramagnetic anisotropy is large. In Sections III and IV. A to D, we have discussed the criteria and examples of anisotropic systems which should serve as guidelines for deciding the cases of high anisotropy. Ratio and spin-isolated nuclei methods which have been used for the calculation of the pseudocontact shifts are approximate and based on assumptions which are not always valid.

A customary method to calculate the pseudocontact shift is to use the g-tensor values. In situations where such ESR data was not available earlier (as is the case for the orbital triplet ground term complexes), the contribution of this term was generally ignored. This is evidently incorrect since such systems usually exhibit very large anisotropy, hence the large contribution of the pseudocontact shift. It should also be emphasized that the ESR experiment is generally done at very low temperatures ($\sim$ 4°K), and resonance is observed from the lowest spin level, although isotropic PMR experiments are done in a temperature range (+60°C to − 100°C) where all the levels within 200 to 300 cm^{-1} are populated. Evidently, this factor should be taken into account in cases where g values are used for the calculation of the pseudocontact shift. This has been adequately illustrated by Horrocks and coworkers (111, 112), and will be discussed below, along with separate discussions on the contact and pseudocontact shifts.

1. Pseudocontact Shift

The usefulness of paramagnetic anisotropy in the calculation of the pseudocontact shift was originally shown by Horrocks and his co-workers. Recently, a number of studies in this direction have been reported which illustrate several aspects of this term. A few examples which highlight these aspects follow.

a. Co(acac)$_2$py$_2$. The bis(pyridine) adduct of cobalt(II) acetylacetonate

$[Co(acac)_2py_2]$ provides a good example to illustrate the application of paramagnetic anisotropy for the calculation of the pseudocontact shift in complexes with an orbital triplet ground state. The cobalt atom is coordinated to the two planer acetylacetonate groups, with the two pyridines bonded axially. The symmetry around the cobalt atom is thus pseudo-octahedral, hence the ground state in the cubic crystal field is $^4T_1$. ESR data on this complex are not available, but the system was the first for which a quantitative separation of the dipolar and contact shifts was attempted long ago by the spin-isolated nuclei method.

Horrocks (113) has measured the paramagnetic anisotropy of this compound at room temperature and estimated the pseudocontact term. The principal susceptibilities are strongly rhombic: $K_x = 9836$, $K_y = 11{,}346$, and $K_z = 7526$ ($\times 10^{-6}$ cm^3 mole^{-1}). Thus Eq. 56 is applicable. The estimated pseudocontact and contact terms of the α, β, and γ pyridine protons and carbons are given in Table X. The large pseudocontact term for all the positions is quite noticeable. For the protons, this contribution is, in fact, larger in magnitude than the observed isotropic shifts, hence its correction completely alters the interpretation. It should also be noted that the dipolar terms for the carbon nuclei are larger than those for the corresponding protons. The values of the pseudocontact shift calculated by the paramagnetic anisotropy method are in agreement with those obtained by the spin-isolated nuclei method.

b. Low-Spin Ferric Heme Proteins. Isotropic proton shift measurements on a large number of low-spin ferric heme proteins have been reported. The ground state of the ferric ion in these proteins is formally a $^2T_2$ in the cubic field, which (of course) splits into three orbital singlets in the rhombic nature of the LF in these complexes. An extensive amount of ESR studies have been reported, thus the g tensors are available for the calculation of the pseudocontact shift. Such calculation has already been done and the values of the pseudocontact shift obtained (114). In two interesting papers, Horrocks and Greenberg (111, 112) illustrated beautifully the shortcoming of such a

TABLE X
Isotropic, Pseudocontact, and Contact Shifts for $Co(acac)_2py_2$ at 293°K (113)

Atoms	$\left(\frac{\Delta H}{H_o}\right)_{iso}$	$\left(\frac{\Delta H}{H_o}\right)_{pcs}$	$\left(\frac{\Delta H}{H_o}\right)_{cs}$
$\alpha - H$	− 32.9	+39.5	− 72.5
$\beta - H$	− 5.0	+18.1	− 23.1
$\gamma - H$	+ 9.4	+15.6	− 6.2
$\alpha - C$	+199	+92.5	+106
$\beta - C$	−229	+35.7	−265
$\gamma - C$	+ 73.8	+28.3	+ 45.5

calculation of the pseudocontact shift from the g values and demonstrated the advantages of the magnetic anisotropy data for this purpose.

Since no paramagnetic anisotropy data have been reported for the low-spin ferric heme proteins, Horrocks and Greenberg calculated the theoretical values of K_x, K_y, and K_z with the help of the experimental g values. The standard method of LF calculations was employed, and second-order Zeeman contributions were included. It was observed that, in some cases (e.g., met-myoglobin cyanide, cytochrome P450), estimation of the pseudocontact shifts directly from the g-tensor anisotropy data without inclusion of the second-order Zeeman-term effects or excited-state contributions led to either an overestimation or underestimation of this contribution by 30 to 40%. It can perhaps be added that the effects of the second-order Zeeman term and low-lying excited states are always included in the susceptibility, but not directly in the g values. This example stresses the necessity of the magnetic anisotropy data for an accurate evaluation of the dipolar shift.

Equation 56 consists of two terms: the axial term $[\frac{1}{2}(K_x + K_y) - K_z]$ and the equatorial term involving the in-plane anisotropy term $(K_y - K_x)$. Most of the low-spin ferric heme proteins show strongly rhombic g tensors; for example, for ferricytochrome C, $g_x = 1.25$, $g_y = 2.25$, and $g_z = 3.15$. Thus the relative importance of the two terms, especially the equatorial term, is interesting to examine. This has been done by Horrocks and Greenberg, and the results of their calculations on ferricytochrome C are given in Table XI. The effect of the equatorial term is very large. This stresses the necessity of taking into account the in-plane anisotropy for the calculation of the pseudocontact shift, wherever it exists.

c. Trivalent Metal Acetylacetonates. Recently, in an interesting paper by Doddrell and Gregson (115), pseudocontact shifts were estimated at room temperature for ^{13}C and ^{1}H in a number of trivalent transition metal complexes of V(III), Cr(III), Mn(III), Fe(III), and Ru(III). As mentioned earlier,

TABLE XI
NMR Data and Calculated Pseudo contact Shifts for Ferricytochrome *e* at 298°K[a]

Protons	$\left(\frac{\Delta H}{H_0}\right)^{b}_{obs}$	$\left(\frac{\Delta H}{H_0}\right)^{ax}_{pcs}$	$\left(\frac{\Delta H}{H_0}\right)^{eq}_{pcs}$	$\left(\frac{\Delta H}{H_0}\right)_{cs}$
CH_3b_1	−31.9	3.9	−2.9	−32.9
CH_3b_2	−27.5	3.9	−2.3	−29.1
CH_3b_3	− 6.8	3.9	2.3	−13.0
CH_3b_4	− 3.8	3.9	2.3	−10.0
Meso H *e*	14.9, 17.1	10.1	1.4	3. 4, 5.6
Imidazole $N_3 - H$	− 4.6	−12.8	0.1	8.1

[a]The shifts are in ppm with respect to TMS (111).
[b]Corrected for diamagnetic shift including the ring current effect.

Cr(III) and Fe(III) acetylacetonates have an orbital singlet as the ground state and the anisotropies are extremely small. The pseudocontact shifts are therefore expected to be very small. On the other hand, the shifts are expected to be quite large in the case of $Ru(acac)_3$ and, to some extent, in $Mn(acac)_3$ In Table XII, some of their experimental and computed results are given. The pseudocontact shift was calculated using Eq. 57, and experimental (in some cases calculated) values of the paramagnetic anisotropy were used. The results in Table XII are quite illustrative. As expected, for $Cr(acac)_3$ and $Fe(acac)_3$, the pseudocontact contribution to the proton shift is very small, whereas in $Ru(acac)_3$ and $Mn(acac)_3$ it is quite large. For $V(acac)_3$, they estimated a nonnegligible dipolar contribution to the isotropic proton shifts, which is contrary to the earlier assumptions (116, 117). The relative contributions of the pseudocontact shifts to ^{13}C and ^{1}H are not uniform. For example, the relative contribution of ^{13}C to the pseudocontact shift in $Ru(acac)_3$ is much larger than that in $V(acac)_3$.

TABLE XII

^{1}H and ^{13}C NMR Parameter for Some Paramagnetic Trivalent Transition Metal Acetylacetonates

Compound	NMR Parameters	$\underline{C}$-H	C-$\underline{H}$	$\underline{C}$-H_3	C$\underline{H}_3$
	$(\Delta H/H_o)_{iso}$	+255.8	−34.6	+328.4	−42.8
$V(acac)_3$	$(\Delta H/H_o)_{pcs}$	+ 11.4	+ 4.7	+ 1.2	+ 1.0
	$(\Delta H/H_o)_{cs}$	+244.4	−39.3	+237.2	−43.8
	$(\Delta H/H_o)_{iso}$			+157	−38
$Cr(acac)_3$	$(\Delta H/H_o)_{pcs}$			0	0
	$(\Delta H/H_o)_{cs}$			+157	−38
	$(\Delta H/H_o)_{iso}$	−298.8	−12.6	− 41.2	−23.0
$Mn(acac)_3$	$(\Delta H/H_o)_{pcs}$			small	small
	$(\Delta H/H_o)_{cs}$			0.41	0.23
	$(\Delta H/H_o)_{iso}$		+32.9	−252.5	−19.6
$Fe(acac)_3$	$(\Delta H/H_o)_{pcs}$	+ 1.0	+ 0.5	0	0
	$(\Delta H/H_o)_{cs}$		+33.4	−252.5	−19.6
	$(\Delta H/H_o)_{iso}$	−42.5	+34.9	+ 48.1	+ 7.6
$Ru(acac)_3$	$(\Delta H/H_o)_{pcs}$	+17.2	+ 7.1	+ 1.8	+ 1.6
	$(\Delta H/H_o)_{cs}$	−59.7	+27.8	+ 46.3	+ 6.0

[a]Shifts are in ppm with respect to TMS (115).

d. Halo-bis(diethyldithiocarbamato)iron(III). All the examples discussed above have acknowledged the importance of pseudocontact shift in complexes with orbital triplet (and, in some cases, orbital singlet) ground terms. Furthermore, these estimates were made only at room temperature, and the effect of the ZFS could not be shown as the splitting was too small [e.g., $Fe(acac)_3$, etc.]. A beautiful example that demonstrated, for the first time, the effect of increasing ZFS on the magnitude and temperature dependence of the pseudocontact shift has recently been provided by Dhingra et al. (118) on the intermediate spin, $S = 3/2$, $Fe(dtc)_2X$ (X = Cl, Br, I), see Section IV.C. They have measured the isotropic proton shifts on all three derivatives of this series between +60° to −60°C. The ground state of the ferric ion in these distorted square pyramidal complexes is an orbital singlet $^4A_2$, hence it is tempting to conclude that the pseudocontact shift would be negligibly small. No detailed ESR data are available (because of line broadening), except for some data reported at 4°K on an analogous derivative Fe(isopropyl dtc)$_2$Cl (82). However, as discussed in Section IV.C, detailed paramagnetic anisotropy data on $Fe(dtc)_2$ X system have been reported between 77° to 290°K (84, 85). Using the paramagnetic anisotropy data (see Table VI and Fig. 32) and the geometric factors calculated from the known crystal structure data at room temperature, the pseudocontact shift was estimated in each case at different temperatures from Eq. 56. The results are summarized in Fig. 44, together with the data at 273°K listed in Table XIII.

Table XIII shows that the pseudocontact shift in this system increases along the series, and it is about 4, 11, and 13% of the total observed shift. At low temperatures this contribution becomes larger, especially for the bromo and iodo derivatives. This result highlights the danger involved in neglecting the pseudocontact shift in cases where the ground state of the metal ion is an orbital singlet. Using the g values available on the isopropyl derivative (82), a calculation of the pseudocontact shift shows that in no case is it more than 2%. This reinforces the conclusion deduced above regarding the usefulness of the paramagnetic anisotropy data for an accurate calculation of this contribution. We will discuss shortly the reason for the above variation of the pseudocontact shift in the $Fe(dtc)_2X$ system.

Figure 44 shows that the pseudocontact shift for the chloro derivative obeys a $1/T$ law very closely, whereas marked deviations from such behavior are observed for the bromo and iodo derivatives. The above variation of the pseudocontact shift along the series and with temperature can be easily understood if we take into account the ZFS of the $^4A_2$ ground state (cf. Table VI), which is also given in Fig. 44 for reference. As explained earlier in Section IV.C, δ is the total ZFS of the $^4A_2$ state, and D and E are the axial and rhombic ZFS parameters. The principal susceptibilities in Eq. 56 can be expressed in terms of these parameters, and the expression for the pseudocontact shift can be rewritten as follows (110):

$$\left(\frac{\Delta H}{H_o}\right)_{pcs} = \frac{1}{6kT}\left[\beta^2 S(S+1)\right]r^{-3}F' + \frac{1}{60k^2T^2}\left[\beta^2 S(S+1)(2S-1)(2S+3)\right]r^{-3}F'' \tag{58}$$

where

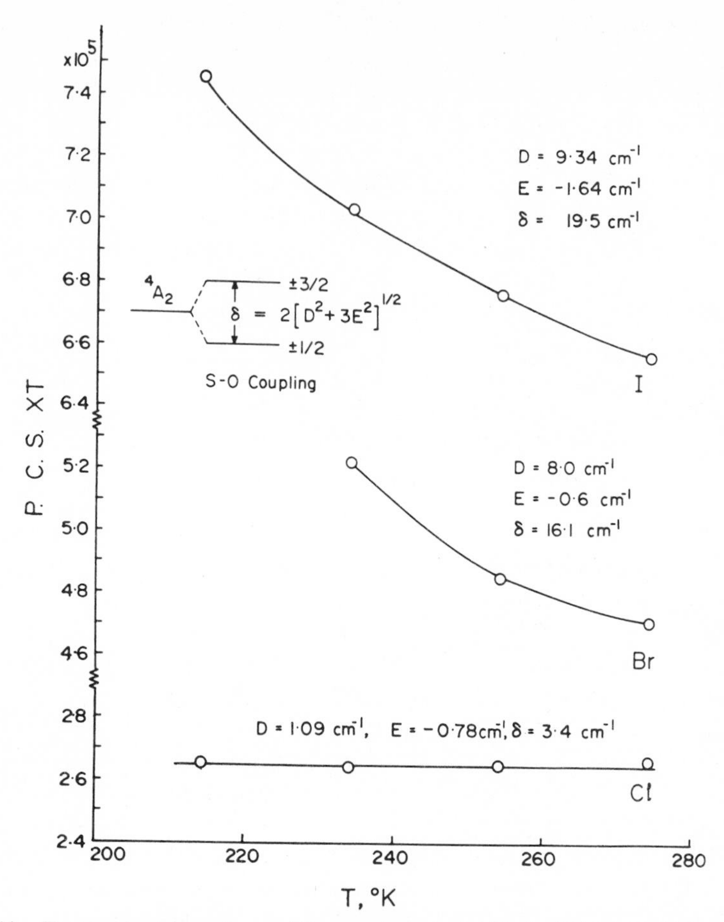

Fig. 44. Temperature dependence of the pseudocontact shifts for $Fe(dtc)_2X$ (118).

$$F' = \left(\frac{2}{3}\right)\left\{g_x^2\left[3\sin^2\theta\cos^2\phi - 1\right] + g_y^2\left[3\sin^2\theta\,\mathrm{Sin}^2\phi - 1\right] + g_z^2\left[3\cos^2\theta - 1\right]\right\}$$

$$F'' = \left(\frac{2}{3}\right)\left\{g_x^2 D_x\left[3\sin^2\theta\cos^2\phi - 1\right] + g_y^2 D_y\left[3\sin^2\theta\sin^2\phi - 1\right] + g_z^2 D_z\left[3\cos^2\theta - 1\right]\right\}$$

and D_x, D_y, and D_z, are the components of the ZFS tensor and are related to D and E. Equation 58 consists of two terms: one varying as $1/T$ and the

TABLE XIII
Isotropic, Pseudocontact, and Contact Shifts for $Fe(dtc)_2X$
(X = Cl, Br, I)

Compound	Isotropic shift, cps	Pseudocontact shift, cps	Contact shift, cps
	1535[b]		
$Fe(dtc)_2Cl$	1049[b]	98	1194
	(1292)		
	1685[b]		
$Fe(dtc)_2Br$	1299[b]	173	1319
	(1492)		
	2000[b]		
$Fe(dtc)_2I$	1720[b]	241	1619
	(1860)		

[a]All the shifts are negative and refer to the CH_2 protons at 273°K (118).
[b]The peak due to the CH_2 protons split into two components. The value given in the parentheses are the mean of these values.

other varying as $1/T^2$. The $1/T$ term depends only on the g values, whereas the $1/T^2$ term depends on the ZFS as well. Evidently, as the ZFS of the $^4A_2$ term increases, the pseudocontact term increases in magnitude and the $1/T^2$ term becomes large enough to cause deviation from the $1/T$ temperature dependence. The values of the total ZFS (δ) for the chloro, bromo, and iodo derivatives are 3.36, 16.1 and 19.5 cm^{-1} respectively. Thus the contirbution of the $1/T^2$ term is expected to increase with the increase in the ZFS from the chloro to the iodo derivative, and this is precisely what is observed in Fig. 44.

2. *Contact Shift*

Once the pseudocontact term is known, the contact term can be easily separated out. Equation 55 gives the relation between the principal susceptibilities and the contact shift, which should enable a more accurate calculation of the hyperfine coupling constant, A_s. However, such an estimate does not appear to have been done. Equation 55 also shows that the temperature dependence of the contact shift should reflect very closely the electronic structure of the metal ion which governs, as we have seen in earlier subsections, the temperature dependence of the susceptibilities. Such a correlation of the contact shift data has recently been shown by Dhingra et al. for the $Fe(dtc)_2X$ system. In Fig. 45, the temperature dependence of the observed and contact shifts is shown. Here again, the contact shift for the chloro derivative obeys $1/T$ dependence very closely, whereas deviations are observed for the bromo and iodo ones. This behavior is in agreement with the increasing ZFS of the $^4A_2$ ground term and can be quantitatively explained on this basis. However, an interesting observation is that the temperature dependence of

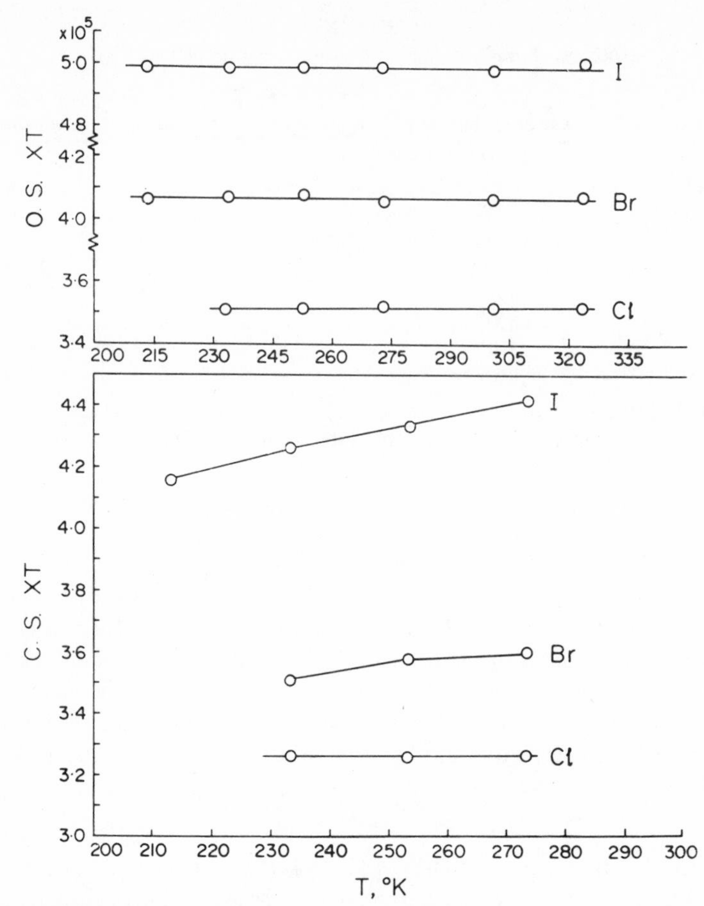

Fig. 45. Temperature dependence of the observed (O.S.) and contact (C.S.) shifts for $Fe(dtc)_2X$ (118).

the observed isotropic shift obeys, in all the derivatives, very closely a $1/T$ temperature dependence, hence this could lead to quite misleading conclusions.

In conclusion, we note that the measurement of paramagnetic anisotropy provides the most accurate and direct method to estimate the pseudocontact shift. A calculation of this contribution from the g values alone could be misleading.

G. Magnetic-Exchange Coupled Systems

In our discussions so far, we have assumed that the magnetic properties of a paramagnetic ion are influenced only by the LFs. In the solid state, however, this is not always true. Magnetic-exchange interactions between the metal ions in the solids affect the magnetic properties significantly, and sometimes they even swamp other effects such as the LF. The exchange interaction is present, even though small, in almost all the so-called paramagnets, and in many cases its presence is dominant only at extremely low temperatures. The extent and manner in which the exchange interaction between the paramagnetic ions influences the magnetic anisotropy depend on the characteristics of the system itself. For example, in a three-dimensional (3-*d*) lattice, the exchange interaction induces an anisotropy in addition to that produced by the LF, below the Neel temperature at which long-range order takes place. On the other hand, in polynuclear cluster complexes, where the long-range order does not normally occur, the anisotropy is not directly affected.

In the area of magnetic-exchange interaction, the interest of the majority of chemists is confined to the study on polynuclear cluster complexes. These complexes provide, in a way, a very convenient system to study the nature and mechanism of the exchange interaction in the absence of complications due to long-range ordering. There is, however, a very striking dearth of single-crystal studies on the cluster complexes, the only detailed study being on some metal dimers. In contrast to this situation, there exist very large and detailed studies on one-, two-, and three-dimensional lattices. These studies have provided the basis for most of our present day understanding of the magnetic-exchange interaction. An exhaustive account of the experimental studies on such extended interacting systems has recently appeared (119). The experimental data are generally analyzed in terms of some models in which the magnetic interaction is often simplified, while the dimensionality of the lattice may be varied. The interaction Hamiltonian may be written as

$$\mathscr{H} = -2J\sum_{i>j}[a(S_i^z S_j^z) + b(S_i^x S_j^x + S_i^y S_j^y)] \tag{59}$$

where the summation is taken over nearest neighboring spins and J is the exchange integral. If $a = b = 1$, we obtain

$$\mathscr{H} = -2J\sum_{i>j}\{S_i^z S_j^z + S_i^x S_j^x + S_i^y S_j^y\} = -2J\sum S_i \cdot S_j \tag{60}$$

This is the Heisenberg model in which the interaction is wholly isotropic. The other extreme situation is obtained by setting $a = 1$ and $b = 0$, which is the case of an anisotropic Ising interaction. The third case, $a = 0$, $b = 1$, is called the *xy* model, or the planar Heisenberg model if one puts the additional requirement that the spins are constrained to lie within the *xy* plane.

The experimental data on polynuclear complexes (which belong to zero dimension) have generally been analyzed in terms of the isotropic Heisenberg model (Eq. 60) and good agreement has been obtained (120). The theory assumes the ground state of the metal ion to be orbitally nondegenerate, the orbital contribution from the excited states being taken through the *g* values.

Recently, the problem of magnetic-exchange interaction between the metal ions with an orbitally degenerate ground state has also been tackled (121, 122). The experimental data on polynuclear complexes are usually limited to the temperature range 77° to 300°K, in which the data are not often sensitive enough to differentiate between various models. Studies on extended systems (e.g., one-, two-, three-dimensional lattices) have often shown the inadequacy of the above models and have added refinements to the above simple pictures. Recently a detailed investigation on some polynuclear copper(II) tetramers down to 1°K has also brought to light the failure of Eq. 60 to account for the temperature dependence of their magnetic susceptibility (123).

1. Polynuclear Complexes

As indicated earlier, the single-crystal magnetic measurements on polynuclear complexes are confined to the studies on dimers of copper(II), cobalt(II), and iron(III). Of these, the most detailed measurements have been reported on the binuclear copper(II) compound, cupric acetate monohydrate. We discuss below the results on these compounds and examine the information obtained from the magnetic anisotropy measurements.

a. Copper(II) Acetate Monohydrate. Copper(II) acetate monohydrate is a compound of central significance in the studies of metal-metal interaction in paramagnetic clusters. Its well-known molecular structure involves a pair of copper atoms, 2.64 Å apart (cf. Cu–Cu distance in metallic copper of 2.65 Å), supported by four acetate bridging groups in an approximate D_{4h} array with two water molecules completing the coordination along the Cu–Cu axis (124). The ground state of the Cu^{2+} ion in such a case is an orbital singlet ($^2B_1$ if the unpaired electron of the Cu^{2+} ion resides in the $d_{x^2-y^2}$ orbital), and an interaction Hamiltonian given by Eq. 60 would give a lower spin singlet ($S = 0$) and an upper spin triplet ($S = 1$) with separation being J (see Fig. 46*b*). The principal molecular susceptibilities are then given by (125)

$$K_{\parallel} = \left(\frac{N\beta^2}{3kT}\right) g_{\parallel}^2 \left[1 + \frac{1}{3}\exp\left(\frac{-J}{kT}\right)\right]^{-1} + \frac{8N\beta^2\kappa^2}{\Delta_{\parallel}}$$
$$K_{\perp} = \left(\frac{N\beta^2}{3kT}\right) g_{\perp}^2 \left[1 + \frac{1}{3}\exp\left(\frac{-J}{kT}\right)\right]^{-1} + \frac{2N\beta^2\kappa^2}{\Delta_{\perp}} \qquad (61)$$

where $\Delta_{\parallel}$ and $\Delta_{\perp}$ are the one-electron excitation energies for the monomer corresponding to $(d_{x^2-y^2}) \leftarrow (d_{xy})$ and $(d_{x^2-y^2}) \leftarrow (d_{xz}, d_{yz})$ transitions, respectively.

The magnetic anisotropy of copper(II) acetate monohydrate has been reported by a number of workers (126–128). Their interpretations revealed serious inconsistencies among themselves and with Eq. 61. Later, a very

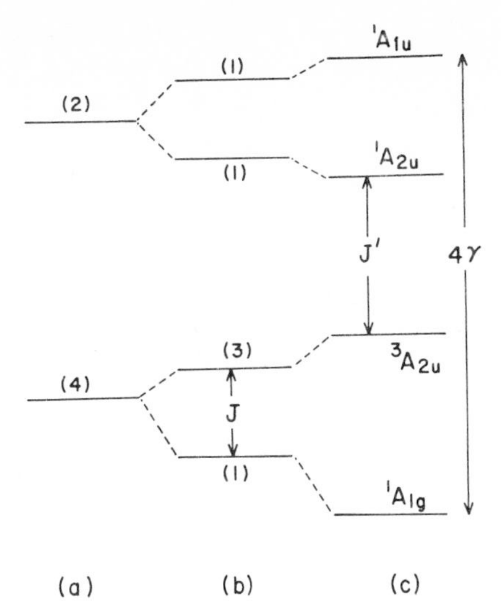

Fig. 46. Schematic energy levels for (*a*) two noninteracting copper atoms; (*b*) Bleaney and Bowers model; (*c*) MO δ-bond model.

careful and detailed study by Gregson, Martin, and Mitra (129) established that all the earlier measurements of magnetic anisotropy of this compound were grossly in error, probably because of erroneous identification of the crystal morphology. The temperature dependence of the molecular anisotropy of this compound, as determined by Gregson et al, is shown in Fig. 47.

The figure shows that the magnetic anisotropy of this dimeric compound follows a temperature dependence which closely resembles that of its average susceptibility. We observe, experimentally, that $K_{\parallel} > K_{\perp}$, which disagrees with an earlier suggestion (130) that the unpaired electron of each Cu^{2+} ion in this compound lies in the d_{z^2} orbital, giving $K_{\perp} > K_{\parallel}$. Furthermore, the percentage anisotropy of this compound is observed to be almost the same as that of "magnetically dilute" copper(II) compounds. This is not surprising since the term involving the effect of the exchange interaction in Eq. 61 gets canceled out in the percentage anisotropy ($\Delta K/\bar{K}$) estimate. The absolute value of the anisotropy is, however, much smaller than the normal copper(II) complexes.

The most outstanding application of the magnetic anisotropy studies on copper acetate monohydrate has been to utilize this result to reinterpret the controversial electronic spectrum of this compound (129). The electronic spectrum of this compound consists of a broad envelope between 10,000 and 20,000 cm^{-1}, with two weaker bands at 11,000 cm^{-1} (z polarized) and 14,000 cm^{-1} (xy polarized). A more intense band is observed at ca. 27,000 cm^{-1}

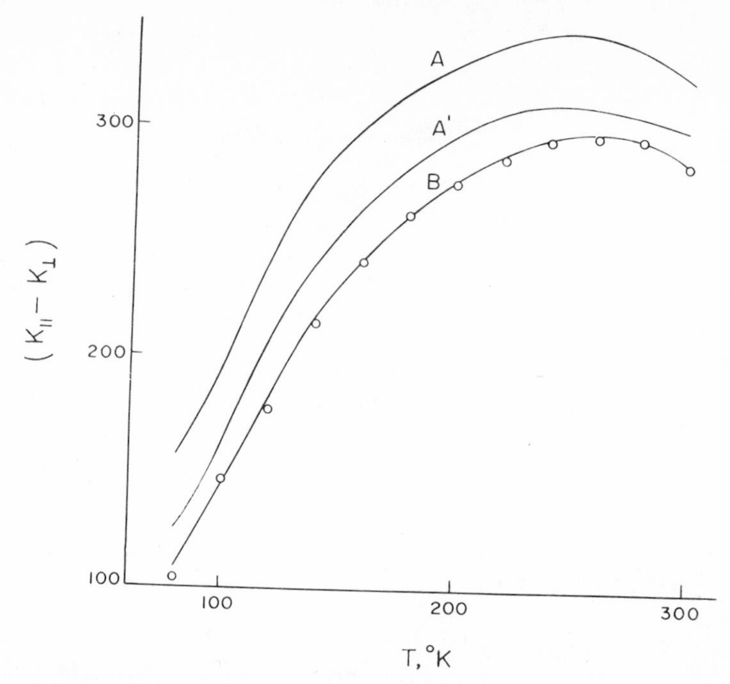

Fig. 47. Temperature dependence of magnetic anisotropy for copper(II) acetate monohydrate: ○, experimental data (129). Full curves are calculated with $g_{\parallel} = 2.34$; $g_{\perp} = 2.06$; $J = -286\ \text{cm}^{-1}$. Curves A and A' have $\Delta_{\parallel} = 11{,}000\ \text{cm}^{-1}$, $\Delta_{\perp} = 14{,}400\ \text{cm}^{-1}$, and $\kappa^2 = 0.8$ and 0.6, respectively. Curve B has $\Delta_{\parallel} = 17{,}000\ \text{cm}^{-1}$, $\Delta_{\perp} = 14{,}400\ \text{cm}^{-1}$, and $\kappa^2 = 0.8$ (129).

(z polarized). The assignment of these bands is rather uncertain and has been widely debated. Two alternative band assignments, as shown below, have been proposed (131, 132):

Band	Proposed assignment: A	B
11,000 cm^{-1}	$(d_{xy}) \longleftarrow (d_{x^2-y^2})$	$(d_{z^2}) \longleftarrow (d_{x^2-y^2})$ and/or $(d_{xy}) \longleftarrow (d_{x^2-y^2})$
14,000 cm^{-1}	$(d_{xz}, d_{yz}) \longleftarrow (d_{x^2-y^2})$	$(d_{xz}, d_{yz}) \longleftarrow (d_{x^2-y^2})$
27,000 cm^{-1}	$(d_{z^2}) \longleftarrow (d_{x^2-y^2})$	?

The origin of the band at 27,000 cm^{-1} is highly disputable and is variously assigned to a charge transfer and double excitation $(d_{z^2}) \longleftarrow (d_{x^2-y^2})$ transition.

In deducing the information about the electronic structure of the copper acetate, Gregson et al. made use of the observation that the calculation of magnetic anisotropy from Eq. 61 implies a marked sensitivity to the anisotropy in the TIP term, $2N\beta^2 (4/\Delta_{\parallel} - 1/\Delta_{\perp}) \kappa^2$. For example, the anisotropy in the TIP term at liquid nitrogen temperature is about 70% of the total anisotropy; even at room temperature, this contribution amounts to about 25%. For this reason it should be possible to assign the LF transitions $(d_{xz}, d_{yz}) \leftarrow (d_{x^2-y^2})$ and $(d_{xy}) \leftarrow (d_{x^2-y^2})$ of this compound from an analysis of its anisotropy data.

In Fig. 47 theoretical curves of Eq. 61 are plotted, together with the experimental data, for different values of κ^2, using the set of energy values A (i.e., $\Delta_{\parallel} = 11{,}000\ \text{cm}^{-1}$ and $\Delta_{\perp} = 14{,}400\ \text{cm}^{-1}$). Constant values of g_i and J have been used. A satisfactory fitting of the experimental data cannot be achieved for acceptable values of κ; a good fit can only be obtained if κ^2 is assumed to be less than 0.5, which is untenable. Figure 47 also includes the theoretical plot of Eq. 61 using the excitation energies normally observed for monomeric copper(II)-oxygen compounds (i.e., $\Delta_{\parallel} = 17{,}000\ \text{cm}^{-1}$ and $\Delta_{\perp} = 14{,}400\ \text{cm}^{-1}$) with the same values of $g_{\parallel}$, $g_{\perp}$, and J. Excellent agreement was obtained when $\kappa^2 = 0.8$, a value typical of copper(II) compounds coordinated by oxygen ligands.

The above analysis of the magnetic anisotropy data indicates that the earlier assignment of the $11{,}000\ \text{cm}^{-1}$ band to the $(d_{xy}) \leftarrow (d_{x^2-y^2})$ transition is no longer tenable, and the alternative assignment B, $(d_{z^2}) \leftarrow (d_{x^2-y^2})$, seems more reasonable. The $(d_{xy}) \leftarrow (d_{x^2-y^2})$ transition is expected to be very weak and, at ca. $17{,}000\ \text{cm}^{-1}$, may be osbcured by the broad envelope of the "copper" band.

There has been some attempt to explain the magnetic behavior of copper acetate on the molecular orbital model (133). In this description, unlike the earlier two-level example (Fig. 46*b*), four terms ${}^1A_{1g}$, ${}^3A_{2u}$, ${}^1A_{2u}$, and ${}^1A_{1g}$, in ascending order of energy are deduced. The energy difference between the bonding (b_{1g}) and antibonding (b_{2u}) MO's (called 2Υ) is now an additional adjustable parameter in this model, and Eq. 61 modifies to the following form (cf. Fig 46c):

$$K_{\parallel} = \frac{N\beta^2}{3kT}\left[\frac{6g_{\parallel}^2}{F} + \frac{48kT}{\Delta_{\parallel}}\right]$$

$$K_{\perp} = \frac{N\beta^2}{3kT}\left[\frac{6g_{\perp}^2}{F} + \frac{12kT}{\Delta_{\perp}}\right] \tag{62}$$

with $F = 3 + [\exp(-J/kT) + \exp(-J'/4kT) + \exp(-2\Upsilon - \tfrac{1}{4}J')/kT]$. However, the additional singlet ${}^1A_{2u}$ lies about $300\ \text{cm}^{-1}$ above the ${}^3A_{2u}$, hence the experimental magnetic susceptibility and anisotropy data in the usual temperature range (80°–300°K) cannot distinguish between these two models, and measurements at much higher temperatures are required to verify the existence of the excited singlet $^1A_{2u}$. Furthermore the magnetic anisotropy data have been unable to contribute directly any new information to the vexed question of the relative contribution to the exchange interaction of the direct Cu–Cu bond and indirect superexchange via the bridging acetate ligands, since Eq. 61 does not depend on the origin of the interaction. However, the electronic structure of this binuclear compound, as deduced from the magnetic anisotropy data, is very similar to that of monomeric, say, cupric sulfate pentahydrate. This suggests that the bond between the two copper atoms in the copper acetate molecule is very weak and thus lends support to the weakly coupled chromophore model of Hansen and Ballhausen (134).

b. $[Fe(salen)]_2O \cdot CH_2Cl_2$. The above ferric salen compound is a prototype of a large family of oxygen-bridged dimeric iron(III) complexes which show appreciable antiferromagnetic magnetic-exchange interaction. The average magnetic moment of these ferric complexes usually lies between 2 to 3 B.M. at room temperature and decreases slowly with the decrease in temperature. It has been observed that the temperature dependence of the magnetic moment of these compounds is rather insensitive to the spin state of the ferric ion, although the high spin state ($S = 5/2$) has often been preferred. Coggon et al. (135) have reported the magnetic anisotropy studies on the above salen compound between 200° to 80°K. Their results show that the anisotropies are extremely small and decrease with decreasing temperature. The small value of the magnetic anisotropy suggests $S = 5/2$ spin state, and its temperature dependence could be fitted to an isotropic value of exchange integral, $J = -87\ \text{cm}^{-1}$. They attributed the observed anisotropy partly to the ZFS of the ground sextet and partly to the anisotropy in g values, which possibly cannot be ascertained uniquely because of several adjustable parameters and uncertainties involved in measuring these extremely low anisotropies.

c. Co(salen). Co(salen) has a binuclear structure with square pyramidal coordination around each cobalt atom. The compound is low spin and the principal magnetic moments at two extreme temperatures are summarized below:

μ_i(B.M.)	300°K	100°K
μ_x	2.54	2.12
μ_y	2.08	1.70
μ_z	1.81	1.51

It is interesting that the magnitude and temperature dependence of the principal magnetic moments of this binuclear compound are very similar to the analogous monomeric low-spin Co(salen)(py) (see Fig. 36). However, no detailed theoretical analysis of the data is available.

2. *Extended Lattices*

As mentioned earlier, very exhaustive experimental studies on one-, two-, and three-dimensional lattices exist. The measurements have been generally reported down to the liquid helium temperatures on single crystals, and results have been interpreted on the basis of an Ising, Heisenberg, or molecular field model. These systems are characterized by long-range ordering, which is theoretically forbidden for the one- and two-dimensional lattices, but allowed for the three-dimensional one. Interchain and interlayer interaction can, of course, cause such ordering, even in one and two dimensions. We discuss below some examples of these systems; a detailed description is available elsewhere (119).

a. Linear Chain. A linear chain can be viewed as a polynuclear system with infinitely long array of metal atoms held up linearly by a ligand system. The ideal situation occurs when these chains are completely isolated so that the interchain interaction can be neglected, and the interaction is only within the chain itself. The theoretical calculations based on the Ising model are available for $S = 1/2, 1, 3/2$ (136–138). Fischer (139) has deduced an expression for the susceptibility in the classical ($S = \infty$) Heisenberg limit, which is given below:

$$\chi = \left[\frac{Ng^2\beta^2S(S+1)}{3kT}\right]\left[\frac{(1+U)}{(1-U)}\right] \tag{63}$$

where $U = \cosh 2JS(S+1)/kT - kT/2JS(S+1)$. This expression has been used for the finite spin. For temperatures $T > 0$, approximate solutions have been derived by Bonner and Fischer (140) by calculating the properties of local rings containing an increasing number of spins and subsequently extrapolating to infinite chain. If $S = 1/2$, an exact solution is possible, which gives the following expressions for the principal susceptibilities:

$$\begin{aligned} K_{\parallel} &= \frac{N\beta^2}{4/J/} g_{\parallel}^2 \left|\frac{J}{kT}\right| \exp\left|\frac{2J}{kT}\right| + \text{TIP} \\ K_{\perp} &= \frac{N\beta^2}{8/J/} g_{\perp}^2 \tanh\left|\frac{J}{kT}\right| + \left|\frac{J}{kT}\right| \text{sech}^2\left|\frac{J}{kT}\right| + \text{TIP} \end{aligned} \tag{64}$$

All these calculations are generally based on the assumption that the interaction is confined to the nearest neighbor only. An elaborate calculation has recently been done for $S = 1/2 \rightarrow 5/2$, taking into account the interaction up to 12 atoms (141). It is observed, however, that the interaction falls off very rapidly after about six atoms.

In Fig. 48, the temperature dependence of the parallel susceptibility of

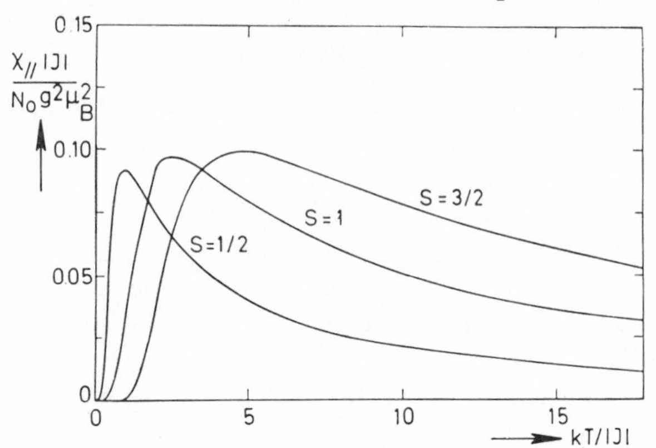

Fig. 48. Dependence on the spin value of the parallel susceptibility of an Ising chain (119).

an ideal linear chain, with $S = 1/2$, 1, 3/2, is shown. Since no long-range ordering exists in these cases, the entropy has to be removed in the short-range-order process. This is reflected in the susceptibility that displays broad maxima. In practice, however, the interaction between chains is always present. An example that shows typical behavior of a practical linear chain is illustrated in Fig. 49 for $CsNiCl_3$ (142). The susceptibility is almost isotropic at

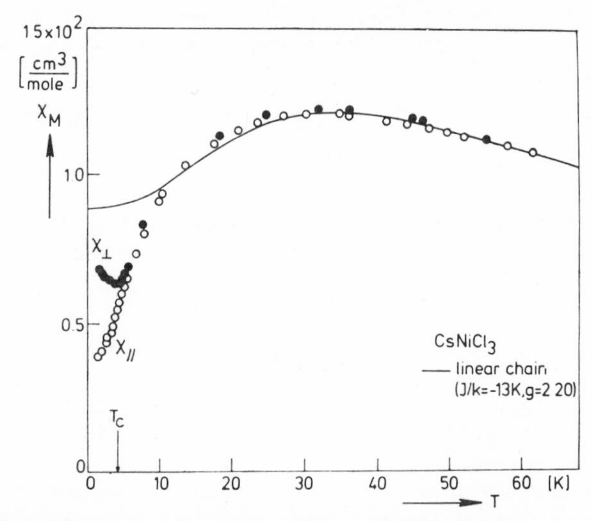

Fig. 49. Susceptibility of $S = 1$ antiferromagnetic Heisenberg chain, $CsNiCl_3$ (142). The full curve is theoretically calculated with $(J/_k) = -13°K$ and $g = 2.20$.

higher temperatures (the anisotropy due to the LF being very small) and passes through a broad maximum, as discussed above. Below about 5°K, long-range ordering takes place and an anisotropy is produced. The Heisenberg model reproduces well the experimental behavior down to ~10°K, below which the theory is inapplicable because of interchain interaction. It should be noted that the anisotropy produced below T_c could partly arise from the ZFS of the nickel(II) ion, which is usually 2 to 3 cm^{-1} and can therefore affect the magnetic properties significantly in this temperature range.

Most of the linear chains conform to the Heisenberg model, although examples of Ising chains are also available. $CsCoCl_3$ is one such example for which $J/_K = -85°$ has been deduced (142). A slight indication of the long-range ordering in the parallel and perpendicular susceptibilities is found around 8°K. Manganese(II) phthalocyanine is another example of an Ising chain (63) with very weak interchain interaction. Its principal susceptibilities conform very well to the ferromagnetic Ising model with $J = +7.6$ cm^{-1}. Copper calcium acetate hexahydrate is another example of a nearly isolated linear chain. The temperature dependence of its principal susceptibilities has been reported down to 1°K and fitted to Eq. 64 with $J \simeq -0.1$ cm^{-1} (143, 144).

b. Layered Structure (two-dimensional lattice). In going from one- to two-dimensional lattices, there arises a basic difference between the Ising model and the Heisenberg and *xy* models. In the case of the ideal chain model there is no transition to long-range order except at $T = 0$ for any type of interaction. The two-dimensional Ising model allows a long-range ordering at finite temperature, but this is not so with the other models. In the isotropic two-dimensional Heisenberg model, the behavior is similar to the linear chain. Accordingly, a broad maximum due to the short-range order effect should be found at higher temperatures, whereas at $T = 0$ the susceptibility should attain a finite value. Calculations on the Ising model have been reported (145–147), but no closed-form theory is available. Generally, high-temperature-series (HTS) expansion and spin-wave theory are used, and they are found to be successful in their respective temperature range. The HTS expansion technique reproduces even the broad maximum in the susceptibility, but it becomes unreliable at low temperatures because of the finite number of terms known in the series. Spin-wave theory has been used to calculate the perpendicular susceptibility for a two-dimensional antiferromagnet, which is given as (119)

$$\chi_{\perp}(0) = \frac{\chi_{\perp}^{0}}{1 + \frac{1}{2}\alpha}\left[1 - \frac{\Delta S(\alpha)}{S} - \frac{e(\alpha)}{(2 + \alpha)ZS}\right] \tag{65}$$

where $\chi_{\perp}^{0}$ (same as $K_{\perp}^{0}$ in our earlier notation) is the susceptibility based on a molecular field model, S is the spin value, Z is the number of nearest neighbors, and $\alpha = H_A/H_E$, H_E being the effective field associated with J, given by $g\beta H_E = 2\pi |J| S$. The anisotropy-dependent quantities, $\Delta S(\alpha)$ and $e(\alpha)$, reflect the effect of zero-point spin deviations on the effective length of the magnetization vector and on the ground state energy, respectively. Thus knowing $\Delta S(\alpha)$ and $e(\alpha)$, one may use the experimental $\chi_{\perp}(0)$ value to determine the exchange constant J.

The $K_2M^{II}F_4$ series provides a good example of a two-dimensional Heisenberg antiferromagnet. A typical set of results on K_2MnF_4 is shown in Fig. 50. Below T_c a large anisotropy is produced. Knowing $(J/_k)$ and α, both $\chi_{\parallel}$ and $\chi_{\perp}$ can be calculated from spin-wave theory, which reproduces well the anisotropy below T_c. The HTS expansion reproduces the broad maximum in susceptibility rather well. It should be noted that the Mn(II) ion having a $^6A_{1g}$ ground state does not have appreciable anisotropy. Copper(II) formate tetrahydrate is another good example with large interlayer separation. The magnetic-exchange interaction is mainly the superexchange-type involving the pi-system of the formate ligands. Its principal susceptibilities in the 77° to 300°K temperature range are reproduced well by the HTS expansion (148). $(C_nH_{2n+1}NH_3)CuX_4$ provides a good example of two-dimensional lattice to study the effects of increasing interlayer separation. By varying n,

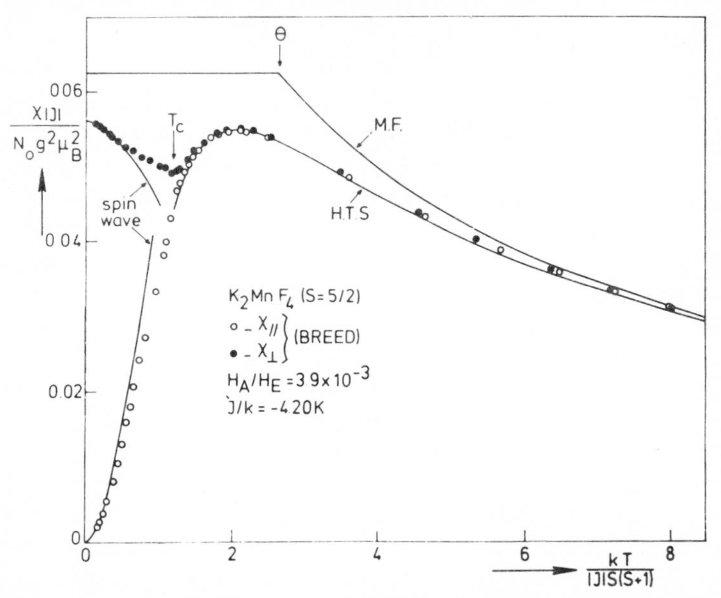

Fig. 50. Parallel and perpendicular susceptibilities of K_2MnF_4, which is an example of a two-dimentional $S = 5/2$ Heisenberg antiferromagnet. The value of J has been obtained by fitting the data to the high-temperature-series expansion (HTS). The experimental data are taken from Breed [*Phys. Letters*, *23*, 181 (1966); *Physica*, *37*, 35 (1967)].

the distance between the Cu(II) ions in the neighboring layers is increased from 9.97°Å ($n = 1$) to 25.8 Å ($n = 10$), while the Cu–Cu separation within the layer ($\sim$ 5.25 Å) is not appreciably changed.

c. Three-Dimensional Lattices. Compounds with three-dimensional lattices exist most abundantly and have been extensively studied. As mentioned earlier, even one- and two-dimensional lattices behave as three-dimensional ones below the temperature (T_c) at which long-range order sets in. Among the classic examples of three-dimensional lattices, attention should be drawn to the perovskite ($KNiF_3$) and rutile (MnF_2) structures.

There is no exact theoretical treatment of three-dimensional Ising and Heisenberg models, although several approximate theories are available. In the temperature region $T \ll T_c$, spin-wave theory gives very good agreement with the experimental results. An interesting point is that unlike the one- and two-dimensional cases, molecular field theory agrees very well with the experiment in the three-dimensional cases.

A very typical result of the magnetic anisotropy measurements on three-dimensional antiferromagnets is provided by MnF_2, which is a good example of an isotropic Heisenberg model (see Fig. 51). At room temperature the anisotropy is only about 0.1% of its average susceptibility, as expected for the $^6A_{1g}$ ground state of the Mn(II) ion. However, a large anisotropy develops below T_c (67.3°K), with the susceptibility along the c axis of this tetragonal crystal decreasing fast with the lowering in temperature (149). This suggests that below the Neél temperature the spins are aligned along the c axis of the crystal, which is in accord with the later neutron diffraction and NMR measurements. Thus the magnetic anisotropy measurement provides a con-

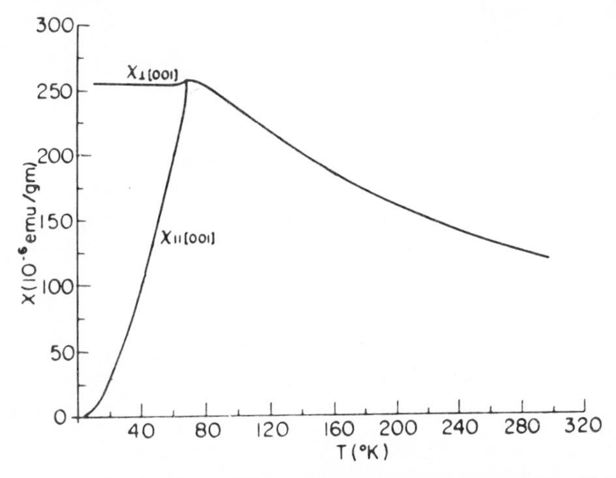

Fig. 51. The behavior of the parallel and perpendicular susceptibilities of MnF_2, which is a typical example of an isotropic three-dimensional antiferromagnet.

venient method to determine the direction of spin alignment in three-dimensional lattices. The anisotropy induced by the exchange interaction below T_c is explained well on the basis of spin-wave theory (150), which gives $J/_k = -0.3°$ along the c axis and $J/_k \simeq 0$ along [100]. Another example that provides close approximation to a three-dimensional nearest-neighbor Heisenberg system is $KNiF_3$, with $T_c = 246°K$. The exchange-interaction constant $(J/_k)$ is deduced to be $(J/_k) = -51°$. Examples of three-dimensional Ising antiferromagnets are found in some rare earth compounds, such as $DyAlO_3$, $DyAl_5O_{12}$, and others.

V. CONCLUDING REMARKS

In the preceding sections we have discussed all the major applications of magnetic anisotropy and highlighted the advantages and limitations of this technique in solving the problems of modern inorganic chemistry. The technique appears very useful and powerful in deducing the nature of ground state properties and ligand fields in a wide variety of transition metal complexes. Very often the information obtained from the magnetic anisotropy studies is found to be comparable to, and sometimes more abundant than, that obtained from other techniques. A point, although stressed on many occasions previously, that needs to be emphasized again is the value of the correct and appropriate analysis of the experimental anisotropy data. Our reanalysis of the anisotropy data on $K_3Fe(oxalate)_3 \cdot 3H_2O$ (cf. Fig. 42) serves as a good illustration for this purpose. We have stressed the desirability of analyzing the data in terms of $\Delta\mu^2$, as this provides a very convenient and sensitive method for the physical understanding of the results. We also suggest that the direct analysis of the anisotropy data should be preferred over the analysis of the principal magnetic moments. Our reanalysis of the data on $VO(acac)_2$ and $K_3Fe(oxalate)_3 \cdot 3H_2O$ again serves as a good example in this regard. While analyzing the experimental anisotropy data, care should be taken about the uniqueness of the LF parameters. The temperature dependence of one molecular anisotropy can at best determine two parameters uniquely; thus caution should be observed when the number of disposable parameters is larger.

It would appear from Section IV that the experimental anisotropy data in most cases are deduced on the assumption of axial symmetry. While this assumption is very convenient for the calculation and analysis of the molecular anisotropies, it may not be valid in many cases. In those cases where the axial (or approximately axial) nature of the LF is known from ESR or other experiments, the above assumption is valid; in other cases, it is always desirable to calculate, if possible, the rhombic molecular anisotropies or susceptibilities. $Fe(dtc)_2X$ provides a good example, where the calculation of rhombic molecular anisotropies has yielded information about the ground state which had earlier remained obscured in the analysis of Mössbauer and other data because of *a priori* assumption of axial symmetry (cf. Section IV.C). For complexes with orbital triplet ground terms or low-lying orbitally degenerate terms, axial symmetry may not be valid and a recalculation of the molecular anisotropies is indicated in such cases.

An obvious limitation of the magnetic anisotropy measurement is its requirement of a single crystal and knowledge of some structural data, in general. Of course, other techniques like ESR, optical spectroscopy, and so on which give similar information, also need a single crystal for a unique analysis of the experimental results. Furthermore, the calculation of molecular anisotropies from the crystalline quantities becomes difficult if the unit cell of the molecule has a large number of differently oriented "magnetically" inequivalent molecules. Fortunately this is a rare occurrence.

Finally, what is the future scope of this technique? With the recognition now that the average magnetic susceptibility is, in general, insensitive to the LF and other parameters, we anticipate that the magnetic anisotropy studies will become increasingly more popular with more extensive applications in chemistry. Already it has assumed an important place in the field of magnetochemistry. We hope that this technique will soon become an integral part of magnetochemistry. With single-crystal x-ray structural studies becoming more popular and simpler, structural data should provide no barrier. It should be noticed that the magnetic anisotropy measurement on systems of chemical interest is almost wholly limited to the liquid nitrogen temperature range, although measurements at lower temperature are expected to give more accurate and new information, in general. The magnetic anisotropy studies below 77°K offer a potentially rich area of investigation in the field of magnetochemistry.

Magnetic anisotropy studies on polynuclear complexes, especially those with orbitally degenerate ground states, would be most valuable to test the validity of different models of exchange interaction. At present, as indicated in Section IV.G, there are very few single-crystal experimental studies available on these interesting systems. We also foresee an increasing application of the magnetic anisotropy study for the calculation of the pseudocontact shift, as it provides the most accurate and direct method for the calculation of this contribution to the isotropic shift. The experimental techniques for the measurement of magnetic anisotropy are now well established, although efforts should be made to avoid the use of fragile quartz fiber and to find some other convanient method of measuring the couple.

Magnetic anisotropy promises to be an important physical technique and will soon become an integral part of magnetochemistry. It is hoped that the present article will provide a definite step in this direction.

Acknowledgment

It is a pleasure to thank my colleagues V. R. Marathe and P. Ganguli for several suggestions and for making available their unpublished results. The author is also indebted to Professor B. Venkataraman for his interest in the present article and to Professor R. L. Martin and Dr. A. K. Gregson for permission to quote some unpublished results. The author wishes to acknowledge long discussions he had several years ago with Professor R. L. Martin on the necessity and scope of the present article. The generous help provided by the drafting section and the typists of our Institute is also thankfully acknowledged.

Reference

1. S. Mitra, *Transition Metal Chem.*, *7*, 183 (1972).
2. W. DeW. Horrocks and D. DeW. Hall, *Coord. Chem. Rev.*, *6*, 147 (1971).
3. K. S. Krishnan and S. Banerjee, *Phil. Trans. Roy. Soc.*, *A234*, 265 (1935).
4. J. W. Stout and M. Griffel, *J. Chem. Phys.*, *18*, 1449 (1950).
5. S. K. Datta, *Indian J. Phys.*, *27*, 155 (1953).
6. D. A. Gordon, *Rev. Sci. Inst.*, *29*, 929 (1958); *J. Phys. Chem.*, *64*, 273 (1960).
7. A. K. Gregson and S. Mitra, *J. Chem. Phys.*, *49*, 3696 (1968).
8. M. Gerloch and P. N. Quested, *J. Chem. Soc. (A)*, 2307 (1971).
9. K. Lonsdale and K. S. Krishnan, *Proc. Roy. Soc. London. A516*, 597 (1936).
10. M. Leela, Ph.D. thesis, University of London. Quoted in Ref. 8.
11. P. Ganguli, V. R. Marathe, and S. Mitra, *Inorg. Chem.*, *14*, 970 (1975).
12. P. Ganguli, V. R. Marathe, and S. Mitra, to be published.
13. K. S. Krishnan, A. Mookherji, and A. Bose, *Phil. Trans. Roy. Soc.*, *A238*, 125 (1939).
14. P. R. Saha, *Indian J. Phys.*, *41*, 628 (1967).
15. H. M. Powell and A. F. Wells, *J. Chem. Soc. (A)*, 359, (1935).
16. A. Bose, S. Mitra, and R. Rai, *Indian J. Phys.*, *39*, 357 (1965).
17. A. Bose, R. Rai, S. Kumar, and S. Mitra, *Physica*, *32*, 1437 (1966).
18. B. N. Figgis, M. Gerloch, and R. Mason, *Proc. Roy. Soc. London*, *A279*, 210 (1964).
19. R. P. Van Stapel, H. G. Beljers, P. F. Bongers, and H. Zijlstra, *J. Chem. Phys.*, *44*, 3719 (1966).
20. S. Mitra, *J. Chem. Phys.*, *49*, 4724 (1968).
21. A. Bose, S. C. Mitra, and S. K. Datta, *Proc. Roy. Soc. London*, *A248*, 153 (1958).
22. M. Gerloch, J. Lewis, and W. R. Smail, *J. Chem. Soc. (A)*, 2434 (1971).
23. A. Bose, S. C. Mitra, and S. K. Datta, *Proc. Roy. Soc. London*, *A239*, 165 (1957).
24. R. C. Marshall and D. W. James, *J. Phys. Chem.*, *78*, 1235 (1974).
25. M. Gerloch and J. R. Miller, *Prog. Inorg. Chem.*, *10*, 1 (1968).
26. B. N. Figgis, *Trans. Faraday Soc.*, *56*, 1553 (1960); *Trans. Faraday Soc.*, *57*, 189, 204 (1961).

27. A. Bose, A. S. Chakravarty, and R. Chatterji, *Proc. Roy. Soc. London, A255*, 145 (1960).
28. H. Kamimura, *J. Phys. Soc. Japan, 11*, 1171 (1956).
29. R. M. Golding, *Applied Wave Mechanics*, Van Nostrand, London, 1969, Chapter 6.
30. F. E. Mabbs and D. J. Machin, *Magnetism and Transition Metal Complexes*, Chapman and Hall, London, 1973, Chapter 5.
31. A. K. Gregson and S. Mitra, *Chem. Phys. Lett., 3*, 392 (1969).
32. B. N. Figgis, J. Lewis, F. E. Mabbs, and G. A. Webb, *J. Chem. Soc. (A)*, 422 (1966).
33. H. S. Jarret, *J. Chem. Phys., 27*, 1298 (1957).
34. B. N. Figgis, M. Gerloch, and R. Mason, *Proc. Roy. Soc. London, A309*, 91 (1969).
35. V. R. Marathe, S. K. Date, and C. R. Kanekar, *Chem. Phys. Lett., 17*, 525 (1972).
36. M. Gerloch, *J. Chem. Soc. (A)*, 2023 (1968).
37. S. Mitra, *Indian J. Pure & Appl. Phys., 2*, 333 (1964).
38. A. Bose, S. Lahiri, and U. S. Ghosh, *J. Phys. Chem. Solids, 26*, 1747 (1965).
39. B. N. Figgis, M. Gerloch, J. Lewis, and R. C. Slade, *J. Chem. Soc. (A)*, 2028 (1968).
40. J. Ferguson, *J. Chem. Phys., 40*, 3406 (1964).
41. A. Bose, A. S. Chakravarty, and R. Chatterji, *Proc. Roy. Soc. London, A261*, 207 (1961).
42. E. Konig and A. S. Chakravarty, *Theoret. Chim. Acta, 9*, 151 (1967).
43. A. Bose and R. Rai, *Indian J. Phys., 39*, 176 (1965).
44. A. K. Gregson and S. Mitra, *Chem. Phys. Lett, 3*, 528 (1969).
45. A. K. Gregson, Ph. D. thesis, Melbourne University, Melbourne, 1971.
46. M. Gerloch, J. Lewis, G. G. Philips, and P. N. Quested, *J. Chem. Soc. (A)*, 1941 (1970).
47. A. K. Gregson and S. Mitra, *Chem. Phys. Lett, 13*, 313 (1972).
48. L. C. Jackson, *Phil. Mag., 4*, 269 (1929).
49. T. Ohtsuka, *J. Phys. Soc. Japan, 14*, 1245 (1959).
50. M. Majumdar and S. K. Datta, *Indian J. Phys., 41*, 590 (1967).
51. R. Ingalls, *Phys. Rev., A133*, 787 (1964).
52. A. Bose, *Indian J. Phys., 22*, 483 (1948).
53. D. Guha Thakurta and D. Mukhopadhyay, *Indian J. Phys., 40*, 69 (1966).
54. A. Bose, A. S. Chakravarty, and R. Chatterji, *Proc. Roy. Soc. London, A261*, 207 (1961).
54a. B. N. Figgis, J. Lewis, F. E. Mabbs, and G. A. Webb, *J. Chem. Soc. (A)*, 442 (1967).
55. H. Montgomery, R. V. Chastain, J. J. Natt, A. M. Witkowska, and E. C. Lingafelter, *Acta Cryst., 22*, 775 (1967).
56. R. Ingalls, K. Ono, and L. Chandler, *Phys. Rev., 172*, 295 (1968).
57. A. Bose, R. Rai, and R. Chatterji, *Proc. Phys. Soc. London, 83*, 959 (1964).
58. J. N. McElearney, R. W. Schwartz, S. Merchant, and R. L. Carlin, *J. Chem. Phys., 55*, 466 (1971).
59. R. L. Carlin and E. G. Terezakis, *J. Chem. Phys., 47*, 4901 (1967).
60. M. Gerloch and R. C. Slade, *J. Chem. Soc. (A)*, 1012, 1022 (1969).
61. S. Lahiry, D. Mukhopadhyay, and D. Ghosh, *Indian J. Phys., 42*, 320 (1968).
62. C. G. Barraclough, R. L. Martin, S. Mitra, and R. C. Sherwood, *J. Chem. Phys., 55*, 1638 (1970).
63. C. G. Barraclough, A. K. Gregson, and S. Mitra, *J. Chem. Phys., 60*, 962 (1974).
64. G. Harris, *Theoret. Chim. Acta, 5*, 369 (1966).
65. V. R. Marathe and S. Mitra, *Ind. J. Pure & Appl. Phys.,14*, 893 (1976).

66. C. G. Barraclough, R. L. Martin, S. Mitra, and R. C. Sherwood, *J. Chem. Phys., 55*, 1643 (1970).
67. D. W. Dale, R. J. P. Williams, C. E. Johnson, and T. L. Thorp, *J. Chem. Phys., 49*, 3441 (1968).
68. R. L. Martin and S. Mitra, *Chem. Phys. Lett., 3*, 183 (1969).
69. A. K. Gregson, R. L. Martin, and S. Mitra, *Chem. Phys. Lett., 5*, 310 (1970).
70. A. K. Gregson, R. L. Martin, and S. Mitra, to be published; see also Ref. 77.
71. R. B. Bentley, F. E. Mabbs, W. R. Smail, M. Gerloch, and J. Lewis, *Chem. Commun.*, 119 (1969); *J. Chem. Soc. (A)*, 3003 (1970).
72. R. J. Fitzgerald and G. R. Brubaker, *Inorg. Chem., 8*, 2265 (1969).
73. R. L. Martin and S. Mitra, *Inorg. Chem., 9*, 182 (1970).
74. D. Kivelson and R. Nieman, *J. Chem. Phys., 35*, 149 (1961).
75. S. E. Harrison and J. M. Assour, *J. Chem. Phys., 40*, 365 (1964).
76. See Ref. 45, Chapter. IV.
77. A. K. Gregson, R. L. Martin, and S. Mitra, *J. Chem. Soc. Dalton*, 1458 (1976).
78. M. H. Valek, W. A. Yearnos, G. Basu, P. K. Hon, and R. L. Belford, *J. Mol. Spectrosc., 37*, 228 (1971).
79. J. Selbin, G. Mans, and D. L. Johnson, *J. Inorg. Nucl. Chem., 29*, 1935 (1967).
80. A. K. Gregson and S. Mitra, *J. Chem. Soc. Dalton*, 1098 (1973).
81. I thank V. R. Marathe for drawing my attention to it.
82. G. E. Chapps, S. W. McCann, H. H. Wickman, and R. C. Sherwood, *J. Chem. Phys., 60*, 990 (1974) and references therein.
83. G. C. Brackett, P. L. Richards, and W. S. Caughy, *J. Chem. Phys., 54*, (1971).
84. P. Ganguli, V. R. Marathe, S. Mitra, and R. L. Martin, *Chem. Phys. Lett., 26*, 529 (1974)
85. P. Ganguli, V. R. Marathe, and S. Mitra, *Inorg. Chem., 14*, 970 (1975).
86. P. Ganguli, V. R. Marathe, and S. Mitra, to be published.
87. R. L. Ake and G. M. Harris Loew, *J. Chem. Phys., 52*, 1098 (1970).
88. M. Gerloch, J. Kohl, J. Lewis, and W. Urland, *J. Chem. Soc. (A)*, 3269, 3283 (1970).
89. K. S. Murray and R. M. Sheahan, *Chem. Phys. Lett., 22*, 406 (1973).
90. M. Weissbluth, *Structure and Bonding, 2*, 1 (1967).
91. J. H. van Vleck and W. G. Penney, *Phils. Mag., 7*, 961 (1934).
92. R. de L. Kornig and C. J. Bouwkamp, *Physica, 6*, 290 (1939).
93. M. H. L. Pryce, *Phys. Rev., 80*, 1107 (1950).
94. H. Watanabe, *Prog. Theoret. Phys., 18*, 405 (1957).
95. M. J. D. Powell, J. R. Gabriel, and D. F. Johnston, *Phys. Rev. Lett., 5*, 145 (1960).
96. J. S. Griffith, *Nature, 180*, 30 (1957); see also *Quantum Aspects of Polypeptides and Polynucleotides*, Biopolymer Symposium 1, Interscience, New York, 1964.
97. M. Kotani, *Prog. Theor. Phys. Suppl., 17*, 41 (1961).
98. M. Gerloch, J. Lewis, and R. Slade, *J. Chem. Soc. (A)*, 1442 (1969).
99. V. R. Marathe and S. Mitra, *Chem. Phys. Lett., 19*, 140 (1973).
100. V. R. Marathe and S. Mitra, *Chem. Phys. Lett., 21*, 62 (1973).
101. D. J. E. Ingram, *Biological and Biochemical Applications of ESR*, Adam Hilger Ltd., London, 1969, p. 239.
102. F. R. Mc Kim, *Proc. Roy. Soc. London, A262*, 287 (1961).
103. H. Venoyama, T. Iizyka, H. Morimoto, and M. Kotani, *Biochim. Biophys. Acta, 160*, 159 (1968).
104. J. Beetlestone, quoted in P. George, *Biopolymers Symposuim, 1*, 45 (1964).

105. P. Ganguli, V. R. Marathe, and S. Mitra, to be published.
106. D. J. Brown, M. Gerloch, and J. Lewis, *Nature, 220*, 256 (1968).
107. M. Gerloch and R. C. Slade, *Ligand Field Parameters*, Cambridge University Press, 1973.
107*a*. K. S. Krishnan and A. Mookherji, phys. Rev., *50*, 800 (1936), *ibid, 54*, 833, 841 (1938).
108. H. M. Mc Connell and R. E. Robertson, *J. Chem. Phys., 27*, 1361 (1958).
109. R. J. Kurland and B. R. McGarvey, *J. Mag. Res., 2*, 286 (1970).
110. B. Bleaney, *J. Mag, Res., 8*, 91 (1972).
111. W. DeW Horrocks, Jr., and E. S. Greenberg, *Biochim. Biophys. Acta, 322*, 38 (1973).
112. W. DeW Horrocks, Jr., and E. S. Greenberg, *Mol. Phys., 27*, 993 (1974).
113. W. DeW Horrocks, Jr., *NMR of Paramagnetic Molecules*, eds. G. N. La Maar, W. DeW Horrocks and R. H. Holm, Eds., Academic, New York, 1973, Chapter 4.
114. R. G. Shulman, S. H. Glarum, and M. Karplus, *J. Mol. Biol., 57*, 93 (1971).
115. D. M. Doddrell and A. K. Gregson, *Chem. Phys., Lett. 29*, 512 (1974).
116. A. Johnson and G. W. Everett, Jr., *J. Am. Chem. Soc., 94*, 1419 (1972).
117. F. Rohrscheid, R. E. Ernst, and R. H. Holm, *Inorg. Chem., 6*, 1315 (1963).
118. M. M. Dhingra, P. Ganguli, V. R. Marathe, S. Mitra, and R. L. Martin, *J. Mag. Res., 20*, 133 (1975).
119. L. J. de Jongh and A. R. Miedema, *Adv. Phys., 23*, 1 (1974).
120. R. L. Martin, *New Parthways in Inorganic Chemistry*, E. A. V. Ebsworth, A. G. Maddock, and A. G. Sharpe, Eds., Cambridge University Press, Cambridge, U. K., 1968, p. 175.
121. C. G. Barraclough and A. K. Gregson, *J. Chem. Soc. Faraday II, 68*, 177 (1972).
122. M. E. Lines, *J. Chem. Phys., 55*, 2977 (1971).
123. A. P. Ginsberg, M. E. Lines, R. L. Martin, and R. C. Sherwood, *J. Chem. Phys., 57*, 1 (1972).
124. J. N. van Niekerk and F. K. Schoening, *Acta Cryst., 6*, 227 (1953).
125. B. Bleaney and K. D. Bowers, *Proc. Roy. Soc. London, A214*, 451 (1952).
126. B. C. Guha, *Proc. Roy. Soc. London, A320*, 473 (1971).
127. A Mookherji and S. C. Mathur, *J. Phys. Soc. Japan, 18*, 977 (1963).
128. A. Bose, R. N. Bagchi, and P. Sengupta, *Indian J. Phys., 42*, 55 (1968).
129. A. K. Gregson, R. L. Martin, and S. Mitra, *Proc. Roy. Soc. London, A320*, 473 (1971).
130. L. S. Foster and C. J. Ballhausen, *Acta Chem. Scand, 16*, 1385 (1962).
131. G. F. Kokoszka, H. C. Allen, and G. Gordon, *J. Chem. Phys., 42*, 3639 (0965).
132. C. W. Reimann, G. F. Kokoszka, and G. Gordon, *Inorg. Chem., 4*, 1082 (1965).
133. R. E. Jotham and S. F. A. Kettle, *J. Chem. Soc. (A)*, 2816 (1969).
134. A. E. Hansen and C. J. Ballhausen, *Trans. Faraday Soc., 61*, 631 (1965).
135. P. Coggon, A. T. McPhail, F. E. Mabbs, and V. N. McLachlan, *J. Chem. Soc. (A)*, 1014 (1971).
136. E. Ising, *Z. Phys., 31*, 253 (1925).
137. T. Obokata and T. Oguchi, *J. Phys. Soc. Japan, 23*, 516 (1967).
138. M. Suzuki, B. Tsujiyama, and S. Katsura, *J. Math. Phys., 8*, 124 (1967).
139. M. E. Fischer, *Am. J. Phys., 32*, 342 (1964).
140. J.C. Bonner and M. E. Fischer, *Phys. Rev., A135*, 640 (1964).
141. C.K. Majumdar, C.S. Jain, and V. Mubai, *Chem. Phys. Lett., 21*, 175 (1973); *Solid State Phys. Sym.*, 1974, Bomday.
142. N. Achiwa, *J. Phys. Soc. Japan, 27*, 561 (1969).

143. A. K. Gregson and S. Mitra, *J. Chem. Phys., 50*, 2021 (1969).
144. J. N. McElearney, D. B. Losee, S. Merchant, and R. L. Carlin, *J. Chem., Phys., 54*, 4585 (1971),
145. L. Onsagar, *Phys. Rev., 65*, 117 (1944); *Nuovo Chim. Suppl., 6*, 261 (1949).
146. M. E. Fischer, *Proc. Roy. Soc. London, A254*, 66 (1960).
147. M. E. Fischer, *Rep. Prog. Phys., 30*, 615 (1967).
148. A. K. Gregson and S. Mitra, *J. Chem. Phys., 51*, 5226 (1969).
149. M. Griffel and J. W. Stout, *J. Chem. Phys., 18*, 1455 (1949).
150. C. Trapp and J. W. Stout, *Phys. Rev. Lett., 10*, 157 (1963).

ADDENDUM

This article was submitted in May 1975. Since then several papers have appeared which show interesting application of single crystal magnetic studies. Some of these results are listed below.

1. Carlin et al. [*J. Am. Chem. Soc., 98*, 685, 3523 (1976)] have reported the principal susceptibilities between 1.5 and 20 °K of $[M^{II}(C_5H_5NO)_6]$-$(ClO_4)_2$ $[M^{II} = Co, Ni]$ where C_5H_5NO is pyridine N-oxide. Both these compounds show very large magnetic anisotropy, although the $M^{II}O_6$ coordination sphere is perfectly octahedral. The origin of this electronic distortion perhaps lies in the total symmetry of the molecule, which is lower. This result is in line with the comments made earlier in Section E. The nickel compound shows a large ZFS of the ground state $^3A_2$. D = 4.35 cm^{-1}. Both the compounds show small antiferromagnetic interaction, the effect being more marked in the nickel analogue.

A direct measurement of paramagnetic anisotropy between 4 and 300 °K has been reported by Mackey and Evans [*J. Chem. Soc. Dalton*, 2004 (1976)] on the analogous $[Ni(pyo)_6]$ $[BF_4]$. The results appear to be similar to $[Ni(pyo)_6]$ $[BF_4]$ with ZFS close to 4 cm^{-1}, and antiferromagnetic exchange interaction being predominant below 20 °K. However, the data above 30 °K as well could not be fitted to a single set of parameters, and the authors suggested a small variation of the trigonal field splitting with temperature.

A single crystal magnetization study on the fluoroborate complex has now been reported [D. J. Mackey, *J. Chem. Soc. Dalton*, 40, 1977] between 2 and 20 °K at magnetic field strengths up to 50 kG. The study reveals that the exchange interaction predominates at low temperatures only in the perpendicular (to the *c*-axis) direction. The magnetization in the parallel direction can be reasonably well reproduced with D = 5 cm^{-1} even without taking exchange interaction into account.

2. An excellent method has recently been used to determine directly the ZFS in $FeSiF_6 \cdot 6H_2O$ by measuring the parallel magnetization in very large magnetic fields up to 400 kG [Varret et al., *Solid State Comm.*, *14*, 17 (1974); *J. Phys. Chem. Solids*, *37*, 257 (1976)]. The method involves detecting the level crossings by measuring the magnetization in intense pulse magnetic fields; in $FeSiF_6 \cdot 6H_2O$ it detects the crossing of $|S_z = 0>$ and $|S_z = +1>$, and $|S_z = 1>$, and $|S_z = 2>$. The experiment gives D = 12.2 cm^{-1}.

3. A direct measurement of paramagnetic anisotropy of $Ni(en)_3(NO_3)$ [en = ethylenediamine] between 4 and 300 °K has been reported [Mackey et al., *J. Chem. Soc. Dalton*, 1515 (1976)]. The data is analyzed in terms of point charge model giving D = -0.65 cm^{-1}. A significant exchange interaction is detected from fitting the anisotropy data below 20 °K as well as from the single crystal magnetization experiments at low temperatures [*Australian J. Chem.*, *30*, 281 (1977)].

4. The theoretical procedure for the application of angular overlap model to magnetic properties of transition metal ions has recently been discussed [Gerloch et al., *J. Chem. Soc. Dalton*, 2443, 2452 (1975)]. The model has been applied to analyze the paramagnetic anisotropies of $M^{II}(py)_4(NCS)_2$- [M^{II} = Fe, Co; py = pyridine] where, it is claimed, the principal molecular susceptibilities do not lie along the bond directions.

Horrocks [*Inorg. Chem.*, *13*, 2775 (1974)] has suggested a theoretical model to analyze the magnetic susceptibilities in terms of orbital energies that are physically more meaningful. The model is applied to analyze the paramagnetic anisotropy data on $Co(acac)_2(py)_2$.

5. Principal crystal and molecular paramagnetic susceptibilities of bis(n-isopropylsalicylideiminato)nickel(II) have been reported between 295 and 20 °K [*J. Chem. Soc.*, *Dalton*, 152 (1977)]. The data are interpreted in terms of angular overlap model.

6. Paramagnetic anisotropy measurement on the Mn(IV) ion in $(NH_4)_6$-$MnMo_9O_{32} \cdot 8H_2O$ single crystals between 4 and 260 °K establishes that the ZFS parameter D is positive and independent of temperature [D. J. Mackey, *Mol. Phys.*, in press (1977)]. Earlier an e.s.r. study on this crystal was unable to give uniquely the sign of D, although it predicted its correct magnitude.

7. Electronic structure of two low-spin Schiff-base planar cobalt(II) complexes has been deduced from the measurement of paramagnetic

anisotropies [Murray and Sheahan, *J. Chem. Soc., Dalton*, 1134 (1976)]. Both these complexes exhibit anisotropies similar to Co(SacSac)$_2$ (see Section IVc, p. 349) and indicate to a d_{z^2} ground state with d_{yz} being very close. It is also observed that presence of a low-lying spin quartet is essential to reproduce the temperature dependence of the principal susceptibilities.

A quantitative interpretation of the data on Co(salen)py (see Section IVc, p. 363) has now been attempted [*J. Chem. Soc., Dalton*, 999 (1976)]. The analysis incorporates spin-orbit coupling between the ground doublet and excited doublet and quartet states, and invokes a small change in the LF with temperature,especially at higher temperatures.

8. The ZFS parameter in the high-spin tetraphenyl porphyrin ferric chloride has been deduced from the measurement of paramagnetic anisotropy between 77 and 300 °K [Behere, Marathe, and Mitra, *J. Am. Chem. Soc.*, in press (1977)]. The experimental results establish D = 5.9 cm^{-1} which is, as expected, about the same as in hemin chloride, but differs appreciably from the value of D deduced for this compound from the average magnetic susceptibility and n.m.r. isotropic proton contact shift studies.

9. Paramagnetic anisotropies of copper(II)tetraphenyl porphyrin have been measured in the liquid nitrogen temperature range [*Aust. J. Chem.*, *28*, 2623 (1975)]. The interpretation of the data on this simple system appears to be complicated as the anisotropy data predicts a very low *g*-anisotropy incompatible with the e.s.r. studies.

10. Measurement of principal susceptibilities of the single crystals of $Cs_3VCl_6 \cdot 3H_2O$ in the 1.5–20 °K temperature range reveals large ZFS D = 8.05 cm^{-1} [Carlin et al., *Inorg. Chem.* *15*,985 (1976)]. This is one of the largest values of D for a V^{3+} ion.

Oxidatively Induced Cleavage of Transition Metal-Carbon Bonds

by Guido W. Daub

Department of Chemistry,
Stanford University,
Stanford, California

INTRODUCTION

The cleavage of transition metal-carbon bonds by electrophilic reagents such as halogen has been known for many years. However, the mechanisms of these reactions have not been discerned. A large body of knowledge (24) for organomercurials suggested an electrophilic attack in some fashion by halogen. One would anticipate retention of configuration at carbon in the analogous reactions of transition metal alkyls. For organomercurials under polar conditions, this is found to be the case. For this reason, the transition metal-carbon cleavage by halogen was believed to be electrophilic in nature.

$$\delta+M \cdots \overset{\delta+}{Br} \text{----} \overset{\delta-}{Br}, \; \delta-C \longrightarrow M^+ \quad C \diagup Br + Br^- \longleftarrow \delta+M \text{------} Br\,\delta- \,/\, \delta-C \text{------} Br\,\delta+$$

Stereochemical studies have shown this not to be the case (35, 15, 23). Whitesides (35) reported that the cleavage proceeded with inversion at carbon for d^6 Fe(II) systems, whereas Johnson (15) and Jensen (23) presented similar findings for d^6 Co(III) complexes.

A large number of recent publications suggest that this cleavage is oxidatively induced. It is becoming clear that oxidation of the metal weakens the metal-carbon bond, resulting in two principal modes of reaction. In its oxidized state, the metal becomes a better leaving group in the sense of nucleophilic substitution at carbon, and the resulting alkyl is subject to SN_2 nucleophilic substitution or, in some cases, solvolysis. Oxidized alkyl-carbonyl complexes exhibit an enhanced tendency to undergo alkyl-to-acyl migratory insertion followed by solvolysis of the resulting highly reactive acyl complex. The general reaction scheme is illustrated in Eq. 1 and 2.

$$M^n - C \xrightarrow{-1e} M^{n+1} - C \xrightarrow{\text{Nuc.}} M^{n-1} + C - \text{Nuc.} \tag{1}$$

$$M^n - C \xrightarrow{-2e} M^{n+2} - C \xrightarrow{\text{Nuc.}} M^n + C - \text{Nuc.} \tag{2}$$

An interesting way of viewing this process is in terms of reactivity patterns. The normal pattern *i* is reversed, allowing nucleophilic attack at the α-carbon atom (pattern *ii*).

(*i*) $M^+ \quad :R^-$

(*ii*) $M:^- \quad R^+$

The evidence for such an electron transfer mechanism involved the use of "nonelectrophilic" oxidants, such as $IrCl_6^{2-}$ and Ce(IV). Johnson reported that treatment of various alkyl-Co(III) complexes with $IrCl_6^{2-}$ (a one-electron outer-sphere oxidant) in the presence of external nucleophiles afforded products derived from the cleavage of the cobalt-carbon bond (2). Reactions run with optically active complexes (at carbon) showed that inversion at carbon occurred. Similar conclusions were drawn when various alkyl Mo(II), W(II), Fe(II), and Co(I) carbonyls were treated with Ce(IV) in methanol to give esters (4). The complexes failed to react with the added nucleophiles in the absence of oxidant.

The transition metal complexes that appear to undergo oxidative cleavage generally fall into three categories: coordinately saturated d^4, d^6, and d^8 systems.

II. SATURATED d^6 SYSTEMS

A. Co(III) Complexes

Much of the research in the area of oxidative cleavage has been done with Co(III). Cobalt(III) compounds are low-spin d^6 complexes with all but the very weakest field ligands. It is suggested (21, 19, 12, 26) that an initial one-electron oxidation of the Co(III) is involved. Oxidized Co(III) species have recently been generated (chemically and electrochemically) by Halpern (19), Costa (12), and Levitin (26). Halpern has characterized this intermediate by EPR techniques; however, it is unclear that the cobalt is actually Co(IV). The metal-leaving group from such an intermediate would be a Co(II) species, a stable and well-characterized oxidation state.

Johnson's initial work was with alkyl [Co(III)(dimethylglyoximato)$_2$] complexes (2, 3, 15). Treatment of the cobalt alkyl **1** with one equivalent of

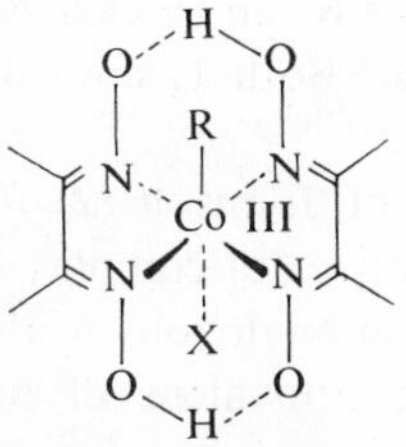

1 R = $C_6H_5CH_2$ X = H_2O
2 R = (+) *sec*-octyl X = pyridine
4 R = Br X = pyridine
8 R = (+) *sec*-butyl X = pyridine

$IrCl_6^{2-}$ in the presence of added Cl^- yielded $C_6H_5CH_2Cl$ (87%). Similarly, the oxidation of complex **1** with one equivalent of ICl in acetic acid with added Cl^- gave a 75% yield of $C_6H_5CH_2Cl$ (2). The reaction of the optically active cobalt alkyl **2** with $IrCl_6^{2-}$ or Br_2 in acetic acid proceeded with inversion, giving modest yields (~30%) of (−) 2-bromooctane (15). Johnson has also used Cl_2 and I_2 to effect the cleavage of complex **1** (3).

Jensen has reported experiments with similar macrocyclic Co(III) alkyls (23). The low-temperature cleavage of complex **3*** with Br_2 in methylene chloride gave low yields (14%) of *trans*-1,4-dibromocyclohexane, which again indicates that inversion is

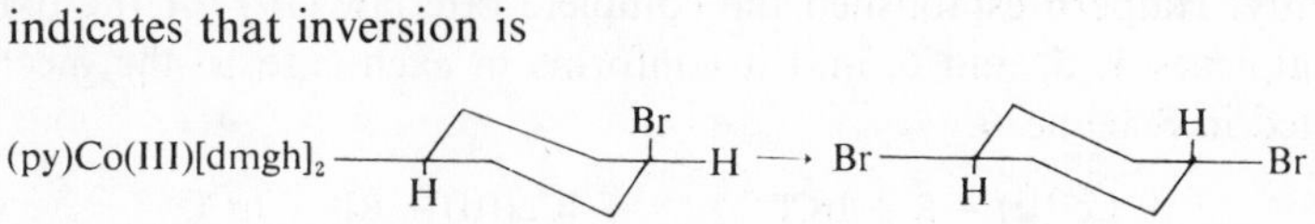

(py)Co(III)[dmgh]$_2$ — H, Br, H → Br, H, H, Br
3

occurring at the α-carbon. The inorganic product was characterized as complex **4**, suggesting the mechanism depicted in Scheme *A*.

*The [dimethylglyoximato]$_2$ macrocycle is abbreviated [dmgh]$_2$. The [*N*, *N*′-ethylene-bis(salicylideneiminato)] macrocycle is abbreviated [salen].

$$\mathrm{Co(III) - R \xrightarrow{Br_2} Co(IV) - R + Br^{\cdot} + Br^-}$$

$$\mathrm{Co(IV) - R \xrightarrow{Br^-} Co(II) + R - Br}$$

Scheme *A*

Halpern (19) observed the oxidation of complexes **1**, **5**, and **6** to the corresponding

$$\mathrm{[dmgh]_2\ Co(III)[X][R] \xrightarrow{-le} [dmgh]_2Co(III)[X][R]^+}$$

5 $R = CH_2CH_2CH_3$ $X = H_2O$

6 $R = CH(CH_3)CH_3$ $X = H_2O$

7 $R = CH_2CH_2CH_2CH_3$ $X = H_2O$

oxidized intermediates. The oxidation can be effected by Ce(IV) or by cyclic voltametric techniques. Levitin (26) reported that the oxidized alkyl intermediate of cobalt alkyl **7** reacts rapidly with added nucleophiles. Jensen (23) also reported that the cobalt-carbon bond in complex **8** can be cleaved with inversion, giving low yields of (−) 2-bromobutane. Both I_2 and Br_2 are effective in this reaction.

Halpern (1) reported results similar to those of Johnson (2) for the cleavage of complex **1** and phenyl-substituted derivatives of complex **1** with $IrCl_6^{2-}$. He presented an electron transfer mechanism analogous to the one in Scheme *A*. In one experiment, the use of one equivalent of oxidant with complex **1** gave **9**, Co(II), and dimethylglyoxime, indicating that in the absence of external nucleophiles, the intermediate decomposes to give alkylated ligand. Johnson (15) observed similar behavior, isolating material he formulated as **10**.

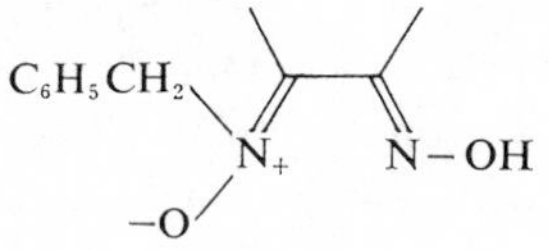

$C_6H_5CH_2$ N+ N−OH −O

9

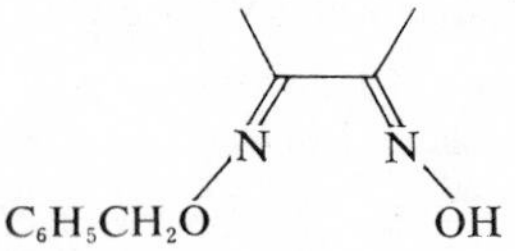

N N $C_6H_5CH_2O$ OH

10

Recently, Halpern established the complete rate law (19) for the oxidation of complexes **1**, **5**, and **6**, and it conforms in each case to the mechanism depicted in Scheme *B*.

$$\mathrm{Co(III) - R + IrCl_6^{2-} \longrightarrow [Co(III) - R]^+ + IrCl_6^{3-}}$$

$$\mathrm{[Co(III) - R]^+ + H_2O \longrightarrow Co(II) + ROH}$$

Scheme *B*

Collman (32) observed the Br_2-induced cleavage of the cobalt-carbon bond in the equatorial cholestanyl Co(III) complex **11**. Yields of 60 to 70% of 3α-bromocholestane were obtained at −78° in methylene chloride.

Similarly, the equatorial complex **12** afforded good yields of 3α-bromocholestane upon Br_2

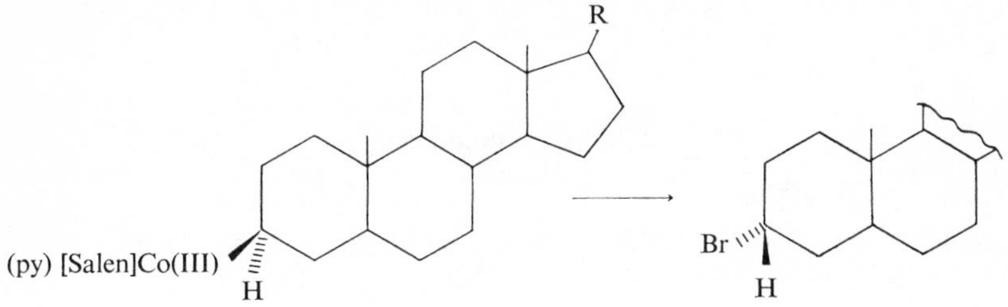

(py) [Salen]Co(III) H R Br H

11

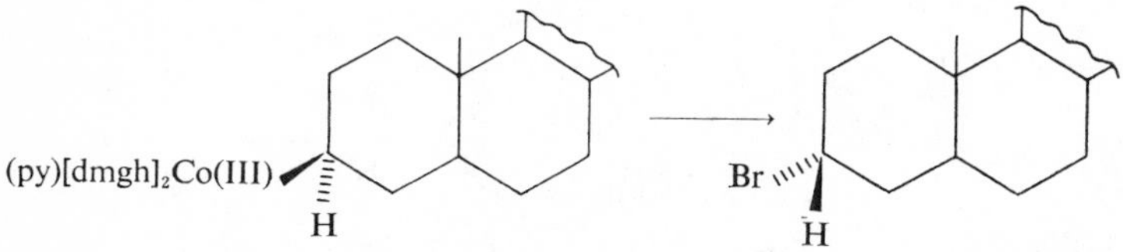

(py)[dmgh]$_2$Co(III) H Br H

12

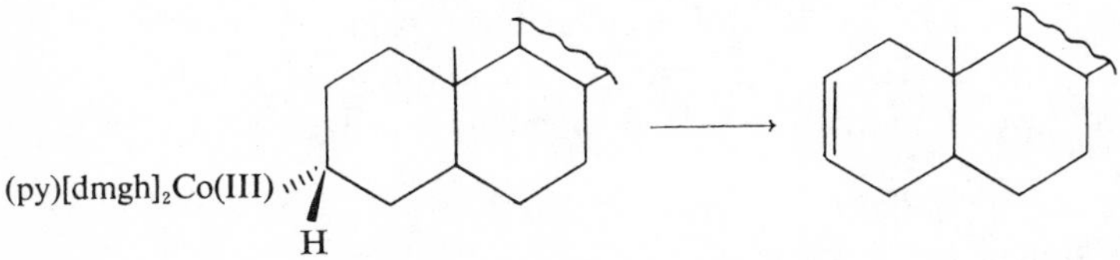

(py)[dmgh]$_2$Co(III) H

13

treatment. The 3α-cholestanyl complex **13** could not be obtained free of the β-epimer, complex **12**; however, oxidation with Br_2 of a mixture enriched in complex **13** gave large amounts of Δ^2-cholestene, in accord with a trans elimination of H^+ and a Co(II) species. The following scheme demonstrates an interesting route for the conversion of the equatorial 3β-cholestanyl bromide to its less stable epimer, 3α-cholestanyl bromide. The Co(I) [salen] · (pyridine) nucleophile

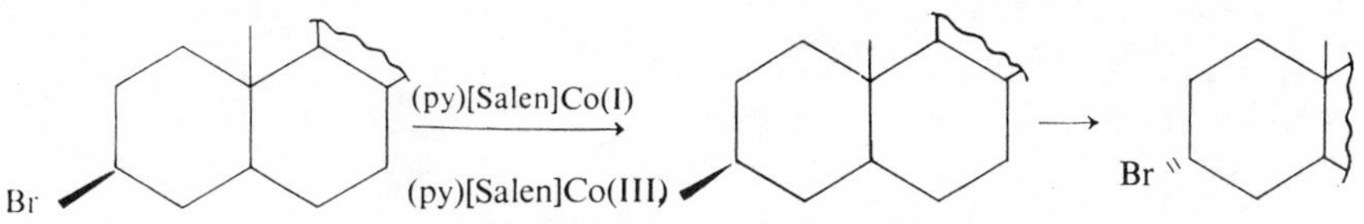

Br (py)[Salen]Co(I) (py)[Salen]Co(III) Br

displaces both bromide and itself until the more stable equatorial complex is formed. This intermediate is then oxidized with Br_2 to give only the contra-thermodynamic α-epimer.

It appears that oxidative attack on Co(III) precedes and facilitates cobalt-alkyl bond cleavage (16). Further studies on electrochemical generation of suspected "Co(IV) intermediates" and their reactions with nucleophiles should firmly establish the generality of this mechanism.

B. Fe(II) Complexes

Iron(II) complexes are generally high-spin systems (13) unless strong ligand fields are present. Hence a one-electron oxidation would give a very stable high-spin d^5 Fe(III) system. The Fe(III) oxidation state is stable and well known. A one-electron oxidation mechanism (Eq. 3) would generate an Fe(I) leaving group, whereas the transfer of two electrons would generate an Fe(II) leaving group (Eq. 4). Little work has been done in this area and experiments should be initiated to clarify this situation.

$$\mathrm{Fe(II)} - \mathrm{R} \xrightarrow{-1e} \mathrm{Fe(III)} - \mathrm{R} \xrightarrow{\mathrm{Nuc.}} \mathrm{Fe(I)} + \mathrm{R} - \mathrm{Nuc.} \quad (3)$$

$$\mathrm{Fe(II)} - \mathrm{R} \xrightarrow{-2e} \mathrm{Fe(IV)} - \mathrm{R} \xrightarrow{\mathrm{Nuc.}} \mathrm{Fe(II)} + \mathrm{R} - \mathrm{Nuc.} \quad (4)$$

Whitesides (7, 35) has used NMR techniques to ascertain that the I_2 or Br_2 cleavage of the iron alkyl **14** in chloroform, pentane, or carbon disulfide proceeds with inversion at carbon, forming the alkyl halide. Whitesides suggested several mechanisms, one of which involved an initial electron transfer.

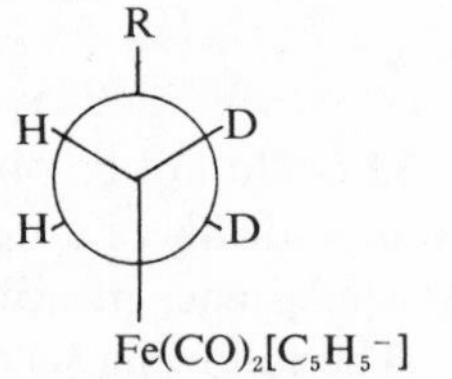

$Fe(CO)_2[C_5H_5^-]$

14 R = *t*-butyl

15 R = C_6H_5

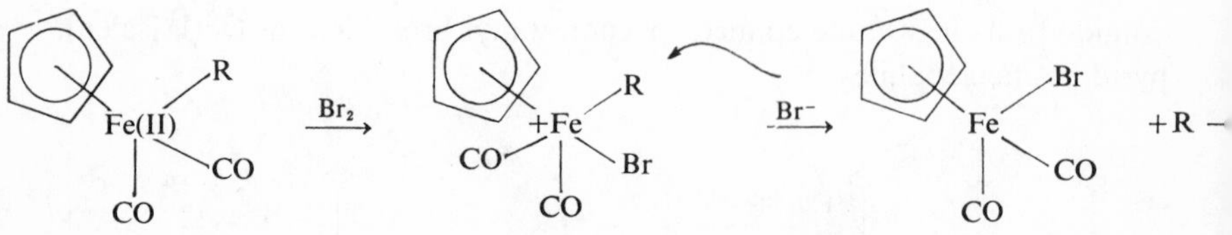

16

The reaction takes a different course in more polar solvents such as alcohols. Whitesides has shown that migratory insertion to the acyl iron occurs prior to the iron-carbon bond rupture. The resultant iron acyl subsequently reacts with the solvent to produce esters. Migration is observed when the iron acyl **14** is treated with $Cl_2/CHCl_3$, implying that the decreased nucleophilicity of the chloride ion over the bromide ion allows favorable competition of the migratory-insertion process. It should be noted that the $Br_2/CHCl_3$ reaction with complex **14** gives only a 50% yield of alkyl bromide (38), allowing for the possibility of some migratory insertion.

Rosenblum (28, 29) and Johnson (4) have shown that various iron alkyls **16** give esters in good yield upon treatment with alcoholic $CuCl_2$ and alcoholic Ce(IV), respectively. This is in accord with the results of Whitesides in polar solvents. The fact that Cu(II) and Ce(IV) are one-electron oxidants leaves open the question of whether an Fe(III) or an Fe(IV) is involved in the cleavage. Certainly, two successive one-electron oxidations are mechanistically plausible; however, there is no *a priori* reason to rule out a one-electron oxidation. Rosenblum reported (29) that $FeCl_3$ (a very weak oxidant) and DDQ could also serve as oxidants, and that the use of an aqueous medium (acetone/H_2O) and oxidant gave rise to carboxylic acids. The fact that $FeCl_3$ was effective in these reactions argues strongly for an initial one-electron oxidation.

Flood (18) has reported evidence which implicates some sort of iron-leaving group. The reaction between the deuterated iron alkyl **16a** and Br_2/CS_2 at 0°, Br_2/CH_2Cl_2 at −78°, or I_2/CS_2 at 0° produces equal amounts of two products resulting from carbon-iron bond cleavage. The expected product, $C_6H_5CH_2CD_2X$, was found to be

16a [CpFe(CO)$_2$$CD_2CH_2C_6H_5$] $\xrightarrow{X_2}$ $C_6H_5CH_2CD_2X + C_6H_5CD_2CH_2X$ (1:1)

stable to reaction conditions. Flood favors an oxidative addition by "Br^+" followed by nucleophilic attack by the aromatic ring, displacing an iron-leaving group.

16a $\xrightarrow{Br_2}$ [Cp(CO)$_2$(Br)Fe^+–CD_2CH_2–C_6H_5] ⟶ [X^- phenonium ion (CD_2–CH_2 bridged, +)] ⟶ products

This mode of aryl participation could explain the results of Baird (36), who reported that retention of configuration occurs in the Br_2 cleavage of the threo iron alkyl **15**. The intermediacy of an erythro phenonium ion would lead to the formation of the threo isomer as the final product.

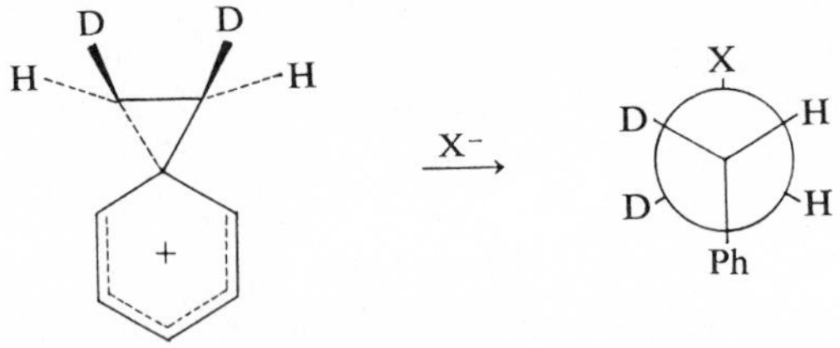

Curiously, he reports no such rearrangement for an analogous Co(III) alkyl **17**, attributing this to the fact that the phenyl group is not an effective enough nucleophile to displace the cobalt-leaving group. The absence of rearranged product might also be attributed to steric interactions of the developing phenonium ion with the macrocyclic ligand.

$$(py)[dmgh]_2Co(III)-CD_2CH_2C_6H_5 \xrightarrow{Br_2} BrCD_2CH_2C_6H_5$$

17

C. W(0) Complexes

Casey (8) has observed the oxidative cleavage of the carbene-tungsten bond in compounds **18** and **19**. Neutral carbenes are known to be electrophilic,

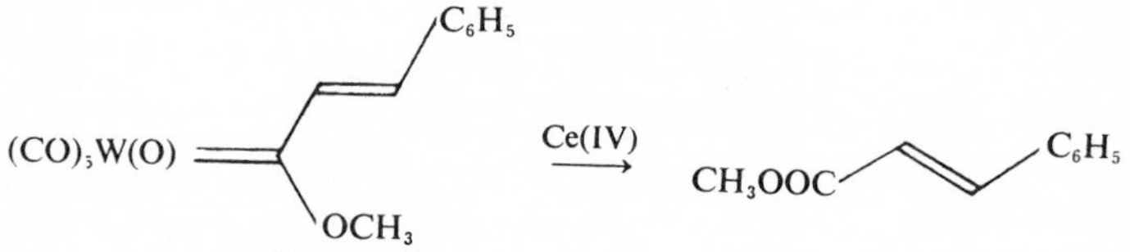

18

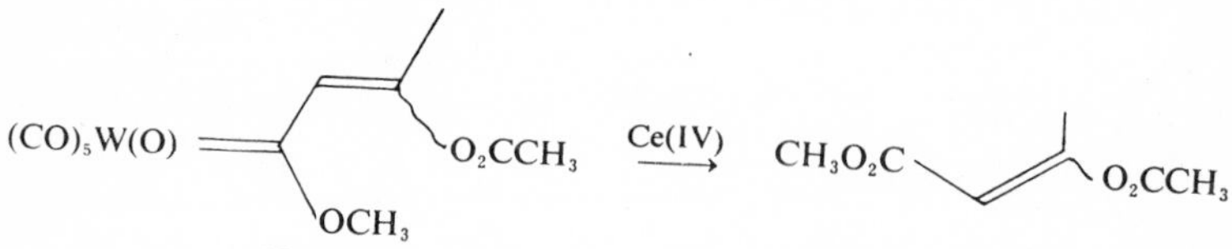

19

and this tendency should be enhanced by an oxidation of the metal. The oxidation of W(0) would certainly facilitate both the nucleophilic attack by water and the subsequent loss of tungsten. It is not clear whether the actual species that is oxidized is the carbene complex, an intermediate such as shown in Eq. 5, or both.

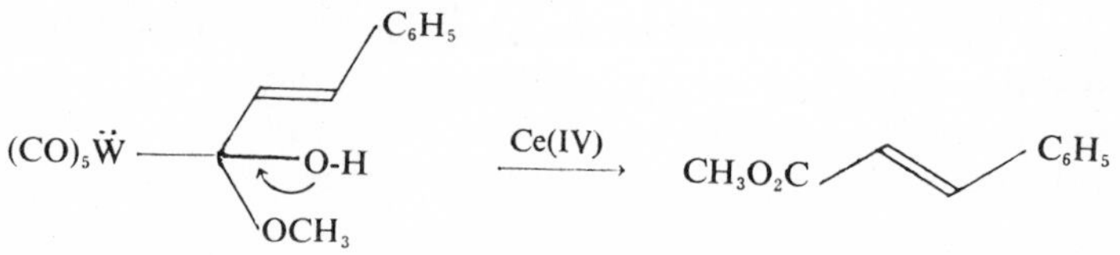

(5)

III. SATURATED d^8 SYSTEMS

A. Fe(0) Complexes

Collman (11, 36) and Takegami (27, 34) have used various oxidants, including I_2, Br_2, Cl_2, and O_2, to cleave the iron acyls **20** and iron alkyls **21** in good yield. Attack by the solvent (water, alcohol, or amine) on the

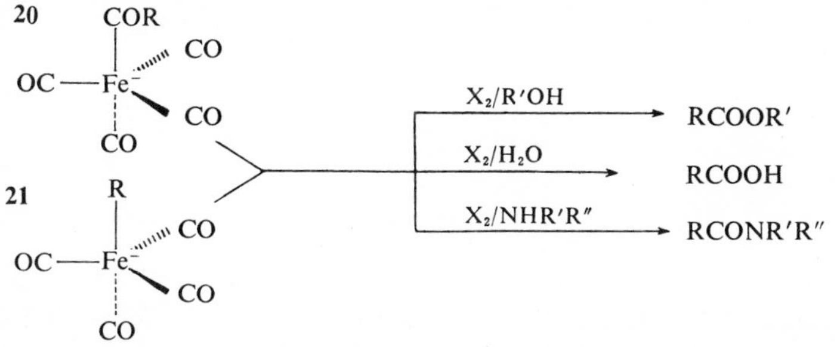

20: $(OC)_3Fe^{-}(COR)(CO)$; 21: $(OC)_3Fe^{-}(R)(CO)$

$X_2/R'OH \rightarrow RCOOR'$

$X_2/H_2O \rightarrow RCOOH$

$X_2/NHR'R'' \rightarrow RCONR'R''$

oxidized iron acyl leads to the production of acids, esters, or amides. The question of whether the initial oxidation is a one or two-electron process is unclear. Molecular oxygen is a one-electron oxidant; however, the iron complexes **20** and **21** are known to possess nucleophilic character (10), suggesting an oxidative addition mechanism of the following type:

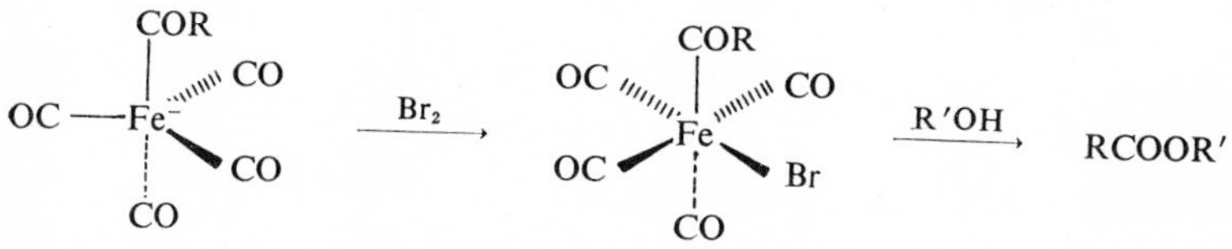

$$(OC)_4Fe^{-}(COR) \xrightarrow{Br_2} (OC)_4Fe(COR)Br \xrightarrow{R'OH} RCOOR'$$

The second step may be a reductive elimination, somewhat in contrast to an external nucleophile's attack on the α-carbon. In any case, oxidative attack of some fashion is occuring, the details of which need to be clarified.

The question of the generation of acyl products from the iron alkyls **21** is more complex. Winter (36) suggested that the migration is facilitated by an oxidation of complex **21** to an Fe(I) or Fe(II) state. In view of the recent reports of Whitesides for d^6 Fe(II) systems (7), the polarity of the solvents may have some bearing on the issue. Winter generally conducted these reactions in polar solvents (THF/HMPA, or NMP) with added alcohol or amine. It is reasonable that the migratory insertion occurs faster once the metal is oxidized, since this weakens the carbon-metal bond. Oxidation also increases the electrophilicity of the carbon monoxides coordinated to the metal, both electronically and by virtue of the decreased back-bonding by the electron-deficient iron (28).

B. Co(I) Complexes

There is one isolated example of a d^8 Co(I) system in Johnson's reports (4). The cobalt acyl **22** was treated with Ce(IV)/CH_3OH to give the corresponding methyl ester.

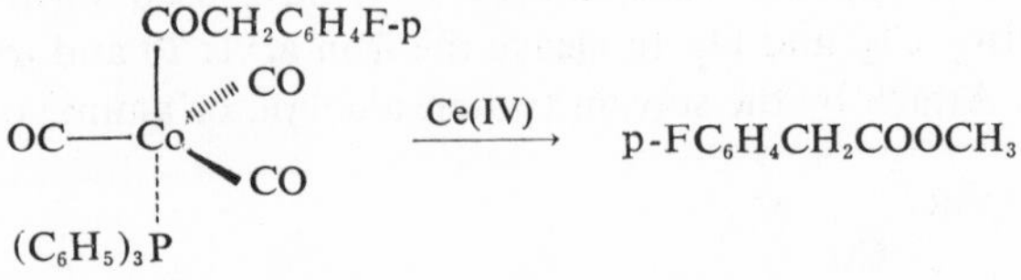

$$\underset{\mathbf{22}}{OC{-}Co(CO)_2(COCH_2C_6H_4F\text{-}p)(P(C_6H_5)_3)} \xrightarrow{Ce(IV)} p\text{-}FC_6H_4CH_2COOCH_3$$

IV. SATURATED d^4 SYSTEMS

A. Mo(II) Complexes

Johnson (4) observed the oxidative cleavage of the Mo(II) alkyl complexes **23** and **24**. Treatment of either complex with Ce(IV) in methanol/LiCl gave quantitative yields of the methyl ester $RCH_2CO_2CH_3$. Certainly, an oxidative attack is occuring since neither complex reacts in the absence of Ce(IV).

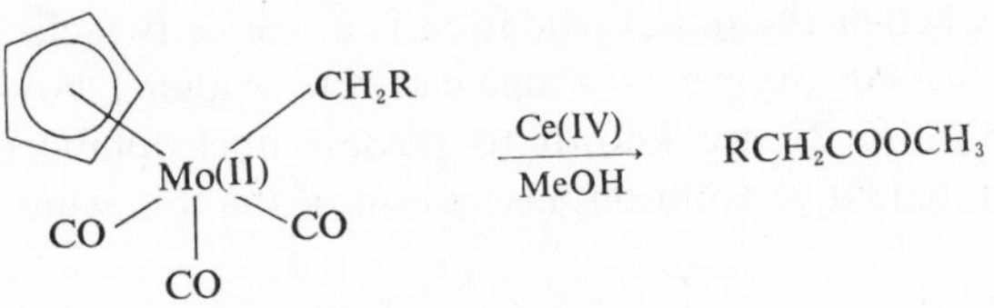

$$CpMo(II)(CO)_3CH_2R \xrightarrow[MeOH]{Ce(IV)} RCH_2COOCH_3$$

23 R = 3-pyridyl

24 R = p-FC_6H_4

Again, the oxidation of the metal and the nature of the solvent account for the migratory insertion prior to carbon-metal bond rupture.

B. W(II) Complexes

Johnson (4) observed results analogous to those of molybdenum for the W(II) complex **25**. Ceric ion oxidation in methanol/LiCl leads to the methyl ester. It is curious that *p*-$FC_6H_4CH_2Cl$ and *p*-$FC_6H_4CH_2OCH_3$ were formed in small amounts;

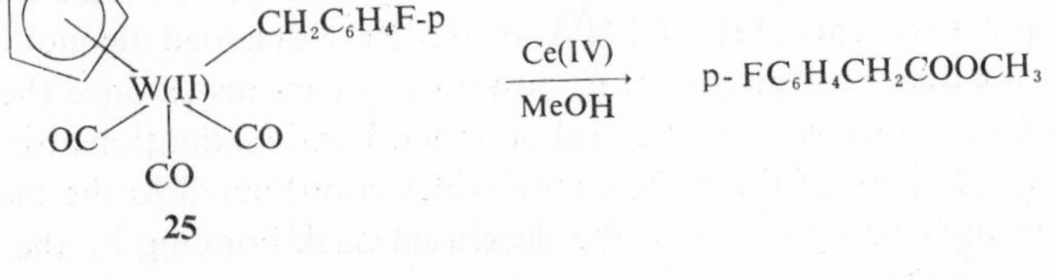

$$\underset{\mathbf{25}}{CpW(II)(CO)_3CH_2C_6H_4F\text{-}p} \xrightarrow[MeOH]{Ce(IV)} p\text{-}FC_6H_4CH_2COOCH_3$$

however, this may simply be a function of the tungsten.

Barnett (5) has reported the iodine cleavage of the W(II) acyl **26**. It is unfortunate that the nature of the organic product was not investigated; however, in view of the above examples, it is reasonable to predict that it is acetic acid or methyl acetate.

$$CpW(II)(CO)_3(COCH_3)\ \textbf{26} \xrightarrow{I_2} CpW(II)(CO)_3I + ?$$

V. OTHER SYSTEMS

A. Cr(III)—An Unsaturated d^3 System

Espenson has reported the bromine cleavage of Cr(III)-alkyl bonds (17). Unfortunately, no stereochemical studies were reported, thus no meaningful conclusions can be drawn. Espenson preferred an electrophilic-type mechanism, and did not mention the possibility of an oxidative process. He

$$R\text{-}Cr(III)(H_2O)_5^{2+} \xrightarrow{Br_2} Cr(III)(H_2O)_6^{3+} + R - Br + Br^-$$

concluded, by employing kinetic arguments, that homolysis to a Cr(II) species and an alkyl radical was not occuring.

B. Polynuclear Cobalt Species

Seyferth has treated the binuclear cobalt acetylene complex **27** and the trinuclear cobalt cluster **28** with oxidants, liberating acetylenes and carboxylic acids (30, 31). Certainly, these decompositions are initiated by some oxidative process.

$$(CO)_3Co(\text{Ar-C}\equiv\text{C-Ar})Co(CO)_3\ \textbf{27} \xrightarrow{Ce(IV)} Ar-C\equiv C-Ar$$

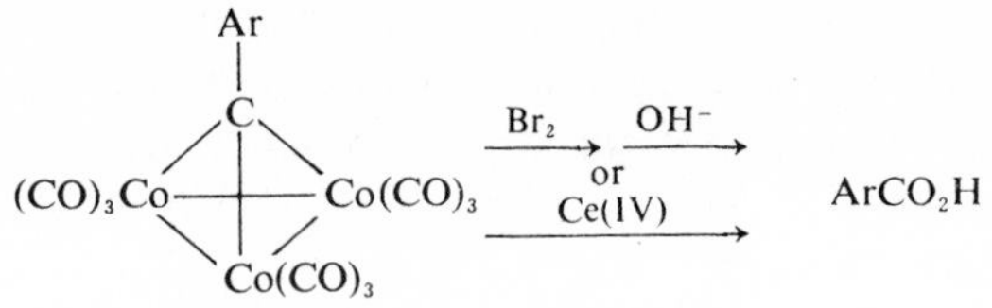

$$(CO)_3Co_3C\text{–}Ar \;(\mathbf{28}) \xrightarrow{Br_2} \xrightarrow{OH^-} \text{ or } \xrightarrow{Ce(IV)} ArCO_2H$$

C. Pd(II) Unsaturated d^8 Systems

Coulson (14) has demonstrated that the unsaturated Pd(II) complex **29** undergoes carbon-palladium bond cleavage upon treatment with halogen, with retention of configuration at carbon. Whitesides (7) has suggested an oxidative addition/reductive elimination sequence (Scheme *C*) to account for the stereochemical observations of Coulson.

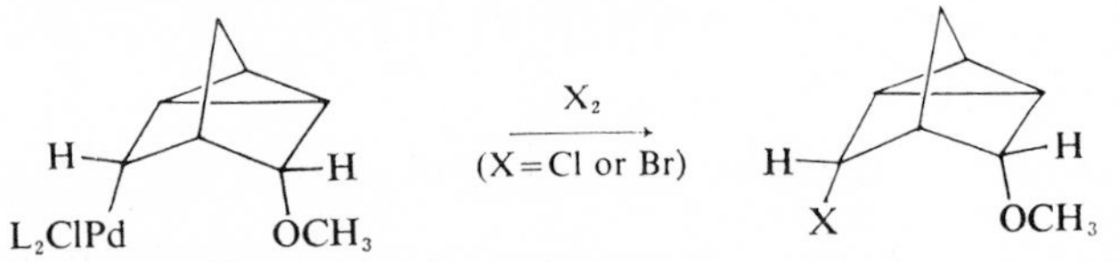

$$\mathbf{29}\ (L_2ClPd,\ H,\ H,\ OCH_3) \xrightarrow[(X=Cl \text{ or } Br)]{X_2} (X,\ H,\ H,\ OCH_3)$$

$$L_2Pd(L)(R) \xrightarrow{X_2} L_2PdX_2(L)(R) \longrightarrow R{-}X + L_2Pd(L)(X)$$

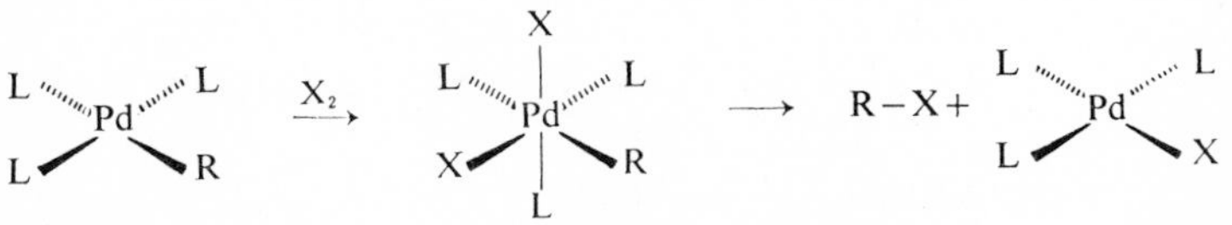

Scheme C

Stille (25, 37) has used bromine in methanol to cleave acyl-palladium bonds in complex **30**. Here again, one can envision that the unsaturated d^8 system

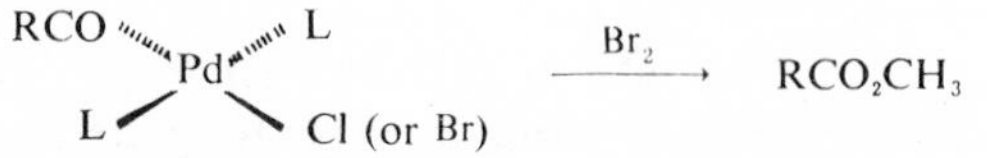

$$\mathbf{30}\ (RCO)Pd(L)_2Cl \text{ (or Br)} \xrightarrow{Br_2} RCO_2CH_3$$

undergoes oxidative addition, followed either by reductive elimination of the acyl halide which subsequently reacts with solvent to give the ester, or by direct attack of the solvent on the oxidized palladium acyl species. There is no reason, however. to rule out a one-electron oxidative process.

D. Zr(IV) Unsaturated d^0 Systems

Schwartz (22) has cleaved carbon-zirconium bonds in complex **31** with

Br_2, I_2, and $C_6H_5ICl_2$ to give the corresponding alkyl halides in good yield (60–90%). He has also treated analogous acyl-zirconium complexes **32** with Br_2/MeOH or H_2O_2/OH^- to give the corresponding esters and acids (50–70%).

$$Cp_2Zr(IV)(R)Cl \ (\mathbf{31}) \xrightarrow{I_2} RI$$

$$Cp_2Zr(IV)(COR)Cl \ (\mathbf{32}) \xrightarrow[MeOH]{NBS} RCOOCH_3$$

The zirconium is formally d^0 Zr(IV), which makes an electron transfer mechanism very unlikely. Schwartz suggests that simple electrophilic attack is taking place on the zirconium-carbon bond.

VI. CONCLUSION

Clearly there is more work to be done before the electron transfer mechanism is to be accepted as general for the oxidative cleavage of transition metal alkyls. Further stereochemical studies are certainly necessary, especially for systems for which there have been no such studies at all. The recent publication by Whitesides (7) is an excellent beginning in this area. The extensive use of outer-sphere nonelectrophilic oxidizing agents should continue to be an important technique, especially for those systems where only halogen has been used to initiate decomposition [Fe(0)]. Nonchemical oxidation techniques, particularly cyclic voltametry (21), should be used widely. Electrochemical methodology may be invaluable in probing the question of whether a one-electron or two-electron oxidation is the initial step. The characterization of oxidized intermediates is a difficult yet necessary task. The work of Halpern (19) and Levitin (26) are efforts in this direction.

Organometallic chemistry is only recently finding broad generalized concepts to organize the body of reactions that are known. Examples of such concepts include oxidative addition, migratory insertion, and reductive elimination. Any generalized scheme for the oxidative cleavage of transition metal-carbon bonds would be a significant contribution to this evolving field.

Acknowledgment

This paper was written in fulfillment of the requirements of Chemistry 255, Stanford University. The author wishes to acknowledge and thank Professor J. P. Collman for helpful suggestions.

Finally the author is grateful to the following for permission to reproduce some of the Figures in this article: Tailor and Francis Ltd., London, for Figures 48–51; the Chemical Society, London, for Figures 11, 33–35 and 38; North-Holland Publishing Co., for Figures 7, 10, 13, and 25; American Instibute of Physics, for Figures 19–23; Academic Press, for Figure 44; Elsevier Scientific Publishing Co., for Figure 39; American Chemical Society, for Figure 32, and the Royal Society, London, for Figure 47.

References

1. P. Abley, E. R. Dockal, and J. Halpern, *J. Am. Chem. Soc., 94*, 659 (1972).
2. S. N. Anderson, D. H. Ballard, J. Z. Chrzastowski, D. Dodd, and M. D. Johnson, *J. Chem. Soc., Chem. Commun.*, 685 (1972).
3. S. N. Anderson, D. H. Ballard, and M. D. Johnson, *J. Chem. Soc., Perkin II*, 311 (1972).
4. S. N. Anderson, C. W. Fong, and M. D. Johnson, *J. Chem. Soc., Chem. Commun.*, 163 (1973).
5. K. W. Barnett, D. L. Beach, S. P. Gaydos, and T. G. Pollman, *J. Organometal. Chem., 69*, 121 (1974).
6. C. A. Bertelo and J. Schwartz, *J. Am. Chem. Soc., 97*, 228 (1975).
7. P. L. Bock, D. J. Boschetto, J. R., Rasmussen, J. P. Demers, and G. M. Whitesides, *J. Am Chem. Soc., 96*, 2814 (1974).
8. C. P. Casey, R. A. Boggs, and R. L. Anderson, *J. Am. Chem. Soc., 94*, 8947 (1972).
9. C. P. Casey and T. J. Burkhardt, *J. Am. Chem. Soc., 94*, 6543 (1972).
10. J. P. Collman, S. R. Winter, and D. R. Clark, *J. Am. Chem. Soc., 94*, 1788 (1972).
11. J. P. Collamn, S. R. Winter, and R. G. Komoto, *J. Am. Chem. Soc., 95*, 249 (1973).
12. G. Costa, A. Puxeddu, and E. Reisenhofer, *J. Chem. Soc. Dalton*, 1519 (1972).
13. F. A. Cotton and G. Wilkinson, *Advanced Inorganic Chemistry, A Comprehensive Text*, 3rd ed., Interscience, New York, 1972.
14. D. R. Coulson, *J. Am. Chem. Soc., 91*, 200 (1969).
15. D. Dodd and M. D. Johnson, *J. Chem. Soc., Chem. Commun.*, 571 (1971).
16. D. Dodd and M. D. Johnson, *J. Organometal. Chem., 69*, 121 (1974).
17. J. H. Espenson and D. A. Williams, *J. Am. Chem. Soc., 96*, 1008 (1974).
18. T. C. Flood and F. J. DiSanti, *J. Chem. Soc., Chem. Commun.*, 18 (1975).
19. J. Halpern, M. S. Chan, J. Hanson, T. S. Roche, and J. A. Topich, *J. Am. Chem. Soc., 97*, 1606 (1975).
20. D. W. Hart and J. Schwartz, *J. Am. Chem. Soc., 96*, 8115 (1974).
21. J. B. Headridge, *Electrochemical Techniques for Inorganic Chemists*, Academic, London, England, 1969, pp. 42–56.
22. F. R. Jensen, V. Madan, and D. H. Buchanan, *J. Am. Chem. Soc., 92*, 1414 (1970).
23. F. R. Jensen, V. Madan, and D. H. Buchanan, *J. Am. Chem. Soc., 93*, 5283 (1971).
24. F. R. Jensen and B. Rickborn, *Electrophilic Substitution of Organomercurials*, McGraw-Hill, New York, 1968, pp. 75–99.

25. K. S. Y. Lau, R. W. Fries, and J. K. Stille, *J. Am. Chem. Soc., 96*, 4983 (1974); see also *J. Am. Chem. Soc., 96*, 5956 (1974).
26. I. Y. Levitin, A. L. Sigan, and M. E. Vol'pin, *Izv. Akad. Nauk. SSSR, Ser. Khim., 23*, 1205 (1974).
27. H. Masada, M. Mizuno, S. Suga, Y. Watanabe, and Y. Takegami, *Bull. Chem. Soc. Japan, 43*, 3824 (1970).
28. K. M. Nicholas and M. Rosenblum, *J. Am. Chem. Soc., 95*, 4983 (1974).
29. M. Rosenblum, *Acc. Chem. Res.*, 122 (1974).
30. D. Seyferth and A. T. Wehman, *J. Am. Chem. Soc., 92*, 5320 (1970).
31. D. Seyferth, G. H. Williams, A. T. Wehman, and M. O. Nestle, *J. Am. Chem. Soc., 97*, 2107 (1975).
32. K. B. Sharpless, Research Report to J. P. Collman, Stanford University, Stanford, California.
33. D. Slack and M. C. Baird, *J. Chem. Soc., Chem. Commun.*, 701 (1974).
34. Y. Takegami, Y. Watanabe, H. Masada, and I. Kanaya, *Bull. Chem. Soc. Japan, 40*, 1456 (1967).
35. G. M. Whitesides and D. J. Boschetto, *J. Am. Chem. Soc., 93*, 1529 (1971).
36. S. R. Winter, Ph.D. dissertation, Stanford University, Stanford, California, 1973.
37. P. K. Wong, K. S. Y. Lau, and J. K. Stille, *J. Am. Chem. Soc., 96*, 5956 (1974).

Subject Index

Cumulative Index, Volumes 1-22